Novel Applications of Carbon Based Nano-Materials

Editors

Swamini Chopra
Assistant Professor - Research
Mechanical Engineering Department
Maharashtra Institute of Technology, Aurangabad, India

Kavita Pande
R & D Chief
Mayur Industries, Nagpur, Maharashtra, India

Vincent Shantha Kumar
Assistant Professor
Department of Mechanical Engineering
BITS Pilani Dubai Campus, Dubai, United Arab Emirates

Jitendra A. Sharma
Structural Engineer Lead
Voetsalpine Group, GL Apeldoorn, Netherlands

CRC Press
Taylor & Francis Group
Boca Raton London New York

CRC Press is an imprint of the
Taylor & Francis Group, an **informa** business
A SCIENCE PUBLISHERS BOOK

First edition published 2023
by CRC Press
6000 Broken Sound Parkway NW, Suite 300, Boca Raton, FL 33487-2742

and by CRC Press
4 Park Square, Milton Park, Abingdon, Oxon, OX14 4RN

Library of Congress Cataloging-in-Publication Data (applied for)

ISBN: 978-1-032-02480-6 (hbk)
ISBN: 978-1-032-02482-0 (pbk)
ISBN: 978-1-003-18354-9 (ebk)

DOI: 10.1201/9781003183549

Typeset in Times New Roman
by Innovative Processors

Acknowledgement

The completion of a task is the result of invaluable contributions, direction, supervision, criticism, suggestions and perpetual encouragement from a large number of people. Over this important commitment, the editors would like to thank CRC Press, Taylor and Francis Group, for encouraging this initiative and for providing the necessary guidance for the fulfillment of this work.

The editors also express a deep sense of gratitude to all the chapter authors, who have opted to contribute their research work in our title *Novel Applications of Carbon Based Nano-Materials*. Their warm support throughout the process has been encouraging for the completion of this endeavor.

The editors express a word of thanks to their employers and institutions for continuous encouragement and moral support throughout this work.

The editors are thankful to their families, friends and acquaintances for their continuous encouragement and unconditional support.

Finally, we believe that God is there all the time with us in the form of wonderful human beings who have helped us in any direct or indirect way possible. So finally we would like to thank the Almighty for providing us with the strength and goodwill for this task.

Preface

Carbon is a prime element used widely in various aspects of human civilization, since it forms strong bonds than any other materials in different forms. In the past few decades, several carbon-based nano-materials have been developed through various synthesis processes. The unique and tunable properties of carbon-based nano-materials enable new technologies for identifying and addressing varied field challenges. This book critically assesses the contributions of carbon-based nano-materials and related case studies in the allied field. The unique morphology and multifaceted properties of Carbon has enabled them to be used in multidisciplinary fields.

This special volume on **"Novel Applications of Carbon Based Nano-Materials"** published by **CRC Press, Taylor and Francis Group** mainly deals with different types of allotropes and forms of Carbon that are widely used in the field of Energy Storage, Waste Water Treatment, Mechanical Engineering and Medical Applications. This book consists of total **21 Chapters** contributed by Experts and Researchers in their respective fields, broadly classified here under **4** different categories.

Part I covers the area of **Photo catalytic and Energy Storage Applications**; where the chapters mainly cover the topics of Carbon Quantum Dots synthesis and application in the photo catalytic applications and mainly projects the super capacitors application for energy storage using carbon based nano-materials.

Part II illustratively discusses the use of carbon based nano-materials for **Mechanical and Engineering Applications**; wherein the chapters mainly deal with the synthesis of the Carbon nanomaterials or their composites which have the Thermal, Aerospace, Tribological and other allied mechanical or Engineering applications.

Part III deals with **Bioscience and Medical Applications** discussing the recent trends in the field of biomedical. The use of nanotechnology against Corona Virus and removal of antibiotics is discussed here. This section also covers scope and challenges for Green synthesis of functional nano particles.

Part IV covers the **Waste Water Treatment and Environmental Applications**; subdivided into chapters which precisely cover the Socio-economic Impact of Waste management especially targeting water treatments. Apart the impacts of carbon-based nanomaterials in biological and allied systems are also discussed in the said section.

    We hope this volume will be very much helpful not only to the researchers in the field of Nanotechnology and Carbon-based nano-materials; but also the students studying in the allied areas. It is our pleasure to offer sincere gratitude to all the authors for their valuable contributions and the reviewers for the improvement of the articles. We extend our thanks to the publisher for creating and making us to avail the platform.

Swamini Chopra
Kavita Pande
Vincent Shantha Kumar
Jitendra A. Sharma

Contents

Section I – Photocatalytic, Energy Storage and Mechanical Engineering Applications

Part I: Photocatalytic and Energy Storage Applications

Part II: Mechanical and Engineering Applications

Section II – Biomedical, Wastewater Treatment and Environmental Applications

Part III: Bioscience and Medical Applications

Part IV: Wastewater Treatment and Environmental Applications

Section I – Photocatalytic, Energy Storage and Mechanical Engineering Applications

Part I: Photocatalytic and Energy Storage Applications

An Overview of Synthesis and Photocatalytic Application of Carbon Quantum Dots-Based Nanocomposites

Pooja Shandilya[1]*, Adesh K. Saini[2], Rishav Sharma[1], Parteek Mandyal[1], Rohit Sharma[1] and Ankita Guleria[1]

[1] School of Advanced Chemical Sciences, Faculty of Sciences, Shoolini University, Solan, Himachal Pradesh - 173212, India
[2] Department of Biotechnology and Central Research Cell, MMEC, Markandeshwar (Deemed to be University), Mullana - 173207, Ambala, Haryana, India

1. Introduction

Nanotechnology plays a predominant role in various scientific domains, including electronics, biosciences, chemical sensors, fuel cells, space, fresh air, fabric, IT and water purification (Tlili et al. 2019). In today's world, everyone acknowledges that water pollution is rising at an alarming rate and needs to be addressed immediately. Different type of physical, chemical or biological pollutants contaminate the water, changing its properties and appearance. Various conventional methods were available to access good quality water for human use, such as boiling, filtration, distillation and chlorination. Nevertheless, these methods are not enough to purify all types of contaminations, including microbes (Hinrichsen et al. 2002, Sharma et al. 2021, Shandilya et al. 2018). Accordingly, quick action is needed to implement green and renewable technologies in the degradation of organic-inorganic contaminants and energy conservation processes. The idea of utilizing solar energy-assisted semiconductor quantum dots (QDs) has become an appropriate approach with huge potential for solving water pollution (Xue et al. 2017). QDs generally comprise II-VI or III-V groups of periodic table elements with sizes ranging between 1-10 nm (Ratnayake et al. 2019). The photocatalytic property of QDs enables the generation of reactive oxidative species (ROS) on its surface responsible for the degradation of pollutants (Chibli et al. 2011). The essential parameters that make QDs more famous are precise control over their size, energy levels and emission color. Several properties such as broad absorption range, narrow emission and wavelength of QDs

Corresponding author: pooja@shooliniuniversity.com, poojashandil03@gmail.com can be directly related to their size (Medintz et al. 2005, Ilaiyaraja et al. 2018). Since a nanoparticle is made up of around 100-10,000 atoms, therefore, QDs are also known as artificial atoms. Photoluminescence (PL) is one of the most specific properties of QDs responsible for its various fluorescent-based application, such as bio-sensing, catalysis, optoelectronic, bio-imaging and drug delivery (Hada et al. 2003, Shandilya 2021, Shandilya et al. 2022a). The use of QDs is preferably compared to organic fluorescent agents due to their high photostability and resistance toward photo-bleaching. Previous reports showed that most conventional semiconductor QDs were fabricated using heavy metals, like CdSe, PbS, PbSe, HgTe, InAs and InP, which furnished harmful impact on the environment and thereby limited their practical implication (Chan et al. 1998, Bruchez et al. 1998, Michalet et al. 2005, Resch-Genger et al. 2008, Shandilya et al. 2020a, Guleria et al. 2021, Rogach et al. 2007, Xu et al. 2010). Hence, metal-free carbon quantum dots could be the better substitute with the most negligible toxicity and high efficiency compared to conventional QDs (Wang et al. 2018).

Carbon, which is the second-most bulk element present in sp, sp^2 and sp^3 hybridization states, also exhibits catenation displaying an unusual ability to form the polymeric carbon chain. Furthermore, based on the dimensions, carbon nanostructures can be categorized into 0D (fullerenes, onion-like carbon structures, carbon quantum dots and nano-diamonds), 1D (carbon nanotubes and nanofibers), 2D (graphene and graphene nanoribbons) and 3D (diamond and graphite) as shown in Figure 1(a). These availabilities in different dimensions introduce carbon-based QDs as fascinating materials to be explored by the scientific community (Sharma et al. 2020, Georgakilas et al. 2015). The CQDs are composed of quasi-spherical and distinct carbon nanomaterials with size <10 nm, whereas GQDs are made up of graphene nanosheets smaller than 100 nm in size (Hu et al. 2016, Bacon et al. 2014). The unique features of CQDs, such as non-expensiveness, high stability, displaying no optical blinking, non-toxicity, chemical inertness, exceptional multiphoton properties and biocompatibility, permit them to be used exceptionally in various photocatalytic fields (Lim et al. 2015). Due to the advantageous features and unique properties of CQDs, considerable improvements have been observed for their photocatalytic applications in the last 10 years (Figure 1b).

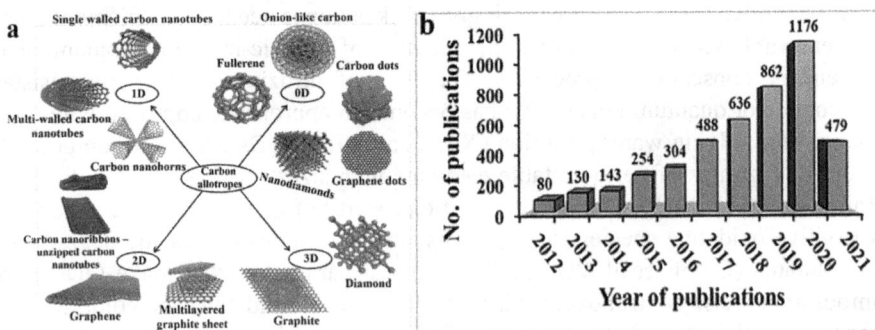

Figure 1: (a) Classification of carbon allotropes according to their dimensions 'reprinted from (Chem. Rev. 2015, 115, 11, 4744–4822) copyright (2015) American Chemical Society)' [19] and (b) Significant growth of carbon quantum dots as photocatalysts from 2012 to 2021.

CQDs were accidentally invented by Xu and co-workers in 2004 while purifying single-walled carbon nanotubes and then in 2006 during laser ablation of graphite and cement (Xu et al. 2004, Sun et al. 2006). The fabrication methods of CQDs can be categorized into two classes: "top-down" and "bottom-up" techniques. CQDs synthesized by break-down of carbonaceous materials are assembled in a top-down approach while polymerizing different aromatic compounds are grouped in a bottom-up approach (Cong et al. 2018). These approaches contain electrochemical carbonization, microwave irradiation, laser ablation, chemical ablation, hydrothermal or solvothermal treatment. So far, apart from photocatalysis, CQDs applications have also been extended to healthcare industries and optoelectronics devices as well (Wang et al. 2014). The increasing attention toward the usage of CQDs in photocatalysis has promising potential to be applied in environmental remediation and solar fuel generation (Wei et al. 2019, Gholipour et al. 2017, Kumar et al. 2018, Wood et al. 2019, Phang et al. 2019, Shandilya et al. 2022b). CQDs have become a new member of the nanocarbon family due to their abundant, inexpensive nature and attractive properties, especially their good optical absorption in the UV region (260-320 nm) showing an absorption band at 270 nm (Baker et al. 2010, Zhou et al. 2007). Their large surface area enhances the adsorption capacities of the nanocomposite, apart from also improving its migration rate and lowering the recombination process of charge carriers (Di et al. 2015, Shandilya et al. 2020b). For better photo efficiency, various synthetic strategies and modifications of CQDs-based nanocomposites have been widely explored. Like doping of carbonaceous material with different inorganic compounds, such as ZnS and ZnO was done to synthesize CZnS-dots and CZnO-dots leading to excellent, bright fluorescence performance (Sun et al. 2008). When CQDs are dissolved into polyvinyl alcohol matrix under UV light at room temperature, the phosphorescence property of CQDs could be observed. The phosphorescence is generated through excited triplet states in aromatic carbonyls above the surface of CQDs (Deng et al. 2013).

Besides, CQDs hold another interesting up-conversion property where low energy photon is converted into high energy photons with lower wavelength contrary to conventional inorganic semiconductor QDs (Jia et al. 2012). This upconversion becomes advantageous for utilizing a high band-gap semiconductor as a photocatalyst. In conventional semiconductor QDs, some part of the energy is liberated as thermal energy, which reduces photoefficiency and light-harvesting capacity. The upconversion photoluminescence (UCPL) mechanism can be well explained by TiO_2.CQDs, which initiate charge carrier generation on TiO_2 surface after solar irradiation (Ke et al. 2017). The photoexcited electron in the conduction band (CB) of CQDs migrates to CB of TiO_2 surface, thereby assisting the formation of superoxide radical anion ($^\cdot O_2^-$), while holes in the valence band (VB) of TiO_2 generate hydroxyl radical ($OH^\cdot$). During the relaxation process, photoexcited electrons in CQDs emit high-energy photons that subsequently activate the TiO_2 surface and generate charge carriers. The TiO_2 exhibits an improved visible-light response after coupling with CQDs due to its UCPL of photogenerated charge carrier. This process improves the photo efficiency of nanohybrids and leads to the degradation of various pollutants and dyes, as shown in Figure 2(a). The UCPL property of CQDs is also advantageous in cellular vision by the two-photon luminescence microscope phenomenon. The scanning electron microscope (SEM) image of CQDs-TiO$_2$ composites shows dispersed particle and microsphere surface in Figure 2(b). It exhibits bright emission in the visible range when irradiated with argon-ion laser (458 nm) or through femtosecond pulsed laser in NIR region (800 nm); it clearly describes that CQDs one photon and the two photons excitations results are similar in Figure 2(c-e) (Cao et al. 2007). However, the UCPL properties were confirmed via a two-photon luminescence spectrum. Li et al. stated that C-dots fabricated through alkali-supported ultrasonic treatment exhibits strong UCPL properties and size-based PL (Li et al. 2011). Various reports indicate that the UCPL phenomenon is similar to ordinary fluorescence and not shown by all CQDs and also nearly all CQDs do not exhibit UCPL property (Zhuo et al. 2012).

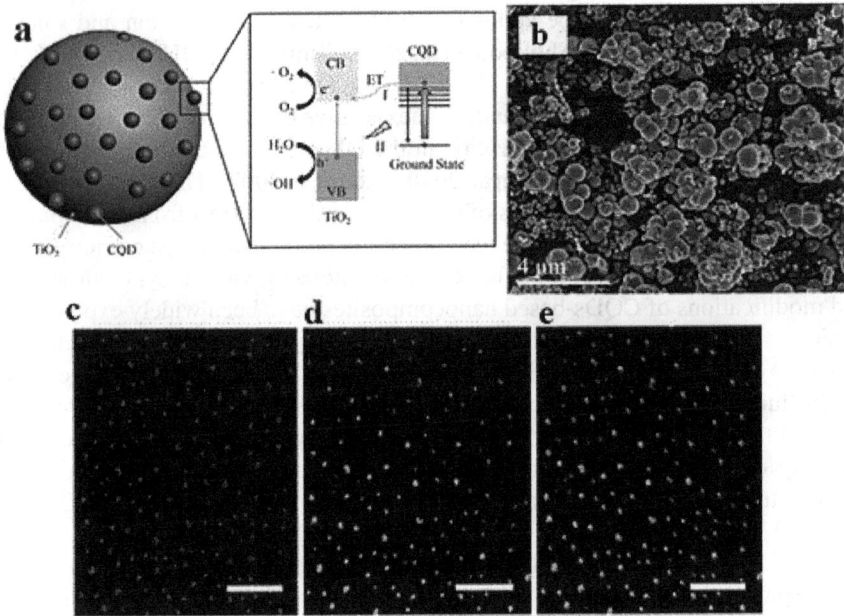

Figure 2: (a) Charge transfer mechanism on CQDs-TiO$_2$, (b) SEM image of CQDs-TiO$_2$ composites 'Reprinted from *Journal of Colloid and Interface Sciences* with permission from Elsevier (License No. 5044141251769)' [38], Luminescence images of CDs, (c) Argon ion laser excitation at 458 nm, (d) Femtosecond pulsed laser excitation at 800 nm and (e) overlay of (c) and (d). 'Reprinted from (J. Am. Chem. Soc. 2007, 129, 37, 11318–11319) copyright (2007) American Chemical Society' [39].

2. Method of Preparation for CQDs Modified Photocatalysts

CQDs are mainly formulated via two methods i.e., 'Top-down' and 'Bottom-up' approaches. The starting material that can be utilized for the synthesis are graphite, activated carbon, nano-diamonds, graphene, graphene oxide, carbon nanotube and carbon fibers. Agglomeration, surface functionalization, size control and uniformity are the significant difficulties that arise during the synthesis. In the top-down method, a large molecule is broken into nano-sized material. Whereas in the bottom-up method, larger structures are formed by massing smaller atoms or molecules shown as in Figure 3(a) (Namdari et al. 2017).

2.1 Top-Down Approach

2.1.a Arc Discharge Method

Here, an inert atmosphere of helium gas at 660 m bar pressure is used to generate electric current at cathode and anode, which is made up of pure graphite rods. An anode with a 3.5 mm hole in diameter and 40 mm deep is filled with carbonaceous materials and catalyst. This method is worked by applying 100 Å electric current and a 30 V potential in-between the electrodes separated by 3 mm distance. The different atomic percentages of catalysts mixture, such as Ni-Y, Co-Y or Ni-Co, are used to enhance the product yield by arc discharge method. In order to obtain a high yield of single-walled carbon nanotubes, Journet and co-workers used a catalyst mixture of 4.2% Ni and 1% Y. They also suggested that the growth mechanism of these nanotubes greatly depends on the kinetics of carbon condensation rather than experimental conditions (Journet et al. 1997). C-dots are fabricated from raw carbon nanotube residue by arc-discharge technique with a quantum yield of 1.6%. In this method, firstly, the crude carbon nanotube soot is oxidized with HNO_3 and after that, the resultant material is separated by an alkaline solution of pH value 8.4. Afterward, separated particles are sterilized that is followed by gel electrophoresis to obtain highly fluorescent CQDs of size 18 nm (Cao et al. 2007). Also, unique PL nanomaterials are synthesized from nitric acid-oxidized carbon nanotubes via the electric arc technique. However, pure carbon nanotube-derived fluorescent nanoparticles are hydrophobic, which confines their distribution in water; whereas, oxidized carbon nanotube-acquired fluorescent nanoparticles are accumulated when dissolved in water laminated with a thin carbon layer that extended their distribution. CQDs synthesized by arc discharge technique have complex compositions and are hard to purify and acquire low quantum yield (Bottini et al. 2006).

2.1.b Laser Ablation Technique

The laser ablation method involves the absorption of high-energy laser pulse by carbon reactants, which evacuates the electrons and creates a high-energy electric field. These high electric fields can produce high repulsive forces, responsible for the break-down of carbon precursors into CQDs (White et al. 2012). It is a simple chemical technique that produces fewer side products at mild temperature and pressure conditions. CQDs could also be prepared by dissolving the carbon nanomaterial in organic solvent (ethanol, acetone, etc.) and subsequently followed by ultrasonication (Li et al. 2010). Laser irradiation is the preferable technique to control the size of CQDs. Moreover, this technique can also be used to modify the surface of CQDs simply by choosing an appropriate solvent during laser irradiation. This technique provides different PL characteristics to fabricated CQDs. Arvind et al. reported nanosecond laser ablation for the synthesis of highly stable CQDs suspension size ranges from 5 to 15 nm by shifting the time of laser irradiation. This work was an ecofriendly and single-step reaction to form CQDs, starting from the graphite crystalline particles spread in water and generated CQDs shown in Figure 3(b). They noticed that diverting the time of laser ablation and optimizing parameters, such as laser wavelength and power, pulse regularity rate and its duration control the size of CQDs. It was observed that fragmentation was not sufficient when the laser beam was kept stationary inside graphitic dispersion. Therefore, the focused beam was line scanned repetitively during the process, which enhanced the interaction of graphitic microcrystal, maximized heat dissipation rate and converted an ample amount of microcrystal into CQDs. Two possible mechanisms, namely Columbic explosion and melting-evaporation, are generally responsible for CQDs production during laser irradiation. In this work, metal evaporation is a dominant mechanism where a highly energetic laser beam is used to produce an excited plasma of ions or electrons along with neutral molecules or atoms. These high energetic plasma species generate a plume of very high pressure and temperature. This plume then extends with time and squeezes the cold liquid medium which results in creating a shockwave at the liquid/plume interface. At the interface, thermalization is initiated and the plume species that are liberated there are energy-caused deacceleration of particles. In the meantime, inward shockwaves are also produced due to the development of reverse pressure gradient. These outward and inward shockwaves cycle continues till the complete thermalization takes place and result in the condensation, nucleation and clustering of plume species, which are followed by fabrication of CQDs.

TEM analysis was used for examining the morphology of prepared CQDs; with a rise in laser irradiation time, the average size of CQDs particles became smaller. The different irradiation times of 40, 60 and 80 minutes are labeled as S40, S60 and S80, respectively. The average particles size obtained is $\sim$ 15, $\sim$ 10 and $\sim$ 4.5 nm as shown in Figure 3(c-e) (Singh et al. 2019). The various top-up and bottom-up approaches have been already employed to synthesize CQDs. Still, most of them bear the pitfalls, such as consuming harsh chemicals, releasing toxic side products, tedious synthetic methods, less control over the size and composition of quantum dots. In contrast to these methods, the synthesis of CQDs by laser ablation method is more recommended due to rapid synthesis, less consumption of harmful chemicals, good control over size, shape and composition, etc. However, the high cost and complicated operation limit the broader applications of this method.

2.1.c Electro-Chemical Approach

The electrochemical approach is an analytical technique that measures a potential, charge or current so as to determine the concentration. In the last few years, many approaches were used for the preparation of quantum dots. However, these methods are pretty difficult to perform; for example, in ultrasonication, potassium ion needs to be eliminated for distillation of potassium intercalation (Lin et al. 2013). The electrochemical method is also used for the preparation of metal sulfide and metal oxide CQDs. In 2018, CQDs of different sizes were formed using the graphite electrode in the presence of EtOH/NaOH solution. Devi et al. explained the electrochemical synthesis of CQDs by using a 5 cm graphite rod with a width $\sim$ 0.2 cm, which was used as cathode and anode dipped in an alkaline solution. Here, this solution was composed of 0.1 M sodium hydroxide, ethanol and water, which performed as an electrolyte. Moreover, a continuous direct current of 50 mA was also attached to both of the electrodes. The mechanism of fabrication is explained by the presence of electrolyte solution which formed hydroxyl radicals and sodium salt of ethanol. The free hydroxyl radical attack on the anode initiated the oxidation process and led to the exfoliation of the graphite rod, which consequently generated CQDs. The progress of the route can be monitored easily by a change in color from colorless to yellowish-red and finally dark red, confirming the preparation of CQDs. Here, the electrolytes used during electrochemical synthesis exhibit a key role. For example, in a controlled experiment, when the above process was carried in the presence of an aqueous medium without ethanol or in the presence of ethanol only, no changes were observed in the reaction that indicated no CQDs formation. In an aqueous solution, a medium to strong exothermic reaction takes place with the hydrolysis of water that produces hydrogen gas but not CQDs. Therefore, the aqueous solution of both NaOH and C_2H_5OH are must needed to successfully accomplish the reaction. By this method, various dimensions of CQDs can be obtained just by reducing the graphite superficial. The formation of these CQDs can be characterized by XRD, as CQDs display broader peaks around 24.3° whereas graphite shows a sharp peak at 25.76°. This shift in peaks from 25.76° to 24.3° is due to a change in interlayer distance in CQDs, which is because of the intercalation of more oxygen functional groups during the exfoliation (Devi et al. 2018). By changing the current density, the amount of CQDs can be controlled. Moreover, the electrochemical method acquires low cost, step method, high output and easy manipulation, which makes it the preferable technique for the production of CQDs (Cao et al. 2007).

2.1.d   Acid Oxidizing Exfoliation

CQDs are prepared by acid oxidizing exfoliation method using H_2SO_4 and HNO_3. GQDs provide the primary application in nanotechnology. Peng et al. synthesized GQDs supported by resin-rich carbon fiber via acid oxidizing exfoliation method (Peng et al. 2012). Different size GQDs were obtained simply by changing the reaction temperature. The blue shift was observed from 330 to 270 nm when the temperature raised from 80 to 120 °C. These results indicate that reaction temperature alters the distribution of emission wavelength as well as modified band-gap value and absorption properties of GQDs. Generally, the GQDs prepared at low temperature formed GQDs showed absorption at a longer wavelength. Additionally, the optical characteristics of GQDs could also be adjusted by changing the size of GQDs attribute to change in density and nature of sp^2 sites present in the structure. In this scenario, the carbon precursor used during the synthesis is also an important factor to predetermine the size, shape and yield of the product. Liu et al. fabricated graphene and GQDs employing graphite as the precursor. Graphite nanoparticles are ideal precursors to fabricate CQDs, GQDs and oxygen functionalities containing GOQDs. The graphite nanoparticles contain numerous graphene layers stacked via pi-pi or Van der Waals interaction. These layers can be exfoliated easily under the acid treatment to obtain GOQDs or GQDs with uniform shape and size. During the process, graphene nanomaterials mixed with N-Methyl-2-pyrrolidone, dimethylformamide, dimethyl

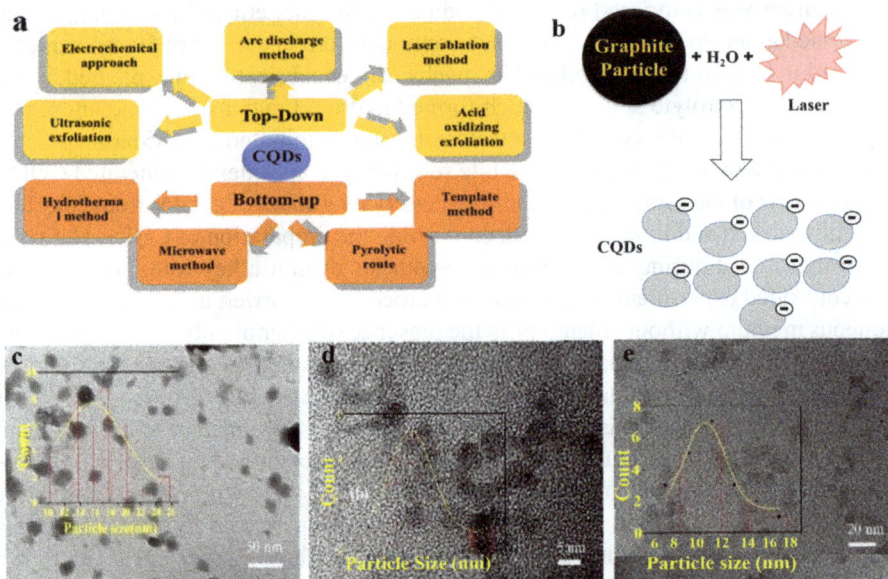

Figure 3: (a) Methods for CQDs preparation, (b) Laser ablation method for CQDs, (c) TEM images S40, (d) S60 and (e) S80 CQDs. 'Reprinted from *Materials Chemistry and Physics* with permission from Elsevier (License No. 5045931374543)' [47].

sulfoxide under ultrasonication and followed by centrifugation to obtain quantum-sized graphene (Liu et al. 2013). However, performing acid exfoliation under a harsh and rigid environment is one of the major disadvantages. Thus, achieving precise particle size becomes a difficult task under such harsh reaction conditions (Bao et al. 2011).

2.1.e   Ultrasonic Exfoliation

Using ultrasonication for CQDs synthesis is more advantageous; it creates high- and low-pressure waves within the liquids, responsible for creating and collapsing vacuum bubbles in the reaction mixture. The continuous genesis of alternate waves inhibits cluster formation of QDs and produces strong hydrodynamic shear forces responsible for their high stability. CQDs/Cu_2O nanostructure was prepared through simple single-step ultrasonic waves by utilizing $C_6H_{12}O_6$ as starting material in basic conditions (Li et al. 2012). The nanostructure exhibits higher productivity due to UCPL of CQDs and the high light reflecting the capability of Cu_2O. Since 53% of the solar spectrum consists of IR radiation, CQDs/Cu_2O photocatalytic system is reported for the first time that is active in the NIR region. The characterization results of SEM and TEM images show spherically shaped colloidal particles as shown in Figure 4(a-c). HRTEM reveals lattice spacing around 0.25 and 0.32 nm for Cu_2O and CQDs that confirms the loading of CQDs on the surface of Cu_2O particles in Figure 4(d). The elemental mapping clearly indicates the presence of Cu, O and C in Figure 4(e-h). These analyzes suggested a spherical-shaped nanostructure that was effective against photocatalytic degradation of methylene blue. Park et al. executed a new environmentally friendly ultrasonic method to synthesize CQDs by utilizing cost-effective waste food. Various steps are involved for the synthesis of CQDs, dehydration, polymerization and carbonization by passivation. Since waste food contains various organic entities based on H-bonding, therefore during ultrasonication dehydration, polymerization and carbonization occur correspondingly that causes a single explode of nucleation. The emerging nuclei then grow on carbon nanoparticles surface via diffusion and form highly functionalized CQDs. In this work, a rotten eatable was mixed with ethanol followed by ultrasonication for 45 minutes at a frequency of 40 kHz; lastly, heavy and agglomerated particles of average size 4.6 nm CQDs were isolated through centrifugation (Park et al. 2014). The procedure of this method is easy to perform and is cost-effective.

2.2 Bottom-Up approach

2.2.a   Hydro-Thermal Method

CQDs are synthesized by hydrothermal carbonization route by using various precursors, including chitosan, banana juice, citric acid and glucose that are sealed and heated at high temperature in a thermal reactor. Hydrothermal is an environmentally friendly, cost-effective and non-toxic route to fabricate carbon-based materials. Shen et al. (2018) formulated CQDs/TiO_2 heterojunction by using glucose and citric acid as starting materials. Here, the CQDs synthesized from glucose and citric acid were denoted by CQDs-G/TiO_2 and CQDs-CA/TiO_2 heterojunction, respectively. A higher photocatalytic degradation of phenol was observed for CQDs-G/TiO_2 compared to CQDs-CA/TiO_2 and bare photocatalyst under UV radiation. The characterization results suggested high crystallinity and charge carrier efficiency of CQDs-G/TiO_2 and TEM images indicated by spherical shape with no apparent aggregation and narrow size distribution for both heterojunctions in Figure 4(i and j). However, the size of CQDs-G (3-6 nm) was found more significant than CQDs-CA (2-4 nm) (Shen et al. 2018). Sarkar et al. synthesized yellow-colored water-soluble CQDs using sucrose and ethanol precursor placed in Teflon autoclave and heated at 175-180°C followed by centrifugation. The PL spectra show a maximum intensity observed around 490 nm, emission generally ascribed to the electron-hole recombination process or quantum size effect (Sarkar et al. 2016). To enhance the photophysical activity of CQDs, surface functionalization and heteroatom doping are two common strategies. Out of these two, surface functionalization is less preferable as the organic molecule or polymer that is used during functionalization occupies some specific functionalized position. Therefore, heteroatom doping is considered a more advantageous approach to improve fluorescence properties of CQDs, which can be done easily by the hydrothermal method. For example, composites like N/Cu-CQDs were prepared by the one-step hydrothermal route and obtain a brown powder of N/Cu-CQDs. The reported nanocomposite revealed high stability, metal ion insensitivity and special fluorescence properties along with a high quantum yield of 50% (Ma et al. 2017). CQDs synthesize by the hydrothermal route do not require any toxic chemicals and are also able to synthesize hydrophilic CQDs.

2.2.b   Microwave Method

The microwave technique is green, cost-effective, easy and highly efficient for constructing CQDs in contrast with the hydrothermal method. This method is less time-consuming due to the usage of a standard microwave oven. In 2012, Qu et al. synthesized fluorescent-dependent CDs by utilizing citric acid and urea in an oven at 700 W (Qu et al. 2012). The synthesized CQDs were used for layering on commercial gauzes, vegetal fibers, mammals' furs, feathers and skins as fluorescent ink. The morphology of CDs was characterized using TEM analysis, displaying spherical shape and well dispersed with 1-5 nm diameter. The two broad peaks at 6.8 and 3.4 Å in XRD patterns confirm the disordered nature of carbon. Raman spectra of CDs display 1,365 and 1,575 cm^{-1}, two broad peaks, which were assigned for D and G bands. The surface functional group were analyzed by FTIR, broad absorption bands obtained at 3,100-3,500 cm^{-1} (O-H) and (N-H), 1,600-1,700 cm^{-1} (C=O) and 1,350-1,460 cm^{-1} (CH_2). Toxicity studies were also conducted by various research groups and suggested no toxicity of CDs. Previous studies suggested that the microwave method is one of the most favorable techniques for preparing CQDs as it provides constant heating and uniform size arrangement of CQDs. Surface passivation reagents are generally needed to construct CDs with good quantum yield without surface passivating agent quantum yield that are comparatively low. Tang et al. used a microwave-assisted hydrothermal approach to synthesize homogeneous and uniform size GQDs. This process brought the advantages of both microwave and hydrothermal processes and also did not require any inorganic additives or surface passivating agents. The quantum yield found in this work was 7-11% even without using a surface passivating agent. In this process, they used glucose as a precursor. Other than glucose, sucrose, fructose or carbohydrates with 1:2:1 proportion of C, H and O can also be used (Tang et al. 2012). The obtained CQDs from the microwave method were highly biocompatible and had excellent potential in biomedical applications (Li et al. 2013).

2.2.c   Pyrolytic Route

The pyrolytic route is an easy and single-step method with high quantum yield, multicolor fluorescent CQDs and soluble in water and organic solvents. This approach has great importance to synthesize tunable and uniform size CQDs through confined pyrolysis of organic solvent. During the synthesis, organic precursors are absorbed on porous nano-reactors through capillary forces. In the second step, the organic precursor undergoes pyrolysis and is converted into carbonaceous matter CQDs, which are then collected from the nano-reactor. The size and structure of CQDs depend upon the texture of the nano-reactor. Porous silica nano-reactor is most commonly used for size-controlled confined pyrolysis due to its variety of texture, easy removal and high thermal stability. CQDs were also synthesized by pyrolysis of EDTA (ethylenediaminetetraacetic acid) at low temperature. EDTA is usually more beneficial as the synthesized CQDs are hydrophilic and soluble in a variety of polar organic solvents. Additionally, pyrolysis of EDTA form highly selective and much more sensitive CQDs used to analyze Hg^{2+} ion and bio-thiols. Since Hg^{2+} completely quenched the fluorescent signal of C-dots due to high selectivity toward CQDs, it can be easily detected. The obtained CQDs emit light with blue fluorescence and PL quantum yield reached 66.03% (Li et al. 2015). In 2012, Guo et al. used a chemical unzipping of epoxy-polystyrene photonic crystal as a precursor to synthesize CDs through a one-step pyrolytic route. In this process, blue, orange and white-emitting quantum dots were generated by altering reaction temperature (Guo et al. 2012). Further, a two-step pyrolytic method can be employed to synthesize chiral CQDs with enantiomeric ligands by surface passivation. The acquired optically active CQDs show photostability, suitable biocompatibility and promising chiroptical nano-pores beneficial in chirality-dependent applications (Rao et al. 2019).

2.2.d   Template Method

For tuning pore size so as to attain narrow size distribution of CQDs template method can be employed; however, it is pretty challenging as it does not entirely regulate the pore size. This technique involves using a silicon sphere that acts as a mesoporous template for the calcination or rubbing to formulate small-sized CQDs. Silicon sphere as a mesoporous template is used to control the size of CQDs. Here, tetra-ethoxy silane was used as a precursor, N-hexadecyl amine as a surfactant (Grün et al. 2000). In 2013, Yang et al. illustrated the fabrication of monodispersed PL CDs via a soft-hard template technique (Yang et al. 2013). In this process, 1, 3, 5-trimethylbenzene, diaminebenzene and pyrene were used as carbon precursors covered by micelles of copolymer Pluronic (P123) silica accompanied by carbonization, elimination of template and passivation to obtain the quantum dots. Oleyl amine-capped CNDs (carbon nano-dots) of different sizes were also synthesized via the soft-template method illustrated by Kwon et al. (2014). After that, intermolecular dehydration of citric acid gave 'polymer-like' intermediates through water evaporation. Finally, the organic-soluble CNDs were obtained through these intermediates' carbonization and further capped via oleylamine molecules. One of the essential properties of this soft-template technique is the size of CNDs which can be managed by changing Oleyl-amine concentration. It acts as a capping agent, affecting the termination rate for the growth of CNDs. Similarly, by changing the concentration of citric acid, the size of CNDs was also regulated. So, two different-sized CNDs were characterized by TEM analysis. Overall, CQDs synthesized under harsh conditions, consume high energy and give low yield in the top-down method. Whereas, low yield bottom-up approaches are broadly utilized because they are simple and one-step syntheses. Still, this method results in identical nanomaterials, having broad size distribution and is time-consuming.

3. Application of CQDs-Based Nanocomposites

Photocatalysis acts as a sensational field that aims to construct an effective nano-catalyst with improved photo efficiency. Photocatalysis, a green technique utilizes solar radiation and is broadly employed for environmental remediation and energy application (Ratnayake et al. 2019). Photocatalysis is a redox process that depends upon band structure and charge carrier's migration on CQDs surface. In photocatalysis, CQDs are preferable due to their tuneable size, which leads to the controlled emissions of light and absorption of different wavelengths of light from the visible to NIR range. In this section, applications of CQDs-based 2D nanocomposites in photodegradation, photocatalytic water splitting and CO_2 photoreduction are discussed in detail. The photocatalytic mechanism and different roles of CQDs for improving overall composites efficiency are also highlighted.

3.1 CQDs in Photodegradation of Pollutant

Previous reports proposed the enhanced photocatalytic performance of CQDs via doping (Hu et al. 2013). Additionally, photoinduced CQDs hold excellent separation efficiency of charge species due to their electron-donating and accepting nature. The up-conversion phenomenon holds a crucial part in photocatalysis, which is meant to break down various pollutants. CQDs hold distinctive up-conversion properties due to oxygen-containing functional groups and quantum confinement effect. The quantum confinement effect contains discrete electronic energy levels due to the confinement of electronic wave function. The basic mechanism of CQDs photocatalysis can be explained in three steps:

(1) Effective photogeneration of the charge carrier is induced when the light of appropriate frequency falls on the semiconducting surface. The photo-excited electrons present in the valence band (VB) move to the conduction band (CB) and generate the charge carriers. Generally, the band-gap of CQDs lies in the range of 2.15-2.89.

(2) Oxidation and reduction reaction is carried out by these photogenerated holes in VB and photoexcited electrons in the CB. Photoexcited electrons in the CB react with the O_2 molecule and converts it into superoxide ($^{\cdot}O_2^{-}$) radical; simultaneously, positively charged holes react with the ^{-}OH ion to generate $OH^{\cdot}$. The redox potential of holes and electrons is +1.0 to +3 V and +0.5 to −1.5 V w.r.t NHE, respectively. The strong oxidation potential of $^{\cdot}O_2^{-}$ and $OH^{\cdot}$ is responsible for diminishing organic and inorganic pollutants into CO_2, H_2O and simpler inorganic species (Wang et al. 2017).

(3) Recombination of charge carrier emits a certain frequency of light in the form of PL (Jing et al. 2005). The PL intensity decreases, the recombination rate also decreases and consequently, the photo-efficacy of photocatalyst increases (Yu et al. 2003, Li et al. 2001). Therefore, significant efforts have been done to decrease the recombination process of charged species.

The catalytic activity of CQDs strongly relies on the shifting and partition capability of photoinduced electrons. Small-size CQDs (1-4 nm) show excellent photocatalytic activity in the NIR region, particularly for the oxidation of alcohol into benzaldehydes with high reproducibility (Li et al. 2013). On the other hand, larger CQDs of size 5-10 nm exhibit light-induced proton characteristics and are utilized in organic transformations in the aqueous system under visible light. Liu et al. illustrated a procedure for environmentally-friendly oxidation of cyclohexane to cyclohexanone by AuNP-CQDs nano-complexes through H_2O_2 act as an oxidant, which had high transformation productivity (63.8%) and selectivity (99.9%) (Liu et al. 2014). In comparison with Fe_2O_3 particles, Fe_2O_3/CQDs nanocomposites hold excellent photocatalytic performance toward gas-phase methanol degradation (Zhang et al. 2012). For example, CQDs/Ag_3PO_4 degraded MO in just 25 minutes, whereas the pristine Ag_3PO_4 takes 55 minutes. In that context, Zhang et al. described a preparation procedure for hydroxide-dispersible fluorescent n-CDs via one-pot ultrasonic interaction (Zhang et al. 2012). Thakur et al. reported efficient tunable photocatalyst $P_{M(Au/Ag)}$@CQDs/TiO_2 nanofibers nanocomposites, enhancing light harnessing ability and generated plenty of active oxygen species, which are responsible for high-performance. The photoreaction mechanism and path of photoinduced degradation of MB and pharmaceutical drugs in Figure 4(m). The morphology and structural analysis of P_M@CQDs was characterized by HRTEM images, displaying the spherical morphology of nanostructure with ~10 nm diameter as shown in Figure 4(k and l) (Thakur et al. 2020, Shandilya et al. 2020c). Large numbers of CQDs nanocomposites discussed for the degradation of various dyes and pollutants degradation are illustrated in Table 1.

3.2 CQDs in Hydrogen Production via Water Splitting

The hydrogen generation through photocatalytic water splitting under solar irradiation utilizes a green energy source other than fossil fuels. Photocatalytic water splitting involves creating holes and electrons and subsequently, generation of the ROS species on the surface of the semiconductor, which serves as the oxidizing and reducing agents for the evolution of O_2 and H_2 gases (Saravanan et al. 2019). This splitting process of H_2O into O_2 and H_2 requires high energy of 237 kJ/mol (Reddy et al. 2020, Monga et al. 2020). The ideal band edge position of the conduction band should be extra negative compared to the redox potential of H^+/H_2 lies at 0 V vs NHE, whereas the valence band must be at a different favorable position than the standard oxidation potential of O_2/H_2O i.e. 1.23 V vs NHE. The complete water splitting process mainly performs by two half-reactions involving the H_2 evolution reaction and O_2 evolution reaction (Fang et al. 2019). The mechanism firstly includes creating electrons and holes through solar irradiation and partition of charges occurs, which are then captured by active surface sites causing water splitting process shown in Figure 4(n) (Monga et al. 2020).

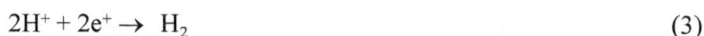

$$H_2O \rightarrow H_2O + 1/2\, O_2 \qquad (1)$$

$$4H^+ + H_2O \rightarrow O_2 + 4H^+ \qquad (2)$$

$$2H^+ + 2e^+ \rightarrow H_2 \qquad (3)$$

The semiconductor materials' electronic structure and electron holes separation play a crucial role in water splitting under light irradiation. Thus, by minimizing the recombination rate in both bulk and the exterior parts of photocatalyst, their efficiency can be improved (Ni et al. 2007, Serpone et al. 2006). So, in this condition, TiO_2 is a preferable semiconductor as its conduction band lies –0.6 eV, which is favorable for the reduction of H^+ to H_2. But due to the wide band-gap of TiO_2 (3.2 eV), it only works in the UV spectrum of sunlight (Serpone et al. 2006). Thus, scientists focus on developing the photocatalyst with optimum band-gap and displaying activity in the visible and NIR region. Nowadays, CQDs-based photocatalysts are utilized in

Table 1: Summary of CQDs-based nanocomposites in photodegradation.

CQDs-based Nanocomposites	Synthetic Method	Precursor	Targeted Pollutant	% degradation	Reference
CQDs-TiO_2-x/rGO	Hydrothermal method	Citric acid, urea, TiO_2/rGO,	Bisphenol A, phenol and Cr (VI)	BPA=82%, Phenol=76%, Cr (VI)=80%	[77]
CQDs @ Pd- SnS_2	Hydrothermal method	Sucrose, H_2PdC_{14} solution	4-nitro phenol	4-nitro phenol=99.7%	[78]
Fe_3O_4/C/TiO_2/N-CQDs	Hydrothermal method	Citric acid, glycine, Fe_3O_4/C	Methylene blue	Methylene blue=95.5%	[79]
CQDs@$CdIn_2S_4$/CdS	Microwave-assisted	Glucose, NaOH $Cd(NO_3)_2$, $In(NO_3)_3$	Methyl orange	95.5%	[80]
CQDs/$PbBiO_2Cl$	Solvent-thermal process	Citric acid, Bi $(NO_3)_3 \cdot 5H_2O$, 1 mmol of $Pb(NO_3)_2$	Ciprofloxacin, tetracycline and bisphenol A	CIP=78.9%, TC=79.8% and BPA=92%	[81]
CdS/CQDs/BiOCl	Facile region-selective deposition process	Glucose, TGA, $CdCl_2 \cdot 2.5H_2O$, CQDs/BiOCl, CdS QDs solution	Phenol	99.5%	[82]
TiO_2/BiOI/CQDs (TBC)	Step-by-step process	KNO_3 solution, titanium isopropoxide, TBC, TiO_2,	Methyl orange	90%	[83]
CQDs@CuS	Hydrothermal treatment	Copper acetate monohydrate, thiourea and ammonium citrate, ethylenediamine	Methylene blue	100%	[84]

$CeO_2:Tm^{3+}$/CQDs/ATP	Facile deposition method	Cerium nitrate hexahydrate, thulium nitrate hexahydrate and Hexamethylenetetramine	Dibenzothiophene	96%	[85]
N, S-CQDs/TiO_2	Hydrothermal process	Tetra-butyl ortho-titanate, Acetylacetone, citric acid, thiourea, EDTA and L-ascorbic acid	Acid red 88	77.29%	[86]

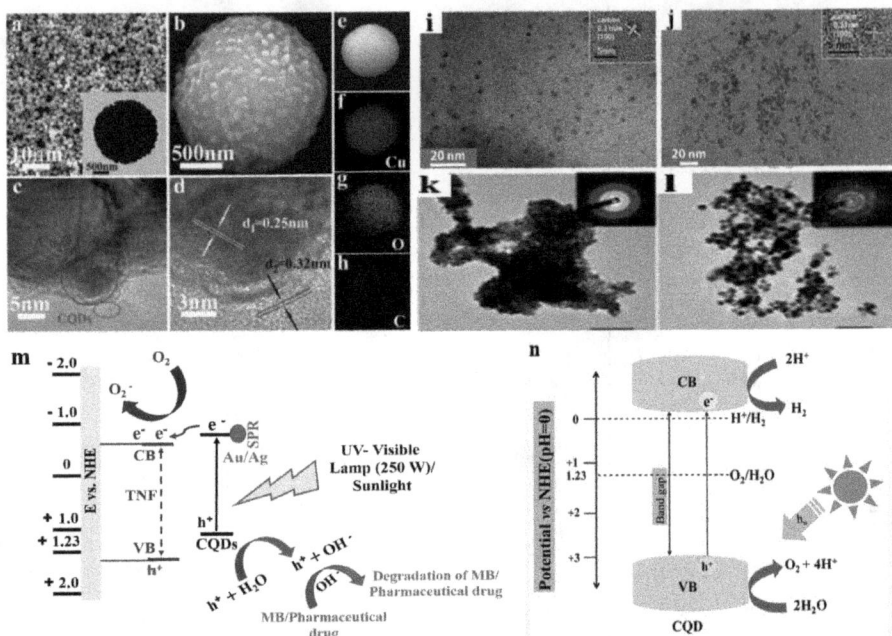

Figure 4: (a–c) SEM and TEM (inset of a), (d) HRTEM images of the CQDs/Cu$_2$O composite, (e) SEM image of CQDs/Cu$_2$O particle for EDS and (f–h) EDS of Cu, O and C. 'Reprinted from *Journal of Materials Chemistry* with permission from (Royal Society of Chemistry) (License No. 1096241-1)' [53]. TEM images of (i) CQDs-CA, (j) CQDs-G and insets show the HRTEM image of each sample. 'Reprinted from *Ceramics International* with permission from Elsevier (License No. 5044151303496)' [55]. HR-TEM micrograph of (k) P$_{Ag}$-CQDs, (l) P$_{Au}$-CQDs, (m) Schematic proposed mechanism for the MB dye and pharmaceutical drug degradation using PM-CQDs/TiO$_2$. 'Reprinted from *RSC Advances* with permission from creative commons' [75] and (n) General mechanism for hydrogen production.

solar water splitting and are quite advantageous in some aspects, such as migration properties, taking electrons, up and down converted PL and making NIR active hybrid photocatalysts. Au/CQDs nanocomposite used hydrogen production, which gives a quantum yield of 56% when placed under sunlight (Mehta et al. 2017). A large number of carbon-based nanocomposites, such as Cu/CQDs, amino-conjugated CQDs, Ag/CQDs/C$_3$N$_4$, PVP-coated CQDs and NiNPs, g-C$_3$N$_4$/graphene quantum dots have been used in hydrogen production (Zhang et al. 2017, Xu et al. 2016, Qin et al. 2017, Virca et al. 2017, Zou et al. 2016). Though CQD is a metal-free photocatalyst, its photocatalytic efficiency was enhanced by using noble metals as co-catalysts. CQDs exhibit lower efficiency in the water-splitting photocatalytic process, which can be tuned by band-gap engineering and surface modification.

3.3 CQDs for CO$_2$ Photoreduction

A large part of human society depends upon the consumption of fossil fuels to fulfill their energy requirements. The burning of fossil fuels generates greenhouse gases, including CO$_2$, which are essentially responsible for environmental pollution and

CNNS/CQDs	Bottom-up method	Photocatalyst=50 mg, Light source=300W Xe arc lamp, Light intensity=100 mW/cm^2	Visible irradiation	H$_2$ production	116.1 µmol/h	[108]
CQDS/CDS	Chemical absorption method	Light source=400 W metal halide lamp, Sacrificial reagent=Na$_2$S and Na$_2$SO$_3$, Room temperature	Visible irradiation	H$_2$ production	309 mmol/g/h	[109]
Bi-CDs/CdS	Pyrolysis method	Photocatalyst=50 mg, Light source=300 W Xe lamp, Sacrificial reagent=Na$_2$SO$_3$, Na$_2$S	Visible irradiation	H$_2$ production	1.77 mmol/g/h	[110]

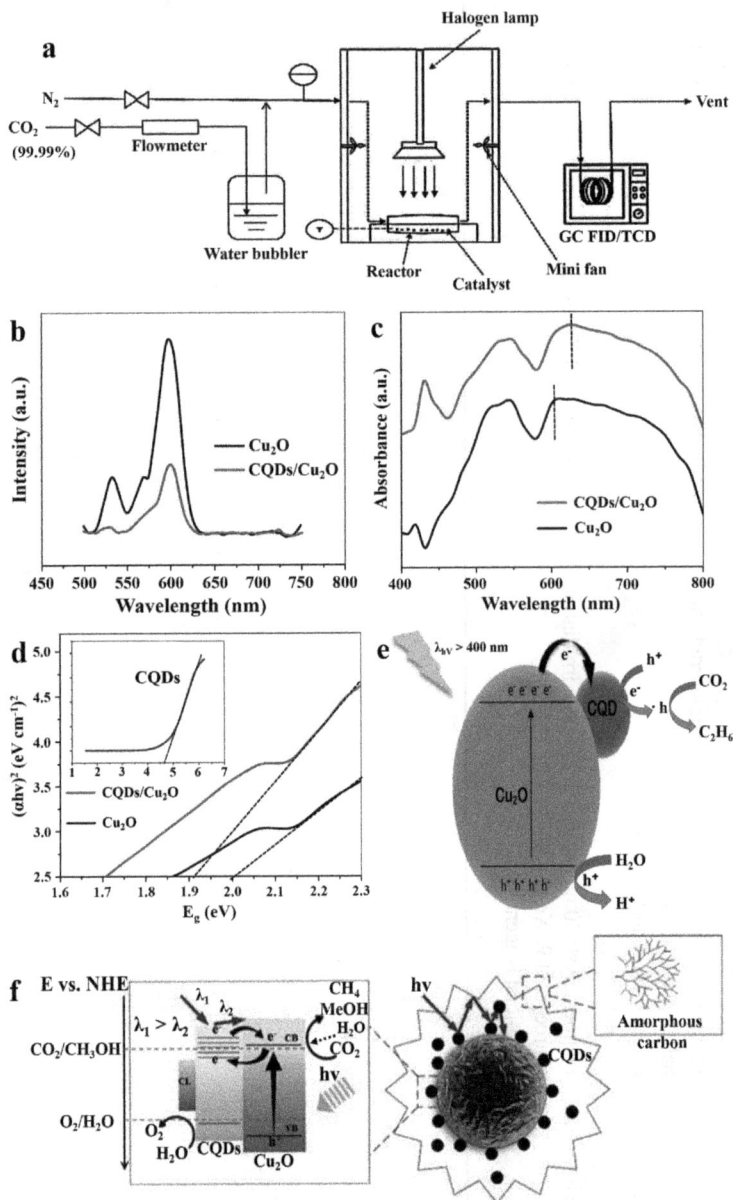

Figure 6: (a) Schematic illustration of system set-up for CO_2 photoreduction, (b) PL spectra of Cu_2O and $CQDs/Cu_2O$ nanocomposite, (c) UV-vis absorption spectra of Cu_2O and $CQDs/Cu_2O$ nanocomposite, (d) Tauc's plot of CQDs (inset), Cu_2O and $CQDs/Cu_2O$ nanocomposite, (e) Mechanism diagram of CO_2 photoreduction of $CQDs/Cu_2O$ photocatalyst. 'Reprinted from *Diamond and Related Materials* with permission from Elsevier (License No. 5044700284102)' [105] and (f) Proposed mechanism of CO_2 photoreduction of $CL@CQDs/Cu_2O$ 'Reprinted from *Chemical Communication* with permission from (Royal Society Chemistry) (License No. 1110671-1)' [106].

In another work carbon layer (CL) covered CQDs incorporated with Cu_2O (CL@ CQDs/Cu_2O) to improve the optical and physical properties toward photocatalytic CO_2 reduction (Li et al. 2019). The nano-catalyst provided high stability (loses only 1.2% yield after three-cycle) and high photoconversion efficiency of CO_2 into methanol (249 μmol/g in 2.5 hours). Figure 6f illustrates that the extra light is reflected and absorbed by the CL@CQDs part, which improved the light-harvesting capacity of the composite. Also, the amorphous CL redirected the up-converted light from CQDs into Cu_2O particles and assisted Cu_2O in producing charge carriers to perform the photoreactions.

4. Conclusion

Nowadays, most reviews based on CQDs concentrate on their description and fabrication techniques; only a few reviews have focused on their application. The framework holds the current progress in synthetic methods, properties, photocatalytic application in water purification, water splitting and CO_2 photoreduction using CQDs-based nanocomposites. CQDs nanohybrids reveal high light-harvesting capacity, infrared responsible UCPL, tunable PL and unique photogenerated electron transport. Selecting a specific fabrication method, surface properties of CQDs could be modified to attain a tunable light emission. In particular, UCPL is an optical phenomenon whereby fluorescence produces a shorter wavelength as compared to the excitation wavelength. This distinctive characteristic is able to rationalize a high-performance CQDs photocatalyst system for the utilization of a broad solar spectrum. The PL of CQDs quenched either electron acceptor or donor molecules, which photogenerated CQDs to give excellent electron transfer properties and the abovementioned properties made CQDs efficient for photocatalytic applications. Although significant modification has been made in the last decades, there are still many issues for further investigation. The photocatalytic activity is troubled by oxygenic groups existing on their surface and also its absorption in the UV region responsible for its limitation. Formulating CQDs with heteroatom doping or surface modification allows them to work even under visible or NIR regions. Accordingly, for diversified applications, a more reliable CQDs-based system can be explored.

References

Aghamali, A., Khosravi, M., Hamishehkar, H., Modirshahla, N. and Behnajady, M.A. 2018. Preparation of novel high performance recoverable and natural sunlight-driven nanocomposite photocatalyst of Fe_3O_4/C/TiO_2/N-CQDs. Mater. Sci. Semicond. Process 87: 142-154. DOI: {doi.org/10.1016/j.mssp.2018.07.018}

Bacon, M., Bradley, S.J. and Nann, T. 2014. Graphene quantum dots. Part. Part. Syst. Charact. 31: 415-428. DOI: {doi.org/10.1002/ppsc.201300252}

Baker, S.N. and Baker, G.A. 2010. Luminescent carbon nanodots: emergent nanolights. Angew. Chem. Int. Ed. 49: 6726-6744. DOI: {doi.org/10.1002/anie.200906623}

Bao, L., Zhang, Z.L., Tian, Z.Q., Zhang, L., Liu, C., Lin, Y., Qi, B. and Pang, D.W. 2011.

Electrochemical tuning of luminescent carbon nanodots: from preparation to luminescence mechanism. Adv. Mater. 23: 5801-5806. DOI: {doi.org/10.1002/adma.201102866}

Barber, J. and Tran, P.D. 2013. From natural to artificial photosynthesis. J. R. Soc. Interface 10: 20120984. DOI: {doi.org/10.1098/rsif.2012.0984}

Bottini, M., Balasubramanian, C., Dawson, M.I., Bergamaschi, A., Bellucci, S. and Mustelin, T. 2006. Isolation and characterization of fluorescent nanoparticles from pristine and oxidized electric arc-produced single-walled carbon nanotubes. J. Phys. Chem. B 110: 831-836. DOI: {doi.org/10.1021/jp055503b}

Bruchez, M., Moronne, M., Gin, P., Weiss, S. and Alivisatos, A.P. 1998. Semiconductor nanocrystals as fluorescent biological labels. Science 281: 2013-2016. DOI: {10.1126/science.281.5385.2013}

Cao, L., Wang, X., Meziani, M.J., Lu, F., Wang, H., Luo, P.G., Lin, Y., Harruff, B.A., Veca, L.M., Murray, D. and Xie, S.Y. 2007. Carbon dots for multiphoton bioimaging. J. Am. Chem. Soc. 129: 11318-11319. DOI: {doi.org/10.1021/ja073527l}

Chan, W.C. and Nie, S. 1998. Quantum dot bioconjugates for ultrasensitive nonisotopic detection. Science 281: 2016-2018. DOI: {10.1126/science.281.5385.2016}

Chibli, H., Carlini, L., Park, S., Dimitrijevic, N.M. and Nadeau, J.L. 2011. Cytotoxicity of InP/ZnS quantum dots related to reactive oxygen species generation. Nanoscale 3: 2552-2559. DOI: {doi.org/10.1039/C1NR10131E}

Cong, S. and Zhao, Z. 2018. Carbon Quantum Dots: A Component of Efficient Visible Light Photocatalysts. InTech. DOI: {dx.doi.org/10.5772/interchopen.70801}

Deng, Y., Zhao, D., Chen, X., Wang, F., Song, H. and Shen, D. 2013. Long lifetime pure organic phosphorescence based on water soluble carbon dots. Chem. Comm. 49: 5751-5753. DOI: {doi.org/10.1039/C3CC42600A}

Devi, N.R., Kumar, T.V. and Sundramoorthy, A.K. 2018. Electrochemically exfoliated carbon quantum dots modified electrodes for detection of dopamine neurotransmitter. J. Electrochem. Soc. 165: G3112. DOI: {doi.org/10.1149/2.0191812jes}

Di, J., Xia, J., Ji, M., Li, H., Xu, H., Li, H. and Chen, R. 2015. The synergistic role of carbon quantum dots for the improved photocatalytic performance of Bi_2MoO_6. Nanoscale 7: 11433-11443. DOI: {doi.org/10.1039/C5NR01350J}

Fang, S., Sun, Z. and Hu, Y.H. 2019. Insights into the thermo-photo catalytic production of hydrogen from water on a low-cost NiO x-Loaded TiO_2 Catalyst. ACS Catal. 9: 5047-5056. DOI: {doi.org/10.1021/acscatal.9b01110}

Georgakilas, V., Perman, J.A., Tucek, J. and Zboril, R. 2015. Broad family of carbon nanoallotropes: classification, chemistry, and applications of fullerenes, carbon dots, nanotubes, graphene, nanodiamonds, and combined superstructures. Chem. Rev. 115: 4744-4822. DOI: {doi.org/10.1021/cr500304f}

Gholipour, M.R., Béland, F. and Do, T.O. 2017. Post-calcined carbon nitride nanosheets as an efficient photocatalyst for hydrogen production under visible light irradiation. ACS Sustain. Chem. Eng. 5: 213-220. DOI: {doi.org/10.1021/acssuschemeng.6b01282}

Gogoi, D., Koyani, R., Golder, A.K. and Peela, N.R. 2020. Enhanced photocatalytic hydrogen evolution using green carbon quantum dots modified 1-D CdS nanowires under visible light irradiation. Sol. Energy 208: 966-977. DOI: {doi.org/10.1016/j.solener.2020.08.061}

Grün, M., Büchel, G., Kumar, D., Schumacher, K., Bidlingmaier, B. and Unger, K.K. 2000. Rational design, tailored synthesis and characterisation of ordered mesoporous silicas in the micron and submicron size range. Stud. Surf. Sci. Catal. 128: 155-165. DOI: {doi.org/10.1016/S0167-2991(00)80019-1}

Guleria, A., Sharma, R. and Shandilya, P. 2021. Photocatalytic and adsorptional removal of heavy metals from contaminated water using nanohybrids. Photocatalysis: Advanced Materials and Reaction Engineering 100: 113-160. DOI: {doi.org/10.21741/9781644901359-4}

Guo, X., Wang, C.F., Yu, Z.Y., Chen, L. and Chen, S. 2012. Facile access to versatile fluorescent carbon dots toward light-emitting diodes. Chem. Comm. 48: 2692-2694. DOI: {doi.org/10.1039/C2CC17769B}

Hada, Y. and Eto, M. 2003. Electronic states in silicon quantum dots: Multivalley artificial atoms. Phys. Rev. B Condens. Matter 68: 155322. DOI: {doi.org/10.1103/PhysRevB.68.155322}

Hinrichsen, D. and Tacio, H. 2002. The coming freshwater crisis is already here. The linkages between population and water. Washington, DC: Woodrow Wilson International Center for Scholars 1-26.

Hu, S. 2016. Tuning optical properties and photocatalytic activities of carbon-based "quantum dots" through their surface groups. Chem. Rec. 16: 219-230. DOI: {doi.org/10.1111/tcr.201500225}

Hu, S., Tian, R., Dong, Y., Yang, J., Liu, J. and Chang, Q. 2013. Modulation and effects of surface groups on PL and photocatalytic activity of carbon dots. Nanoscale 5: 11665-11671. DOI: {doi.org/10.1039/C3NR03893A}

Huang, J., Li, L., Chen, J., Ma, F. and Yu, Y. 2020. Broad spectrum response flower spherical-like composites CQDs@ CdIn$_2$S$_4$/CdS modified by CQDs with up-conversion property for photocatalytic degradation and water splitting. Int. J. Hydrogen Energy 45: 1822-1836. DOI: {doi.org/10.1016/j.ijhydene.2019.11.078}

Ilaiyaraja, N., Fathima, S.J. and Khanum, F. 2018. Quantum dots: a novel fluorescent probe for bioimaging and drug delivery applications. Nanomedicines: 529-563. DOI: {doi.org/10.1016/B978-0-12-813661-4.00012-2}

Jia, X., Li, J. and Wang, E. 2012. One-pot green synthesis of optically pH-sensitive carbon dots with upconversion luminescence. Nanoscale 4: 5572-5575. DOI: {doi.org/10.1039/C2NR31319G}

Jing, L., Yuan, F., Hou, H., Xin, B., Cai, W. and Fu, H. 2005. Relationships of surface oxygen vacancies with PL and photocatalytic performance of ZnO nanoparticles. Sci. China, Ser. B: Chem. 48: 25-30. DOI: {doi.org/10.1007/BF02990909}

Journet, C., Maser, W.K., Bernier, P., Loiseau, A., de La Chapelle, M.L., Lefrant, D.S., Deniard, P., Lee, R. and Fischer, J.E. 1997. Large-scale production of single-walled carbon nanotubes by the electric-arc technique. Nature 388: 756-758. DOI: {doi.org/10.1038/41972}

Ke, J., Li, X., Zhao, Q., Liu, B., Liu, S. and Wang, S. 2017. Upconversion carbon quantum dots as visible light responsive component for efficient enhancement of photocatalytic performance. J. Colloid Interface Sci. 496: 425-433. DOI: {doi.org/10.1016/j.jcis.2017.01.121}

Kong, X.Y., Tan, W.L., Ng, B.J., Chai, S.P. and Mohamed, A.R. 2017. Harnessing Vis–NIR broad spectrum for photocatalytic CO$_2$ reduction over carbon quantum dots-decorated ultrathin Bi$_2$WO$_6$ nanosheets. Nano Res. 10: 1720-1731. DOI: {doi.org/10.1007/s12274-017-1435-4}

Kulandaivalu, T., Rashid, S.A., Sabli, N. and Tan, T.L. 2019. Visible light assisted photocatalytic reduction of CO$_2$ to ethane using CQDs/Cu$_2$O nanocomposite photocatalyst. Diam Relat Mater 91: 64-73. DOI: {doi.org/10.1016/j.diamond.2018.11.002}

Kumar, A., Prajapati, P.K., Pal, U. and Jain, S.L. 2018. Ternary rGO/InVO$_4$/Fe$_2$O$_3$ Z-scheme heterostructured photocatalyst for CO$_2$ reduction under visible light irradiation. ACS Sustain. Chem. Eng. 6: 8201-8211. DOI: {doi.org/10.1021/acssuschemeng.7b04872}

Kwon, W., Lee, G., Do, S., Joo, T. and Rhee, S.W. 2014. Size-controlled soft-template synthesis of carbon nanodots toward versatile photoactive materials. Small 10: 506-513. DOI: {doi.org/10.1002/smll.201301770}

Li, H., Liu, R., Lian, S., Liu, Y., Huang, H. and Kang, Z. 2013. Near-infrared light controlled photocatalytic activity of carbon quantum dots for highly selective oxidation reaction. Nanoscale 5: 3289-3297. DOI: {doi.org/10.1039/C3NR00092C}

Li, H., Liu, R., Liu, Y., Huang, H., Yu, H., Ming, H., Lian, S., Lee, S.T. and Kang, Z. 2012. Carbon quantum dots/Cu_2O composites with protruding nanostructures and their highly efficient (near) infrared photocatalytic behaviour. J. Mater. Chem. 22: 17470-17475. DOI: {doi.org/10.1039/C2JM32827E}

Li, H., He, X., Liu, Y., Huang, H., Lian, S., Lee, S.T. and Kang, Z. 2011. One-step ultrasonic synthesis of water-soluble carbon nanoparticles with excellent photoluminescent properties. Carbon 49: 605-609. DOI: {doi.org/10.1016/j.carbon.2010.10.004}

Li, H., He, X., Kang, Z., Huang, H., Liu, Y., Liu, J., Lian, S., Tsang, C.H.A., Yang, X. and Lee, S.T. 2010. Water-soluble fluorescent carbon quantum dots and photocatalyst design. Angew. Chemie 122: 4532-4536. DOI: {doi.org/10.1002/ange.200906154}

Li, H., Deng, Y., Liu, Y., Zeng, X., Wiley, D. and Huang, J. 2019. Carbon quantum dots and carbon layer double protected cuprous oxide for efficient visible light CO_2 reduction. Chem. Comm 55: 4419-4422. DOI: {doi.org/10.1039/C9CC00830F}

Li, K., Peng, B. and Peng, T. 2016. Recent advances in heterogeneous photocatalytic CO_2 conversion to solar fuels. ACS Catal. 6: 7485-7527. DOI: {doi.org/10.1021/acscatal.6b02089}

Li, K., Su, F.Y. and Zhang, W.D. 2016. Modification of g-C_3N_4 nanosheets by carbon quantum dots for highly efficient photocatalytic generation of hydrogen. Appl. Surf. Sci. 375: 110-117. DOI: {doi.org/10.1016/j.apsusc.2016.03.025}

Li, X., Chang, J., Xu, F., Wang, X., Lang, Y., Gao, Z., Wu, D. and Jiang, K. 2015. Pyrolytic synthesis of carbon quantum dots, and their PL properties. Res. Chem. Intermed. 41: 813-819. DOI: {doi.org/10.1007/s11164-013-1233-x}

Li, X., Ma, S., Qian, H., Zhang, Y., Zuo, S. and Yao, C. 2019. Upconversion nanocomposite CeO_2: Tm^{3+}/attapulgite intermediated by carbon quantum dots for photocatalytic desulfurization. Powder Technol. 351: 38-45. DOI: {doi.org/10.1016/j.powtec.2019.04.004}

Li, X.Z., Li, F.B., Yang, C.L. and Ge, W.K. 2001. Photocatalytic activity of WOx-TiO_2 under visible light irradiation. J. Photochem. Photobiol. A 141: 209-217. DOI: {doi.org/10.1016/S1010-6030(01)00446-4}

Lim, S.Y., Shen, W. and Gao, Z. 2015. Carbon quantum dots and their applications. Chem. Soc. Rev. 44: 362-381. DOI: {doi.org/10.1039/C4CS00269E}

Lin, L., Xu, Y., Zhang, S., Ross, I.M., Ong, A.C. and Allwood, D.A. 2013. Fabrication of luminescent monolayered tungsten dichalcogenides quantum dots with giant spin-valley coupling. ACS Nano 7: 8214-8223. DOI: {doi.org/10.1021/nn403682r}

Lingampalli, S.R., Ayyub, M.M. and Rao, C.N.R. 2017. Recent progress in the photocatalytic reduction of carbon dioxide. ACS Omega 2: 2740–2748. DOI: {doi.org/10.1021/acsomega.7b00721}

Liu, F., Jang, M.H., Ha, H.D., Kim, J.H., Cho, Y.H. and Seo, T.S. 2013. Facile synthetic method for pristine graphene quantum dots and graphene oxide quantum dots: origin of blue and green luminescence. Adv. Mater. 25: 3657-3662. DOI: {doi.org/10.1002/adma.201300233}

Liu, M., Wang, R., Liu, B., Guo, F. and Tian, L. 2019. Carbon quantum dots@ Pd-SnS_2 nanocomposite: the role of CQDs@Pd nanoclusters in enhancing photocatalytic reduction of aromatic nitro compounds. J. Colloid Interface Sci. 555: 423-430. DOI: {doi.org/10.1016/j.jcis.2019.08.002}

Liu, R., Huang, H., Li, H., Liu, Y., Zhong, J., Li, Y., Zhang, S. and Kang, Z. 2014. Metal nanoparticle/carbon quantum dot composite as a photocatalyst for high-efficiency cyclohexane oxidation. ACS Catal. 4: 328-336. DOI: {doi.org/10.1021/cs400913h}

Ma, Y., Cen, Y., Sohail, M., Xu, G., Wei, F., Shi, M., Xu, X., Song, Y., Ma, Y. and Hu, Q. 2017. A ratiometric fluorescence universal platform based on N, Cu codoped carbon dots to detect metabolites participating in H_2O_2-generation reactions. ACS Appl. Mater. Interfaces 9: 33011-33019. DOI: {doi.org/10.1021/acsami.7b10548}

Medintz, I.L., Uyeda, H.T., Goldman, E.R. and Mattoussi, H. 2005. Quantum dot bioconjugates for imaging, labelling and sensing. Nat. Mater 4: 435-446. DOI: {doi.org/10.1038/nmat1390}

Mehta, A., Pooja, D., Thakur, A. and Basu, S. 2017. Enhanced photocatalytic water splitting by gold carbon dot core shell nanocatalyst under visible/sunlight. New J. Chem. 41: 4573-4581. DOI: {doi.org/10.1039/C7NJ00933J}

Michalet, X., Pinaud, F.F., Bentolila, L.A., Tsay, J.M., Doose, S.J.J.L., Li, J.J., Sundaresan, G., Wu, A.M., Gambhir, S.S. and Weiss, S. 2005. Quantum dots for live cells, in vivo imaging, and diagnostics. Science 307: 538-544. DOI: {10.1126/science.1104274}

Monga, D., Ilager, D., Shetti, N.P., Basu, S. and Aminabhavi, T.M. 2020. 2D/2D heterojunction of MoS_2/g-C_3N_4 nanoflowers for enhanced visible-light-driven photocatalytic and electrochemical degradation of organic pollutants. J. Environ. Manage. 274: 111208. DOI: {doi.org/10.1016/j.jenvman.2020.111208}

Namdari, P., Negahdari, B. and Eatemadi, A. 2017. Synthesis, properties and biomedical applications of carbon-based quantum dots: an updated review. Biomed. Pharmacother. 87: 209-222. DOI: {doi.org/10.1016/j.biopha.2016.12.108}

Ni, M., Leung, M.K., Leung, D.Y. and Sumathy, K. 2007. A review and recent developments in photocatalytic water-splitting using TiO_2 for hydrogen production. Renew. Sust. Energ. Rev. 11: 401-425. DOI: {doi.org/10.1016/j.rser.2005.01.009}

Ong, W.J., Putri, L.K., Tan, Y.C., Tan, L.L., Li, N., Ng, Y.H., Wen, X. and Chai, S.P. 2017. Unravelling charge carrier dynamics in protonated g-C_3N_4 interfaced with carbon nanodots as co-catalysts toward enhanced photocatalytic CO_2 reduction: a combined experimental and first-principles DFT study. Nano Res. 10: 1673-1696. DOI: {doi.org/10.1007/s12274-016-1391-4}

Pan, J., Liu, J., Zuo, S., Khan, U.A., Yu, Y. and Li, B. 2018. Structure of Z-scheme CdS/CQDs/BiOCl heterojunction with enhanced photocatalytic activity for environmental pollutant elimination. Appl. Surf. Sci. 444: 177-186. DOI: {doi.org/10.1016/j.apsusc.2018.01.189}

Park, S.Y., Lee, H.U., Park, E.S., Lee, S.C., Lee, J.W., Jeong, S.W., Kim, C.H., Lee, Y.C., Huh, Y.S. and Lee, J. 2014. Photoluminescent green carbon nanodots from food-waste-derived sources: large-scale synthesis, properties, and biomedical applications. ACS Appl. Mater. Interfaces 6: 3365-3370. DOI: {doi.org/10.1021/am500159p}

Peng, J., Gao, W., Gupta, B.K., Liu, Z., Romero-Aburto, R., Ge, L., Song, L., Alemany, L.B., Zhan, X., Gao, G. and Vithayathil, S.A. 2012. Graphene quantum dots derived from carbon fibers. Nano Lett. 12: 844-849. DOI: {doi.org/10.1021/nl2038979}

Phang, S.J. and Tan, L.L. 2019. Recent advances in carbon quantum dot (CQDs)-based two dimensional materials for photocatalytic applications. Catal. Sci. Technol. 9: 5882-5905. DOI: {doi.org/10.1039/C9CY01452G}

Qin, J. and Zeng, H. 2017. Photocatalysts fabricated by depositing plasmonic Ag nanoparticles on carbon quantum dots/graphitic carbon nitride for broad spectrum photocatalytic hydrogen generation. Appl. Catal. B: Environ. 209: 161-173. DOI: {doi.org/10.1016/j.apcatb.2017.03.005}

Qu, S., Wang, X., Lu, Q., Liu, X. and Wang, L. 2012. A biocompatible fluorescent ink based on water-soluble luminescent carbon nanodots. Angew. Chemie 124: 12381-12384. DOI: {doi.org/10.1002/ange.201206791}

Qu, X., Yi, Y., Qiao, F., Liu, M., Wang, X., Yang, R., Meng, H., Shi, L. and Du, F. 2018. TiO_2/BiOI/CQDs: enhanced photocatalytic properties under visible-light irradiation. Ceram. Int. 44: 1348-1355. DOI: {doi.org/10.1016/j.ceramint.2017.08.185}

Rahbar, M., Mehrzad, M., Behpour, M., Mohammadi-Aghdam, S. and Ashrafi, M. 2019. S,

N co-doped carbon quantum dots/TiO_2 nanocomposite as highly efficient visible light photocatalyst. Nanotechnology 30: 505702. DOI: {doi.org/10.1088/1361-6528/ab40dc}

Rao, X., Yuan, M., Jiang, H., Li, L. and Liu, Z. 2019. A universal strategy to obtain chiroptical carbon quantum dots through the optically active surface passivation procedure. New J. Chem. 43: 13735-13740. DOI: {doi.org/10.1039/C9NJ03434J}

Ratnayake, S.P., Mantilaka, M.M.M.G.P.G., Sandaruwan, C., Dahanayake, D., Murugan, E., Kumar, S., Amaratunga, G.A.J. and de Silva, K.N. 2019. Carbon quantum dots-decorated nano-zirconia: a highly efficient photocatalyst. Applied Catalysis A: GEN. 570: 23-30. DOI: {doi.org/10.1016/j.apcata.2018.10.022}

Reddy, C.V., Reddy, K.R., Shetti, N.P., Shim, J., Aminabhavi, T.M. and Dionysiou, D.D. 2020. Hetero-nanostructured metal oxide-based hybrid photocatalysts for enhanced photoelectrochemical water splitting – A review. Int. J. Hydrog. Energy 45: 18331-18347. DOI: {doi.org/10.1016/j.ijhydene.2019.02.109}

Resch-Genger, U., Grabolle, M., Cavaliere-Jaricot, S., Nitschke, R. and Nann, T. 2008. Quantum dots versus organic dyes as fluorescent labels. Nat. Methods 5: 763. DOI: {doi.org/10.1038/nmeth.1248}

Rogach, A.L., Eychmüller, A., Hickey, S.G. and Kershaw, S.V. 2007. Infrared-emitting colloidal nanocrystals: synthesis, assembly, spectroscopy, and applications. Small 3: 536-557. DOI: {doi.org/10.1002/smll.200600625}

Sahu, S., Liu, Y., Wang, P., Bunker, C.E., Fernando, K.S., Lewis, W.K., Guliants, E.A., Yang, F., Wang, J. and Sun, Y.P. 2014. Visible-light photoconversion of carbon dioxide into organic acids in an aqueous solution of carbon dots. Langmuir 30: 8631-8636. DOI: {doi.org/10.1021/la5010209}

Saravanan, K.A., Prabu, N., Sasidharan, M. and Maduraiveeran, G. 2019. Nitrogen-self doped activated carbon nanosheets derived from peanut shells for enhanced hydrogen evolution reaction. Appl. Surf. Sci. 489: 725-733. DOI: {doi.org/10.1016/j.apsusc.2019.06.040}

Sarkar, S., Banerjee, D., Ghorai, U.K. and Chattopadhyay, K.K. 2016. Hydrothermal synthesis of carbon quantum dots and study of its photoluminecence property. International Conference on Microelectronics, Computing and Communications: 1-3. DOI: {doi.org/10.1109/MicroCom.2016.7522521}

Serpone, N. 2006. Is the band gap of pristine TiO_2 narrowed by anion- and cation-doping of titanium dioxide in year-generation photocatalysts? J. Phys. Chem. B 110: 24287-24293. DOI: {doi.org/10.1021/jp065659r}

Shandilya, P., Mittal, D., Soni, M., Raizada, P., Lim, J.H., Jeong, D.Y., Dewedi, R.P., Saini, A.K. and Singh, P. 2018. Islanding of $EuVO_4$ on high-dispersed fluorine doped few layered graphene sheets for efficient photocatalytic mineralization of phenolic compounds and bacterial disinfection. J. Taiwan Inst. Chem. Eng. 93: 528-542. DOI: {https://doi.org/10.1016/j.jtice.2018.08.034}

Shandilya, P., Sudhaik, A., Raizada, P., Hosseini-Bandegharaei, A., Singh, P., Rahmani-Sani, A., Thakur, V. and Saini, A.K. 2020a. Synthesis of Eu^{3+}-doped ZnO/Bi_2O_3 heterojunction photocatalyst on graphene oxide sheets for visible light-assisted degradation of 2,4-dimethyl phenol and bacteria killing. Solid State Sciences, 102: 106164. DOI: {https://doi.org/10.1016/j.solidstatesciences.2020.106164}

Shandilya, P., Raizada, P. and Singh, P. 2020b. Photocatalytic degradation of Azo dyes in water. *In:* Water Pollution and Remediation: Photocatalysis 119-146. DOI: {doi.org/10.1007/978-3-030-54723-3_4}

Shandilya, P., Raizada, P., Sudhaik, A., Saini, A., Saini, R. and Singh, P. 2020c. Metal and carbon quantum dot photocatalysts for water purification. *In:* Water Pollution and Remediation: Photocatalysis 81-118. DOI: {doi.org/10.1007/978-3-030-54723-3_3}

Shandilya, P., Sambyal, S., Sharma, R., Kumar, A. and Dai-Viet, N.V. 2021. Recent advancement on ferrite based heterojunction for photocatalytic degradation of organic pollutants: a review. Materials Research Foundations 112: 121-161. DOI: {https://doi. org/10.21741/9781644901595-3}

Shandilya, P., Mandyal, P., Kumar, V. and Sillanpää, M. 2022a. Properties, synthesis, and recent advancement in photocatalytic applications of graphdiyne: a review. Sep. Purif. Technol. 281: 119825. DOI: {https://doi.org/10.1016/j.seppur.2021.119825}

Shandilya, P., Sambyal, S., Sharma, R., Mandyal, P. and Fang, B. 2022b. Properties, optimized morphologies, and advanced strategies for photocatalytic applications of WO3 based photocatalysts. J. Hazard. Mater. 128218. DOI: {https://doi.org/10.1016/j. jhazmat.2022.128218}

Sharma, R., Arizaga, G.G.C., Saini, A.K. and Shandilya, P. 2021. Layered double hydroxide as multifunctional materials for environmental remediation: from chemical pollutants to microorganisms. Sustain Mater. and Tech. 29: e00319. DOI: {https://doi.org/10.1016/j. susmat.2021.e00319}

Sharma, V., Getahun, T., Verma, M., Villa, A. and Gupta, N. 2020. Carbon based catalysts for the hydrodeoxygenation of lignin and related molecules: A powerful tool for the generation of non-petroleum chemical products including hydrocarbons. Renew. Sust. Energ. Rev. 133: 110280. DOI: {doi.org/10.1016/j.rser.2020.110280}

Shen, T., Wang, Q., Guo, Z., Kuang, J. and Cao, W. 2018. Hydrothermal synthesis of carbon quantum dots using different precursors and their combination with TiO_2 for enhanced photocatalytic activity. Ceram. Int. 44: 11828-11834. DOI: {doi.org/10.1016/j. ceramint.2018.03.271}

Sheng, Y., Yi, D., Qingsong, H., Ting, W., Ming, L., Yong, C., Yifan, S., Jun, D., Bi, W., Xia, J. and Huaming, L. 2019. CQDs modified $PbBiO_2Cl$ nanosheets with improved molecular oxygen activation ability for photodegradation of organic contaminants. J. Photochem. Photobiol. A 382: 111921. DOI: {doi.org/10.1016/j.jphotochem.2019.111921}

Singh, A., Mohapatra, P.K., Kalyanasundaram, D. and Kumar, S. 2019. Self-functionalized ultrastable water suspension of luminescent carbon quantum dots. Mater. Chem. Phys. 225: 23-27. DOI: {doi.org/10.1016/j.matchemphys.2018.12.031}

Sun, Y.P., Zhou, B., Lin, Y., Wang, W., Fernando, K.S., Pathak, P., Meziani, M.J., Harruff, B.A., Wang, X., Wang, H. and Luo, P.G. 2006. Quantum-sized carbon dots for bright and colorful PL. J. Am. Chem. Soc. 128: 7756-7757. DOI: {doi.org/10.1021/ja062677d}

Sun, Y.P., Wang, X., Lu, F., Cao, L., Meziani, M.J., Luo, P.G., Gu, L. and Veca, L.M. 2008. Doped carbon nanoparticles as a new platform for highly photoluminescent dots. J. Phys. Chem. C. 112: 18295-18298. DOI: {doi.org/10.1021/jp8076485}

Tang, L., Ji, R., Cao, X., Lin, J., Jiang, H., Li, X., Teng, K.S., Luk, C.M., Zeng, S., Hao, J. and Lau, S.P. 2012. Deep ultraviolet PL of water-soluble self-passivated graphene quantum dots. ACS Nano 6: 5102-5110. DOI: {doi.org/10.1021/nn300760g}

Thakur, A., Kumar, P., Kaur, D., Devunuri, N., Sinha, R.K. and Devi, P. 2020. TiO_2 nanofibres decorated with green-synthesized PAu/Ag@CQDs for the efficient photocatalytic degradation of organic dyes and pharmaceutical drugs. RSC Adv. 10: 8941-8948. DOI: {doi.org/10.1039/C9RA10804A}

Tlili, I. and Alkanhal, T.A. 2019. Nanotechnology for water purification: electrospun nanofibrous membrane in water and wastewater treatment. J. Water Reuse and Desalination 9: 232-248. DOI: {doi.org/10.2166/wrd.2019.057}

Virca, C.N., Winter, H.M., Goforth, A.M., Mackiewicz, M.R. and McCormick, T.M. 2017. Photocatalytic water reduction using a polymer coated carbon quantum dot sensitizer and a nickel nanoparticle catalyst. Nanotechnology 28: 195402. DOI: {doi.org/10.1088/1361-6528/aa6ae3}

Wang, R., Lu, K.Q., Zhang, F., Tang, Z.R. and Xu, Y.J. 2018. 3D carbon quantum dots/ graphene aerogel as a metal-free catalyst for enhanced photosensitization efficiency. Appl. Catal. B Environ. 233: 11-18. DOI: {doi.org/10.1016/j.apcatb.2018.03.108}

Wang, R., Lu, K.Q., Tang, Z.R. and Xu, Y.J. 2017. Recent progress in carbon quantum dots: synthesis, properties and applications in photocatalysis. J. Mater. Chem. A 5: 3717-3734. DOI: {doi.org/10.1039/C6TA08660H}

Wang, X., Li, L., Fu, Z. and Cui, F. 2018. Carbon quantum dots decorated CuS nanocomposite for effective degradation of methylene blue and antibacterial performance. J. Mol. Liq. 268: 578-586. DOI: {doi.org/10.1016/j.molliq.2018.07.086}

Wang, Y. and Hu, A. 2014. Carbon quantum dots: synthesis, properties and applications. J. Mater. Chem. 2: 6921-6939. DOI: {doi.org/10.1039/C4TC00988F}

Wang, Y., Chen, J., Liu, L., Xi, X., Li, Y., Geng, Z., Jiang, G. and Zhao, Z. 2019. Novel metal doped carbon quantum dots/CdS composites for efficient photocatalytic hydrogen evolution. Nanoscale 11: 1618-1625. DOI: {doi.org/10.1039/C8NR05807E}

Wei, Y., Zhu, Y. and Jiang, Y. 2019. Photocatalytic self-cleaning carbon nitride nanotube intercalated reduced graphene oxide membranes for enhanced water purification. Chem. Eng. J. 356: 915-925. DOI: {doi.org/10.1016/j.cej.2018.09.108}

White, C.W. (ed.). 2012. Laser and Electron Beam Processing of Materials. Elsevier.

Wood, D., Shaw, S., Cawte, T., Shanen, E. and Van Heyst, B. 2019. An overview of photocatalyst immobilization methods for air pollution remediation. Chem. Eng. J. 123490. DOI: {doi.org/10.1016/j.cej.2019.123490}

Xu, L., Yang, L., Bai, X., Du, X., Wang, Y. and Jin, P. 2019. Persulfate activation towards organic decomposition and Cr (VI) reduction achieved by a novel CQDs-TiO$_2$–x/rGO nanocomposite. Chem. Eng. Technol. 373: 238-250. DOI: {doi.org/10.1016/j.cej.2019.05.028}

Xu, M., Deng, G., Liu, S., Chen, S., Cui, D., Yang, L. and Wang, Q. 2010. Free cadmium ions released from CdTe-based nanoparticles and their cytotoxicity on Phaeodactylum tricornutum. Metallomics 2: 469-473. DOI: {doi.org/10.1039/c005387m}

Xu, X., Ray, R., Gu, Y., Ploehn, H.J., Gearheart, L., Raker, K. and Scrivens, W.A. 2004. Electrophoretic analysis and purification of fluorescent single-walled carbon nanotube fragments. J. Am. Chem. Soc. 126: 12736-12737. DOI: {doi.org/10.1021/ja040082h}

Xu, X., Bao, Z., Zhou, G., Zeng, H. and Hu, J. 2016. Enriching photoelectrons via three transition channels in amino-conjugated carbon quantum dots to boost photocatalytic hydrogen generation. ACS Appl. Mater. Interfaces 8: 14118-14124. DOI: {doi.org/10.1021/acsami.6b02961}

Xue, X.Y., Cheng, R., Shi, L., Ma, Z. and Zheng, X. 2017. Nanomaterials for water pollution monitoring and remediation. Environ. Chem. Lett. 15: 23-27. DOI: {doi.org/10.1007/s10311-016-0595-x}

Yang, Y., Wu, D., Han, S., Hu, P. and Liu, R. 2013. Bottom-up fabrication of photoluminescent carbon dots with uniform morphology via a soft–hard template approach. Chem. Comm. 49: 4920-4922. DOI: {doi.org/10.1039/C3CC38815H}

Yu, H., Zhao, Y., Zhou, C., Shang, L., Peng, Y., Cao, Y., Wu, L.Z., Tung, C.H. and Zhang, T. 2014. Carbon quantum dots/TiO$_2$ composites for efficient photocatalytic hydrogen evolution. J. Mater. Chem. A 2: 3344-3351. DOI: {doi.org/10.1039/C3TA14108J}

Yu, J.G., Yu, H.G., Cheng, B., Zhao, X.J., Yu, J.C. and Ho, W.K. 2003. The effect of calcination temperature on the surface microstructure and photocatalytic activity of TiO$_2$ thin films prepared by liquid phase deposition. J. Phys. Chem. B 107: 13871-13879. DOI: {doi.org/10.1021/jp036158y}

Zhang, H., Ming, H., Lian, S., Huang, H., Li, H., Zhang, L., Liu, Y., Kang, Z. and Lee, S.T. 2011. Fe$_2$O$_3$/carbon quantum dots complex photocatalysts and their enhanced photocatalytic activity under visible light. Dalton Trans 40: 10822-10825. DOI: {doi.org/10.1039/C1DT11147G}

Zhang, H., Huang, H., Ming, H., Li, H., Zhang, L., Liu, Y. and Kang, Z. 2012. Carbon quantum dots/Ag$_3$PO$_4$ complex photocatalysts with enhanced photocatalytic activity and stability under visible light. J. Mater. Chem. 22: 10501-10506. DOI: {doi.org/10.1039/C2JM30703K}

Zhang, P., Song, T., Wang, T. and Zeng, H. 2017. In-situ synthesis of Cu nanoparticles hybridized with carbon quantum dots as a broad spectrum photocatalyst for improvement of photocatalytic H_2 evolution. Appl. Catal. B: Environ. 206: 328-335. DOI: {doi.org/10.1016/j.apcatb.2017.01.051}

Zhou, J., Booker, C., Li, R., Zhou, X., Sham, T.K., Sun, X. and Ding, Z. 2007. An electrochemical avenue to blue luminescent nanocrystals from multiwalled carbon nanotubes (MWCNTs). J. Am. Chem. Soc. 129: 744-745. DOI: {doi.org/10.1021/ja0669070}

Zhuo, S., Shao, M. and Lee, S.T. 2012. Upconversion and downconversion fluorescent graphene quantum dots: ultrasonic preparation and photocatalysis. ACS Nano 6: 1059-1064. DOI: {doi.org/10.1021/nn2040395}

Zou, J.P., Wang, L.C., Luo, J., Nie, Y.C., Xing, Q.J., Luo, X.B., Du, H.M., Luo, S.L. and Suib, S.L. 2016. Synthesis and efficient visible light photocatalytic H_2 evolution of a metal-free g-C_3N_4/graphene quantum dots hybrid photocatalyst. Appl. Catal. B: Environ. 193: 103-109. DOI: {doi.org/10.1016/j.apcatb.2016.04.017}

Pentagraphene: Structure, Properties, and Electronic Device Applications

Khurshed A. Shah[1,2]*, M. Shunaid Parvaiz[1,3] and Mubashir Qayoom[2]

[1] Department of Physics, S.P. College, Cluster University, Srinagar, J&K - 190001, India
[2] Department of Nanotechnology, University of Kashmir, Srinagar, J&K - 190006, India
[3] Department of Physics, University of Kashmir, Srinagar, J&K - 190006, India

1. Introduction

Carbon presents itself in many forms, from diamond and graphite to graphene, nanotube, and fullerenes. As we know, hexagons are the primary building blocks of many of these materials, except for C20 fullerene, carbon structures that are made exclusively of pentagons are not known. It is known to all that many of the exotic properties of carbon are associated with their unique structures. Some fundamental questions that arise are whether it is possible to have materials made exclusively of carbon pentagons and if so, will they be stable and have unusual properties? Based on this curiosity, we present this chapter on pentagraphene, which is made up of only carbon pentagons and resembling the Cairo pentagonal tiling pattern. We try to understand its structure, stability, properties, tubes, and electronic device applications. The study shows that pentagraphene has superiority over graphene in terms of its negative Poisson's ratio, large bandgap, and ultrahigh mechanical strength.

Carbon is one of the most interesting elements, having a number of allotropes, like graphite, diamond, fullerene, carbon nanotube, C60, and graphene (Kroto et al. 1985). After the successful synthesis of graphene in the year 2014, significant research has been conducted in characterizing carbon-based nanomaterials (Shah et al. 2016, Shah et al. 2016, Shah et al. 2019, Zwanenburg et al. 2009). A large number of 2D carbon allotropes are being studied. Although some of the polymorphs, such as graphdiyne, are metastable compared with graphene, they have been successfully synthesized (Li et al. 2014). Some of the 2D allotropes of carbon are being researched for their interesting properties that are an improvement on graphenes, such as anisotropic Dirac cones, ferromagnetic nature, high catalytic behavior, and high superconductivity, due to the high density of states at the Fermi level (Malko

---

Corresponding author: drkhursheda@gmail.com

et al. 2012, Mina et al. 2013, Terrones at al. 2000). These results reveal that the novel properties of carbon allotropes are related to the topological arrangement of carbon atoms and thus highlight the importance of structure-property relationships (Xu et al. 2014).

Carbon nanostructures are based on two important structural motifs, namely pentagons and hexagons. The hexagon is the building block of zero-dimensional nanoflakes, nanotube, graphene, graphite, and metallic carbon phases (Bucknum et al. 1994, Zhang et al. 2013, Omachi et al. 2013). Carbon materials composed of pentagons are rarely found. Carbon pentagons are generally known as topological defects or geometrical frustrations as stated in the well-known 'isolated pentagon rule' (IPR) for fullerenes, where pentagons must be separated from each other by neighboring hexagons to reduce the steric stress (Deza et al. 2000, Tan et al. 2009). For example, C60 is made up of 12 pentagons separated by 20 hexagons resembling the shape of a soccer ball, which explains the IPR rule perfectly. In some examples, the existence of carbon pentagons is also accompanied by carbon heptagons, which are separated from one another (Deza et al. 2000). The synthesis of pentagon-based C20 cage has inspired many researchers and considerable research has been done to stabilize pentagon-based and non-IPR carbon materials of different dimensionalities (Prinzbach et al. 2000). Some non-IPR fullerenes have also been experimentally synthesized (Tan et al. 2009). During the growth of 2D carbon sheets, researchers have confirmed that a 'pentagon first' mechanism has been brought into play in order to transform sp carbon chains to sp2 carbon rings (Wang et al. 2011). Therefore, pursuing the idea of building a pentagraphene carbon sheet using fused pentagons is considered as a structural motivation. The dynamical, thermal, and mechanical stability of such a structure has been confirmed in a series of publications (Yagmurcukardes et al. 2015, Yu et al. 2015, Xu et al. 2015, Cranford et al. 2016, Lopez-Bezanilla et al. 2015, Stauber et al. 2016, Rajbanshi et al. 2016). Zhang and coworkers (Zhang et al. 2013) have reported the thermal and mechanical stability of pentagon-based carbon nanotubes (CNTs). Additionally, Avramov et al. (2015) have reported the binding energy per atom for (n, n) pentagraphene-based carbon nanotubes (CNTs), confirming that the (7,7) pentagraphene-based CNT is the most energetically stable structure.

Pentagraphene is an attractive material for optoelectronic applications because it is a semiconductor with a quasi-direct bandgap. Additionally, it has a reduced thermal conductivity as compared with graphene and has an auxetic material, which means it has a negative Poisson's ratio. Furthermore, many properties of pentagraphene can be tuned by doping and functionalization; therefore it will be an exciting material in the coming times. Even though pentagraphene has not been synthesized experimentally, the theoretical analysis points out many intriguing properties that are worth discussing in more detail. Most strikingly in terms of applications is the largely predicted bandgap of 3.25 eV, which makes it a potential candidate for blue absorption/emission.

Prediction of either the existence of a material or the properties of a material can be decided from 'density functional theory' (DFT) and 'molecular dynamics' (MD) based calculations. DFT is the most successful ab initio/first principle calculation, which is based on the quantum mechanical description of the material and governed by the laws of condensed matter physics (Hasnip et al. 2014, Hohenberg et al. 1964, Kohn et al. 1965). The underlying principle is that "the total energy of the system is a unique function of the electron density" (Hohenberg et al. 1964, Kohn et al. 1965). DFT calculations are based on the 'Kohn-Sham (KS)' method which is an independent particle method for many-body electron problems. The framework of DFT is based on computing the ground state energy and charge density of the many-body system of the electron by solving a set of Schrödinger-like equations, representing a fictitious but simplified system of non-interacting electrons moving in an effective potential [11, 12]. Effective potential consists of the external potential, Coulomb potential, and Exchange correlation potential. The Exchange correlation potential is a measure of the quantum mechanical exchange and correlation of the particles (Ghosh et al. 2017). Exchange correlation potential plays a critical role in DFT calculation. The ground state energy of the material is identified from the DFT calculation. Molecular Dynamics (MD) simulation calculates the dynamics for systems for which the interatomic forces are derived from DFT-based calculations.

The possibility of the existence of pentagraphene and its properties were predicted (Kresse et al. 1996) with the help of DFT-based MD simulations performed using the Vienna Ab initio Simulation Package (VASP). Interactions between ion cores and valence electrons were treated with a projector augmented wave (PAW) basis set (Bliichl et al. 1994). Proper choice of 'exchange-correlation' potential function is important for predicting material properties using DFT-based calculations. Zhang et al. (2015) used generalized gradient approximation (GGA) with Perdew-Burke-Ernzerhof exchange-correlation potential for most of the calculations. Since electronic band structure calculations demand high precision, Zhang et al. (2015) used hybrid Heyd-Scuseria-Ernzerhof exchange-correlation potential (Heyd et al. 2006) only for calculation of electronic band structure of pentagraphene.

Therefore, keeping in view its extraordinary properties, like dynamical and mechanical stability, high-temperature stability, unique atomic configuration, negative Poisson's ratio, large bandgap, and high strength, the pentagraphene proves to be an exciting candidate for novel devices and applications. In this chapter, we first discuss the lattice structure of pentagraphene. Then we introduce the stability of the pentagonal structure and present its various attributes. After that, we discuss the mechanical, electrical, and optical properties of pentagraphene in detail followed by various futuristic device applications.

2. Pentagraphene Lattice Structure

Although proposed theoretically by Zhang et al. in the year 2014 using single layer exfoliation from T12-carbon, pentagraphene (PG) was recently predicted to exist from DFT-based calculations (Zhang et al. 2015). PG with six carbon atoms per unit cell displays a tetragonal lattice structure with lattice parameters optimized as a = b = 3.64 Å. PG carbon atoms exhibit hybridization of two types, viz. sp3 and sp2, which are distinguished customarily as C1 and C2, respectively (Zhang et al. 2015). Figure 1 illustrates the optimized structure of pentagraphene in the top and side view. PG appears like a Cairo pentagonal tiling when viewed from the top. The bond (carbon-carbon) lengths in pentagraphene are 1.55 Å (C1–C2) and 1.34 Å (C2–C2), showing

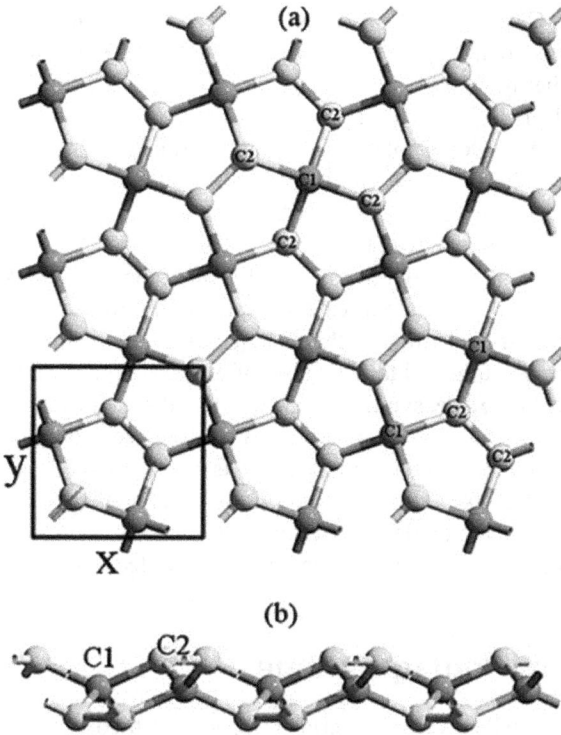

Figure 1: Schematic showing the (a) top view and (b) side view of
the pentagraphene structure.

pronounced single and double bond character, respectively. These bond lengths, however, differ from the bond (carbon-carbon) lengths in diamond and graphite/graphene, which are 1.54 and 1.42 Å, respectively. The bonding characteristics in PG, thus, are dissimilar to well-known diamond and graphite/graphene structures. In pentagraphene, the bond angle (hC2–C1–C2) is 134.2°, indicating the distorted sp3 C1 atoms character. As we know, sp3 carbon bonds exhibit a tetrahedral structure; PG is not truly planner in nature (Einollahzadeh et al. 2016). A buckling of 0.6 Å is detected upon viewing PG from the side. PG is, therefore, a two-dimensional sheet with 1.2 Å thickness. The structure of pentagraphene appears like a 'multi-decker sandwich', where the C1 atoms are sandwiched between the C2 (sp2 hybridized C) atoms. It is interesting to note that the structures of pentagraphene and experimentally identified layered silver azide AgN3 (Zhang et al. 2015) are similar.

The Cairo pentagonal patterning, in its regular form, is characterized by four bonds forming the pentagon, three having the distance 'a' and one has the distance 'b', where b = (3 − 1) a. Six atoms of carbon constitute the unit cell and is defined by two lattice vectors a1 and a2, where a1 = a (3, 3) and a2 = a (3, − 3). The quadratic unit cell length, a1 = a2 = 6a, is obtained from first-principle studies to be 3.64 Å, which calculates a = 1.49 Å and b = 1.09 Å.

The real distances, from first-principle calculations, turn out to be slightly different, i.e., C1–C2 = 1.55 Å and b = C2–C2 = 1.34 Å. There are two C1-atoms and four C2-atoms, denoting the carbon atoms with sp3 and sp2-hybridization, respectively. Furthermore, the arrangement of the atoms is in three horizontal planes, where the central plane is formed by the C1-atoms and two of the four C2-atoms form the upper and the lower planes. The total distance between the C1 and C2-atomic planes is h = 0.6 Å, which yields the 2D-distance as a = C1–C2 = 1.43 Å and b = C2–C2 = 1.34 Å. The bonds of C1 atoms, both horizontal and vertical, of length *b* connect the C2-atoms, whereas the bonds, of projected length *a*, connect the C1 with the C2-atoms. A slight difference of 1% between the two bonds connecting the C1–C2 atoms that form a pentagon exists but is negligible. Another assumption is that the C1–C2 distance is similar to the distance 'a' of the regular Cairo patterning. The square denotes the unit cell that consists of two C1-atoms and four C2-atoms.

Finally referring to the symmetry group, the three-dimensional (3D) lattice consists of the S4 and D2d as point group and full space group, respectively. The latter contains symmetry elements mentioned as one C2 axis perpendicular to the a1 − a2 plane, two C2′ axes along the direction perpendicular to C2, two improper S4 axes, and bisecting the angles formed by the C2′ axes and two dihedral planes. For strictly 2D lattice, the symmetry elements are doubled while going from D2d to D4h full space group.

3. Stability of Pentagraphene

The first question—whenever a material is proposed theoretically—that arises is concerning its stability. The study of stability encompasses ground state energy identification from the total energy calculation using DFT as well as the system behavior in other perturbations, like mechanical, thermal, etc. Ab initio MD simulation in the DFT framework is essential in studying the perturbation behavior of the system.

3.1 Mechanical Stability

As pentagraphene structure is simulated as a fixed supercell using the molecular dynamics approach, the assessment of the effect on structural stability by lattice distortion becomes necessary. The linear elastic constants of a stable crystal have to conform to the Born-Huang criteria in order to guarantee the positive-definiteness of strain energy following lattice distortion (Ding et al. 2013). To determine the pentagraphene mechanical stability, the energy change due to the in-plane strain has been calculated by the researchers. For a 2D sheet, using the standard Voigt notation, i.e., 1-*xx*, 2-*yy*, and 6-*xy*, the elastic strain energy per unit area can be stated as (Andrew et al. 2012),

$$U(\varepsilon) = \frac{1}{2} C_{11}\varepsilon_{xx}^2 + \frac{1}{2} C_{22}\varepsilon_{yy}^2 + C_{12}\varepsilon_{xx}\varepsilon_{yy} + 2C_{66}\varepsilon_{xy}^2 \tag{1}$$

where C_{11}, C_{22}, C_{12}, and C_{66} are components of the elastic modulus tensor, conforming to the second partial derivative of strain energy with respect to strain. By fitting the energy curves corresponding to uniaxial and equibiaxial strains, the elastic constants can be derived. For a mechanically stable 2D sheet (Ding et al. 2013), the elastic constants must satisfy $C_{11} C_{22} - C_{12} > 0$ and $C_{66} > 0$. As pentagraphene shows tetragonal symmetry, therefore $C_{11} = C_{22}$. Thus, in this case, the constants only need to satisfy $C_{11} > |C_{12}|$ and $C_{66} > 0$. The calculated elastic constants satisfy the uniaxial and equibiaxial strains with positive C_{66}, confirming that the pentagraphene sheet is mechanically stable.

3.2 Thermal Stability

The PG thermal stability is inspected by using canonical ensemble and carrying out ab initio MD simulations. To decrease the constraint of periodic boundary condition and explore probable structure reconstruction, the 2D sheet is simulated by (4 × 4), (6 × 6), and (4√2 × 4√2) R45° supercells, respectively. After heating at room temperature (300 K) for 6 ps with a time step of 1 fs, no structure reconstruction is found to have occurred in any of the cases. Moreover, the PG sheet can endure high temperatures (as high as 1,000 K), inferring that this 2D carbon phase, on the potential energy surface (PES) of elemental carbon, is separated from other local minima by high-energy barriers. The consequence of point defects or rim atoms on the stability of the PG sheet is also examined by introducing mono- and di-vacancies, Stone-Wales-like defect, ad-atoms, and edge atoms.

3.3 Energetic Stability

Total energy calculations are made to examine the thermodynamic stability of pentagraphene. Although this phase is metastable, in comparison to graphene and formerly reported 2D carbon allotropes (Li et al. 2014, Terrones et al. 2000, Deza et al. 2000) due to its violation of the IPR, it is more stable than some nanoporous carbon phases such as 3D T-carbon (Sheng et al. 2011), 2D α-graphyne (Autreto et al. 2015), and (3, 12)-carbon sheets. Pentagraphene is energetically superior over some experimentally recognized carbon nanostructures, such as the smallest fullerene, C20, and the smallest carbon nanotube, inferring that 2D pentagraphene sheets could be synthesized. Although pentagraphene shares the structural theme of fused pentagons with C20 cage, unlike the highly curved C20 cage where distorted sp2 hybridization is exhibited by all of the C atoms resulting in large strain energy, in pentagraphene the inception of sp3 hybridization lowers the fused carbon pentagons curvature and hence releases the strain partially.

3.4 Dynamic Stability

The researchers have also studied the lattice dynamics of pentagraphene by calculating the phonon dispersion. The presence of no imaginary modes in the whole Brillouin zone approves that pentagraphene is stable dynamically. Like graphene phonons (Marianetti et al. 2010, Si et al. 2012), three acoustic modes are also found in pentagraphene phonon spectrum. Linear dispersion occurs near the Γ point in in-plane longitudinal and in-plane transverse modes, while quadratic dispersion is found in out-of-plane (ZA) mode as q approaches 0. In the long-wavelength region, the quadratic ZA mode is closely linked with the lattice heat capacity and bending rigidity of the nanosheet. A significant phonon gap is observed in the phonon spectra. Detailed analysis of the atom-resolved phonon density of states (PhDOS) reveals that the double bonds amongst the C2 atoms (sp2 hybridized) are predominant in the dispersion-less high-frequency modes, which is quite alike to the previously reported sp2-sp3 hybrid carbon structure phonon modes (Bucknum et al. 1994, Zhang et al. 2013).

4. Properties of Pentagraphene

Validation of stability of the material appeals to curiosity vis-à-vis the properties of pentagraphene. Researches are in progress in studying the thermal, electronic, mechanical, optical, and magnetic properties of pentagraphene.

4.1 Mechanical Properties

The elastic strain per unit area in a 2D material is expressed in Equation 1 above. Young's modulus of a material is defined as (Lee et al. 2008),

$$E = (C_{11}^2 - C_{12}^2)/C_{11} \tag{2}$$

The value of Young's modulus in pentagraphene is found to be 263.8 GPa nm, whereas the corresponding value is 345 GPa nm in the case of graphene (Lee et al. 2008). The value of Young's modulus in pentagraphene is comparable to that of h-BN monolayer (Andrew et al. 2012). Another important parameter that articulates the mechanical property of a material is Poisson's ratio. It is defined as the negative ratio of the transverse strain to the corresponding axial strain. In the case of most of the solids, this ratio is positive since under the uniaxial compression solids expand in the transverse direction. Although the continuum mechanics theory does not discard the possibility of the existence of negative Poisson's ratio (NPR) in a stable linear elastic material, it is rarely found in nature. Very few artificial materials are characterized with NPR. Materials with NPR, known as 'auxetic materials', are of great scientific and technological interest (Jiang et al. 2014, Greaves et al. 2011). The major area of application of 'auxetic materials' is in the biomedical field as a material for structure/muscle/alignment anchors, surgical implant, prosthetic materials, dilator to open up blood vessels during heart surgery, etc. Another application field of auxetic materials is piezoelectric sensors and actuators. Auxetic foams and fibers find application in filters and woven structures. Interestingly, pentagraphene exhibits a negative Poisson's ratio viz., $v_{12} = v_{21} = \dfrac{C_{12}}{C_{11}} = -0.068$ since C_{12} is negative for PG (Zhang et al. 2015).

The NPR characteristics of PG have been testified comprehensively for both of the cases i.e., the lateral response in the x-direction when the tensile strain is applied in the y-direction and vice versa. Pentagraphene as a nano-auxetic material may have multiple applications in micro-mechanical and micro-electromechanical systems (MEMS). In addition to in-plane stiffness, ideal strength is also an important mechanical property of a nanomaterial. The strain at maximum stress is 21% beyond which the mechanical failure starts. However, the critical strain is less than 21%

for pentagraphene since phonon instability i.e., phonon softening induced by Kohn anomaly occurs before the stress reaches its maximum (Marianetti et al. 2010). It is found that in the case of pentagraphene, phonon softening arises when the equibiaxial strain reaches 17.2%, whereas the corresponding value is 14.8% for graphene. Thus, the ideal strength of PG is higher than that of graphene. At the critical strained condition, the single bond lengths between C1 and C2 atoms are found to be 1.77 Å, which is comparable with the experimentally and theoretically reported longest C–C bond length. Beyond critical strength, the structure fracture starts through the breaking of the bonds between C1 and C2 atoms.

4.2 Electrical Properties

From the computation of band structure using Scuseria-Ernzerhof potential functional, pentagraphene is found to be an indirect bandgap semiconductor with a bandgap of 3.25 eV (Zhang et al. 2015). As per the electronic band structure, the conduction band minimum lies on the M–C path, whereas the valence band maximum (VBM) is located on the C–X path. However, there exists a sub-VBM on the M–C path which has an energy value that is very close to that of true VBM (Zhang et al. 2015). Thus, pentagraphene can also be considered as a 'quasi-direct-bandgap' semiconductor. Analysis of partial DOS reveals that the major contribution of electronic states near the Fermi level is due to the sp2 hybridized C2 atoms. In pentagraphene, the pz orbitals of sp3-hybridized C2 atoms are spatially separated by the sp-hybridized C1 atoms. Thus, full electron delocalization is hindered, which gives rise to a finite bandgap. According to Zhang et al., there is a possibility to achieve superconductivity in pentagraphene through a hole doping since there are dispersionless, partially degenerate valance bands, which leads to a high total DOS near the Fermi level (Zhang et al. 2015, Savini et al. 2010).

4.3 Optical Properties

The DFT optical spectrum shows that there are three marked peaks in the low-energy region. The lowest peak (~2.45 eV) is due to transitions near the bandgap; it is dominated by contributions from a region around the Δ line, where the conduction and valence band states have an energy difference ~2.5 eV. The next peak (~3 eV) has contributions from the Σ line across the BZ, where many transitions are allowed due to the low symmetry. The last and higher peak in this energy window (~3.8 eV) can be related to the Δ line, where several allowed transitions due to the low symmetry are present in this line, similar to those from the Σ direction and the peak near 3 eV. Also, the Γ point does not have a major weight in the peak; its height is due to other low-symmetry points involving the local maxima and minima of the valence bands. Additional symmetry analysis gives another interesting property related to the optical response of PG. Due to the group of the wave vector at the high symmetry points X and M of the BZ, a selection rule is framed that forbids transitions from the valence to the conduction band at these two points. This is because of the different parities of the irreducible representations that are coupled by the momentum operator, giving a direct product that does not contain the invariant irreducible representation of group 36. Moving away from these points, the restriction is relaxed; the shoulder at ~4 eV has contributions of transitions from the Σ and Y lines around M and neighboring low-symmetry points.

4.4 Thermal Properties

Thermal conductivity is an important physical parameter that assesses the heat dissipation ability of material and the high-temperature application possibility. Joule heating is a major problem in electronic devices, which affects the performance, reliability, and lifetime of the device. Moreover, the lower heat capacity of nano-electronic devices is a critical issue. Hence, the device material should be thermally highly conductive so that the generated heat during device operation can dissipate fast. On the other hand, thermoelectric devices demand lower thermal conductivity. Although graphene is characterized by ultrahigh lattice thermal conductivity (3,000-5,800 W/mK), the absence of bandgap limits its application in most electronic devices, including the transistor. From the solution of linearized phonon 'Boltzman transport equation' (BTE) with first-principle calculation, lattice thermal conductivity of PG is found to be 645 W/mK at room temperature, which is significantly lower in comparison to that of graphene. The presence of more scattering channels in buckled pentagraphene structure may be the possible reason behind the significantly lower thermal conductivity of PG in comparison to that of graphene (Wang et al. 2016). It is found that the lattice thermal conductivity of stacked PG does not depend on the number of layers present. This characteristic of PG is also in sharp contrast to that of graphene. In metals, the thermal conductivity is mainly due to electrons, while in semiconductors and insulators, phonons play a significant role in determining the resultant thermal conductivity.

From MD simulations using the Tersoff interatomic potential, the intrinsic thermal conductivity of PG is found to be 167 W/mK (Xu et al, 2015). In PG, the acoustic phonon lifetimes are 20 ps and 1-10 ps for lower (1-5 THz) and higher (5-15 THz) frequency ranges, respectively, which are comparable to that of the graphene. However, the remarkably lower thermal conductivity of PG, in comparison to graphene, arises from the lower phonon group velocities and fewer collective phonon excitations in PG.

In pentagraphene, the phonon scattering rate is proportional to temperature and consequently, the thermal conductivity is inversely proportional to the temperature (Xu et al. 2015). From non-equilibrium MD simulations using the ReaxFF force field, the thermal conductivity of infinitely long pentagraphene is found to be 112.35 W/mK (Zhang et al. 2017). The cause behind the difference in results is due to the method of simulation and the type of interatomic force field employed. Researchers are still in search of appropriate potential for pentagraphene (Winczewski et al. 2018).

5. Pentagraphene Nanotubes

Graphene can be rolled into one-dimensional carbon nanotubes (CNTs), which was discovered by Iijima with experiment methods (Iijima 1991). A great deal of research focuses on the properties of CNTs due to their superior performance on mechanical, electrical, and thermal properties (Zhang et al. 2010, Wong et al. 1997,

Wong et al. 2010, Bao et al. 2004, Anandatheertha et al. 2010, Vodenitcharova et al. 2003, Giannopoulos et al. 2008). Similarly, pentagraphene can be also rolled into a pentagraphene nanotube (PGNT) as shown in Figure 2a. Due to the 2D bulking structure, PGNTs have rough inner and outer surfaces. Combined with its pentaring atomic configuration, PGNT might exhibit special electrical and mechanical properties, different from those of CNTs. Some pioneer works have been carried out to illustrate this issue. Zhang et al. (2015) have reported that PGNTs are semiconductor materials regardless of their chirality. Structural transitions of (4,4), (5,5), and (6,6) PGNTs have been found under the compressive strains, and increasing strain may result in the semiconductor-metal transitions (Wang et al. 2017).

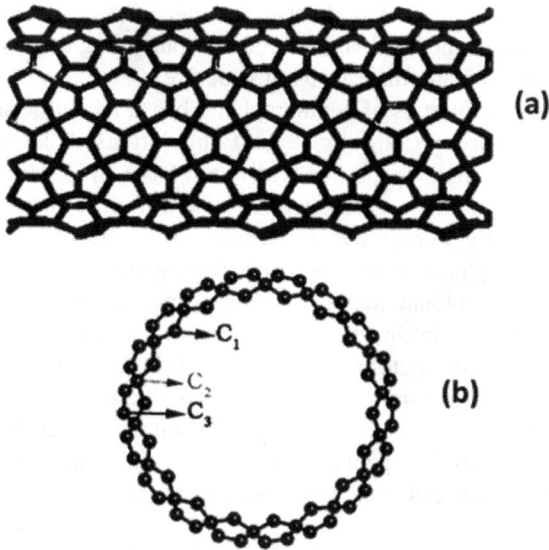

Figure 2: (a) Schematic of the pentagraphene nanotube (PGNT) structure. (b) Cross-section view of PGNT structure showing C1, C2, and C3 atoms.

A PGNT is generated by rolling up the 2D pentagraphene along the C_{nm} direction, thus the chiral vector (n, m) is defined to demonstrate the chirality of the tubes. The orientation of PGNT could be characterized by the equation (Wang et al. 2017),

$$C_{nm} = na_1 + ma_2 \tag{3}$$

where a_1 and a_2 are the unit vectors. The integers n and m denote the number of unit vectors a_1 and a_2 along with the two vector directions in the five-membered crystal lattice. For instance, when the integer $n = m = 9$, the model is named as (9, 9) PGNT. The PGNT models with DFT calculation have been proved stable when integer $n = m$, while the $(n, 0)$ models could not converge to the stable nanotubes. This situation has been certified by the previous investigation of Zhang et al. (2015).

Due to the bulking structure of the 2D pentagraphene, PGNT acts as a geometry structure with three layers i.e., the inner layer, the median layer, and the outer layer. Taking the particular case of (9,9) PGNT in Figure 2b, three kinds of carbon atoms can be distinguished and marked in the diagram, named C1, C2, and C3. The inner layer of PGNT is composed of sp2 hybridized C1 atoms; the median layer consists of sp3 hybridized C2 atoms; the outer layer consists of sp2 hybridized C3 atoms. For the pristine (9,9) PGNT without applying strain, its bond length of C1–C1, C1–C2, C2–C3, and C3–C3 is 1.318 Å, 1.503 Å, 1.555 Å, and 1.359 Å, respectively.

The bandgap of PGNTs ranges from 2.008 to 2.731 eV, thus all of the PGNTs act as direct bandgap semiconductors. As the diameter of the inner layer increases from 5.69 Å (for (4,4) PGNT) to 12.10 Å (for (8,8) PGNT), its bandgap fluctuates with the tube size. Then, the bandgap decreases rapidly with the diameter increase. When the inner layer size decreases to 20.22 Å (for (13,13) PGNT), the values change slowly. This might be due to a curvature effect. The small-size tubes like (4,4) PGNT have a highly curved performance. As the diameter increases, a lower curvature is constructed that reduces the bandgap. Therefore, controlling the diameter of PGNTs is a possible method to manipulate its band structure.

The mechanical properties of pentagraphene-based nanotube structure (PGNT) have been acquired through large-scale MD simulations. It is found that PGNT has comparable mechanical properties with that of the CNT. However, unlike the brittle CNT, PGNT exhibits large extensibility (with failure strain exceeding 60%) and behaves plastically during tensile deformation. The plasticity is inherently originated from the phase transition from pentagonal to polygonal (including trigonal, tetragonal, hexagonal, heptagonal, and octagonal) carbon rings. The plastic characteristic of PGNTs is intrinsic with tube-diameter and strain-rate independence. Additionally, within the ultimate temperature (T < 1100 K), tensile-deformed PGNTs manifest similar phase transitions with an approximate transition ratio from the pentagon to the hexagon. For a given PGNT, the total energy required to achieve the maximum transition ratio from pentagon to hexagon is a constant and independent of the contribution between tensile strain energy and kinetic energy.

6. Futuristic Electronic Device Applications

To understand the futuristic device applications of pentagraphene, recently we have modeled a saw-tooth pentagraphene nanoribbon (SPGNR) based on a two-probe system attached with two (4,0) zig-zag CNT electrodes using the Atomistic Toolkit (ATK) (Quantum ATK 2019) (version P-2019.03) software and its graphical interface virtual nano-lab, as shown in Figure 3 (Parvaiz et al. 2020).

CNT electrodes have been used in previous research work designed to study the electronic properties of various devices (Baughman et al. 2002). In order to understand the behavior of the PGNR-based system for comparative studies, we substitutionally doped the scattering region of the device with four atoms of boron and nitrogen, both homogeneously and heterogeneously. In all simulated models, both electrodes consist of 16 atoms each, and the central scattering consists of 120 atoms. The models used in the study were simulated using Huckel parameters with an electrode temperature of 300 K. The Density mesh cut-off was set to 55 Hartee. The set of k-points were chosen as 1, 1, 125 for appropriate sampling.

All of the structures were fully relaxed in order to make the residual forces on each atom smaller than 0.05 eV/A. In order to achieve the requisite results, we

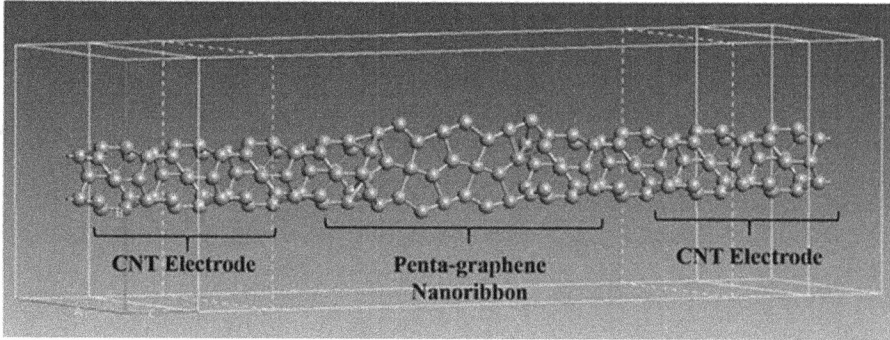

Figure 3: The computational model of SPGNR with CNT electrodes (the figure has been taken from our research article, reference 66).

analyzed the projected device density of states (PDDoS) and transmission spectra of all models and plotted the I-V curves and conductance curves under different bias voltages. Figure 4 shows the projected PDDoS of the proposed models.

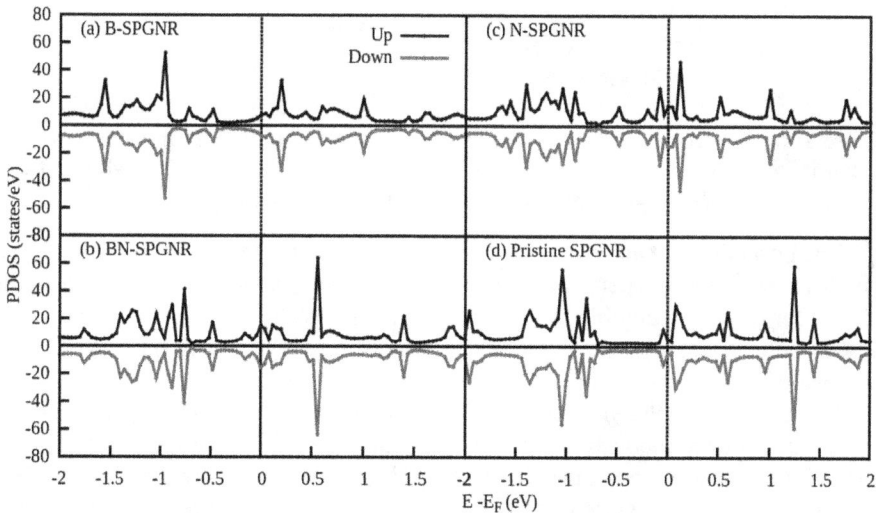

Figure 4: Projected Device Density of States (PDDoS) for (a) B-SPGNR, (b) BN-SPGNR, (c) N- SPGNR and (d) Pristine SPGNR (the figure has been taken from our research article, reference 66).

In the case of pristine SPGNR, the states of C atoms are almost non-existent in the energy zone with a similar trend in all models doped with B. In the case of N-SPGNR, new electronic states of N atoms are observed in the energy zone corresponding to the bandgap of the pristine SPGNR, and thereby reducing the bandgap. This in turn also results in the change of the band structure leading to a high current in N-SPGNR at low applied voltages. It is observed that the alteration in band structure is mainly dependent upon the doped element and not on the electronic distribution from the dopants (Berdiyorov et al. 2016). Figure 5(a) shows the I–V curves of pristine, boron-doped, nitrogen-doped, and heterogeneously boron-nitrogen doped systems, respectively. The results show that the pristine SPGNR model produces the maximum current at higher applied voltages, while the nitrogen-doped model produces a high current at lower applied voltages. Using CNT electrodes results in a remarkable increase in current at higher bias voltages in pristine SPGNR, as compared to other reports (Tien et al. 2019).

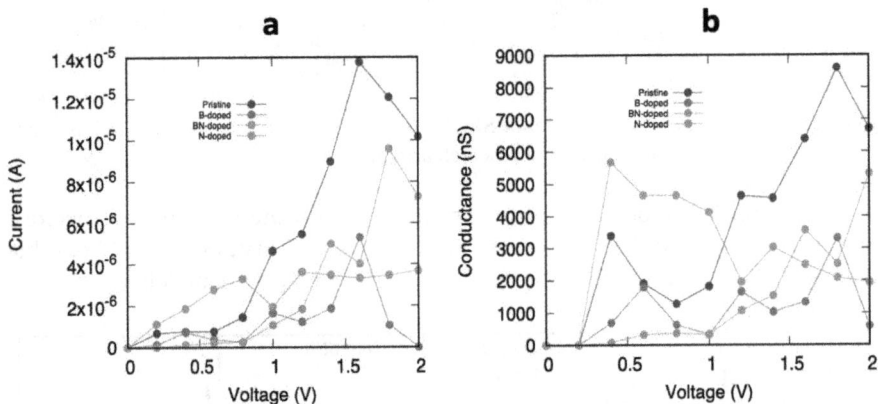

Figure 5: (a) I–V curves of Pristine SPGNR, B-SPGNR, BN-SPGNR, and N-SPGNR. (b) Conductance curves of Pristine SPGNR, B-SPGNR, BN-SPGNR, and N-SPGNR (the figure has been taken from our research article, reference 66).

The possible mechanism for this behavior is due to the fact that the CNT exhibits a ballistic transport property and acts as an efficient charge injecting electrode in this model due to negligible contact resistance associated with the C–C bonds of the CNT electrodes and the SPGNR scattering region. At higher applied voltages, the charge transfer from the CNT electrodes is so high that a bandgap reduction takes place in the SPGNR, resulting in high output current. For pristine boron-doped and heterogeneously boron-nitrogen doped models, the current changes slightly as the bias voltage increases from 0 V to 0.6 V. The current behavior of the pristine pentagraphene system starts to increase at a bias voltage of 0.8 V, reaches its maximum at 1.6 V and then decreases afterward. In the boron-doped model, the first peak appears at 1.0 V followed by a sharp increase in current reaching its maximum value at 1.6 V and then decreasing to its minimum at 2.0 V. A similar trend is observed in the BN-doped model, where the current starts to increase at 0.8 V, forming one peak at 1.4 V and then reaching up to the highest value at 1.8 V. However, the current intensity in these three models is high with a peak intensity of 13.774 µA in the pristine model. On the other hand, the current intensity of the nitrogen-doped model increases linearly with the bias voltage from 0 V to 0.8 V; the current then decreases at 1.0 V and subsequently increases again at 1.2 V and remains almost constant with the rest of the bias voltage. This variation of current with the bias voltage is due to the difference in the electronic scattering of different dopant atoms. Nitrogen has five outer electrons, which means that it will have free electrons in the doped pentagraphene nanoribbon configuration; this, in turn, results in a high electron density, thereby explaining the high value of current at low bias.

Figure 5(b) gives a comparison between the conductance of the proposed models, showing that the nitrogen-doped model has a higher conductance at low voltages while the pristine model shows increased conductance at higher bias voltages. It is important to mention here that pentagraphene is an indirect bandgap semiconductor with a bandgap of 3.25 eV and the electronic states near the Fermi level originate from the sp^2 hybridized carbon atoms. The sp^3 hybridized carbon atoms spatially separate the pz orbitals of the sp^2 hybridized carbon atoms, which screens the full electron delocalization leading to a finite bandgap. The non-dispersive partially degenerate valence bands give a high density of states near the Fermi level, thereby leading to high conductance.

Therefore, by analyzing the electronic and transport characteristics of pristine and doped pentagraphene nanoribbons with CNT electrodes, we have observed that using CNT electrodes results in a remarkable increase in current at higher bias voltages in pristine SPGNR as compared to other cited reports. Doping also significantly affects the I–V characteristics and transmission spectra of the two probe systems. Specifically, the operating voltage of the nitrogen-doped model is significantly reduced as compared to the pristine, homogenously boron-doped and heterogeneously boron-nitrogen doped models. Moreover, the results show that the magnitude of the current in the pristine pentagraphene model is maximum at higher bias voltages. The behavior of the computed transmission spectrum completely matches the I–V characteristics. The small but perceptible NDR behavior is also observed in doped pentagraphene models, especially in the nitrogen-doped model.

7. Summary

Pentagraphene is a two-dimensional carbon allotrope that has been proposed theoretically very recently. Although its practical synthesis/implementation is yet a crucial challenge, some of its unique electrical, thermal, mechanical, magnetic, and optical properties highlighting promising application in nano-electronic devices have motivated researchers to study the material thoroughly. The zero bandgap characteristic of graphene limits its application in various cases, whereas PG is a quasi-direct bandgap semiconductor with a bandgap of 3.25 eV. It can withstand a temperature as high as 1,000 K. Thermal conductivity of PG is significantly lower in comparison to graphene, which enhances the possibility of using PG in thermoelectric devices. The mechanical strength of PG also outperforms graphene. There would have an enormous application of PG in the biomedical and sensor field as nano-auxetic materials. However, despite such a wide spectrum of interesting properties, the practical realization of pristine PG is a crucial challenge. Judicious chemical functionalization could be a possible solution for achieving a stable PG structure. Functionalized PG is shown to be more stable than its pristine counterpart suggesting the possibility of synthesis of PG functionalized state as a precursor. Interestingly, PG was proposed to be synthesized by hydrogen intercalation from T12-carbon in the pioneering work on PG.

The electronic, mechanical, thermal, optical, and magnetic properties of PG can be significantly tailored by judicious chemical functionalization. The mechanical strength of PG was found to improve significantly upon chemical functionalization, especially for hydrogen and hydroxyl groups. The thermal conductivity of PG dramatically increases upon hydrogenation which is in contrast to graphene, where hydrogenation reduces thermal conductivity. The possibility of tuning thermal conductivity by varying the percentage of coverage of functional groups widens the application of PG in thermoelectric devices. Bandgap tailoring of PG is also possible by suitable chemical functionalization, which is of huge interest for nano-electronic devices. Magnetization can also be induced in PG by hydrogen functionalization which finds application in next-generation spintronic devices. Mainly the effects of three types of functionalization of PG have been studied so far. There are many others in the queue to be studied. However, the successful synthesis of PG with desired stability and predicted properties is still a crucial challenge.

References

Anandatheertha, S., Narayana, G., Gopalakrishnan, S. and Rao, P.S. 2010. High accuracy curve fits for chirality, length and diameter dependent initial modulus of single walled carbon nanotubes. Physica E 43: 252–255.

Andrew, R.C., Mapasha, R.E., Ukpong, A.M. and Chetty, N. 2012. Mechanical properties of graphene and boronitrene. Phys. Rev. B 85: 125428.

Autreto, P. and Galvao, D.S. 2015. Site Dependent Hydrogenation in Graphynes: A Fully Atomistic Molecular Dynamics Investigation. MRS Online Proceedings Library Archive 1726.

Avramov. P., Demin, V., Luo, M., Choi, C.H., Sorokin, P.B., Yakobson, B. and Chernozatonskii, L. 2015. Translation symmetry breakdown in low-dimensional lattices of pentagonal rings. J. Phys. Chem. Lett. 6: 4525–4531.

Bao, W.X., Zhu, C.C. and Cui, W.Z. 2004. Simulation of Young's modulus of single-walled carbon nanotubes by molecular dynamics. Phys. B 352: 156–163.

Baughman, R.H., Zakhidov, A.A. and de Heer, W.A. 2002. Carbon nanotubes – the route toward applications. Science 297: 787–792.

Berdiyorov, G., Dixit, G. and Madjet, M. 2016. Band gap engineering in pentagraphene by substitutional doping: first-principles calculations. J. Phys. Condensed Matter 28: 475001–475010.

Bliichl, P.E. 1994. Projector augmented-wave method. Phys. Rev. B: Condens. Matter 50: 17953–17979.

Bucknum, M.J. and Hoffmann, R.A. 1994. Hypothetical dense 3,4-connected carbon net and related B_2C and CN_2 nets built from 1,4-cyclohexadienoid units. J. Am. Chem. Soc. 116: 11456–11464.

Cranford, S.W. 2016. When is 6 less than 5? Penta- to hexa-graphene transition. Carbon 96: 421–428.

Deza, M., Fowler, P.W., Shtogrin, M. and Vietze, K. 2000. Pentaheptite modifications of the graphite sheet. J. Chem. Inf. Comput. Sci. 40: 1325–1332.

Ding, Y. and Wang, Y. 2013. Density functional theory study of the silicene-like SiX and XSi3 (X = B, C, N, Al, P) honeycomb lattices: The various buckled structures and versatile electronic properties. J. Phys. Chem C 117: 18266–18278.

Einollahzadeh, H., Dariani, R.S. and Fazeli, S.M. 2016. Computing the band structure and energy gap of pentagraphene by using DFT and G_0W_0 approximations. Solid State Commun. 229: 1–4.

Ghosh, K., Rahaman, H. and Bhattacharyya, P. 2017. Potentiality of density-functional theory in analyzing the devices containing graphene-crystalline solid interfaces: a review. IEEE Trans. Electron Dev. 64: 4738–4745.

Giannopoulos, G.I., Kakavas, P.A. and Anifantis, N.K. 2008. Evaluation of the effective mechanical properties of single walled carbon nanotubes using a spring based finite element approach. Comput. Mater. Sci. 41: 561–569.

Greaves, G.N., Greer, A.L., Lakes, R.S. and Rouxel, T. 2011. Poisson's ratio and modern materials. Nat. Mater. 10: 823–837.

Hasnip, P.J., Refson, K., Probert, M.I.J., Yates, J.R., Clark, S.J. and Pickard, C.J. 2014. Density functional theory in the solid state. Philos. Trans. Royal Soc. Math. Phys. Eng. Sci. 372.

Heyd, J., Scuseria, G.E. and Ernzerhof, M. 2006. Erratum: hybrid functionals based on a screened Coulomb potential. J. Chem. Phys. 124: 219906.

Hohenberg, P. and Kohn, W. 1964. Inhomogeneous electron gas. Phys. Rev. 136: B864–B887.

Iijima, S. 1991. Helical microtubules of graphitic carbon. Nature 354: 56–58.

Jiang, J.W. and Park, H.S. 2014. Negative Poisson's ratio in single-layer black phosphorus. Nat. Commun. 5: 4727.

Kohn, W. and Sham, L.J. 1965. Self-consistent equations including exchange and correlation effects. Phys. Rev. 140: A1133–A1138.

Kresse, G. and Furthmuller, J. 1996. Efficient iterative schemes for abinitio total-energy calculations using a plane-wave basis set. Phys. Rev. B Condens. Matt. 54: 11169–11186.

Kroto, H.W., Heath, J.R., O'Brien, S.C., Curl, R.F. and Smalley, R.E. 1985. C60: Buckminsterfullerene. Nature 318: 162–165.

Lee, C., Wei, X., Kysar, J.W. and Hone, J. 2008. Measurement of the elastic properties and intrinsic strength of monolayer graphene. Science 321: 385–388.

Li, Y., Xu, L., Liu, H. and Li, Y. 2014. Graphdiyne and graphyne: from theoretical predictions to practical construction. Chem. Soc. Rev. 43: 2572–2586.

Lopez-Bezanilla, A. and Littlewood, P.B. 2015. σ–π-Band inversion in a novel two-dimensional material. J. Phys. Chem. 119: 19469–19474.

Malko, D., Neiss, C., Vines, F. and Gorling, A. 2012. Competition for graphene: Graphynes with direction-dependent dirac cones. Phys. Rev. Lett. 108: 086804.

Marianetti, C.A. and Yevick, H.G. 2010. Failure mechanisms of graphene under tension. Phys. Rev. Lett. 105: 245502.

Mina, M. and Susumu, O. 2013. Two-dimensional sp^2 carbon network of fused pentagons: all carbon ferromagnetic sheet. Appl. Phys. Express 6: 095101.

Omachi, H., Nakayama, T., Takahashi, E., Segawa, Y. and Itami, K. 2013. A grossly warped nanographene and the consequences of multiple odd-membered-ring defects. Nat. Chem. 5: 739–744.

Parvaiz, M.S., Shah, K.A., Dar, G.N., Choudry, S., Farinre, O. and Misra, P. 2020. Electronic transport in pentagraphene nanoribbon devices using carbon nanotube electrodes: a computational study. Nanosystems: Physics, Chemistry, Mathematics 11: 176–182.

Prinzbach, H., Weiler, A., Landenberger, P., Wahl, Fabian, Worth, J., Scott, L.T., Gelmont, M., Olevano, D. and Issendorf, B.V. 2000. Gas-phase production and photoelectron spectroscopy of the smallest fullerene, C20. Nature 407: 60–63.

Quantum ATK version P-2019.03, Synopsis Quantum ATK (www.synopsys.com/silicon/quantumatk.html).

Rajbanshi, B., Sarkar, S., Mandal, B. and Sarkar, P. 2016. Energetic and electronic structure of pentagraphene nanoribbons. Carbon 100: 118–125.

Savini, G., Ferrari, A.C. and Giustino, F. 2010. First-principles prediction of doped graphene as a high-temperature electron-phonon superconductor. Phys. Rev. Lett. 105: 037002.

Shah, K.A. and Parvaiz, M.S. 2016. Computational comparative study of substitutional, endo and exo BN co-doped single walled carbon nanotube system. Superlattices and Microstructures 93: 234–241.

Shah, K.A. and Parvaiz, M.S. 2016. Negative differential resistance in BN co-doped coaxial carbon nanotube field effect transistor. Superlattices and Microstructures 100: 375–380.

Shah, K.A., Parvaiz, M.S. and Dar, G.N. 2019. Photocurrent in single walled carbon nanotubes. Phys. Lett. A 383: 2207–2212.

Sheng, X.L., Yan, Q.B., Ye, F., Zheng, Q.R. and Su, G. 2011. T-carbon: a novel carbon allotrope. Phys. Rev. Lett. 106: 155703.

Si, C., Duan, W., Liu, Z. and Liu, F. 2012. Electronic strengthening of graphene by charge doping. Phys. Rev. Lett. 109: 226802.

Stauber, T., Beltran, J.I. and Schliemann, J. 2016. Tight-binding approach to pentagraphene. Sci. Rep. 6: 22672–22680.

Tan, Y.Z., Xie, S.Y., Huang, R.B. and Zheng, L.S. 2009. The stabilization of fused-pentagon fullerene molecules. Nat. Chem. 1: 450–460.

Terrones, H., Terrones, M., Hernandez, E., Grobert, N., Charlier, J.C. and Ajayan, P.M. 2000. New metallic allotropes of planar and tubular carbon. Phys. Rev. Lett. 84: 1716–1719.

Tien, N.T., Thao, P.T.B., Phuc, V.T. and Ahuja, R. 2019. Electronic and transport features of sawtooth pentagraphene nanoribbons via substitutional doping. Physica E: Low-dimensional Systems and Nanostructure 114: 113572.

Vodenitcharova, T. and Zhang, L.C. 2003. Effective wall thickness of a single-walled carbon nanotube. Phys. Rev. B 68: 165401.

Wang, Y., Page, A.J., Nishimoto, Y., Qian, H.J., Morokuma, K. and Irle, S. 2011. Template effect in the competition between haeckelite and graphene growth on Ni(111): quantum chemical molecular dynamics simulations. J. Am. Chem. Soc. 133: 18837–18842.

Wang, Z., Cao, X., Qiao, C., Zhang, R.J., Zheng, Y.X., Chen, L.Y., Wang, S.Y., Wang, C.Z., Ho, K.M., Fan, Y.J., Jin, B.Y. and Su, W.S. 2017. Novel pentagraphene nanotubes: strain induced structural and semiconductor–metal transitions. Nanoscale 9: 19310–19317.

Wang, F.Q., Yu, J., Wang, Q., Kawazoe, Y. and Jena, P. 2016. Lattice thermal conductivity of pentagraphene. Carbon 105: 424–429.

Winczewski, S., Shaheen, M.Y. and Rybicki, J. 2018. Interatomic potential suitable for the modeling of pentagraphene: molecular statics/molecular dynamics studies. Carbon 126: 165–175.

Wong, C.H. 2010. Elastic properties of imperfect single-walled carbon nanotubes under axial tension. Comput. Mater. Sci. 49: 143–147.

Wong, E.W., Sheehan, P.E. and Lieber, C.M. 1997. Nanobeam mechanics: elasticity, strength, and toughness of nanorods and nanotubes. Science 277: 1971–1975.

Xu, L.C., Wang, R.Z., Miao, M.S., Wei, X.L., Chen, Y.P., Yan, H., Lau, W.M., Liu, L.M. and Ma, Y.M. 2014. Two dimensional dirac carbon allotropes from graphene. Nanoscale 6: 1113–1118.

Xu, W., Zhang, G. and Li, B. 2015. Thermal conductivity of pentagraphene from molecular dynamics study. J. Chem. Phys. 143: 154703–154706.

Yagmurcukardes, M., Sahin, H., Kang, J., Torun, E., Peeters, F.M. and Senger, R.T. 2015. Pentagonal monolayer crystals of carbon, boronnitride, and silver azide. J. Appl. Phys. 118: 104303–104306.

Yu, Z.G. and Zhang, Y.W. 2015. A comparative density functional study on electrical properties of layered pentagraphene. J. Appl. Phys., 118: 165706.

Zhang, S., Wang, Q., Chen, X. and Jena, P. 2013. Stable three-dimensional metallic carbon with interlocking hexagons. Proc. Natl. Acad. Sci. USA 110: 18809–18813.

Zhang, Y.Y., Wang, C.M. and Xiang, Y. 2010. A molecular dynamics investigation of the torsional responses of defective single-walled carbon nanotubes. Carbon 48: 4100–4108.

Zhang, S., Zhou, J., Wanga, Q., Chend, X., Kawazoef, Y. and Jenac, P. 2015. Pentagraphene: a new carbon allotrope. PNAS 112: 2372–2377.

Zhang, Y.Y., Pei, Q.X., Cheng, Y., Zhang, Y.W. and Zhang, X. 2017. Thermal conductivity of pentagraphene: the role of chemical functionalization. Comput. Mater. Sci. 137: 195–200.

Zwanenburg, F.A., VanderMast, D.W., Heersche, H.B., Kouwenhoven, L.P. and Bakkers, E.P.A.M. 2009. Electric field control of magnetoresistance in InP nanowires with ferromagnetic contacts. Nano Lett. 9: 2704–2709.

A Review of Carbon Dots – A Versatile Carbon Nanomaterial

Jayanta Sarmah Boruah[1,2], Ankita Deb[1], Jahnabi Gogoi[1], Kabyashree Phukan[1], Neelam Gogoi[3] and Devasish Chowdhury[1]*

[1] Material Nanochemistry Laboratory, Physical Sciences Division, Institute of Advanced Study in Science and Technology, Paschim Boragaon, Garchuk, Guwahati - 781035, India
[2] Department of Chemistry, Cotton University, Assam, India
[3] School of Life and Environmental Sciences, School of Chemistry, The University of Sydney, Camperdown, NSW 2006, Australia

1. Introduction

Different types of nanomaterials derived from the carbon family, such as fullerenes, functionalized graphene, carbon nanotubes, and fluorescent carbon nanoparticles, have found their use in a wide range of applications. Among them, carbon dots (CDs) have gained immense popularity in the research community for their ease of synthetic procedure with minimal cost, unique and tunable photoluminescence properties, easy modification, and chemical inertness (Baker and Baker 2010, He et al. 2017, Zheng et al. 2014, Zhu et al. 2015). CDs can be defined as carbon nanoparticles with all three dimensions below the 10 nm range. It is, therefore, also denoted as a zero-dimensional (0-D) nanomaterial. The discovery of CDs dated back to 2004 when single-walled carbon nanotubes obtained through preparative electrophoresis were purified (Xu et al. 2004) and then again in 2006, they were obtained via laser ablation of graphite powder and cement (Sun et al. 2006). Unlike the semiconductor quantum dots (e.g., CdTe and CdSe) having issues of toxicity that limits their applications in the biological sector, CDs, on the other hand, have been widely used in the field of nanomedicine and theragnostic applications because of their excellent biocompatibility, low cytotoxicity, use of non-toxic carbon precursors (biodegradable), and good aqueous solubility (Lim et al. 2006, Bhartiyaa et al. 2016, Wang and Qiu 2016). Therefore, quantum dots, mostly from carbon, are becoming attractive for people working on nanotechnology owing to significant behaviors with (a) broad absorption spectra, (b) very narrow emission spectra, (c) long fluorescence lifetime (Murray et al. 1993, Sharma and Das 2019). CDs are also known to possess

Corresponding author: devasish@iasst.gov.in

electron acceptor and donor properties, endowing them to explore optoelectronics, catalysis, and sensors (Yuan et al. 2019, Sharma and Das 2019). Apart from small organic molecules and polymers, many natural sources, like plant extract, biodegradable materials eggshells, banana, orange, spinach, sugarcane, papaya, pomegranate, ginger, rose flower, rice, glucose, etc., have been employed to prepare CDs. The utilization of microwaves to prepare CDs is also related to green chemistry. Nowadays, nanocomposite preparation via immobilization of CDs through proper functionalization is getting attention for target-specific applications. Our group has developed a number of hybrid nanocomposite products out of CDs, like film, beads, gel, etc., for many applications (Gogoi et al. 2015, Gogoi and Chowdhury 2014, Baruah and Chowdhury 2016).

Herein, this chapter gives an overview of the different synthetic protocols, photoluminescent properties, and applications in drug delivery, bio-imaging, biosensors, chemosensors, optronics, and catalysis (Figure 1).

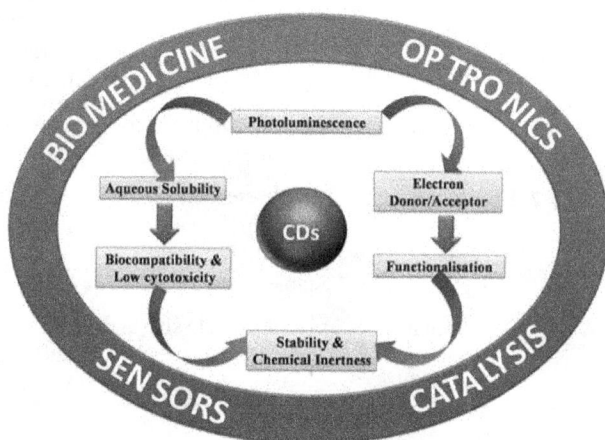

Figure 1: Pictorial description of carbon dots with their unique properties responsible for different potential applications.

2. Synthesis and Properties

2.1 Preparation of CDs

The preparation of CDs can be attained through two synthetic pathways: (A) Bottom-up approach and (B) Top-down approach. The bottom-up approach is based on a chemical reaction between small molecule precursors to form different nano-sized dots. In the top-down approach, the large carbon cluster structure is cleaved into smaller segments, resulting in CDs formation (Zuo et al. 2015). The different synthetic strategy of synthesis of carbon dots and pros and cons of each technique is tabulated in Table 1.

(A) Bottom-Up Approach: The bottom-up methodology of CD synthesis is based on a different strategy. These include hydrothermal, microwave irradiation, and pyrolysis methods.

Table 1: Characteristics of the different synthetic methods used for the preparation of CDs

Types of Synthetic Pathways	Sub-classes	Advantages	Disadvantages
A. Bottom-Up	1. Hydrothermal	Simple operation, controllability, favourable for mass production	Poor control over sizes, contain many impurities
	2. Microwave irradiation	Low cost, convenient and rapid, distinct morphology of product	Particle size distribution not uniform, and poor control over sizes
	3. Pyrolysis	Facile, high yield, repeatability	CDs formation and the separation of small molecules of raw materials become difficult
B. Top-Down	1. Laser Ablation	Simple, surface states tuneable	Low yield
	2. Electrochemical	Single-step, size and nanostructure are controllable, and stable	Time-consuming, expensive
	3. Arc discharge	Small particle size and large oxygen content	Complex procedure and contain impurities

Hydrothermal Method: In this methodology, a solution of an organic precursor is sealed in a Teflon-lined autoclave (hydrothermal reactor) where the reaction occurs at a high temperature (100-200 °C) in the absence of air (Anwar et al. 2019). In the process, the organic moieties form carbogenic cores finally turn into CDs. This carbonization method is promising for the synthesis of CDs due to its low cost, environment-friendly, non-toxic, and easy operational technique. However, this method is time-consuming. There are many reports of CDs synthesis using this method. For example, Liu et al. reported a one-step synthesis of amino-functionalized fluorescent CDs by hydrothermal carbonization of chitosan at 180 °C for 12 hours and was used as novel bioimaging agents (Yang et al. 2012).

Microwave Irradiation Method: This strategy is an augmented method to the hydrothermal one, where the microwave is used instead of heat to synthesize CDs. Localized heating via microwave favors carbonization to yield CDs with distinct morphology (Anwar et al. 2019). This technique has gained importance due to its rapidity in synthesizing CDs and its low cost and eco-friendly nature. As such, the procedure involves dissolving the source material in a solvent and subjecting it to microwave treatment at an optimized temperature. The resulting CDs can then be separated and purified. For example, chitosan CD was fabricated from chitosan hydrogel through microwave irradiation and resulted in blue fluorescent CD, thus showing shifting of emission spectra with different metal ions (Chowdhury et al. 2012).

Pyrolysis Method: Pyrolysis is a facile method to synthesize CDs from organic compounds carried out at a very high temperature. This leads to the formation of black carbon materials from which the CDs are separated and purified. Our group has prepared CDs from CTC (Crush, tear, and curl) tea origin of the state of Assam, India, via pyrolysis and has been successful in fabricating nanocomposite chitosan hydrogel film in comparison to pure chitosan hydrogel film (Konwar et al. 2015). The nanocomposite film has shown superior properties. Similarly, other nanocomposite hydrogel films composed of alginate biopolymer immobilized with tea CDs were fabricated using different bivalent and trivalent ions as cross-linker to study the structure-property relationship of the films (Konwar and Chowdhury 2015).

(B) Top-Down Approach: The top-down strategy to synthesize CDs includes laser ablation, electrochemical oxidation, and arc discharge method.

Laser Ablation Method: The laser ablation technique is used for making CDs of varied sizes. In the process, complex organic macromolecules are exposed under laser radiation operated in CW or pulsed mode. Accordingly, nanosized carbon particles get detached from the larger molecular structures (Gayen et al. 2019). The report showed the synthesis of photoluminescent CDs of ~3 nm size by laser irradiation technique from carbon glassy particles in the presence of polyethylene glycol 200 (Doñate-Buendia et al. 2018). CDs, thus prepared, are applied in bioimaging for cancer epithelial human cells.

Electrochemical Method: As the name suggests, the process of synthesizing CDs comprises electrolytes and electrodes. The electrochemical oxidation method is known for synthesizing ultrapure CDs from larger molecular matter, like carbon nanotube, graphite, and carbon fiber, where they are used as an electrode in the presence of proper electrolytes (Gayen et al. 2019). The first reported synthesis of CDs was from multi-walled carbon nanotubes in the presence of tetrabutylammonium perchlorate as the electrolyte (Zhou et al. 2007).

Arc Discharge Method: The synthesis of CDs by arc discharge method occurred accidentally when Xu and his coworkers separated the nanoparticles from the products when they prepared the single-walled carbon nanotubes by arc discharge (Xu et al. 2004). The fluorescent CDs were again arc discharged by treating with fuming nitric acid to improve the hydrophilicity of nanomaterials. Then, NaOH solution (pH = 8.4) was added to remove the impurities to obtain three fluorescent CDs with different molecular weights after electrophoretic separation. These three groups of fluorescent CDs emitted blue-green, yellow, and orange fluorescence, respectively.

Generally, bottom-up methods are used to prepare CDs at a laboratory scale. The as-prepared CDs are then subjected to various characterization techniques, such as DLS (dynamic light scattering) measurement, TEM image, UV-Vis spectrum, and photoluminescence spectra. Figure 2 is a representative example of a CD obtained from aloe vera displaying the different techniques used to understand the formation of the synthesized CD.

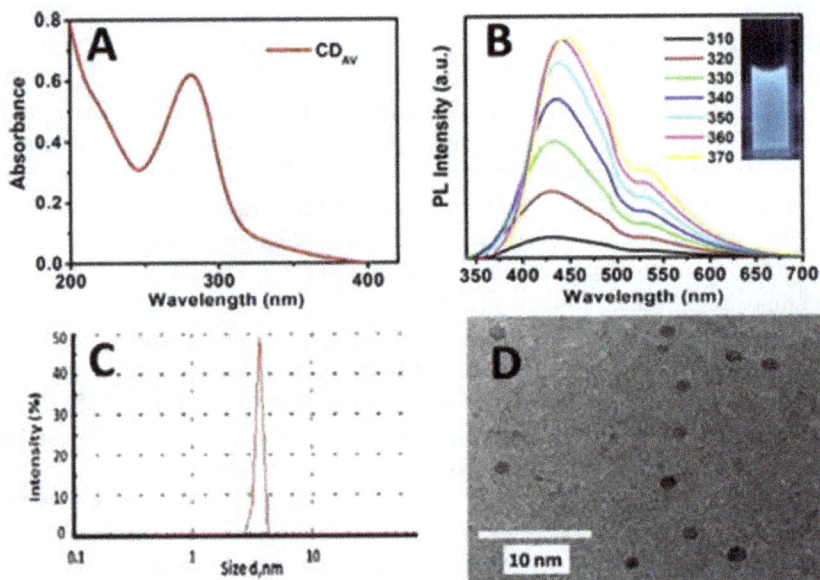

Figure 2: (A) UV-Vis spectrum, (B) PL spectra at the different excitation wavelength, (C) DLS size, and (D) TEM image of Aloe vera CD. The inset of Figure 2B shows the blue fluorescence of CDs under UV light. (Reproduced from reference (Deb, Saikia and Chowdhury 2019) copyright @ American Chemical Society).

2.2 Properties of Carbon Dot

The properties of CDs can be broadly divided into two parts: physical and optical properties. In fact, these properties are being exploited depending on the kind of application to be demonstrated (Zhang et al. 2018). The physical properties are utilized for catalytic studies and energy-related applications, whereas the optical properties are used for biological applications.

2.2.a Physical Properties

Structure and Components: As reported by Sun et al. (2006), CDs are quasi-spherical nanoparticles usually <10 nm in diameter. They are generally amorphous and occasionally show nanocrystalline with sp^2 carbon clusters. Some reports showed the formation of a diamond-like structure with sp^3 carbon network (Hu et al. 2009). The graphitic in-plane lattice spacing by HRTEM and XRD measurements could be found to be 0.18-0.24 nm, which is corresponding to different diffraction planes. Depending upon the precursor and the synthetic protocol, doping with different hetero-atoms, introducing functional groups, and surface passivating, the CD surface can be attained to modulate and play with the photophysical and photochemical properties.

Photoinduced Electron Transfer: Photoinduced electron transfer (PET) is an excited state electron transfer process by which an excited electron is transferred from donor to acceptor. CDs can serve as both electron donors and acceptors

(Liang et al. 2016). They typically demonstrate fluorescence emission; however, the addition of electron acceptors causes quenching of fluorescence intensity because of photoinduced electron transfer between the CDs and the electron acceptor (Liang et al. 2016). Photoinduced electron transfer is important in CD-sensitized materials and CD-based photocatalysts (Wang et al. 2017).

2.2.b   Optical Properties

Absorbance: CDs generally show optical absorbance in the UV region (260-320 nm) with a weak absorption tail extending in the visible region (Baker and Baker 2010). The absorption maxima at the UV region signify π-π* transition arising due to the C=C bonds and n-π* transitions of C=O, C-N, or C-S bonds present. The absorption of the CD can be modulated by surface passivating agents or by the addition of functionalities on the CD surface. Studies have shown that the presence of carbonyl and amino functionalities on the CD shifts the band maxima to longer wavelengths in the UV-Vis spectra due to the variations in HOMO-LUMO energy levels of CDs (Li et al. 2015).

Photoluminescence: The PL of CDs is a fascinating feature in terms of both fundamental and application perspectives. The unique feature of PL of CDs is that they exhibit excitation wavelength-dependent emission behavior (Zhu et al. 2015). The emission peaks of CDs are usually broad with large Stokes shifts compared with the emission of organic dyes. The occurrence of such behavior may be attributed to the wide distribution of differently sized nanoparticles, various emissive traps on the CD surface along with different surface chemistry, or other unresolved mechanisms (Wang et al. 2013). There are reports of CDs displaying excitation wavelength-independent emission behavior, which is assumed to occur due to uniformly sized dots forming along with similar surface chemistry (Wang et al. 2013). Our group has also synthesized dual emissive CDs from a biological molecule by tuning its electronic environment (Konwar et al. 2019).

The mechanistic explanation behind the fluorescence nature of CDs can be categorized into three parts: (1) quantum size effect in CDs, (2) surface state in CDs, and (3) molecule state and carbon-core state in CDs (Liu et al. 2019). As the carbon particle decreases in size to reach the nano-level (below 10 nm), the particle becomes smaller than the exciton Bohr radius of the electron in the dot. This is termed the quantum confinement effect. Therefore, quantization of the energy into discrete levels in the conduction and valence band takes place (Liu et al. 2019). The emission energy depends on the radius of the particle so that as the particle gets small, both the excitation and emission spectrum shift to shorter wavelengths. Lin's group prepared three CDs with bright red, green, and blue fluorescence, respectively, under single UV-light excitation (Jiang et al. 2015). These three CDs have similar chemical compositions but different particle size distributions. Therefore, the change in fluorescence emission of these CDs was assumed to result from the quantum confinement effect.

The surface state of CDs includes the degree of surface oxidation and surface functional groups. The higher the degree of surface oxidation, the greater is the number of surface defects. These defects act as capture centers for excitons, and the radiation from the recombination of trapped excitons causes surface-state-related fluorescence emission. Also, the increase in the oxygen content on the CDs' surface reduces the bandgap; that is, the red-shifted emission could be achieved from the increased degree of surface oxidation (Liu et al. 2019).

Different functional groups on the CD surface give rise to diverse fluorophores or energy levels in CDs, which may result in a series of emissive traps. The functional groups on the CDs' surface can tune the fluorescence of CDs by influencing the emission centers. Baruah et al. (2014) demonstrated reversible on/off switching of fluorescence by esterifying the surface functional groups of the CDs.

The molecule state is the PL center formed solely by an organic fluorophore connected to the surface or interior of the carbon backbone and can exhibit PL emission directly (Zhu et al. 2015). The molecule state is the emerging PL center for a type of CD prepared by small-molecule carbonization (bottom-up synthesis).

Up-conversion Photoluminescence (UCPL): Up-conversion is another optical phenomenon where the fluorescence emission wavelength is shorter than the excitation wavelength. This UCPL property can otherwise be termed as multi-photon emission where two or more long-wavelength photons are simultaneously absorbed by the luminescent material to reach the excited state, which leads to the release of a photon at the ground state having a wavelength less than the excitation wavelength (anti-Stokes type emission) (Anwar et al. 2019). The UCPL property avails opportunities for *in vivo* bioimaging applications because bioimaging requires a longer wavelength to penetrate deep into tissue without any damage, reduced background auto-fluorescence, and low photon-induced toxicity due to the highly-localized nonlinear photon adsorption process (Zheng et al. 2015). This process is advantageous compared to that of single-photon fluorescence, wherein its detection speed is fast and likely to tissue damage due to higher excitation energy (Anwar et al. 2019). Cui et al. (2015) prepared green photoluminescent CDs from ammonium citrate and ammonia. These CDs were used for HeLa cell imaging via the UCPL phenomenon.

Aggregation-Induced Emission (AIE): CDs generally display fluorescence emission in solution or dispersed form. However, there are few reports of CDs demonstrating aggregation-induced emission enhancement effects. Such effects result from the restriction of intramolecular rotations and vibrations due to aggregation of the CDs (Jia et al. 2012). The AIE CDs are sensitive to the surrounding environmental changes, such as pH and solvent, and hence are used in biological and chemical sensing. The AIE performance of CDs enables the design of photosensitizing/photo-thermal probes for photodynamic therapy/photo-thermal therapy that would permit the dense packing in the LED without the impedance of the PL efficiency (Anwar et al. 2019). Recently, researchers from our group carried out the synthesis of CDs in an organic medium from a novel bile acid hydrazone-based organogel in THF/water mixture (Gogoi et al. 2017). The CDs exhibited the AIE phenomenon and could successfully detect cholesterol.

Phosphorescence: Phosphorescence is a spin forbidden process and refers to a radiative transition from the triplet state (T_1) to the singlet ground state (S_0). The phosphorescence lifetime is, therefore, longer than the fluorescence lifetime. CDs are reported to exhibit phosphorescence, monitored by steady-state PL spectroscopy and time-resolved PL spectroscopy. Deng et al. (2013) synthesized CDs from disodium salt of EDTA (ethylene diamine tetraacetic acid) in polyvinyl alcohol (PVA) by forming a CD-PVA composite. The composite obtained a phosphorescence peak at 500 nm, having a lifetime of 380 ms by exiting the sample with 325 nm. It has been assumed that the phosphorescence originated from the triplet excited states of aromatic carbonyls on the surface of the CDs, wherein the PVA molecules can effectively protect the triplet excited state from being quenched by rigidifying these groups with hydrogen bonding and oxygen. Apart from using in anti-counterfeiting applications, the room-temperature phosphorescent material can also be used in chemical and biological sensing and time-resolved imaging because of the water-soluble and biocompatible nature of the material (Anwar et al. 2019).

Chemiluminescence (CL): CL can be described as a process of emitting photons from the excited states to the ground states owing to a chemical reaction. CL has similar properties of fluorescence. The only difference is that the excitation energy of fluorescence comes from the excitation of external light sources, whereas the excitation energy of CL is produced from chemical reactions. That is why CL shows high sensitivity due to the absence of an external light source (Wang et al. 2019). The CL analysis of CDs is still at its infancy stage. Lin et al. (2012) first reported CL of CDs where an intense CL was obtained in the presence of oxidants, such as $KMnO_4$ and cerium (IV) ion. Electron paramagnetic resonance revealed that the oxidants could inject holes into CDs. This process increased the population of holes in the CDs and accelerated electron-hole annihilation, resulting in the release of energy in the form of CL emissions.

Electrochemiluminescence (ECL): ECL is a phenomenon where chemical substances produced on the electrode surface go through electron transfer reactions to generate light emissions. ECL is the result of the combination of electrochemistry and chemiluminescence. According to the source of free radicals, the ECL mechanism is categorized into two types, one the annihilation pathway and the other is the co-reaction pathway (Zhao et al. 2020). Yang and co-workers reported the first investigation of CDs on ECL property in 2009. The CDs were prepared through a simple microwave reaction using poly (ethylene glycol) (PEG-200) and saccharide as the carbon sources. Due to its high sensitivity, the ECL technique on CDs could be effectively used for single-cell analysis and biosensing applications.

3. Applications

3.1 Sensing

Out of the many applications, CDs have found a great candidate in themselves for sensing applications. Because of the advantageous properties of low cost, easy preparation methods, fluorescence, biocompatibility, higher photostability, and excitation-dependent emissions, CDs are very popular for biosensing and chemical sensing applications. CDs take advantage of the changes in the fluorescence intensity that is induced by the different analyte systems for sensing the analytes. The changes in fluorescence intensity may occur through different mechanisms, such as resonance energy transfer, inner filter effect, photo-induced electron, and charge transfer (Sharma and Das 2019). The inter sp^2 hybridized carbon atoms at the core of the

CDs have different functional groups, like hydroxyl, carbonyl, carboxylate, amino, and epoxy on their surface. These functional groups in the surface of the CDs make them easier for functionalization with desired recognition elements, like antibodies and aptamers, through derivatization that helps in attaining selectivity and specificity toward analytes (Xu et al. 2019).

3.1.a.   Chemosensing

CDs are utilized for sensing different types of metal ions, such as Hg^{2+}, Ag^+, Cu^{2+}, Fe^{3+} etc. These metal ions interact with the surface functional groups of the CDs or the functionalized CDs lead to the formation of new electron-hole recombination through any of the energy transfer routes, resulting in changes in the fluorescence intensity. This property is utilized in sensing applications by CDs. CDs are highly utilized for the sensing of toxic mercury ions. In their work by Xu et al. (2018), they prepared novel manganese doped CDs (Mn-CDs) with a quantum yield of 54%. Using these Mn-CDs as fluorescent probes, they obtained a very sensitive and selective detection of Hg^{2+} with a detection limit in nM. Previously, Xu et al. (2015) synthesized sulfur and nitrogen co-doped semicrystalline CDs (S, N-CDs) from sodium citrate and sulfamide. Later, they used these S, N-CDs as probes for the detection of Hg^{2+} in the range of 2-5,000 nM. This system exhibited excellent detection capability with a detection limit of 100 pM. Zhang et al. (2014) used highly luminescent nitrogen-doped CDs (N-CQDs) for the detection of Hg^{2+} ions. They synthesized the CDs from folic acid via a simple economical hydrothermal strategy as the carbon and nitrogen source. This carbon dot system proved to be an effective fluorescent sensing platform for the selective detection of Hg^{2+} ions with a detection limit of 0.23 µM. A turn-off-on fluorescence sensor was reported by Xavier et al. (2018) in their report, where they constructed an environmentally benign and cost-efficient cysteine fluorescence sensor. This sensor was constructed on the basis of nitrogen-doped CDs, where citron fruit extract and human urine waste acted as the carbon and nitrogen sources, respectively. Initially, the fluorescence of these CDs was quenched with Hg^{2+} and owing to the competitive binding of Hg^{2+} with cysteine, the turn-off sensor changed to a turn-on sensor for cysteine. The nitrogen-doped CDs-Hg^{2+} system under optimized conditions had a detection limit of 40 nM and for cysteine detection, it showed a wide linear range of 1-10 µM. Sun's group reported a fluorescent detection of Hg^{2+} ions with a detection limit of 0.23 nM via fluorescence quenching strategy induced by the Hg^{2+} ions into the carbon dot system. They reported the detection with the help of the carbon dot system where the carbon source was low-cost wastes of pomelo peel. This simple, green, and economical hydrothermal method of synthesis resulted in the formation of water-soluble fluorescent carbon nanoparticles with a quantum yield of approximately 6.9% (Lu et al. 2012). Apart from Hg^{2+}, detection of methylmercury is also very important. Bendicho's group reported a fluorescent assay for the detection of methylmercury using a carbon dot system where D-Fructose acted as the carbon source. The CDs were synthesized *in situ* via an ultrasound-assisted method. With the application of high-intensity sonication, simultaneous synthesis of fluorescent CDs along with the detection of the target analyte was achieved in a single step where the fluorescence quenching was observed in presence of methylmercury. The assay used low amounts of organic precursors, such as fructose and polyethylene glycol, and a detection limit of 5.9 nM was obtained (Costas-Mora et al. 2014). CD-based detection and differentiation of inorganic and organic sulfur were reported by Majumdar et al. (2016a) in their work, where they derived CDs from paper ash. A conjugate system of paper CDs with Au^{3+} was developed, which functioned as a fluorescence turn-on and turn-off sensor for the detection of organic and inorganic sulfur in analytes, respectively. This carbon dot system had real-time applications as it was found effective for the detection and differentiation of inorganic and organic sulfur in milk samples. Li et al. (2011) demonstrated for the first time the detection of Ag^+ ions using fluorescent CDs from carbon soot. A cheap and effective fluorescent sensing platform was constructed using carbon soot by lighting a candle. A detection limit, as low as 500 Pm, was obtained with high selectivity. Furthermore, they also used this sensing platform for Ag^+ detection in real samples. Simple and eco-friendly N-doped CDs were fabricated from black soya beans as reported by Jia et al (2019) with a quantum yield of 38.7 ± 0.64%. The synthesized N-CDs possessed responsive properties toward free radical and Fe^{3+} ions, where fluorescence quenching was induced by Fe^{3+} ions in the NCDs and thus making it a suitable candidate for the Fe^{3+} sensor. A fluorescence turn-off sensor for detection of Fe^{3+} ions was reported by Xu et al. (2016), wherein they synthesized copper doped carbon dots via a single-step hydrothermal process. After the optimization of the synthesis conditions 9.81% quantum yield was obtained and fast detection of Fe^{3+} in human blood serum with a detection limit of 1 Nm was obtained in the range of 0.001-200 Mm. Tabaraki and his co-worker synthesized by microwave-assisted method and designed a turnoff fluorescence sensor for Pb^{2+}. They used Doehlert experimental design for optimizing the important parameters, like pH, the concentration of carbon dots, CCDs, and time. And as such the maximum turn-off was observed at a pH 3 with a carbon dot concentration of CCDs 8 mg ML^{-1} and a response time of 5 minutes. The limit of detection was 0.02 μM in the linear range of 0.05-25 μM (Tabaraki and Abdi 2020). Meanwhile, the application of CD for the detection of some hazardous chemicals present in different cosmetics items is getting interesting. For that, our group has developed a CD incorporated fluorescent film to sense para-aminobenzoic acid (PABA), benzophenone, hydroquinone, and propylparaben, which are considered as red-listed chemicals (Deb et al. 2019). The respective sensing pattern of PABA and Stern-Volmer non-linear plot by the conjugate is represented in Figure 3. Moreover, detection of herbicide in the soil can also be achieved by CDs treatment. CDs prepared from polyethylene glycol, decorated with calcium ion have been used to detect Trifluralin (a known toxic herbicide) which uses the PL property of CD (Gogoi and Chowdhury 2020a). Detection of another herbicide (pretilachlor) using CDs synthesized from natural product extract (water hyacinth) with the help of phosphoric acid is an addition to the soil testing application (Deka et al. 2019). Apart from sensing and detection of different types of chemicals, a green sulfonated CDs-chitosan hybrid hydrogel nanocomposite film was applied for water softening applications for removal of calcium and magnesium ions from water (Baruah et al. 2016).

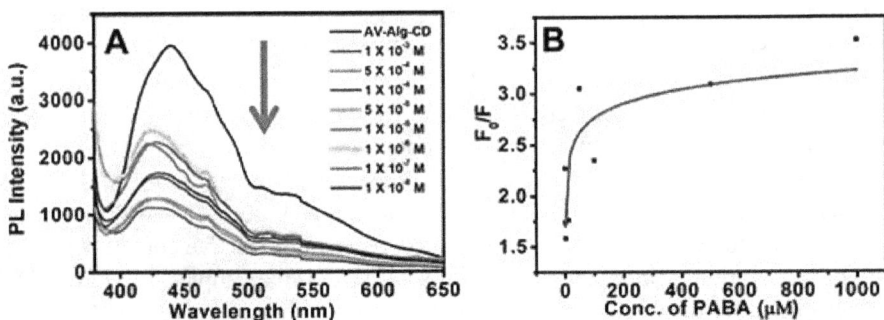

Figure 3: (A) PL emission spectra of AV-Alg-CD$_{AV}$ nano-bioconjugate film in the presence of different concentrations of PABA at an excitation wavelength of 370 nm and (B) corresponding Stern-Volmer non-linear plot. (Reproduced from reference (Deb, Saikia and Chowdhury 2019) copyright @ American Chemical Society).

3.1.b   Biosensing

CDs have also been employed for sensing a variety of other small molecules like drugs, vitamins, pesticides, and other contaminants apart from different types of metal ions. In some cases, the analyte acted as a fluorescence enhancer and in others as fluorescence quenchers. In a work, Li et al. (2015) synthesized high-intensity fluorescent CDs from polyethylene glycol of different molecular weights (PEG 200, PEG 600, and PEG 800) via a low-cost electrolysis method. The CDs where PEG 600 was used as the carbon source resulted in the highest quantum yield of 38%. Coupled with tyrosinase, these hybrid fluorescent CDs acted as a fast, efficient, and stable sensing probe for the detection of levodopa (L-DOPA). The limit of detection for L-DOPA was found to be as low as 9.0×10^{-8} mol L^{-1} with a wide linear range from 3.17×10^{-4} mol L^{-1} to 3.11×10^{-7} mol L^{-1}. Wang et al. (2015) in their work reported a novel thermally reduced blue luminescent CD-based FRET sensor for vitamin B$_{12}$ determination in an aqueous solution. The CDs were synthesized by esterification reactions by carbonization of citric acid and then thermally reduced by a thermogravimetric analyzer. Vit-B$_{12}$ were detected using these CDs with concentrations ranging from 1 to 12 µg mL^{-1}, and the detection limit was as low as 0.1 µg mL^{-1}. Qian and his co-workers designed a label-free real-time fluorometric assay for the sensitive detection of alkaline phosphatase (ALP). Copper ion-assisted aggregation and disaggregation of CDs was the basis of the sensor that resulted in a special fluorescence off-on-off process. The carboxyl groups present in the surface of the CDs resulted in aggregation, which caused fluorescence quenching by copper ions. And the competitive interaction among carboxyl, copper ions, and PPi enabled disaggregation-induced fluorescence enhancement. This first report on quantitative and sensitive detection of ALP activity in real-time had a limit of detection of 1.1 U/L, which is sensitive enough for real-time determination of ALP level in human serum (Qian et al. 2015). Huang et al. (2015) developed a fluorescent nano-sensor for selective detection of DNA using the CDs as the reference fluorophore and the ethidium bromide as a specific organic fluorescent dye toward DNA. The microwave irradiation technique was used for synthesizing the fluorescent CDs. The fluorescence of the CDs was quenched in presence of EB and with the addition of the DNA, the fluorescence of the CDs enhanced significantly. The detection limit of DNA was found to be 0.47 µM under the optimum conditions in a linear range of 1.0 µM ~ 100 µM with a correlation coefficient of 0.9999. Apart from detecting different types of small molecules and biological molecules, CDs have also been used for detecting cells and bacteria. In their study by Wang et al. (2015), a novel fluorescent probe for sensitive and quantitative detection of Salmonella typhimurium was proposed by a combination of CDs and aptamers. Citric acid was used as the carbon source for the preparation of CDs via the hydrothermal method. CDs-aptamer complexes were obtained by combining amino acid-modified aptamers to the carboxyl-modified carbon dot surface. By modifying the aptamer, different bacteria can be detected using this method. In this case, under optimized conditions, *Salmonella typhimurium* was detected, and a linear relationship between the concentration of *Salmonella typhimurium* and fluorescence intensity was obtained in the range of 10^3-10^5 cfu mL^{-1} and the limit of detection was found to be 50 cfu mL^{-1} (Wang et al. 2015). However, the use of CDs for the detection of biological toxins available in our body has been successful when the biomimicking system (e.g. vesicles) is attached with the CDs. Uric acid is such a toxin that can be absorbed by a novel nanocomposite prepared from palmitic acid/cholesterol mixture CD and Palmitic vesicles. The change in optical properties is the indication for the successful detection of uric acid (Boruah and Chowdhury 2020). A therapeutic agent, retinoic acid, is known for its use by pregnant women. But, studies reveal that retinoic acid can cause the malformation of embryos and so it is indeed essential to find out or sense the presence of retinoic acid in a medium. To solve the purpose, CDs derived from chitosan hydrogel have been employed to exclusively sense the chemical with the help of the PL phenomenon (Majumdar et al. 2018).

3.2. Catalysis (Photocatalysis, Electrocatalysis, and Other Catalysis)

CDs are known as the best carbon nanomaterials for catalysis purposes because of their strong and broad optical absorption, high chemical stability, and fast electron transfer processes. Further improvement in their catalytic activity can be accomplished through composite formation utilizing functional nanomaterials, such as inorganic nanostructures, biomaterials, polymers, and so on (Ryplida et al. 2019, Zhu et al. 2018, Bui and Park 2016, Mandani et al. 2017). CDs work not only as efficient catalysts but also assist in catalyst design to promote broad-spectrum absorption, electron-hole separation, and to stabilize photolysis semiconductors (Liu et al. 2014). CDs based on nano-enzymes have been developed with excellent peroxidase-like activity (Sun et al. 2015). The photo-catalysis by CD has been intensively explored by many scientists (Sharma et al. 2018, Zhang et al. 2018, Qian et al. 2018). CDs have already been involved as efficient catalysts for various reactions, including 'click reaction', photooxidation reaction, ring-opening reactions, aldol condensation, and so on (Han et al. 2017, Liu et al. 2017, 2013, 2015). The higher catalytic performance of CD-based composite catalysts over corresponding single form is induced by the process of efficient electron transfer at the interface of CDs and other nanomaterials.

Those composite catalysts are also being used in photocatalysis, peroxidase-like catalysis, electrocatalysis, Fenton-like catalysis, and chemical catalysis.

The limitation of conventional photocatalyst, available in the market (TiO_2, $BiPO_4$, Bi_2O_3, etc.), in terms of weak light absorbance, wide bandgap, and fast charge recombination has made it necessary to design CD-based composite photocatalysts that can enhance photocatalytic performance through the photogenerated charge transfer occurring on the conjugated π-structure of CDs. Eventually, CDs become electron acceptors after photoexcitation (Chen et al. 2020). The efficiency of photocatalysis is regulated by promoting the separation of photogenerated electrons and holes. It has been noted that nanoscale heterojunctions with an intimate interface found in the composite system can offer electron-hole pairs separation. In the CD-based composite photocatalysts, the chance of recombination of photogenerated carriers is retarded and hence acts as a better photocatalyst. Huang's group developed a hybrid of CDs with a carbon layer (CL) covering Cu_2O catalysts for the photocatalytic conversion of CO_2. The new catalyst could protect the Cu_2O from photo-corrosion under light irradiation, thereby its efficiency is improved. The photo-induced holes transport to CDs and are consumed by H_2O to yield O_2, whereas the electrons get involved in the reduction of CO_2 (adsorbed on Cu_2O) to produce methane and methanol. Interestingly, the CO_2 conversion is higher for composite catalysts than the Cu_2O alone and hence the methanol yield (Li et al. 2019). Another example of photocatalytic reduction by CD-based composites is the reduction of hexavalent chromium (Cr^{6+}) to trivalent chromium (Cr^{3+}) reducing its toxicity. The carboxyl and hydroxyl functional groups present on the CDs can adsorb Cr^{6+} by capturing photo-induced electrons and preventing electron-hole recombination (Zhang et al. 2018). Another group developed a photocatalytic H_2 production system in an aqueous solution with CDs as photosensitizers along with molecular Ni catalysts (Martindale et al. 2015).

CDs have emerged as a replacement for enzyme peroxidase due to the high cost in enzyme preparation and availability of active sites on CD surfaces (Ge et al. 2014). The oxidation of 3′,5,5′-tetramethylbenzidine (TMB) in presence of the CDs is attained due to electron transfer from lone-pair electrons in TMB amino groups to the CDs. In the presence of H_2O_2, electron transfer from the CDs to H_2O_2 takes place and increases the oxidation rate of TMB (Shi et al. 2011). Moreover, Majumdar et al. (2018) prepared iron oxide nanoparticle/CD composites as effective catalysts for the cyclo-oxidative tandem synthesis of quinazolinones using alcohols with higher yield (94%), which was 57% when only Fe_3O_4 nanoparticle was used. For clean energy generation electrochemically, different noble-metallic catalysts (Pt, Ir) have been used so far worldwide. The H_2 and O_2 production through water splitting is also a renewable clean energy source. The hydrogen evolution reaction (HER) and oxygen evolution reaction (OER) need catalysts to reduce the overpotentials and improve the energy conversion (Merki and Hu 2011, Reier et al. 2012). Meanwhile, the scarcity and high costs of noble metal catalysts led to the incorporation of CD into the picture. CD/NiFe-LDH nanoplate composites are such types of electrocatalysts for the OER in alkaline electrolytes (Tang et al. 2013) MoS_2/GQD composites with zero bandgaps were employed for the electrochemical HER. The bandgap of the pure MoS_2 monolayer was 1.76 eV and hence the composite was found to show higher electrical conductivity (Guo et al. 2017).

The transition metal ions catalyzed H_2O_2 decomposition, known as Fenton-like reactions, possess some limitations, and to overcome those, people have attempted conjugation of CD. For instance, a catalyst of CDs and dodecylbenzenesulfonate-layered double hydroxides (DBS-LDHs) was reported. The composite catalysts could decompose acidified H_2O_2 to produce •OH radicals (Zhang et al. 2014).

The application of CD-based composites in chemical catalysis is limited to only some particular known reactions. Few of them are oxidation reaction (Wu et al. 2015), reduction reaction (Yang et al. 2020a, Yang et al. 2020b), Suzuki-Miyaura reaction (Bayan and Karak 2017), biodiesel production (Macina et al. 2019), ring-opening reaction (Li et al. 2015), and Maillard reaction (Wei et al. 2014) esterification, Beckmann rearrangement, and aldol condensation (Li et al. 2014). The gold nanoparticle/graphene quantum dot (AuNP/GQD) composites used in Veratryl alcohol oxidation had the potential to prevent the AuNPs from aggregation. The alcohol was absorbed on the composites through π–π stacking and got oxidized by the singlet oxygen generated by GQDs with H_2O_2 under acidic conditions. In an alkaline medium, both AuNPs and GQDs with H_2O_2 generate superoxide anion and singlet oxygen which in turn lead to alcohol oxidation with more than 99% yield (Wu et al. 2015).

3.3 Optronics (Dye-Sensitized Solar Cell, Organic Solar Cell, Supercapacitor, and LED)

It is interesting to note that the chemical structure of CD having a chance for electron conduction through a functional group or its active site leads to their use in various optro-electronic applications. To mention a few, conducting textile material produced from tea CDs with reduced graphene oxide is one of the significant contributions. The electrical conductivity of the material attained an optimum value, which is similar to already reported graphitic conducting materials (Konwar et al. 2017). Photo-responsive materials are a great candidate for developing optoelectronic devices. CD-doped photoresponsive nanocomposite materials can be tuned according to the light conditions for developing optoelectronic devices that show different conductivity in the absence and presence of UV light (Gogoi et al. 2020b). Also, the photosensitive materials are found to behave differently in interaction with chiral CDs derived from chiral precursors, and they have been found to affect the optical properties of these photo-switchable materials, thereby opening up new avenues in using these materials in optoelectronics (Deka and Chowdhury 2017). Dye-sensitized solar cells (DSCs) utilize the diversity of organic dyes for solar energy harvesting. But the photobleaching of organic dyes, the high cost, and the toxicity of ruthenium-containing dyes affect their use in general practices. Due to the tunable optical and electronic properties of CDs, they have been used in DSC too as a sensitizer to absorb sunlight (Mirtchev et al. 2012). The limitation of charge recombination of photogenerated electrons taking place in DSC could be overcome by a CD-bridged dye/semiconductor complex system proposed by Lee et al. leading to efficient photoelectric conversion systems (Ma et al. 2013). It is noted that doping of CDs into the dye/semiconductor complex enhanced the photoelectric conversion efficiency by ~7 times.

Organic solar cells (OSCs) are another platform to utilize solar energy to convert into electricity. For example, the P3HT:PCBM based solar cell is only associated with the sunlight of 480 to 650 nm, which is a small part of visible light (380-780 nm). It is unable to harvest the remaining portions (380-480 nm and 650-780 nm). Hence, to use maximum solar energy, CDs filled with polysiloxane composite were employed which showed emission in the 400-650 nm range under excitation of 320-450 nm. Its PL spectra coincided with the harvesting spectrum (480–650 nm) and unused light spectrum (380–480 nm) of the bulk heterojunction (BHJ) solar cell. (Huang et al. 2014).

The involvement of CDs as excellent electrode materials for supercapacitors is also gaining attention. Decoration of CDs with RuO_2 gave rise to a remarkable electrochemical performance that could offer fast charge transportation and ionic motion during the charge-discharge process (Zhu et al. 2013). The CDs-based ionic liquid can function as a conductive agent and binder in an activated carbon electrode with better supercapacitor performance due to the tendency of the electrolyte to wet on the composite functionalized activated carbon electrode (Wei et al. 2013).

The stable light-emitting characteristic of CDs is the primary reason for use in light-emitting devices (LED). Even, the broad and bright visible light of nitrogen-rich CDs under UV illumination makes their entry in phosphor applications (Kwon et al. 2013). White LEDs consisting of the CDs embedded PMMA matrix film as the color-converting phosphors were also reported. CD-based LEDs with a driving current-controlled color change have been investigated where a CD emissive layer was sandwiched between an organic hole transport layer and an organic or inorganic electron transport layer. Multicolor emission from CDs will be visible when the device structure and applied voltage are tuned. The electron transport layer materials and the thickness of the electrode will also influence the emission behaviors of CD (Zhang et al. 2013).

3.4 Biomedicine (Bio-imaging and Drug Delivery)

3.4.a Bio-imaging

Fluorescent nature of CDs owing to many factors like small size, functionality, etc., has led to their use in different imaging areas. For example, broad emission spectra of CDs have been applied in image-guided tracking of nanoparticles by *in vivo* experiments. A type of CD was synthesized from hydrothermal treatment of dandelion as the carbon source and ethane diamine (EDA) as a passivating agent and further conjugated the CD with folic acid. Receptor-mediated endocytosis facilitates the uptake of that conjugate (FA-CDs) into HepG-2 cells. *In vitro* experiments demonstrated that the FA-CDs could accurately recognize positive FR (FR++/FR+) cancer cells in different cell mixtures of MCF-7/HepG-2 cells and HepG-2/PC-12 cells and could markedly show the elevated expression level of folic acid receptors on the surface of the cancer cells that were detected via confocal laser scanning microscopy. The quantum yield was found to be 13.9%, and this covalent conjugation of FA with CDs successfully differentiated cancer cells from normal cells. Therefore, in addition to cancer detection, prognosis, and individualized treatment, CDs could also act as an effective tool for accurately differentiating cancer cells from normal cells (Zhao et al. 2017). However, another study revealed the enhancement of the quantum yield of nascent CDs utilizing folic acid as a sole precursor via a hydrothermal method, up to 94.5% in water, which was more than most of the organic fluorescent dyes. The resulting CDs procured high photostability, suitable biocompatibility, and magnificent photoluminescent activity. The remaining residuals of folic acid assisted detection of folate receptor-mediated cellular uptake by CDs, thereby showing great potential in targeted drug delivery and bioimaging studies (Liu et al. 2018). Other CDs prepared from Malus floribunda (*M. floribunda*) fruits showed fluorescent quantum yield (18%). In an aqueous medium, this work has led to the facile detection of ferric ions by a fluorescent method.

The biocompatibility of CDs was checked in human colon cancer (HCT-116) cells and showed no cytotoxicity, thereby acting as good fluorescent probes for *in vitro* imaging of HCT-116 cells. In fact, the CDs also exhibited multicolor fluorescence imaging when internalized into the whole body (Atchudan et al. 2020). A cost-effective one-pot synthesis of Hf-doped CDs (Hafnium-doped CDs) from citric acid (CA), thiourea (TU), and ruthenium chloride ($HfCl_4$) was proposed by the pyrolysis method (Su et al. 2020). Due to their outstanding full-color emission, good biosafety, and water solubility (80 mg mL^{-1}), Hf-CDs can be efficiently utilized for *in vivo* CT/FI (fluorescence imaging) dual-model imaging of orthotopic liver tumors with high tumor targeting specificity, effective renal clearance capability, and excellent stability. Concentration-dependent multicolor photoluminescence behavior was also observed from the CDs synthesized from citric acid as a raw material and propylene diamine as a passivation agent (Zhang et al. 2020) that allowed efficient bioimaging at different concentrations in HeLa cells. Results obtained from this study revealed that multicolor luminescence exhibited blue, cyan, green, and red light under 330-380 nm, 380-420 nm, 450-490 nm, and 510-560 nm, respectively. However, at low concentrations, green and red light could not be emitted. The intensity gradually increased with increasing concentration. This characteristic facilitates the contribution of imaging by estimating the changes in concentration of N-CDs in cells and thereby promotes the tracking of cell proliferation.

3.4.b Drug Delivery

CD-assisted drug delivery is considered to be a successful field of research, particularly due to its fluorescent nature and great biocompatibility to be applied in living cells. Biopolymeric hydrogels have an issue of uncontrolled drug release. The incorporation of CDs in fabricating nanocomposite hydrogels has proved to be efficient in preventing the burst release of drugs before reaching the targeted site. The nano-engineered materials can also be designed in such a way that the drug releases only at a given specific stimulus. One such work of pH-responsive sustained drug delivery was carried out by Gogoi et al. (2014) where calcium alginate beads coated with CDs prepared from chitosan gel (CA-CD) showed maximum drug release at pH 1, enabling its use in the gastrointestinal tract where the pH is low. Taking the same system and using garlic extract as the model drug, Majumdar et al. (2016b) demonstrated the drug release occurs according to the pathogen concentration (MRSA) whose growth depended on the pH of the medium. The drug got released only when there was an increase in MRSA concentration, understood by the decrease of pH of the medium (Figure 4). Another recent work of pH-responsive drug delivery was reported by our group where a jute CD incorporated cotton patch (Jt-CD-CP) acted as a drug delivery vehicle. The cotton patch released the drug more at lower pH (pH 5) than at higher pH (pH 7), enabling its use in cuts/wounds where there is a lowering in pH during any pathogenic bacterial infections (Deb et al. 2020). The enhanced antimicrobial activity of extracellular metabolite derived from actinobacteria, conjugated with a novel CD, synthesized from Propolis (produced by a honey bee) was an interesting finding for modification of drug to make it more active compared to available antibiotic rifampicin (Chowdhury et al. 2020).

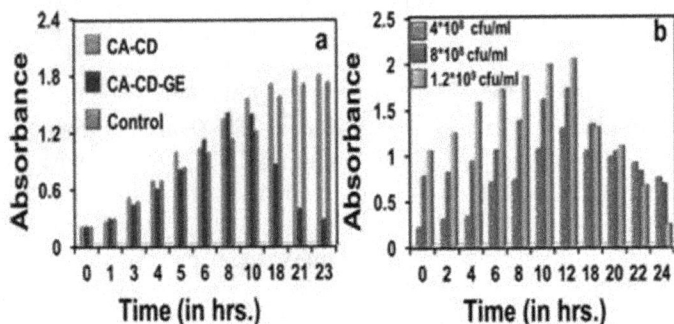

Figure 4: (a) UV analysis of cell death with pH change at different time intervals by CD beads (CA-CD), drug-loaded CD (CA-CD-GE) beads, and control and (b) concentration of cell vs. cell death assay at different time intervals. (Reproduced from reference (Majumdar et al. 2016b) copyright @ American Chemical Society).

Another study stated that hyaluronic acid conjugated polyethylenimine CDs could carry doxorubicin drugs for targeted imaging, drug delivery through electrostatic self-assembly (Gao et al. 2017). These DOX-loaded polyethyleneimine CDs successfully could distinguish NIH-3T3 cells (non-cancerous cells) from cancerous HeLa cells due to overexpression of CD44 marker on the surface of cancerous HeLa cells and found high toxicity from the analysis of MTT assay. Moreover, CDs from *Daucus carota* subsp. *sativus* (carrot) roots extract was employed for pH-sensitive drug delivery purposes. The mitomycin drug was loaded on the CDs via hydrogen bonding. It was noted that in the acidic tumor microenvironment (pH ~6.8) the hydrogen bond becomes weak and releases mitomycin. The good biocompatibility and small size of CDs have led to possessing a strong affinity toward cancer cells which results in the incorporation of mitomycin CDs by Bacillus subtilis cells to a great extent (D'souza et al. 2018). Synthesis of polydopamine-folate CDs reported by Phan et al. (2019) showed image-guided photothermal therapy in the treatment of prostate cancer cells. By using photo features of polydopamine-folate CDs, exhibited outstanding photothermal effect with robust blue fluorescence emission, at

808 nm laser treatment that leads to complete irradiation of the prostate cancer cells. Hemin/CDs prepared by Yang et al. (2019) was found to have a synergistic effect of photothermal and photodynamic therapies aided by the fluorescence resonance energy transfer effect in the treatment of cancer with the improvement of temperature to 26 °C under laser irradiation which resulted in 90% death of cancer cells after 10 minutes laser treatment. Due to several advantages, cationic CDs have received more attention in gene delivery (Salem et al. 2003, Dou et al. 2015). An observation was reported by Cao and coworkers in which the preparation of dual-functional cationic CDs through single-step microwave-assisted pyrolysis of glucose and arginine for the application of both self-imaging and a non-viral gene delivery (Cao et al. 2018). Condensation of cationic CDs with model gene plasmid SOX9 developed NPs (10-30 nm) which exhibited various suitable properties. The MTT assay of CDs/plasmid SOX9 NPs in mouse embryonic fibroblasts showed very little cytotoxicity compared to Lipofectamine 2000 and PEI. CDs/plasmid SOX9 NPs also exhibited good tunable fluorescence and dual functions of plasmid SOX9 delivery into MEFs with intracellular tracking of the NPs at great efficiency. Pandit et al. (2019) synthesized CDs functionalized with an amino acid with good tuneable properties. Because of the functionalization of the surface, these prepared CDs possessed protein sensing and other targeted drug delivery, including aptamer, monoclonal antibody, and various conjugations with good photoresponsive characteristics without any cytotoxicity. The functionalization of CDs with amino acid proline can be successfully engineered with folic acid. Another report has been investigated by Jiao and co-workers (2019) where FA (folic acid) functionalized on the surface of CDs and made ligand receptor-based drug delivery. The uptake of folate functionalized CDs was measured by the concentration of FR (folic acid receptor) positive cancerous MCF-7 cells (FR^{++}) and HepG-2 cells (FR^{+}) via receptor-mediated endocytosis, which was analyzed by a comparative study with non-cancerous FR-negative PC-12 cells (FR-).

4. Conclusion

Since the advent of carbon quantum dots or carbon dots (CDs), as it is simply referred to, it has significantly attracted the attention of the scientific community working on the development of carbon-based nanomaterials. It can be prepared by microwave irradiation, pyrolysis technique, laser ablation method, electrochemical method, arc discharge method, etc. However, most of the techniques of synthesis of CDs are simple with less sophistication leading to an easy synthetic protocol to synthesize CDs. Moreover, it is also easy to functionalize the CDs to tune their properties. The CDs possess exciting photophysical and photochemical properties. The photoluminescence properties especially up-conversion photoluminescence, Aggregation-Induced Emission (AIE), Chemiluminescence, and Electrochemiluminescence has been utilized extensively for varied applications. It has been used for developing chemical and biological sensors. Similar to other quantum dots, its bandgap too can be tuned and hence has been used in fabricating several optoelectronic devices. As CDs are less toxic, it has found applications in the biomedical field. CDs have been used extensively in bioimaging as well as in drug delivery. CDs have also become the material of choice for catalysis purposes because of their strong and broad optical absorption, high chemical stability, and fast electron transfer processes. The popularity, significance, and attraction toward the use of carbon dots can be reflected by the fact that in the last eight years, there has been a more than 600 times increase in the number of publications on carbon dots.

References

Anwar, S., Ding, H., Xu, M., Hu, X., Li, Z., Wang, J., Liu, L., Jiang, L., Wang, D., Dong, C., Yan, M., Wang, Q. and Bi, H. 2019. Recent advances in synthesis, optical properties, and biomedical applications of carbon dots. ACS Appl. Bio. Mater. 2: 2317-2338. DOI: 10.1021/acsabm.9b00112

Atchudan, R., Edison, T.N.J.I., Perumal, S., Muthuchamy, N. and Lee, Y.R. 2020. Eco-friendly synthesis of tunable fluorescent carbon nanodots from Malus floribunda for sensors and multicolor bioimaging. J. Photochem. Photobiol. A: Chemistry 390: 112336. DOI: 10.1016/j.jphotochem.2019.112336.

Baker, S.N. and Baker, G.A. 2010. Luminescent carbon nanodots: emergent nanolights. Angew. Chem. Int. Ed 49: 6726-6744. DOI: 10.1002/anie.200906623

Baruah, U., Deka, M.J. and Chowdhury, D. 2014. Reversible on/off switching of fluorescence via esterification of carbon dots. RSC Adv. 4: 36917-36922. DOI: 10.1039/c4ra04734f

Baruah, U., Konwar, A. and Chowdhury, D. 2016. Sulphonated carbon dots-chitosan hybrid hydrogel nanocomposite as an efficient ion-exchange film for Ca^{2+} and Mg^{2+} removal. Nanoscale 8: 8542-8546. DOI: 10.1039.C6NR01129B

Baruah, U. and Chowdhury, D. 2016. Functionalized graphene oxide quantum dot–PVA hydrogel: a colorimetric sensor for Fe^{2+}, Co^{2+} and Cu^{2+} ions. Nanotechnology 27: 145501-145516. DOI: 10.1088/0957-4484/27/14/145501

Bayan, R. and Karak, N. 2017. Photo-assisted synthesis of a Pd–Ag@CQD nanohybrid and its catalytic efficiency in promoting the Suzuki–Miyaura cross-coupling reaction under ligand-free and ambient conditions. ACS Omega 2: 8868–8876. DOI: 10.1021/acsomega.7b01504

Bhartiyaa, P., Singha, A., Kumara, H., Jaina, T., Singha, B.K. and Dutta, P.K. 2016. Carbon dots: chemistry, properties and applications. J. Indian Chem. Soc. 93: 1-8.

Boruah, J.S. and Chowdhury., D. 2020. Palmitic acid-carbon dot hybrid vesicles for absorption of uric acid. Appl. Nanosci. 10: 2207-2218. DOI: 10.1007/s13204-020-01374-2

Bui, T.T. and Park, S.Y. 2016. A carbon dot–hemoglobin complex-based biosensor for cholesterol detection. Green Chem. 18: 4245-4253. DOI: 10.1039/C6GC00507A

Cao, X., Wang, J., Deng, W., Chen, J., Wang, Y., Zhou, J. Du, P., Xu, W., Wang, Q., Wang, Q., Yu, Q., Spector, M., Yu, J. and Xu, X. 2018. Photoluminescent cationic carbon dots as efficient non-viral delivery of plasmid SOX9 and chondrogenesis of fibroblasts. Sci. Rep. 8: 7057. DOI: 10.1038/s41598-018-25330-x.

Chen, B.B., Liu, M.L. and Huang, C.Z. 2020. Carbon dot-based composites for catalytic applications. Green Chem. 22: 4034-4054. DOI: 10.1039/D0GC01014F.

Chowdhury, D., Gogoi, N. and Majumdar, G. 2012. Fluorescent carbon dots obtained from chitosan gel. RSC Adv. 2: 12156-12159. DOI: 10.1039/c2ra21705h.

Chowdhury, D., Majumdar, S. and Thakur, D. 2020. Actinobacteria mediated synthesis of bio-conjugate of carbon dot with enhanced biological activity. Appl. Nanosci. 10: 2199-2206. DOI: 10.1007/s13204-020-01392-0.

Costas-Mora, I., Romero, V., Lavilla, I. and Bendicho, C. 2014. In situ building of a nanoprobe based on fluorescent carbon dots for methylmercury detection. Anal. Chem. 86: 4536-4543. DOI: 10.1021/ac500517h.

Cui, Y., Zhang, C., Sun, L., Hu, Z. and Liu, X. 2015. Simple and efficient synthesis of strongly green fluorescent carbon dots with upconversion property for direct cell imaging. Part. Part. Syst. Charact. 32: 542-546. DOI: 10.1002/ppsc.201400221.

Deng, Y.H., Zhao, D.X., Chen, X., Wang, F., Song, H. and Shen, D.Z. 2013. Long lifetime pure organic phosphorescence based on water soluble carbon dots. Chem. Commun. 49: 5751-5753. DOI: 10.1039/C3CC42600A.

Deb, A., Saikia, R. and Chowdhury, D. 2019. Nano-bioconjugate film from aloevera to detect hazardous chemicals used in cosmetics. ACS Omega 4: 20394-20401. DOI: 10.1021/acsomega.9b03280.

Deb, A., Konwar, A. and Chowdhury, D. 2020. pH-responsive hybrid jute carbon dot-cotton patch. ACS Sustainable Chem. Eng. 8: 7394-7402. DOI: 10.1021/acssuschemeng.0c01221.

Deka, M.J., Dutta, P., Sarma, S., Medhi, O.K., Talukdar, N.C. and Chowdhury, D. 2019. Carbon dots derived from water hyacinth and their application as a sensor for pretilachlor. Heliyon 5: e01985. DOI: 10.1016/j.heliyon.2019.e01985.

Deka, M.J. and Chowdhury, D. 2017. Chiral carbon dots and their effect on the optical properties of photosensitizers. RSC Adv. 7: 53057-53063. DOI: 10.1039/C7RA10611D.

Doñate-Buendia, C., Torres-Mendieta, R., Pyatenko, A., Falomir, E., Fernández-Alonso, M. and Mínguez-Vega, G. 2018. Fabrication by laser irradiation in a continuous flow jet of carbon quantum dots for fluorescence imaging. ACS Omega 3: 2735-2742. DOI: 10.1021/acsomega.7b02082.

Dou, Q., Fang, X., Jiang, S., Chee, P.L., Lee, T.-C. and Loh, X.J. 2015. Multi-functional fluorescent carbon dots with antibacterial and gene delivery properties. RSC Adv. 5: 46817-46822. DOI: 10.1039/C5RA07968C.

D'souza, S.L., Chettiar, S.S., Koduru, J.R. and Kailasa, S.K. 2018. Synthesis of fluorescent carbon dots using Daucus carota subsp. sativus roots for mitomycin drug delivery. Opt. Int. J. Light Electron Opt. 158: 893-900. DOI: 10.1016/j.ijleo.2017.12.200.

Gayen, B., Palchoudhury, S. and Chowdhury, J. 2019. Carbon dots: a mystic star in the world of nanoscience. J. Nanomater 2019: 1-19. DOI: 10.1155/2019/3451307

Gao, N., Yang, W., Nie, H., Gong, Y., Jing, J., Gao, L. and Zhang, X. 2017. Turn-on theranostic fluorescent nanoprobe by electrostatic self-assembly of carbon dots with doxorubicin for targeted cancer cell imaging, in vivo hyaluronidase analysis, and targeted drug delivery. Biosens. Bioelectron. 96: 300-307. DOI: 10.1016/j.bios.2017.05.019.

Ge, J., Lan, M., Zhou, B., Liu, W., Guo, L., Wang, H., Jia, Q., Niu, G., Huang, X., Zhou, H., Meng, X., Wang, P., Lee, C.S., Zhang, W. and Han, X. 2014. A graphene quantum dot photodynamic therapy agent with high singlet oxygen generation. Nat. Commun. 5: 4596. DOI: 10.1038/ncomms5596

Gogoi, N. and Chowdhury, D. 2014. Novel carbon dot coated alginate beads with superior stability, swelling and pH responsive drug delivery. J. Mater. Chem. B 2: 4089–4099. DOI: 10.1039/c3tb21835j

Gogoi, N., Barooah, M., Majumdar, G. and Chowdhury, D. 2015. Carbon dots rooted agarose hydrogel hybrid platform for optical detection and separation of heavy metal ions. ACS Appl. Mater. Interfaces 7: 3058-3067. DOI: 10.1021/am506558d.

Gogoi, N., Agarwal, D.S., Sehgal, A., Chowdhury, D. and Sakhuja, R. 2017. One-pot synthesis of carbon nanodots in an organic medium with Aggregation-Induced Emission Enhancement (AIEE): a rationale for "Enzyme-Free" detection of cholesterol. ACS Omega 2: 3816-3827. DOI: 10.1021/acsomega.7b00643

Gogoi, J. and Chowdhury, D. 2020a. Calcium-modified carbon dots derived from polyethylene glycol: fluorescence-based detection of Trifluralin herbicide. J. Mater. 55: 11597-11608. DOI: 10.1007/s10853-020-04839-5

Gogoi, J., Shishodia, S. and Chowdhury, D. 2020b. Tunable electrical properties of carbon dot doped photo-responsive azobenzene-clay nanocomposites. RSC Adv. 10: 37545-37554. DOI: 10.1039/D0RA07386E.

Guo, B., Yu, K., Li, H., Qi, R., Zhang, Y., Song, H., Tang, Z., Zhu, Z. and Chen, M. 2017. Coral-shaped MoS_2 decorated with graphene quantum dots performing as a highly active electrocatalyst for hydrogen evolution reaction. ACS Appl. Mater. Interfaces 9: 3653-3660. DOI: 10.1021/acsami.6b14035.

Han, Y.Z., Huang, H., Zhang, H.C., Liu, Y., Han, X., Liu, R., Li, H. and Kang, Z. 2017. Carbon quantum dots with photo enhanced hydrogen-bond catalytic activity in aldol condensations. ACS Catal. 4: 781-787. DOI: 10.1021/cs401118x

He, J., He, Y., Chen, Y., Lei, B., Zhuang, J., Xiao, Y., Liang, Y., Zheng, M., Zhang, H. and Liu, Y. 2017. Solid-state carbon dots with red fluorescence and efficient construction of dual-fluorescence morphologies. Small 13: 1700075. DOI: 10.1002/smll.201700075

Hu, S.L., Niu, K.Y., Sun, J., Yang, J., Zhao, N.Q. and Du, X.W. 2009. One-step synthesis of fluorescent carbon nanoparticles by laser irradiation. J. Mater. Chem. 19: 484-488.

Huang, J.J., Zhong, Z.F., Rong, M.Z., Zhou, X., Chen, X.D. and Zhang, M.Q. 2014. An easy approach of preparing strongly luminescent carbon dots and their polymer based composites for enhancing solar cell efficiency. Carbon 70: 190-198. DOI: 10.1016/j.carbon.2013.12.092

Huang, S., Wang, L., Zhu, F., Su, W., Sheng, J., Huang, C. and Xiao, Q. 2015. A ratiometric nanosensor based on fluorescent carbon dots for label-free and highly selective recognition of DNA. RSC Adv. 5 (55): 44587-44597. DOI:10.1039/C5RA05519A.

Jia, X.F., Lia, J. and Wang, E.K. 2012. One-pot green synthesis of optically pH-sensitive carbon dots with upconversion luminescence. Nanoscale 4: 5572-5575. DOI: 10.1039/C2NR31319G.

Jia, J., Lin, B., Gao, Y., Jiao, Y., Li, L., Dong, C. and Shuang, S. 2019. Highly luminescent N-doped carbon dots from black soya beans for free radical scavenging, Fe^{3+} sensing and cellular imaging. Spectrochim. Acta Part A. Mol. Biomol. Spectrosc. 211: 363-372. DOI: 10.1016/j.saa.2018.12.034.

Jiang, K., Sun, S., Zhang, L., Lu, Y., Wu, A., Cai, C. and Lin, H. 2015. Red, green, and blue luminescence by carbon dots: full-color emission tuning and multicolor cellular imaging. Angew. Chem. Int. Ed. 127: 5450-5453. DOI: 10.1002/ange.201501193.

Jiao, Y., Sun, H., Jia, Y., Liu, Y., Gao, Y., Xian, M., Shuang, S. and Dong, C. 2019. Functionalized fluorescent carbon nanoparticles for sensitively targeted of folate-receptor-positive cancer cells. Microchem. J. 146: 464-470. DOI: 10.1016/j.microc.2019.01.003.

Konwar, A., Deb, A., Kar, A. and Chowdhury, D. 2019. Dual emission carbon dots from carotenoids: converting a single emission to dual emission. Luminescence 34: 790-795. DOI: 10.1002/bio.3685

Konwar, A., Baruah, U., Deka, M.J., Hussain, A.A., Haque, S.R., Pal, A.R. and Chowdhury, D. 2017. Tea-carbon dots-reduced graphene oxide: an efficient conducting coating material for fabrication of an e-textile. ACS Sustainable Chem. Eng. 5: 11645-11651. DOI: 10.1021/acssuschemeng.7b03021.

Konwar, A., Gogoi, N., Majumdar, G. and Chowdhury, D. 2015. Green chitosan–carbon dots nanocomposite hydrogel film with superior properties. Carbohydr. Polym. 115: 238-245. DOI: 10.1016/j.carbpol.2014.08.021

Konwar, A. and Chowdhury, D. 2015. Property relationship of alginate and alginate–carbon dot nanocomposites with bivalent and trivalent cross-linker ions. RSC Adv. 5: 2864-2870. DOI: 10.1039/c5ra09887d

Kwon, W., Do, S., Lee, J., Hwang, S., Kim, J.K. and Rhee, S.W. 2013. Freestanding luminescent films of nitrogen-rich carbon nanodots toward large-scale phosphor-based white-light-emitting devices. Chem. Mater. 25: 1893-1899. DOI: 10.1021/cm400517g.

Li, Y., Shu, H., Niu, X. and Wang, J. 2015. Electronic and optical properties of edge-functionalized graphene quantum dots and the underlying mechanism. J. Phys. Chem. C 119: 24950-24957. DOI: 10.1021/acs.jpcc.5b05935

Li, H., Zhai, J. and Sun, X. 2011. Sensitive and selective detection of silver(I) ion in aqueous solution using carbon nanoparticles as a cheap, effective fluorescent sensing platform. Langmuir 27: 4305-4308. DOI: 10.1021/la200052t.

Li, H., Liu, R., Kong, W., Liu, J., Liu, Y., Zhou, L., Zhang, X., Lee, S.T. and Kang, Z. 2014. Carbon quantum dots with photo-generated proton property as efficient visible light controlled acid catalyst. Nanoscale 6: 867-873. DOI: 10.1039/C3NR03996.

Li, H., Liu, J., Guo, S., Zhang, Y., Huang, H., Liu, Y. and Kang, Z. 2015. Carbon dots from PEG for highly sensitive detection of levodopa. J. Mater. Chem. B 3: 2378-2387. DOI: 10.1039/C4TB01983K.

Li, H., Sun, C., Ali, M., Zhou, F., Zhang, X. and MacFarlane, D.R. 2015. Sulfated carbon quantum dots as efficient visible-light switchable acid catalysts for room-temperature ring-opening reactions. Angew. Chem. Int. Ed. 54: 8420-8424. DOI: 10.1002/anie.201501698.

Li, H., Deng, Y., Liu, Y., Zeng, X., Wiley, D. and Huang, J. 2019. Carbon quantum dots and carbon layer double protected cuprous oxide for efficient visible light CO_2 reduction. Chem. Commun. 55: 4419-4422. DOI: 10.1039/C9CC00830F.

Li, H.T., Liu, R.H., Lian, S.Y., Liu, Y., Huang, H. and Kang, Z.H. 2013. Near-infrared light controlled photocatalytic activity of carbon quantum dots for highly selective oxidation reaction. Nanoscale 5: 3289-3297. DOI: 10.1039/C3NR00092C.

Liang, Z., Kang, M., Payne, G.F., Wang, X. and Sun, R. 2016. Probing energy and electron transfer mechanisms in fluorescence quenching of biomass carbon quantum dots. ACS Appl. Mater. Interfaces 8(27): 17478-17488. DOI: 10.1021/acsami.6b04826.

Lim, S.F., Riehn, R., Ryu, R.S., Khanarian, N., Tung, C., Tank, D. and Austin, R.H. 2006. In vivo and scanning electron microscopy imaging of upconverting nanophosphors in caenorhabditis elegans. Nano Lett. 6(2): 169-174. DOI: 10.1021/nl0519175.

Lin, Z., Xue, W., Chen, H. and Lin, J.M. 2012. Classical oxidant induced chemiluminescence of fluorescent carbon dots. Chem. Commun. 48: 1051-1053. DOI: 10.1039/C1CC15290D.

Liu, M.L., Chen, B.B., Li, C.M. and Huang, C.Z. 2019. Carbon dots: synthesis, formation mechanism, fluorescence origin and sensing applications. Green Chem. 21: 449-472. DOI: 10.1039/c8gc02736f.

Liu, Z.X., Chen, B.B., Liu, M.L., Zou, H.Y. and Huang, C.Z. 2017. Cu(I)-doped carbon quantum dots with zigzag edge structures for highly efficient catalysis of azide–alkyne cycloadditions. Green Chem. 19: 1494-1498. DOI: 10.1039/C6GC03288E.

Liu, R.H., Huang, H., Li, H.T., Liu, Y., Zhong, J., Li, Y.Y., Zhang, S. and Kang, Z. 2014. Metal nanoparticle/carbon quantum dot composite as a photocatalyst for high-efficiency cyclohexane oxidation. ACS Catal. 4: 328-336. DOI: 10.1021/cs400913h.

Liu, H., Li, Z., Sun, Y., Geng, X., Hu, Y., Meng, H., Ge, J. and Qu, L. 2018. Synthesis of luminescent carbon dots with ultrahigh quantum yield and inherent folate receptor positive cancer cell targetability. Sci. Rep. 8: 1-8. DOI: 10.1038/s41598-018-19373-3.

Lu, W., Qin, X., Liu, S., Chang, G., Zhang, Y., Luo, Y., Asiri, A.M., Al-Youbi, A.O. and Sun, X. 2012. Economical, green synthesis of fluorescent carbon nanoparticles and their use as probes for rapid, sensitive, and selective detection of mercury (II) ions. Anal. Chem. 84: 5351-5357. DOI: 10.1021/ac3007939.

Ma, Z., Zhang, Y.L., Wang, L., Ming, H., Li, H., Zhang, X., Wang, F., Liu, Y., Kang, Z. and Lee, S.T. 2013. Bioinspired photoelectric conversion system based on carbon-quantum-dot-doped dye–semiconductor complex. ACS Appl. Mater. Interfaces. 5: 5080-5084. DOI: 10.1021/am400930h.

Macina, A., Medeiros, T.V.D. and Naccache, R. 2019. A carbon dot-catalyzed transesterification reaction for the production of biodiesel. J. Mater. Chem. A 7: 23794-23802. DOI: 10.1039/ C9TA05245C.

Majumdar, S., Baruah, U., Majumdar, G., Thakur, D. and Chowdhury, D. 2016a. Paper carbon dot based fluorescence sensor for distinction of organic and inorganic sulphur in analytes. RSC Adv. 6: 57327-57334. DOI:10.1039/C6RA07476F.

Majumdar, S., Krishnatreya, G., Gogoi, N., Thakur, D. and Chowdhury, D. 2016b. Carbon-dot-coated alginate beads as a smart stimuli-responsive drug delivery system. ACS Appl. Mater. Interfaces 8: 34179-34184. DOI: 10.1021/acsami.6b10914

Majumdar, B., Sarma, D., Jain, S. and Sarma, T.K. 2018. One-pot magnetic iron oxide–carbon nanodot composite-catalyzed cyclooxidative aqueous tandem synthesis of quinazolinones in the presence of *tert*-butyl hydroperoxide. ACS Omega 3: 13711-13719. DOI: 10.1021/ acsomega.8b01794

Majumdar, S., Bhattacharjee, T., Thakur, D. and Chowdhury, D. 2018. Carbon dot based fluorescence sensor for retinoic acid. Chemistry Select. 3: 673-677. DOI: 10.1002/ slct.201702458

Mandani, S., Majee, P., Sharma, B., Sarma, D., Thakur, N., Nayak, D. and Sarma, T.K. 2017. Carbon dots as nanodispersants for multiwalled carbon nanotubes: reduced cytotoxicity and metal nanoparticle functionalization. Langmuir 33: 7622-7632.

Martindale, B.C., Hutton, G.A., Caputo, C.A. and Reisner, E. 2015. Solar hydrogen production using carbon quantum dots and a molecular nickel catalyst. J. Am. Chem. Soc. 137: 6018-6025. DOI: 10.1021/jacs.5b01650

Merki, D. and Hu, X. 2011. Recent developments of molybdenum and tungsten sulfides as hydrogen evolution catalysts. Energy Environ. Sci. 4: 3878-3888. DOI: 10.1039/ C1EE01970H.

Mirtchev, P., Henderson, E.J., Soheilnia, N., Yip, M. and Ozin, G.A. 2012. Solution phase synthesis of carbon quantum dots as sensitizers for nanocrystalline TiO_2 solar cells. J. Mater. Chem. 22(4): 1265-1269. DOI: 10.1039/C1JM14112K

Murray, C.B., Norris, D.B. and Bawendi, M.G. 1993. Synthesis and characterization of nearly monodisperse CdE (E = sulfur, selenium, tellurium) semiconductor nanocrystallites. J. Am. Chem. Soc. 115: 8706.

Pandit, S., Behera, P., Sahoo, J. and De, M. 2019. In situ synthesis of amino acid functionalized carbon dots with tunable properties and their biological applications. ACS Appl. Biomater. 2: 3393-3403. DOI: 10.1021/acsabm.9b00374.

Phan, L.M.T., Gul, A.R., Le, T.N., Kim, M.W., Kailasa, S.K., Oh, K.T. and Park, T.J. 2019. One-pot synthesis of carbon dots with intrinsic folic acid for synergistic imaging-guided photothermal therapy of prostate cancer cells. Biomater. Sci. 7: 5187-5196. DOI: 10.1039/ C9BM01228A.

Qian, Z. S., Chai, L.J., Huang, Y.Y., Tang, C., Shen, J.J., Chen, J.R. and Feng, H. 2015. A real-time fluorescent assay for the detection of alkaline phosphatase activity based on carbon quantum dots. Biosens. Bioelectron. 675-680. DOI: 10.1016/j.bios.2015.01.068.

Qian, J., Yan, J., Shen, C., Xi, F., Dong, X. and Liu, J. 2018. Graphene quantum dots-assisted exfoliation of graphitic carbon nitride to prepare metal-free zero-dimensional/two-dimensional composite photocatalysts. J. Mater. Sci. 53: 12103-12114. DOI: 10.1007/ s10853-018-2509-8

Reier, T., Oezaslan, M. and Strasser, P. 2012. Electrocatalytic Oxygen Evolution Reaction (OER) on Ru, Ir, and Pt catalysts: a comparative study of nanoparticles and bulk materials. ACS Catal. 2: 1765-1772. DOI: 10.1021/cs3003098

Ryplida, B., Lee, G., In, I. and Park, S.Y. 2019. Zwitterionic carbon dot-encapsulating pH-responsive mesoporous silica nanoparticles for NIR light-triggered photothermal therapy through pH-controllable release. Biomater. Sci. 7: 2600-2610. DOI: 10.1039/ C9BM00160C.

Salem, A.K., Searson, P.C. and Leong, K.W. 2003. Multifunctional nanorods for gene delivery. Nat. Mater. 2: 668-671. DOI: 10.1038/nmat974.

Sharma, A. and Das, J. 2019. Small molecules derived carbon dots: synthesis and applications in sensing, catalysis, imaging, and biomedicine. J. Nanobiotechnology 17: 1-24. DOI: 10.1186/s12951-019-0525-8

Sharma, S., Umar, A., Mehta, S.K., Ibhadon, A.O. and Kansal, S.K. 2018. Solar light driven photocatalytic degradation of levofloxacin using TiO_2/carbon-dot nanocomposites. New J. Chem. 42: 7445-7456. DOI: 10.1039/C7NJ05118B

Shi, W.B., Wang, Q.L., Long, Y.J., Cheng, Z.L., Chen, S.H., Zheng, H.Z. and Huang, Y. 2011. Carbon nanodots as peroxidase mimetics and their applications to glucose detection. Chem. Commun. 47: 6695-6697. DOI: 10.1039/C1CC11943E.

Su, Y., Liu, S., Guan, Y., Xie, Z., Zheng, M. and Jing, X. 2020. Renal clearable Hafnium-doped carbon dots for CT/Fluorescence imaging of orthotopic liver cancer. Biomaterials 255: 120110. DOI: 10.1016/j.biomaterials.2020.120110.

Sun, Y.P., Zhou, B., Lin, Y., Wang, W., Fernando, K.A.H., Pathak, P., Meziani, J.M., Harruff, B.A., Wang, X., Wang, H., Luo, P.G., Yang, H., Kose, M.E., Chen, B., Veca, L.M. and Xie, S.Y. 2006. Quantum-sized carbon dots for bright and colorful photoluminescence. J. Am. Chem. Soc. 128: 7756-7757. DOI: 10.1021/ja062677d

Sun, H.J., Zhao, A.D., Gao, N., Li, K., Ren, J.S. and Qu, X.G. 2015. Deciphering a nanocarbon-based artificial peroxidase: chemical identification of the catalytically active and substrate-binding sites on graphene quantum dots. Angew. Chem. Int. Ed. 54: 7176-7180. DOI: 10.1002/anie.201500626

Tabaraki, R. and Abdi, O. 2020. Fluorescent sensing of Pb^{2+} by microwave-assisted synthesized N-doped carbon dots: application of response surface methodology and Doehlert design. J. Iran. Chem. Soc. 17: 839-846. DOI: 10.1007/s13738-019-01815-y.

Tang, D., Liu, J., Wu, X.Y., Liu, R.H., Han, X., Han, Y., Huang, H., Liu, Y. and Kang, Z. 2014. Carbon quantum dot/NiFe layered double-hydroxide composite as a highly efficient electrocatalyst for water oxidation. ACS Appl. Mater. Interfaces 6: 7918-7925. DOI: 10.1021/am501256x.

Wang, J. and Qiu, J. 2016. A review of carbon dots in biological applications. J. Mater. Sci. 51: 4728-4738. DOI: 10.1007/s10853-016-9797-7

Wang, R., Lu, K.Q., Tang, Z.R. and Xu, Y.J. 2017. Recent progress in carbon quantum dots: synthesis, properties and applications in photocatalysis. J. Mater. Chem. A 5: 3717-3734. DOI: 10.1039/C6TA08660H

Wang, Y.Y., Li, Y., Yan, Y., Xu, J., Guan, B., Wang, Q., Li, J. and Yu, J. 2013. Luminescent carbon dots in a new magnesium aluminophosphate zeolite. Chem. Commun. 49: 9006-9008. DOI: 10.1039/C3CC43375G

Wang, Y., Dong, L., Xiong, R. and Hu, A. 2013. Practical access to bandgap-like N-doped carbon dots with dual emission unzipped from PAN@PMMA core–shell nanoparticles. J. Mater. Chem. C 1: 7731-7735. DOI: 10.1039/C3TC30949E

Wang, D.M., Lin, K.L. and Huang, C. 2019. Carbon dots-involved chemiluminescence: recent advances and developments. Luminescence 34: 4-22. DOI: 10.1002/bio.3570

Wang, R., Xu, Y., Zhang, T. and Jiang, Y. 2015. Rapid and sensitive detection of Salmonella typhimurium using aptamer conjugated carbon dots as fluorescence probe. Anal. Methods 7(5): 1701-1706. DOI 10.1039/C4AY02880E.

Wang, J., Wei, J., Su, S. and Qiu, J. 2015. Novel fluorescence resonance energy transfer optical sensors for vitamin B_{12} detection using thermally reduced carbon dots. New J. Chem 39(1): 501-507. DOI: 10.1039/c4nj00538d.

Wei, W., Xu, C., Wu, L., Wang, J., Ren, J. and Qu, X. 2014. Non-enzymatic-browning-reaction: a versatile route for production of nitrogen-doped carbon dots with tunable multicolor luminescent display. Sci. Rep. 4: 3564-3567. DOI: 10.1038/srep03564.

Wei,Y., Liu, H., Jin, Y., Cai, K., Li, H., Liu, Y., Kang, Z. and Zhang, Q. 2013. Carbon nanoparticle ionic liquid functionalized activated carbon hybrid electrode for efficiency enhancement in supercapacitors. New J. Chem. 37(4): 886-889. DOI: doi.org/10.1039/C3NJ41151F.

Wu, X., Guo, S. and Zhang, J. 2015. Selective oxidation of veratryl alcohol with composites of Au nanoparticles and graphene quantum dots as catalysts. Chem. Commun. 51: 6318-6321. DOI: 10.1039/C5CC00061K

Xavier, S.J., Siva, G., Annaraj, J., Kim, A.R., Yoo, D.J. and Kumar, G.G. 2018. Sensitive and selective turn-off-on fluorescence detection of Hg^{2+} and cysteine using nitrogen doped carbon nanodots derived from citron and urine. Sens. Actuators B. Chem. 259: 1133-1143. DOI: 10.1016/j.snb.2017.12.046.

Xu, X., Ray, R., Gu, Y., Ploehn, H.J., Gearheart, L., Raker, K. and Scrivens, W.A. 2004. Electrophoretic analysis and purification of fluorescent single-walled carbon nanotube fragments. J. Am. Chem. Soc. 126(40): 12736-12737. DOI: 10.1021/ja040082h

Xu, D., Lin, Q. and Chang, H.T. 2019. Recent advances and sensing applications of carbon dots. Small Methods 4(4): 1900387. DOI: 10.1002/smtd.201900387.

Xu, Q., Liu, Y., Gao, C., Wei, J., Zhou, H., Chen, Y., Dong, C., Sreeprasad, T.S., Li, N. and Xia, Z. 2015. Synthesis, mechanistic investigation, and application of photoluminescent sulfur and nitrogen co-doped carbon dots. J. Mater. Chem. C 3(38): 9885-9893. DOI: 10.1039/C5TC01912E.

Xu, Q., Wei, J., Wang, J., Liu, L., Li, N., Chen, Y., Gao, C., Zhang, W. and Sreeprased, T.S. 2016. Facile synthesis of copper doped carbon dots and their application as "turn-off" fluorescent probe in the detection of Fe^{3+} ion. RSC Adv. 6(34): 28745-28750. DOI: 10.1039/C5RA27658F.

Xu, Q., Su, R., Chen, Y., Sreenivasan, S.T., Li, N., Zheng, X., Zhu, J., Pan, H., Li, W., Xu, C., Xia, Z. and Dai, L. 2018. Metal charge transfer doped carbon dots with reversibly switchable, ultra-high quantum yield photoluminescence. ACS Appl. Nano Mater. 1(4): 1886-1893. DOI: 10.1021/acsanm.8b00277.

Yang, W., Wei, B., Yang, Z. and Sheng, L. 2019. Facile synthesis of novel carbon-dots/hemin nanoplatforms for synergistic photo-thermal and photo-dynamic therapies. J. Inorg. Biochem. 193: 166-172. DOI: 10.1016/j.jinorgbio.2019.01.018.

Yang, Y., Cui, J., Zheng, M., Hu, C., Tan, S., Xiao, Y., Yanga, Q. and Liu, Y. 2012. One-step synthesis of amino-functionalized fluorescent carbon nanoparticles by hydrothermal carbonization of chitosan. Chem. Commun. 48: 380-382. DOI: 10.1039/C1CC15678K

Yang, Y., Zhu, W., Shi, B. and Lü, C. 2020a. Construction of a thermo-responsive polymer brush decorated Fe_3O_4@catechol-formaldehyde resin core–shell nanosphere stabilized carbon dots/PdNP nanohybrid and its application as an efficient catalyst. J. Mater. Chem. A 8(7): 4017-4029. DOI: 10.1039/C9TA12614G.

Yang, Y., Duan, H., Xia, S. and Lü, C. 2020b. Construction of a thermo-responsive copolymer-stabilized Fe_3O_4@CD@PdNP hybrid and its application in catalytic reduction. Polym. Chem. 11: 1177-1187. DOI: 10.1039/C9PY01529A.

Yuan, T., Meng, T. and He, P. 2019. Carbon quantum dots: an emerging material for optoelectronic applications. J. Mater. Chem. C 7: 6820-6835. DOI: 10.1039/C9TC01730E

Zhang, J., Liu, X., Wang, X., Mu, L., Yuan, M., Liu, B. and Shi, H.J. 2018. Carbon dots-decorated $Na_2W_4O_{13}$ composite with WO_3 for highly efficient photocatalytic antibacterial activity. Hazard. Mater. 359: 1-8. DOI: 10.1016/j.jhazmat.2018.06.072J.

Zhang, Y.H., Xu, M.J., Li, H., Ge, H. and Bian, Z.F. 2018. The enhanced photoreduction of Cr(VI) to Cr(III) using carbon dots coupled TiO_2 mesocrystals. Appl. Catal., B 226: 213-219. DOI: 10.1016/j.apcatb.2017.12.053

Zhang, X., Zhang, Y., Wang, Y., Kalytchuk, S., Kershaw, S.V., Wang, Y., Wang, P., Zhang, T., Zhao, Y., Zhang, H., Cui, T., Wang, Y., Zhao//, J., Yu, W.W.// and Rogach, A.L. 2013.

Color-switchable electroluminescence of carbon dot light-emitting diodes. ACS Nano 7(12): 11234-11241. DOI: 10.1021/nn405017q.

Zhang, M., Yao, Q., Guan, W., Lu, C. and Lin, J.M. 2014. Layered double hydroxide-supported carbon dots as an efficient heterogeneous fenton-like catalyst for generation of hydroxyl radicals. J. Phys.Chem. C 118: 10441-10447. DOI: 10.1021/jp5012268.

Zhang, R. and Chen, W. 2014. Nitrogen-doped carbon quantum dots: facile synthesis and application as a "turn-off" fluorescent probe for detection of Hg^{2+} ions. Biosens. Bioelectron. 55: 83-90. DOI: 10.1016/j.bios.2013.11.074.

Zhang, X., Jiang, M. and Niu, N. 2018. Natural-product derived carbon dots: from natural products to functional materials. Chem. Sus. Chem. 11: 11-24. DOI: 10.1002/cssc.201701847

Zhang, Y., Zhang, X., Shi, Y., Sun, C., Zhou, N. and Wen, H. 2020. The synthesis and functional study of multicolor nitrogen-doped carbon dots for live cell nuclear imaging. Molecules 25(2): 306. DOI:10.3390/molecules25020306.

Zhao, Q., Song, W., Zhao, B. and Yang, B. 2020. Spectroscopic studies of the optical properties of carbon dots: recent advances and future prospects. Mater. Chem. Front. 4: 472-488. DOI: 10.1039/c9qm00592g.

Zhao, X., Zhang, J., Shi, L.L., Xian, M., Dong, C. and Shuang, S. 2017. Folic acid-conjugated carbon dots as green fluorescent probes based on cellular targeting imaging for recognizing cancer cells. RSC Adv. 7: 42159-42167. DOI: 10.1039/c7ra07002k

Zheng, X.T., Ananthanarayanan, A., Luo, K.Q. and Chen, P. 2015. Glowing graphene quantum dots and carbon dots: properties, syntheses, and biological applications. Small 11(14): 1620-1636. DOI: 10.1002/smll.201402648

Zhou, J., Booker, C., Li, R., Zhou, X., Sham, T.K., Sun, X. and Ding, Z. 2007. An electrochemical avenue to blue luminescent nanocrystals from multiwalled carbon nanotubes (MWCNTs). J. Am. Chem. Soc. 129(4): 744-745. DOI: 10.1021/ja0669070

Zhu, H., Wang, X., Li, Y., Wang, Z., Yang, F. and Yang, X. 2009. Microwave synthesis of fluorescent carbon nanoparticles with electrochemiluminescence properties. Chem. Commun. 5118-5120. DOI: 10.1039/B907612C

Zhu, S., Song, Y., Zhao, X., Shao, J., Zhang, J. and Yang, B. 2015. The photoluminescence mechanism in carbon dots (graphene quantum dots, carbon nanodots, and polymer dots): current state and future perspective. Nano Res. 8(2): 355-381. DOI: 10.1007/s12274-014-0644-3

Zhu, Y., Ji, X., Pan, C., Sun, Q., Song, W., Fang, L., Chen, Q. and Banks, C.E. 2013. A carbon quantum dot decorated RuO_2 network: outstanding super capacitances under ultrafast charge and discharge. Energy Environ. Sci. 6(12): 3665-3675. DOI: 10.1039/C3EE41776J

Zhu, Q.D., Zhang, L.H., Vliet, K.V., Miserez, A. and Holten-Andersen, N. 2018. White light-emitting multistimuli-responsive hydrogels with lanthanides and carbon dots. ACS Appl. Mater. Interfaces 10(12): 10409-10418. DOI:10.1021/acsami.7b17016

Zuo, J., Jiang, T., Zhao, X., Xiong, X., Xiao, S. and Zhu, Z. 2015. Preparation and application of fluorescent carbon dots. J. Nanomater. 1-13. DOI: 10.1155/2015/78786.

Carbon-Based Photocatalytic Nanomaterials for Clean Fuel Production

Saikumar Manchala[1]* and **Jaison Jeevanandam**[2]

[1] Department of Chemistry, Malla Reddy Engineering College (Autonomous), Maisammaguda, Dhulapally, Secunderabad - 500100, Telangana, India
[2] CQM-Centro de Química da Madeira, MMRG, Universidade da Madeira, Campus da Penteada, 9020-105 Funchal, Portugal

1. Introduction

In view of the rapid rising of the world population and development of science and technology, people of developing countries are urging for better living standards which subsequently resulted in overusage of traditional energy sources (Tonda et al. 2017). Furthermore, it has led to the rise in the cost of energy and depletion in the availability of traditional fuels. The emission of hazardous pollutants during the combustion of these fuels is not only causing the greenhouse effect and climate change but is also affecting the health of living beings (Ulmer et al. 2019). In this regard, H_2 is considered a clean fuel. So, the development of appropriate methods for the generation of H_2 is necessary. Photocatalytic water splitting can mimic natural photosynthesis, which involves the conversion of sunlight energy into chemical energy in the form of H_2 fuel, and it is an attractive avenue for clean energy generation and environmental safety (Chen et al. 2010, Wang et al. 2019b). Unfortunately, the performance of the presently developed photocatalysts does not meet the demands of large-scale practical applications. Therefore, the development of highly efficient visible light-active photocatalysts for renewable energy generation and environmental protection is still being sought.

Energy is a fundamental and necessary prerequisite demand for the development of modern society. One of the greatest challenges of the twenty-first century is sufficient energy supply, particularly for transport and household applications. The primary sources of energy are non-renewable fuels. Current overall energy utilization is quickly expanding but the amount of non-renewable fuels is limited, and it is

Corresponding author: smartsai@student.nitw.ac.in

approximated that the globe will need twice its current energy supply by the middle of the century and thrice by the end of the century. Unfortunately, the uncontrollable hazardous emissions from the utilization of non-renewable fuels will lead to adverse effects on the environment and human life (Sahaym et al. 2008, Balat et al. 2010, Dai et al. 2012). Present clean energy generation and environmental related issues are topping the list of global challenges in our society due to the exponential progress of the human population and intensification of economic development. Therefore, the development of alternative pollution-free supplies and clean energy technologies for green remediation in the sustainable growth of human society is an urgent need (Lewis et al. 2006, Chang et al. 2016, Tian et al. 2016).

Recently, the significance of hydrogen (H_2) energy has become more popularized and identified, particularly for its use in fuel cells (Nizovskii et al. 2015). The H_2 can be generated through various renewable sources (Manchala et al. 2020a). Compared to all renewable sources, such as wind, solar, geo- and hydrothermal, utilization of sunlight energy for the generation of H_2 from water splitting offers a prominent eco-friendly method at ambient conditions (Hosseini et al. 2016). At present, non-renewable technologies such as steam-reforming of natural gas and methane and coal-gasification are the major sources of H_2 generation on an industrial scale and hence the development of new and renewable energy technologies is in great demand (Nabgan et al. 2017). Mainly, H_2 can be produced by two energy resources, renewable and non-renewable. If one is concerned about environmental issues, H_2 can be produced from water using natural sunlight energy that has the advantage over the other resources as this method emit environment-friendly gases as the final product (Hosseini et al. 2016, Huang et al. 2017a). Therefore, photocatalytic water splitting using a semiconductor catalyst is an interesting and prominent method for solving energy and environmental issues (Manchala et al. 2018). The H_2 generation from water splitting using catalysts and sunlight is not new. For about 45 years, water splitting has been established and investigated in the areas of photochemistry, electrochemistry, catalysis, organic and inorganic chemistry. Since the Honda-Fujishima effect was studied using a TiO_2 semiconductor electrode (Li et al. 2016, Yang et al. 2017). Although the reported catalyst is able to photo-decompose water into H_2 and O_2 in a stoichiometric amount with a reasonable activity, it is yet to be seen. Here, sustained photo-decomposition of water has been achieved under conditions where photogenerated electrons and holes are separated for maximum photoreaction yield.

2. Reason and Advantages of Hydrogen as an Energy Carrier

The present energy and transport framework, which is based predominantly on fossil fuels, such as gasoline, natural gas, and diesel, cannot be considered sustainable. There is a growing interest in H_2 as an energy carrier and world over researchers are associated with making it an industrially accessible source of energy. The following features depict that H_2 is the best energy source:

1. Abundant element on the earth.
2. Lightest element.
3. Provides clean energy.
4. Stored readily compared to electricity.
5. Can be generated from water.
6. Decentralized production is possible.
7. Energy storage in remote zones.
8. Very proficient when utilized in energy units like fuel cells.
9. High energy per unit mass (Table 1).

Table 1: The heat of combustion of fuels with air

Fuel	Energy (k cal g^{-1})
Hydrogen	34.0
Coal	7.8
Gasoline	10.3-8.4
Oil	9.4
Wood	4.2

3. Challenges of Photocatalysis

However, photocatalytic water-splitting technology has been considered as a sustainable strategy for clean energy generation, still it experiences some major problems as follows:

(i) Photocatalysts absorb the visible light spectrum of sunlight for the water splitting technique is limited.

(ii) Suffers from fast recombination.

(iii) Poor surface-interface reactions between photocatalysts and the adsorbed species in suspension.

Therefore, the current research in the field of photocatalytic water splitting mainly focuses on the search for active photocatalytic materials with bandgap corresponding to the energy of visible light for efficient utilization of sunlight, which contributes 5% UV and 45% visible light. Recently, significant efforts have also been made to overcome the above-mentioned problems for modified semiconductor photocatalysts that are capable of using visible light, including metal ion doping, non-metallic element doping, creating oxygen vacancies, dye sensitization, and coupling with metal or other semiconductors.

4. Benefits and Roles of Carbon Nanomaterials in Carbon-Based Photocatalysis

The development of carbon-based nanostructured materials has been selected as one of the best methods to extend the photoresponse of materials into the visible region. Carbon nanomaterials, especially graphene, CQDs, and CNs, are commonly used as supporting materials to anchor semiconductor photocatalysts and noble-metal nanoparticles (Cao et al. 2016). This strategy is imputed to the various advantages of these carbon nanomaterials. They are shown in Figures 1 and 2.

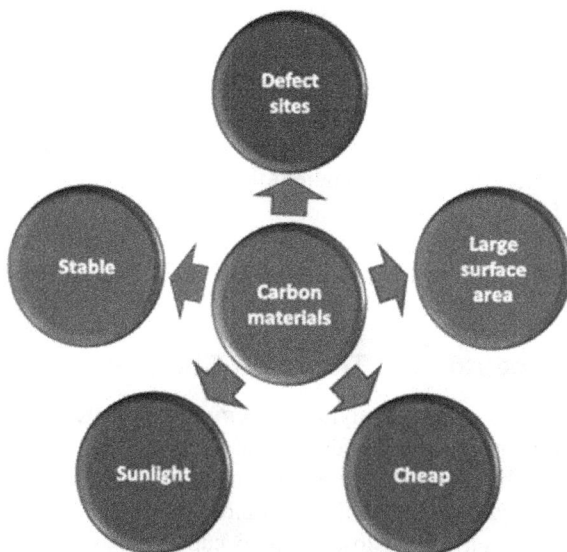

Figure 1: Advantages of carbon nanomaterials for serving as supporting materials in photocatalytic water splitting.

Figure 2: Roles of carbon nanomaterials for H_2 generation from photocatalytic water splitting.

4.1 Supporting Material for Enhanced Structural Stability

(1) These materials have a large surface area and provide a structure on which nanoparticles can be immobilized and distributed.

(2) These materials have oxygen-containing groups and defect sites, which are able of providing numerous nucleation sites for the anchor and growth of noble metal nanoparticles.

(3) These materials are thermally inert and chemically stable, thus can maintain their structure and properties throughout the photocatalytic reaction.

(4) Carbon is the fourth abundant element on the earth; nowadays, graphene, CQDs, and CNs can be easily and economically prepared on a large scale.

(5) These materials are lightweight, which is essential for a supporting material (Li et al. 2011, Cao et al. 2016).

4.2 Electron Acceptor and Transport Channel

Usually single-segment photocatalytic systems are suffering from a high recombination rate of photo-generated excitons without moving to the surface of the catalyst and/or further to reactive species, which directly affects the efficiency of photocatalytic H_2 generation. The high work function and high electrical conductive nature of carbon nanomaterials allow them to capture photo-generated electrons from the conduction band of semiconductor photocatalysts (Zhang et al. 2010, Dong et al. 2012). These captured electrons by carbon nanomaterials are then quickly transferred to the reactive species. In this manner, the carbon nanomaterials serve as excellent electron acceptors and transport channels to increase the lifetime of photo-generated excitons and thus result in efficient photocatalytic H_2 generation (Tan et al. 2012, Xie et al. 2013, Yue et al. 2015).

4.3 Increasing Adsorption and Active Sites

Carbon nanomaterials can enhance the surface areas of anchored semiconductor or noble metal nanoparticles, improve the dispersion for more adsorption and increase the active sites for photocatalytic reaction due to their large specific surface area. Theoretical results have demonstrated that carbon nanomaterials can promote the adsorption and splitting of H_2O molecules, which is beneficial for H_2 generation. This is because photocatalytic reactions occur not only on the interface between carbon nanomaterials and semiconductors but also on the surface of carbon nanomaterials (Ambrosetti et al. 2011, Ma et al. 2011, Chen et al. 2013, Xie et al. 2013, Byeon et al. 2014).

4.4 Co-catalyst

The introduction of co-catalysts into the photocatalytic systems is a well-known method to improve charge separation, lowering the activation energy, reducing the over-potential, and providing abundant catalytic sites for efficient H_2 evolution. Usually, noble metals, including Pt, Au and Pd, have displayed significant H_2 production when utilized as co-catalysts because of their higher work function when compared to that of semiconductor photocatalysts (Yang et al. 2013, Ran et al. 2014).

However, it is a fact that noble metals are expensive and rare. Therefore, an alternative solution for noble metal co-catalysts with low cost is highly advisable for the designing of efficient and economical photocatalytic systems. In this regard, carbon nanomaterials are good candidates and have been widely explored as co-catalysts for photocatalytic H_2 generation (Li et al. 2014a, Lang et al. 2015, Manchala et al. 2019c); additionally, the carbon materials can provide some more additional features compared to that of noble metal co-catalysts, such as extended light absorption and provide more adsorption sites due to the large surface area (Gan et al. 2014).

4.5 Photocatalyst

Theoretical and experimental studies have also described that carbon nanomaterials can be used as photocatalysts for H_2 evolution due to the semiconductor nature of GO and RGO (Manchala et al. 2019b, Manchala et al. 2020b). While the semiconducting carbon nanomaterials can have more negative LUMO positions than the reduction potential of H^+/H_2 (Meng et al. 2013, Yeh et al. 2013).

4.6 Photosensitization

Carbon nanomaterials can also exhibit dye-like characteristics and thus could act as a photo-sensitizer to extend the light-harvesting property of a semiconductor photocatalyst with additional photo-generated electrons (Manchala et al. 2019a). When the LUMO of carbon nanomaterials such as graphene is more negative than the CB of the integrated semiconductor like ZnS so that the photo-generated electrons in graphene can be transferred to the CB of the ZnS and is then involved in H_2 generation (Zhang et al. 2011, Wang et al. 2014).

4.7 Bandgap Narrowing Effect

Theoretical (Li et al. 2014b, Yuan et al. 2015) and experimental studies revealed (Shaban et al. 2008, Sun et al. 2014a, Sun et al. 2014b) that carbon nanomaterials can form specific chemical bonding (e.g., metal-O-C bonds) with the semiconductor photocatalysts. Such strong interactions between them would lead to a bandgap narrowing effect on the semiconductor photocatalysts, which further results in the efficient photocatalytic H_2-generation performance.

5. Different Types of Carbon-based Photocatalytic Nanomaterials

In recent times, there are numerous carbon-based photocatalytic nanomaterials that are fabricated for solar fuel production. These carbon nanomaterials can be classified into graphene, CNTs, CQDs, CNs, CNFs, fullerenes, nanocarbon, and activated carbon nanoparticles as displayed in Figure 3.

5.1 Graphene Nanostructures

Graphene is a two-dimensional nanostructure that can be obtained from carbon materials, where the six carbon atoms are arranged in a hexagonal shape and these

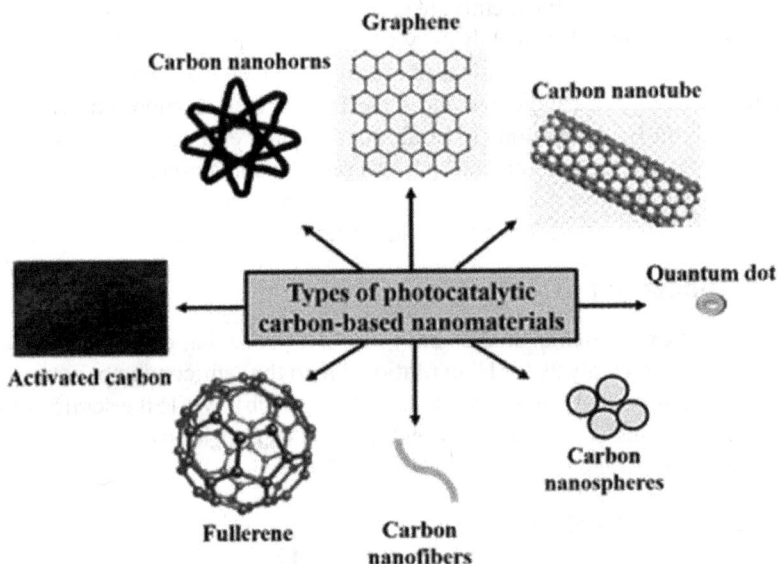

Figure 3: Distinct carbon nanomaterials for photocatalytic applications.

hexagonal structures are arranged horizontally. It is noteworthy that individual graphene does not possess photocatalytic activity, whereas it acts as a support material to improve the photocatalytic property of other carbon or metallic nanoparticles (Jeevanandam et al. 2019, Mateo et al. 2018). Similarly, Safajou et al. (2017) reported the synthesis of novel graphene-palladium-titanium dioxide nanocomposites via combinatorial photo-deposition and hydrothermal approaches. In this study, titanium oxide was fabricated into both nanosized particles and wires, and its photocatalytic ability to degrade toxic Rhodamine B dye was evaluated under ultraviolet light irradiation. The results revealed that the composite with nanowires possesses enhanced photocatalytic property due to the surface area of titanium dioxide substrate (Safajou et al. 2017).

Likewise, Mahdiani et al. (2018) grafted copper-hexaferrite nanoparticles on the surface of CNTs and graphene to improve their photocatalytic activity. In this study, copper-hexaferrite nanostructures were synthesized using amino acids via the sol-gel method. These nanostructures were grafted on graphene and CNT surfaces via a pre-graphenization approach followed by sonication and microwave irradiation. The novel nanostructure was evaluated to degrade toxic organic pollutant dyes, such as erythrosine, methylene white, and methylene blue under ultraviolet radiation. The results showed that the copper-hexaferrite nanostructure in graphene possesses the ability to improve electrochemical properties, shift absorption edge to higher energy, and elevate photocatalytic activity to degrade toxic dyes compared to CNT nanocomposite. The study revealed that the photocatalytic activity is due to the high surface of graphene and elevates bandgap via Burstein effect for enhanced photocatalytic activity under UV irradiation (Mahdiani et al. 2018). Furthermore, Pu et al. (2017) dispersed titanium dioxide nanoparticles on the surface of graphene via a facile *in-situ* design strategy for the effective photocatalytic Rhodamine 6G degradation. The results of the study showed that the 10 wt.% of graphene-nanoparticle combination has degraded toxic Rhodamine 6G dye after visible light irradiation for up to 120 minutes. The study emphasized that the graphene-nanoparticle composite shows 3.5 times higher photocatalytic activity compared to base titanium dioxide particles (Pu et al. 2017). Furthermore, Raizada et al. (2019) listed several graphenes coupled zinc oxide nanostructures with the potential photocatalytic ability for effective wastewater treatment. The authors revealed that the graphene-based nanomaterials and reduced graphene oxide that are coupled with highly porous zinc oxide nanoparticles possess enhanced ability to photo-catalytically reduce the contamination in wastewater via recombination of electron-hole pair with the visible light adsorption capability (Raizada et al. 2019).

5.2 Carbon Nanotubes

Similar to graphene, CNTs also attracted researchers in the field of photocatalysts. CNTs are one-dimensional carbon nanostructures with exclusive properties due to their quantum confinement abilities and high surface to volume ratio. Single and multi-walled are the two main classes of CNTs, which were classified based on the number of one-dimensional carbon structures that can be present by folding graphene layers (Dresselhaus et al. 2000). Free CNTs possess high mechanical strength (Zhu et al. 2018) and are used to prepare nanocomposites for improving the efficiency of the photocatalytic nanomaterials (Bazli et al. 2019). Sapkota et al. (2019) fabricated a novel zinc oxide-single walled CNTs (SWCNTs) nanocomposites that can efficiently degrade persistent organic water pollutant dye via solar light-driven photocatalytic effect. In this study, the authors developed a one-pot-two-chemical re-crystallization technique and thermal decomposition for the fabrication of nanocomposite. The results revealed that the nanocomposite is highly effective in degrading toxic methylene blue dye under the influence of sun (visible light) irradiation compared to pristine SWCNTs and zinc oxide nanorods with the reusability of up to five cycles (Sapkota et al. 2019). Likewise, Payan et al. (2018) synthesized titanate nanotube with SWCNTs as porous nanocomposite to photo-catalytically degrade toxic 4-chlorophenol dye under the influence of UV and solar irradiation. The two-step hydrothermal process was utilized in the fabrication of porous nanocomposite with tubular structures that are uniformly interwoven with each other. The results showed that the titanate nanotubes loaded with 10% and 5% of SWCNT under UV and solar irradiation, respectively, possess significant photocatalytic activity compared to free nanotubes. Also, the study showed that there is an 8% decrement in the stability of the nanocomposite after four cycles, which is low compared to other nanomaterials (Payan et al. 2018). Furthermore, Chatterjee et al. (2017) reported the fabrication of nanocomposite with polyaniline and SWCNTs to perform as a potential photocatalytic agent for the effective degradation of methyl orange and Rose Bengal under solar (visible) light irradiation. The nanocomposite was prepared via aniline monomer polymerization with SWCNT and sulfo-salicylic acid in *in-situ* conditions. The results showed that 2 wt.% of the nanocomposite degrades Rose Bengal and methyl orange with a degradation efficiency of 95.91% and 90.34%, within 10 and 30 minutes, respectively (Chatterjee et al. 2017). Moreover, titanium dioxide-SWCNTs (Ling et al. 2016), titanium oxide shell in SWCNT-fullerodendrone coaxial nanowires (Kurniawan et al. 2017), and ZnO-SWCNTs (Ngai et al. 2016) are the other recent nanocomposites that are synthesized as nanocomposites for the effective photocatalytic activity to degrade toxic pollutants in the presence of a light source (UV or sunlight).

Multi-walled CNTs (MWCNTs) are easier to fabricate compared to SWCNTs due to which they are widely used in photocatalytic pollutant degradation applications. Bellamkonda et al. (2019) fabricated a novel, highly stable, and active nanohybrid using MWCNTs, graphene, and titanium dioxide particles to be useful as a potential photocatalytic agent for water splitting and H_2 production application. The study emphasized that the photocatalytic activity of the nanohybrid is better than graphene-titanium dioxide and free anatase titanium dioxide nanoparticles. Furthermore, the H_2 production rate of nanohybrid is identified to be 29 mmol/h/g under solar visible light irradiation, which is 8-fold higher than the commercial titanium dioxide with 14.6% of estimated solar energy conversion efficiency (Bellamkonda et al. 2019). Besides, Hasan et al. (2020) reported the synthesis of pristine MWCNTs, which are coupled with beta-cyclodextrin and polyaniline hybrid to photo-catalytically degrade crystal violet via advanced oxidative degradation. The study showed that the optimized values such as 120 minutes of radiation time, pH 5 of dye solution, 15 mg of photocatalyst dose, and the dye concentration of 135 mg/L have led to the dye degradation of 93.54-95.39% (Hasan et al. 2020). Also, Gopannagari et al. (2018) demonstrated the influence of surface-functionalized MWCNTs with cadmium sulfide nanohybrids is beneficial for the enhanced photocatalytic production of H_2. The surface functionalization of CNT-cadmium sulfide nanorods with ascorbic acid, amine, and the sulfonic group has been identified to increase H_2 production. Furthermore, platinum incorporated ascorbic acid-functionalized MWCNTs-cadmium sulfide rod nanohybrid can lead to 120.1 mmol/h/g of the highest H_2 production rate, which is 49-fold higher than the pure cadmium sulfide (Gopannagari et al. 2018). Likewise, porous MWCNTs-titanium dioxide (Zouzelka et al. 2016), magnetic MWCNTs-cerium dioxide (Feng et al. 2019), and electrospun MWCNTs-bismuth vanadium oxide (Ye et al. 2019) are the recently synthesized novel nanosized composites that are utilized for the enhanced photocatalytic degradation of pollutants as well as H_2 production.

5.3 Quantum Dots

Quantum dots are another significant carbon-based nanoparticles, which is zero-dimensional in structure with effective photocatalytic ability due to their unique quantum confinement (Wang et al. 2017, Li et al. 2019). Xie et al. (2018) synthesized nanocomposites using tin oxide and graphene quantum dots to be beneficial as a potential photocatalytic agent for the effective oxidation of nitric oxide under visible light irradiation. The study demonstrated that the nanocomposite with 1% graphene quantum dot possesses the highest photocatalytic ability to degrade 57% of the indoor nitric oxide with 5% of nitrogen oxide generated fraction selectivity (Xie et al. 2018). Similarly, Zou et al. (2016) synthesized novel graphene quantum dots hybridized with metal-free graphitic-carbon nitride to be useful as a visible-light-driven photocatalytic agent for the evolution of H_2. The study showed that the nanohybrid with graphene quantum dots were fabricated using melamine, urea, and dicyandiamide as a precursor. The results revealed that the quantum dot-based nanohybrid, prepared using urea, can lead to 2.18 mmol/h/g of H_2 production rate, which is 2.16 times of simple graphitic carbon nitride with 5.25% of quantum efficiency at 420 nm (Zou et al. 2016). Likewise, Lei et al. (2017) demonstrated the formation of graphene quantum dots that are strongly coupled with cadmium sulfide to form nanohybrids, which has been revealed to possess the enhanced photocatalytic ability for H_2 production. The study showed that the nanohybrid with 1 wt.% of graphene quantum dots can lead to superior H_2 production of 95.4 μmol/h upon visible light irradiation, which is 2.7 times higher than free cadmium sulfide nanoparticles with 4.2% of quantum efficiency at 420 nm (Lei et al. 2017). Besides, phosphorus dopants (Qian et al. 2018), zinc oxide (Kumar et al. 2018), graphitic carbon nitride nanorods (Yuan et al. 2019), nitrogen and sulfur co-doped with graphene and titanium dioxide (Tian et al. 2017) were also combined with graphene quantum dots to form nanocomposites in recent time to exhibit enhanced photocatalytic property.

Recently, Lin et al. (2019) demonstrated the fabrication of novel ternary heterostructure photocatalyst using graphitic carbon nitride quantum dots and nitrogen-doped CQDs that are co-decorated with bismuth vanadium oxide microspheres. These novel heterostructures were identified to exhibit enhanced ability to degrade Rhodamine B dye and tetracycline antibiotics to purify wastewater under the illumination of visible light. The results showed that 98% of both the pollutants were degraded by the heterostructures after 100 minutes of visible light irradiation via enhanced photocatalytic and effective adsorption-desorption equilibrium (Lin et al. 2019). Furthermore, Wang et al. (2019) synthesized sulfur-doped CQDs that are loaded in hollow tubular graphitic carbon nitride as an effective photocatalytic agent. These quantum dot-based photocatalysts exhibited enhanced ability to inhibit *Escherichia coli* bacteria as well as degrade toxic tetracycline in the wastewater under the influence of visible light. The results also showed that the novel quantum dot-based photocatalysts possess excellent performance to degrade 99% of bacteria in water with a reaction rate of 0.0293/minute against tetracycline (Wang et al. 2019a). Furthermore, Huang et al. (2017) reported the synthesis of novel zinc-iron oxide-CQDs to be beneficial as a potential photocatalyst for the purification of air. In this study, the 15 vol.% of quantum dot nanocomposite has been identified to possess eight times superior photocurrent transient response compared to zinc-iron oxide nanoparticles. These photocatalytic nanocomposites were further identified to be able to remove nitrous oxide from the air with high selectivity toward the formation of nitrate via visible light irradiation (Huang et al. 2017b). Moreover, ultrathin graphitic carbon nitride (Wang et al. 2018b), copper nanoparticles (Zhang et al. 2017), silver iodide-zinc oxide-phosphorus doped with graphitic carbon nitride (Hasija et al. 2019), and porous graphene-graphitic carbon nitride nanosheet aerogels (He et al. 2018) were combined with CQDs recently to exhibit significant photocatalytic activity.

5.4 Carbon Nanospheres

Carbon nanospheres are three-dimensional carbon-based nanostructures that are commonly blended with metal or other catalytic nanoparticles to improve their photocatalytic property (Gyulavári et al. 2019). Singhal et al. (2018) deposited silver over zinc oxide-CNs via a chemical precipitation-assisted self-assembly approach, where pyrolysis of benzene was utilized for carbon nanosphere fabrication. The study showed that the zinc oxide-CNs possess the ability to reduce ~85.6% of the toxic methylene blue dye after the irradiation of UV light for 25 minutes. The deposition of silver on the carbon nanosphere nanocomposite has improved the photocatalytic dye degradation to ~95% after 15 minutes of UV light irradiation with high stability for up to five cycles (Singhal et al. 2018). Furthermore, Luo et al. (2020) reported the fabrication of amide-functionalized CNs with superior adsorption performance via liquid phase reaction. The study showed that the amide-functionalized CNs have increased 6.7 times of adsorption capacity compared to non-functionalized nanospheres. Also, eosin Y dye-sensitized amide-functionalized CNs along with platinum co-catalyst are reported to exhibit a high H_2 generation activity of 607.4 μmol in 2 hours with a quantum efficiency of 32.9% at 430 nm, which is 14.3 times higher than free platinum and CNs (Luo et al. 2020a). Furthermore, Liu et al. (2020) recently demonstrated that CNs modified with graphitic carbon nitride and doped with nitrogen can exhibit excellent photocatalytic activity. In this study, the hydrothermal method was used to load mono-dispersed nitrogen-doped CNs onto the surface of graphitic carbon nitride for the effective photocatalytic degradation of a toxic chemical named sulfachlorpyridazine. The results revealed that the toxic chemical was removed by a factor of 4.7 and 3.2 under the influence of UV and visible light, respectively, by the nanosphere composite via visible light harvesting and charge mobility mechanisms (Liu et al. 2020a). Besides, Chen et al. (2018) reported a novel fabrication of hollow CNs blended as a composite with copper-titanium dioxide that is coated with surface mounted metal oxide framework via liquid-phase epitaxial layer by layer immersion approach. The resultant novel nanocomposite structure was hollow, uniform, thin, and homogenous with enhanced photocatalytic H_2 production ability under simulated sunlight irradiation and superior recyclability (Chen et al. 2018). In addition, Bakos et al. (2016) synthesized a unique core-shell structure of carbon nanosphere and titanium dioxide composite via an atomic layer deposition approach for photocatalytic applications (Bakos et al. 2016). Recently, Gupta et al. (2019) synthesized CNs using saccharides, such as glucose (polysaccharide), sucrose (disaccharide), and xylose (monosaccharide). The resultant nanospheres are amorphous with a bandgap of 2.2-3.4 eV, which makes them a suitable photocatalytic agent for the organic waste and methylene blue dye degradation under the influence of UV light with 0.3713/hour of rate constant (Gupta et al. 2019).

5.5 Carbon Nanofibers

Apart from other carbon structures, CNFs are recently introduced as a significant photocatalyst for several specific applications. Pant et al. (2016) fabricated a silver-zinc oxide photocatalyst that is anchored on the surface of CNFs. The novel nanocomposite was synthesized via electrospinning and hydrothermal approach.

The resultant nanocomposite was identified to degrade ~95% of methylene blue dye under the irradiation of UV light for 60 minutes. Also, the study showed that the photocatalytic activity is due to the 45% of light adsorption by the nanocomposite, whereas free pristine zinc oxide particles exhibited only ~2% of adsorption (Pant et al. 2016). Likewise, Fernandes et al. (2019) recently demonstrated that the zinc oxide embedded in CNFs can be utilized as a potential photocatalyst for selective alcohol oxidation application. The study confirmed that the addition of zinc oxide with CNFs increased their surface area, which eventually elevated their photocatalytic property. It can be noted that the zinc oxide with 10% of CNFs increased the reaction rate of vanillin formation from vanillyl alcohol by a factor of 2.5 compared to free zinc oxide. It has also been identified that these photocatalytic nanocomposites possess enhanced ability in the synthesis of other aromatic aldehydes, such as piperonal, anisaldehyde, and benzaldehyde (Fernandes et al. 2019). Similarly, bismuth sub-carbonate-molybdenum disulfide-CNF nanocomposites were formed in a recent study via electro-spinning, hydrothermal, and ultra-sonication approaches. The resultant nanocomposite exhibited enhanced photocatalytic activity for the reduction of toxic nitrous oxide in the air. The results showed that a low concentration of 60 PPB (parts per billion) of the nanocomposite can lead to nitrous oxide degradation with 68% of efficiency under visible light irradiation (Hu et al. 2017). Moreover, Han et al. (2016) fabricated novel graphitic carbon nitride-nitrogen-rich CNFs for the enhanced production of H_2 via a photocatalytic approach without co-catalysts. The study has led to the preparation of an interconnected mesoporous framework of graphitic-carbon nitride nanofibers that are merged with *in-situ* incorporated nitrogen-rich carbon, the resultant nanofiber composite exhibited an enhanced H_2 evolution rate of 16,885 μmol/h/g with 14.3% of quantum efficiency at 420 nm (visible light region) without any co-catalysts (Han et al. 2016). In addition, one-dimensional CNFs-titanium dioxide core-shell nanocomposite (Zhang et al. 2018), zinc oxide-titanium dioxide-CNF (Pant et al. 2020), cellulose-derived CNF-bismuth oxide bromide composites (Geng et al. 2018), and foxtail-like molybdenum disulfide on one-dimensional CNFs (Liang et al. 2020) are widely synthesized in recent times to exhibit significant photocatalytic activity.

5.6 Fullerenes

Fullerenes are zero-dimensional nanostructures that are formed by arranging 60 carbon atoms in a spherical morphology (Coro et al. 2016). In recent times, fullerenes were combined with nanoparticles as composites to enhance their photocatalytic property (Pan et al. 2020). Tahir et al. (2018) recently identified the significant role of fullerene to improve the photocatalytic, and the H_2 evolution ability of tungsten oxide. In this study, tungsten oxide-fullerene composite was prepared via a hydrothermal approach for the effective photocatalytic and H_2 production ability using visible light irradiation. It has been identified that the presence of 4% fullerene with tungsten oxide has significantly improved the photocatalytic performance of the composite by improving their surface area, extending the visible light region, and inhibiting recombination losses (Bilal Tahir et al. 2018). Recently, Regulska et al. (2019) demonstrated the synthesis of fullerene functionalized with zinc porphyrin for the enhanced titania sensitization via a visible light-mediated photocatalytic approach. The study showed that the functionalized fullerene possesses enhanced photocatalytic activity to reduce pollutants, such as methylene blue and phenol present in natural rivers and wastewater under visible light irradiation (Regulska et al. 2019). Furthermore, Shahzad et al. (2019) modified the performance of heterogenous tungsten oxide-fullerene-nickel boride-nickel hydroxide composite for enhanced H_2 production. In this study, needle-like nanostructures of nickel boride-nickel hydroxide were integrated into the space between fullerene and tungsten oxide interface. The resultant nanostructure with 1.5% of nickel boride-nickel hydroxide as co-catalyst has led to the H_2 evolution rate of up to 1,578 µmol/h/g, which is 9.6 times higher than the pure photocatalyst (Shahzad et al. 2019). Moreover, Qi et al. (2016) enhanced the photocatalytic performance of titanium dioxide in the anatase crystal structure by modifying them using fullerene. In this study, a solvothermal approach was utilized to modify titanium dioxide by the addition of 2 wt.% fullerenes, which has narrowed the bandgap of the composite and included an additional state of doping between conduction and valence band that has eventually improved their photocatalytic property (Qi et al. 2016). Besides, Martínez-Agramunt and Peris (2019) demonstrated the photocatalytic activity of palladium metallosquare, which is encapsulated in fullerenes for the enhanced generation of singlet oxygen production. In this study, the fullerene composite was performed as a photosensitizer, where it oxidized a series of acyclic and cyclic alkenes at room temperature by generating singlet oxygen by using atmospheric oxygen under the influence of visible light and atmospheric pressure (Martínez-Agramunt et al. 2019).

5.7 Nanocarbon

Nanocarbons are random-shaped nanosized carbon particles that can be blended as composite with other nanoparticles, especially graphitic carbon nitride, to enhance their photocatalytic property (He et al. 2020). Liu et al. (2020) investigated the surface chemistry-dependent photocatalytic activity of graphitic carbon nitride particles that are modified with distinct nanocarbons. The study showed that the nanocarbons in graphitic carbon nitride particles are highly beneficial in the elimination of sulfachloropyridzine under the irradiation of UV light compared to free graphitic carbon nitride (Liu et al. 2020b). Likewise, Wen et al. (2017) utilized a unique approach to fabricate multiple heterojunctions with graphitic-carbon nitride, carbons, and nickel sulfide for the improved production of H_2 via photocatalytic property. In this study, the facile precipitation method was used to fabricate the heterojunctions, which were identified to possess trifunctional roles due to the presence of nanocarbons, such as CNTs, graphite, reduced graphene oxide, acetylene black, and carbon black. Also, the study emphasized that the heterojunction with 0.5% carbon black and 0.5% of acetylene black can lead to highest H_2 evolution rate of 366.4 and 297.7 µmol/g/h, which is 3.17 and 2.57 times higher than heterojunction without nanocarbon (Wen et al. 2017). Similarly, Murugesan et al. (2017) utilized a novel biosynthesis approach to fabricate lanthanum oxide doped nano carbon hollow sphere as a nanocarbon and evaluated its photocatalytic property. In this work, the 40-70 nm sized nanocarbon was prepared via essential oils extracted from the *Ricinus communis* plant using low-temperature direct pyrolysis and multi-metal catalyst prepared from the stem of *Alternanthera sessilis*. These biosynthesized nanocarbon was doped with visible light active lanthanum, and they have exhibited an excellent photocatalytic degradation ability against methylene blue dye by absorbing solar radiation (Murugesan et al. 2017). Additionally, titanium dioxide nanostructures (Reddy et al. 2020), bismuth oxide chloride nanosheets (Shi et al. 2018), molybdenum diselenide (Ren et al. 2017), and self-doped iron oxide from waste printer ink are blended distinct nanocarbons (Saini et al. 2019) to improve their significant photocatalytic effect.

5.8 Activated Carbon

Activated carbons are the recent addition of carbon materials that are included with photocatalytic nanoparticles as composites to improve their photocatalytic efficiency. Activated carbon is a carbon form that is processed to have a low volume of pores with an enhanced surface area to improve their adsorption and chemical reactions, which are available in two forms, such as powdered and granular activated carbon (Johnson 2014). Khan et al. (2019) synthesized a novel manganese dioxide nanocomposite with activated carbon to enhance its UV and visible light-mediated photocatalytic property. In this study, the chemical reduction approach was used to fabricate the cylindrical and spherical as well as dispersed nanocomposite and their photocatalytic ability was evaluated against Congo red dye. The results showed that the nanocomposite with activated carbon possesses enhanced ability to photo-catalytically degrade 98.53% of Congo red dye within 5 minutes of UV irradiation, which is less than free manganese dioxide nanoparticles (66.57%). Meanwhile, the nanocomposite was also proven to possess enhanced photocatalytic activity to reduce 94.21% of Congo red under the influence of visible light, which is less than free manganese dioxide nanoparticles (56.78%) (Khan et al. 2019). Likewise, Aber et al. (2019) synthesized a novel granular activated carbon that is pretreated with ultra-sonication and immobilized with zinc oxide nanoparticles for enhancing its photocatalytic regeneration. The resultant activated carbon particle, which was saturated in reactive red 43 was identified to possess enhanced adsorption capacity to 83% via photocatalysis and higher surface area (Aber et al. 2019). Similarly, Luo et al. (2020) assembled zinc oxide nanorod arrays on the surface of activated carbon fibers to improve their photocatalytic pollutant degradation ability. In this study, the zinc oxide nanorod arrays were synthesized via a sol-gel approach followed by a hydrothermal method to assemble them on the activated carbon fiber surface, which was utilized photo-catalytically to degrade toxic methylene blue dye. The results showed that the novel activated carbon fiber composite exhibited 77.5% of photocatalytic degradation against methylene blue dye with 55% of mineralization under the influence of UV light and can reuse up to five types with excellent stability (Luo et al. 2020b). Moreover, Eshaghi and Moradi (2018) reported that the iron-doped titanium dioxide nanoparticles that are loaded on the surface of activated carbon possess enhanced photocatalytic property. In this study, the composite with activated carbon-containing 35-70 nm sized homogenous titanium dioxide was synthesized via a sol-gel approach, and their photocatalytic activity was evaluated against reactive red 198 dye. The results revealed that the visible light irradiation of the nanocomposite with high light adsorption capacity has led to the degradation of the dye and is proposed to be beneficial to decolorize them in textile effluents (Eshaghi et al. 2018). Furthermore, tungsten oxide as dopants (Tahir et al. 2020), titanium dioxide-silver composite (Jiang et al. 2018), biosynthesized tin oxide (Begum et al. 2018), and aluminum nanoparticles with *Moringa oleifera* gum (Velu et al. 2020) were blended as a composite with activated carbon to improve the efficiency of their photocatalytic activity.

5.9 Other Carbon Forms

Apart from the above-mentioned carbon forms, there are certain peculiar carbon-based nanocomposites, which include charcoal, carbon nanohorns, and graphene nanoribbons. Recently, Selvin et al. (2018) fabricated a unique nanocomposite using zinc oxide, activated charcoal, and polyaniline to exhibit enhanced photocatalytic degradation properties. A simple precipitation approach was used to fabricate these nanocomposites, which has been identified to possess the enhanced photocatalytic ability to degrade toxic Rhodamine B dye upon visible light irradiation (Selvin et al. 2018). Besides, Chandraboss et al. (2016) fabricated a novel bismuth-doped titanium dioxide nanocomposite that is supported on the surface of activated charcoal. This nanocomposite was identified to eliminate methylene blue dye from the contaminated water under the influence of solar light irradiation for 100 minutes due to the decrement in the bandgap of composite with the presence of activated carbon (Chandraboss et al. 2016). Further, Wang et al. (2018) utilized charcoal extracted from the bamboo plant and blended it with titanium dioxide as a composite to improve their photocatalytic activity. The study showed that the bio-charcoal-titanium dioxide nanocomposite can potentially degrade 97% of methylene blue dye upon visible light irradiation for 240 minutes, which remained up to 75% even after four times of re-usage (Wang et al. 2018a). Moreover, Reddy et al. (2021) fabricated an exclusive electrochemical supercapacitor using carbon nanohorns and titanium dioxide nanoflower and utilized them as photocatalytic heterogeneous catalysts for the enhanced production of H_2. In this study, the catalyst was prepared via a one-pot solvothermal approach, which has photo-catalytically degraded 90% of methyl orange and methylene blue dye under the influence of solar radiation and produced about 4,500 µmol/g/h of H_2 (Ramesh Reddy et al. 2021). Furthermore, Kumari et al. (2019) synthesized novel carbon nanohorns in titanium dioxide nanohybrid and evaluated their pre- and post-oxidation capability to enhance photocatalytic H_2 production. In this study, the arc discharge approach was used to synthesize carbon nanohorns, whereas the nanohybrid was fabricated via the wet impregnation method followed by thermal oxidation. The study showed that the solver energy conversion efficiency of the nanohybrid to form H_2 is 6.73%, which was much greater than commercial nanoparticles (0.7%) (M et al. 2019). Recently, Xia et al. (2019) blended graphene nanoribbons with cadmium selenide nanoparticles as a composite material to enable them as a potential visible, light-driven photocatalyst for the enhanced production of H_2. The results showed that the nanocomposite with 2 wt.% of graphene nanoribbons possess enhanced photocatalytic ability to produce 1.89 mmol/h/g of H_2

evolution rate with 19.3% of quantum efficiency, which is 3.7 times higher than pristine cadmium selenide nanoparticles (Xia et al. 2019). It is evident from all these studies that carbon-based nanomaterials can be used to produce nanocomposites with other catalytic nanoparticles to improve their photocatalytic efficiency.

6. General Mechanism of Semiconductor Photocatalysis

In general, a semiconductor photocatalytic cycle comprises three steps. When the incident light energy (ultraviolet or visible) is equal to or larger than that of a bandgap, excitons (electrons and holes) are generated in the conduction and valence bands (CB and VB), respectively [step (i)]. The excited electron-hole pairs can recombine, dissipating the input energy in the form of heat or emitted light with no chemical effect [step (ii)]. However, if the excitons migrate to the semiconductor photocatalyst surface without recombination, they can participate in various redox (both oxidation and reduction) reactions with adsorbed species (protons and hydroxyl ions) to generate H_2 and O_2, respectively [step (iii)] (Tang et al. 2008, Maeda 2011). The main principle involved in overall water splitting is displayed in Figure 4. The reactions involved in overall water splitting are represented in equations 1.1-1.3.

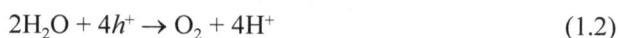

$$2H^+ + 2e^- \rightarrow H_2 \qquad (1.1)$$

$$2H_2O + 4h^+ \rightarrow O_2 + 4H^+ \qquad (1.2)$$

The overall reaction can be represented as:

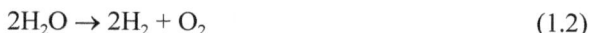

$$2H_2O \rightarrow 2H_2 + O_2 \qquad (1.2)$$

Also, electrons transfer between the noble-metal NPs and carbon nanomaterials is the further step that can occur in carbon-based photocatalytic water splitting. In this way, the lifetime of photo-induced excitons can be increased. Finally, it results in efficient photocatalytic water splitting.

Although the basic physics involved in the separation of the charge carriers varies with different applications and also surface-electronic structure of photocatalyst, it is accepted that the interfacial redox reactions of generated excitons are the primary reactions that are responsible for the positive photocatalytic effect (Herrmann 1999, Anandan et al. 2010). The basic demands of the semiconductor photocatalytic materials are: (i) wide bandgap and (ii) suitable band energy potential, i.e. the top level of the VB shall be more positive than the oxidation potential of O_2/H_2O (1.23 V vs. NHE) while the bottom level of the CB shall be more negative than the reduction potential of H^+/H_2 (0.0 V vs. NHE) as displayed in Figure 4. In addition, (iii) catalytic active sites are for surface-interface reactions. In recent years, heterogeneous photocatalysts have been studied enormously due to their potential use in energy generation (Kudo et al. 2009), environmental remediation (Muhd Julkapli et al. 2014), and organic transformations (Hu et al. 2018).

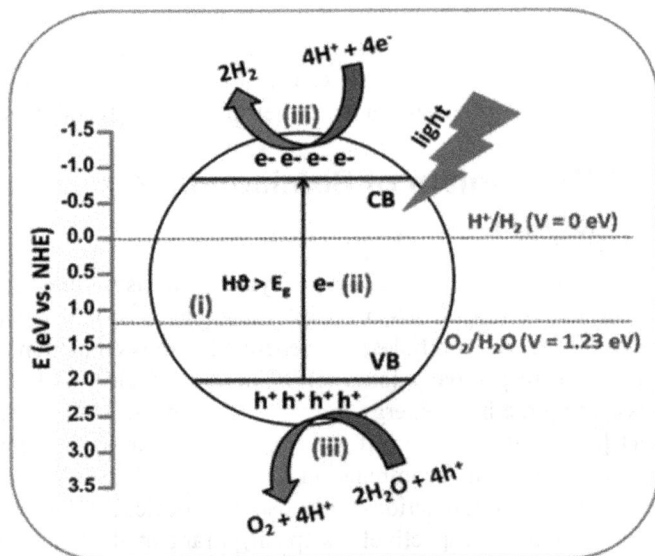

Figure 4: Principle involved in the semiconductor-mediated overall water splitting.

7. Future Perspective

The successfully developed highly efficient visible-light active nanostructured semiconductor materials can be potentially applied for photocatalytic water splitting for H_2 fuel generation, photocatalytic conversion of CO_2 to energy-rich hydrocarbon fuels, and photodecomposition of other organic substances for environmental remediation. Furthermore, it would be interesting to work on the formation of carbon-based photocatalysts with unique nanoarchitecture to fulfill the needs of practical application. More importantly, a cost-effective carbon-based photocatalyst facilitates its re-utilization and also opens the possibility of working in a continuous regime. Finally, carbon-based ternary photocatalytic systems can be very active in the visible light region, the possibility of using natural sunlight as an irradiation source can result in both environmental and economic advantages.

Conflict of Interests

There are no conflicts of interest.

Acknowledgments

Author Saikumar M. thanks to Dr. Vishnu Shanker, Associate Professor, Department of Chemistry, NIT-Warangal, Telangana and Prof. M.V. Shankar, Yogi Vemana University, Department of Materials Science and Nanotechnology, Kadapa, Andhra Pradesh for their continuous support.

References

Aber, S., Tajdid Khajeh, R. and Khataee, A. 2019. Application of immobilized ZnO nanoparticles for the photocatalytic regeneration of ultrasound pretreated-granular activated carbon. Ultrason. Sonchem. 58: 104685.

Ambrosetti, A. and Silvestrelli, P.L. 2011. Adsorption of rare-gas atoms and water on graphite and graphene by van der Waals-corrected density functional theory. J. Phys. Chem. C 115: 3695-3702.

Anandan, S., Ohashi, N. and Miyauchi, M. 2010. ZnO-based visible-light photocatalyst: Band-gap engineering and multi-electron reduction by co-catalyst. Appl. Catal., B: Environmental 100: 502-509.

Bakos, L.P., Justh, N., Hernádi, K., Kiss, G., Réti, B., Erdélyi, Z., Parditka, B. and Szilágyi, I.M. 2016. Core-shell carbon nanosphere-TiO$_2$ composite and hollow TiO$_2$ nanospheres prepared by atomic layer deposition. J. Phys.: Conf. Ser. 764: 012005.

Balat, H. and Kırtay, E. 2010. Hydrogen from biomass – present scenario and future prospects. Int. J. Hydrogen Energy 35: 7416-7426.

Bazli, L., Siavashi, M. and Shiravi, Á. 2019. A review of carbon nanotube/TiO$_2$ composite prepared via sol-gel method. J. Compos. Compd. 1: 1-9.

Begum, S. and Ahmaruzzaman, M. 2018. Biogenic synthesis of SnO$_2$/activated carbon nanocomposite and its application as photocatalyst in the degradation of naproxen. Appl. Surf. Sci. 449: 780-789.

Bellamkonda, S., Thangavel, N., Hafeez, H.Y., Neppolian, B. and Ranga Rao, G. 2019. Highly active and stable multi-walled carbon nanotubes-graphene-TiO$_2$ nanohybrid: an efficient non-noble metal photocatalyst for water splitting. Catal. Today 321-322: 120-127.

Bilal Tahir, M., Nabi, G., Rafique, M. and Khalid, N.R. 2018. Role of fullerene to improve the WO$_3$ performance for photocatalytic applications and hydrogen evolution. International Int. J. Energy Res. 42: 4783-4789.

Byeon, J.H. and Kim, J.-W. 2014. Ambient plasma synthesis of TiO$_2$@ graphite oxide nanocomposites for efficient photocatalytic hydrogenation. J. Mater. Chem. A 2: 6939-6944.

Cao, S. and Yu, J. 2016. Carbon-based H$_2$-production photocatalytic materials. J. Photochem. Photobiol., C: Photochemistry Reviews 27: 72-99.

Chandraboss, V.L., Kamalakkannan, J. and Senthilvelan, S. 2016. Synthesis of activated charcoal supported Bi-doped TiO$_2$ nanocomposite under solar light irradiation for enhanced photocatalytic activity. Appl. Surf. Sci. 387: 944-956.

Chang, X., Wang, T. and Gong, J. 2016. CO$_2$ photo-reduction: insights into CO$_2$ activation and reaction on surfaces of photocatalysts. Energy Environ. Sci. 9: 2177-2196.

Chatterjee, M.J., Ghosh, A., Mondal, A. and Banerjee, D. 2017. Polyaniline–single walled carbon nanotube composite – a photocatalyst to degrade rose bengal and methyl orange dyes under visible-light illumination. RSC Adv. 7: 36403-36415.

Chen, D., Zhang, H., Liu, Y. and Li, J. 2013. Graphene and its derivatives for the development of solar cells, photoelectrochemical, and photocatalytic applications. Energy Environ. Sci. 6: 1362-1387.

Chen, H., Gu, Z.-G., Mirza, S., Zhang, S.-H. and Zhang, J. 2018. Hollow Cu-TiO$_2$/C nanospheres derived from a Ti precursor encapsulated MOF coating for efficient photocatalytic hydrogen evolution. J. Mater. Chem. A 6: 7175-7181.

Chen, X., Shen, S., Guo, L. and Mao, S.S. 2010. Semiconductor-based photocatalytic hydrogen generation. Chemical Reviews 110: 6503-6570.

Coro, J., Suárez, M., Silva, L.S.R., Eguiluz, K.I.B. and Salazar-Banda, G.R. 2016. Fullerene applications in fuel cells: a review. Int. J. Hydrogen Energy 41: 17944-17959.

Dai, L., Chang, D.W., Baek, J.B. and Lu, W. 2012. Carbon nanomaterials for advanced energy conversion and storage. Small 8: 1130-1166.

Dong, C., Li, X., Jin, P., Zhao, W., Chu, J. and Qi, J. 2012. Intersubunit electron transfer (IET) in quantum dots/graphene complex: what features does IET endow the complex with? J. Phys. Chem. C 116: 15833-15838.

Dresselhaus, M.S., Dresselhaus, G., Eklund, P.C. and Rao, A.M. 2000. Carbon Nanotubes. The Physics of Fullerene-Based and Fullerene-Related Materials. W. Andreoni. Dordrecht, Springer Netherlands: 331-379.

Eshaghi, A. and Moradi, H. 2018. Optical and photocatalytic properties of the Fe-doped TiO_2 nanoparticles loaded on the activated carbon. Adv. Powder Technol. 29: 1879-1885.

Feng, K., Song, B., Li, X., Liao, F. and Gong, J. 2019. Enhanced photocatalytic performance of magnetic multi-walled carbon nanotubes/cerium dioxide nanocomposite. Ecotoxicol. Environ. Saf. 171: 587-593.

Fernandes, R.A., Sampaio, M.J., Da Silva, E.S., Serp, P., Faria, J.L. and Silva, C.G. 2019. Synthesis of selected aromatic aldehydes under UV-LED irradiation over a hybrid photocatalyst of carbon nanofibers and zinc oxide. Catal. Today 328: 286-292.

Gan, Z., Wu, X., Meng, M., Zhu, X., Yang, L. and Chu, P.K. 2014. Photothermal contribution to enhanced photocatalytic performance of graphene-based nanocomposites. ACS Nano 8: 9304-9310.

Geng, A., Meng, L., Han, J., Zhong, Q., Li, M., Han, S., Mei, C., Xu, L., Tan, L. and Gan, L. 2018. Highly efficient visible-light photocatalyst based on cellulose derived carbon nanofiber/BiOBr composites. Cellulose 25: 4133-4144.

Gopannagari, M., Kumar, D.P., Park, H., Kim, E.H., Bhavani, P., Reddy, D.A. and Kim, T.K. 2018. Influence of surface-functionalized multi-walled carbon nanotubes on CdS nanohybrids for effective photocatalytic hydrogen production. Appl. Catal., B: Environmental 236: 294-303.

Gupta, A., Kour, R. and Brar, L.K. 2019. Facile synthesis of carbon nanospheres from saccharides for photocatalytic applications. SN Appl. Sci. 1: 1169.

Gyulavári, T., Veréb, G., Pap, Z., Réti, B., Baan, K., Todea, M., Magyari, K., Szilágyi, I.M. and Hernadi, K. 2019. Utilization of carbon nanospheres in photocatalyst production: from composites to highly active hollow structures. Materials 12: 2537.

Han, Q., Wang, B., Gao, J. and Qu, L. 2016. Graphitic carbon nitride/nitrogen-rich carbon nanofibers: highly efficient photocatalytic hydrogen evolution without cocatalysts. Angewandte Chemie 128: 11007-11011.

Hasan, I., Walia, S., Alharbi, K.H., Khanjer, M.A., Alsalme, A. and Khan, R.A. 2020. Multi-walled carbon nanotube coupled β-Cyclodextrin/PANI hybrid photocatalyst for advance oxidative degradation of crystal violet. J. Mol. Liq. 317: 114216.

Hasija, V., Sudhaik, A., Raizada, P., Hosseini-Bandegharaei, A. and Singh, P. 2019. Carbon quantum dots supported AgI/ZnO/phosphorus doped graphitic carbon nitride as Z-scheme photocatalyst for efficient photodegradation of 2,4-dinitrophenol. J. Environ. Chem. Eng. 7: 103272.

He, B., Feng, M., Chen, X. and Sun, J. 2021. Multidimensional (0D-3D Functional Nanocarbon: Promising Material to Strengthen the Photocatalytic Activity of Graphitic Carbon Nitride. Green Energy & Environ. 6(6): 823-845.

He, H., Huang, L., Zhong, Z. and Tan, S. 2018. Constructing three-dimensional porous graphene-carbon quantum dots/g-C_3N_4 nanosheet aerogel metal-free photocatalyst with enhanced photocatalytic activity. Appl. Surf. Sci. 441: 285-294.

Herrmann, J.-M. 1999. Heterogeneous photocatalysis: fundamentals and applications to the removal of various types of aqueous pollutants. Catal. Today 53: 115-129.

Hosseini, S.E. and Wahid, M.A. 2016. Hydrogen production from renewable and sustainable energy resources: promising green energy carrier for clean development. Renewable Sustainable Energy Rev. 57: 850-866.

Hu, J., Chen, D., Li, N., Xu, Q., Li, H., He, J. and Lu, J. 2017. In situ fabrication of $Bi_2O_2CO_3$/ MoS_2 on carbon nanofibers for efficient photocatalytic removal of NO under visible-light irradiation. Appl. Catal., B: Environmental 217: 224-231.

Hu, Z., Quan, H., Chen, Z., Shao, Y. and Li, D. 2018. New insight into an efficient visible light-driven photocatalytic organic transformation over CdS/TiO_2 photocatalysts. Photochem. Photobiol. Sci. 17: 51-59.

Huang, J., Jiang, Y., Li, G., Xue, C. and Guo, W. 2017a. Hetero-structural $NiTiO_3/TiO_2$ nanotubes for efficient photocatalytic hydrogen generation. Renewable Energy 111: 410-415.

Huang, Y., Liang, Y., Rao, Y., Zhu, D., Cao, J.-j., Shen, Z., Ho, W. and Lee, S.C. 2017b. Environment-friendly carbon quantum dots/$ZnFe_2O_4$ photocatalysts: characterization, biocompatibility, and mechanisms for NO removal. Environ. Sci. Technol. 51: 2924-2933.

Jagannatham, M., Berkmans, A. J., Haridoss, P., Reddy, L., & Shankar, M. V. (2019). Influence of pre-oxidation, versus post-oxidation of carbon nanohorns in TiO2 nanohybrid for enhanced photocatalytic hydrogen production. Materials Research Bulletin, 109: 34-40.

Jeevanandam, J., Chan, Y.S., Pan, S. and Danquah, M.K. 2019. Metal oxide nanocomposites: Cytotoxicity and targeted drug delivery applications. Hybrid nanocomposites: fundamentals, synthesis and applications, Pan Stanford Publishing: 111-147.

Jiang, Z., Zhang, X., Yuan, Z., Chen, J., Huang, B., Dionysiou, D.D. and Yang, G. 2018. Enhanced photocatalytic CO_2 reduction via the synergistic effect between Ag and activated carbon in TiO_2/AC-Ag ternary composite. Chem. Eng. J. 348: 592-598.

Johnson, C. 2014. 2.4 - Advances in Pretreatment and Clarification Technologies. Comprehensive Water Quality and Purification. S. Ahuja. Waltham, Elsevier: 60-74.

Khan, I., Sadiq, M., Khan, I. and Saeed, K. 2019. Manganese dioxide nanoparticles/activated carbon composite as efficient UV and visible-light photocatalyst. Environ. Sci. Pollut. Res. 26: 5140-5154.

Kudo, A. and Miseki, Y. 2009. Heterogeneous photocatalyst materials for water splitting. Chem. Soc. Rev. 38: 253-278.

Kumar, S., Dhiman, A., Sudhagar, P. and Krishnan, V. 2018. ZnO-graphene quantum dots heterojunctions for natural sunlight-driven photocatalytic environmental remediation. Appl. Surf. Sci. 447: 802-815.

Kumari Mamatha, M., Jagannatham, M., Berkmans Joseph, A., Haridoss Prathap., Reddy Lakshmana, N. and Shankar, M.V. 2019 Influence of pre-oxidation, versus post-oxidation of carbon nanohorns in TiO_2 nanohybrid for enhanced photocatalytic hydrogen production. Mater. Res. Bull. 109: 34-40.

Kurniawan, K., Tajima, T., Kubo, Y., Miyake, H., Kurashige, W., Negishi, Y. and Takaguchi, Y. 2017. Incorporating a TiO_x shell in single-walled carbon nanotube/fullerodendron coaxial nanowires: increasing the photocatalytic evolution of H_2 from water under irradiation with visible light. RSC Adv. 7: 31767-31770.

Lang, D., Shen, T. and Xiang, Q. 2015. Roles of MoS_2 and graphene as cocatalysts in the enhanced visible-light photocatalytic H_2 production activity of multiarmed CdS nanorods. ChemCatChem 7: 943-951.

Lei, Y., Yang, C., Hou, J., Wang, F., Min, S., Ma, X., Jin, Z., Xu, J., Lu, G. and Huang, K.-W. 2017. Strongly coupled CdS/graphene quantum dots nanohybrids for highly efficient photocatalytic hydrogen evolution: unraveling the essential roles of graphene quantum dots. Appl. Catal., B: Environmental 216: 59-69.

Lewis, N.S. and Nocera, D.G. 2006. Powering the planet: chemical challenges in solar energy utilization. Proc. Natl. Acad. Sci. 103: 15729-15735.

Li, M., Chen, T., Gooding, J.J. and Liu, J. 2019. Review of carbon and graphene quantum dots for sensing. ACS Sens. 4: 1732-1748.

Li, Q., Cui, C., Meng, H. and Yu, J. 2014a. Visible-light photocatalytic hydrogen production activity of $ZnIn_2S_4$ microspheres using carbon quantum dots and platinum as dual co-catalysts. Chem.: Asian J. 9: 1766-1770.

Li, Q., Guo, B., Yu, J., Ran, J., Zhang, B., Yan, H. and Gong, J.R. 2011. Highly efficient visible-light-driven photocatalytic hydrogen production of CdS-cluster-decorated graphene nanosheets. J. Am. Chem. Soc. 133: 10878-10884.

Li, X., Dai, Y., Ma, Y., Han, S. and Huang, B. 2014b. Graphene/$gC_3 N_4$ bilayer: considerable band gap opening and effective band structure engineering. Phys. Chem. Chem. Phys. 16: 4230-4235.

Li, X., Hao, X., Abudula, A. and Guan, G. 2016. Nanostructured catalysts for electrochemical water splitting: current state and prospects. J. Mater. Chem. A 4: 11973-12000.

Liang, H., Bai, J., Xu, T. and Li, C. 2020. Controllable growth of foxtail-like MoS_2 on one-dimensional carbon nanofibers with enhanced photocatalytic activity. Vacuum 172: 109059.

Lin, X., Liu, C., Wang, J., Yang, S., Shi, J. and Hong, Y. 2019. Graphitic carbon nitride quantum dots and nitrogen-doped carbon quantum dots co-decorated with $BiVO_4$ microspheres: a ternary heterostructure photocatalyst for water purification. Sep. Purif. Technol. 226: 117-127.

Ling, L., Wang, C., Ni, M. and Shang, C. 2016. Enhanced photocatalytic activity of TiO_2/single-walled carbon nanotube (SWCNT) composites under UV-A irradiation. Sep. Purif. Technol. 169: 273-278.

Liu, Q., Tian, H., Dai, Z., Sun, H., Liu, J., Ao, Z., Wang, S., Han, C. and Liu, S. 2020a. Nitrogen-doped carbon nanospheres-modified graphitic carbon nitride with outstanding photocatalytic activity. Nano-Micro Lett. 12: 24.

Liu, Q., Zhou, L., Gao, J., Wang, S., Liu, L. and Liu, S. 2020b. Surface chemistry-dependent activity and comparative investigation on the enhanced photocatalytic performance of graphitic carbon nitride modified with various nanocarbons. J. Colloid Interface Sci. 569: 12-21.

Luo, D., Wang, X., Zhang, Z., Gao, D., Liu, Z. and Chen, J. 2020a. Enhancement of photocatalytic hydrogen evolution from dye–sensitized amide–functionalized carbon nanospheres by superior adsorption performance. Int. J. Hydrogen Energy 45: 30375-30386.

Luo, S., Liu, C., Zhou, S., Li, W., Ma, C., Liu, S., Yin, W., Heeres, H.J., Zheng, W., Seshan, K. and He, S. 2020b. ZnO nanorod arrays assembled on activated carbon fibers for photocatalytic degradation: characteristics and synergistic effects. Chemosphere 261: 127731.

Ma, J., Michaelides, A., Alfe, D., Schimka, L., Kresse, G. and Wang, E. 2011. Adsorption and diffusion of water on graphene from first principles. Phys. Rev. B 84: 033402.

Maeda, K. 2011. Photocatalytic water splitting using semiconductor particles: history and recent developments. J. Photochem. Photobiol., C: Photochemistry Reviews 12: 237-268.

Mahdiani, M., Soofivand, F., Ansari, F. and Salavati-Niasari, M. 2018. Grafting of $CuFe_{12}O_{19}$ nanoparticles on CNT and graphene: eco-friendly synthesis, characterization and photocatalytic activity. J. Cleaner Prod. 176: 1185-1197.

Manchala, S., Nagappagari, L.R., Venkatakrishnan, S.M. and Shanker, V. 2018. Facile synthesis of noble-metal free polygonal Zn_2TiO_4 nanostructures for highly efficient photocatalytic hydrogen evolution under solar light irradiation. Int. J. Hydrogen Energy 43: 13145-13157.

Manchala, S., Nagappagari, L.R., Venkatakrishnan, S.M. and Shanker, V. 2019a. Solar-light harvesting bimetallic Ag/Au decorated graphene plasmonic system with efficient photoelectrochemical performance for the enhanced water reduction process. ACS Appl. Nano Mater. 2: 4782-4792.

Manchala, S., Tandava, V., Jampaiah, D., Bhargava, S.K. and Shanker, V. 2019b. Novel and highly efficient strategy for the green synthesis of soluble graphene by aqueous polyphenol extracts of Eucalyptus bark and its applications in high-performance supercapacitors. ACS Sustainable Chem. Eng. 7: 11612-11620.

Manchala, S., Gandamalla, A., Vempuluru, N.R., Venkatakrishnan, S.M. and Shanker, V. 2020a. High potential and robust ternary LaFeO$_3$/CdS/carbon quantum dots nanocomposite for photocatalytic H$_2$ evolution under sunlight illumination. J. Colloid Interface Sci. 583: 255-266.

Manchala, S., Tandava, V., Jampaiah, D., Bhargava, S.K. and Shanker, V. 2020b. A novel strategy for sustainable synthesis of soluble-graphene by a herb delphinium denudatum root extract for use as light-weight supercapacitors. Chemistry Select 5: 2701-2709.

Manchala, S., Tandava, V., Nagappagari, L.R., Venkatakrishnan, S.M., Jampaiah, D., Sabri, Y.M., Bhargava, S.K. and Shanker, V. 2019c. Fabrication of a novel ZnIn$_2$S$_4$/gC$_3$ N$_4$/graphene ternary nanocomposite with enhanced charge separation for efficient photocatalytic H$_2$ evolution under solar light illumination. Photochem. Photobiol. Sci. 18: 2952-2964.

Martínez-Agramunt, V. and Peris, E. 2019. Photocatalytic properties of a palladium metallosquare with encapsulated fullerenes via singlet oxygen generation. Inorg. Chem. 58: 11836-11842.

Mateo, D., Albero, J. and García, H. 2018. Graphene supported NiO/Ni nanoparticles as efficient photocatalyst for gas phase CO$_2$ reduction with hydrogen. Appl. Catal., B: Environmental 224: 563-571.

Meng, F., Li, J., Cushing, S.K., Zhi, M. and Wu, N. 2013. Solar hydrogen generation by nanoscale p-n junction of p-type molybdenum disulfide/n-type nitrogen-doped reduced graphene oxide. J. Am. Chem. Soc. 135: 10286-10289.

Muhd Julkapli, N., Bagheri, S. and Bee Abd Hamid, S. 2014. Recent advances in heterogeneous photocatalytic decolorization of synthetic dyes. The Sci. World J. 2014: 25. Article ID 692307.

Murugesan, B., Sivakumar, A., Loganathan, A. and Sivakumar, P. 2017. Synthesis and photocatalytic studies of lanthanum oxide doped nano carbon hollow spheres. J. Taiwan Inst. Chem. Eng. 71: 364-372.

Nabgan, W., Abdullah, T.A.T., Mat, R., Nabgan, B., Gambo, Y., Ibrahim, M., Ahmad, A., Jalil, A.A., Triwahyono, S. and Saeh, I. 2017. Renewable hydrogen production from bio-oil derivative via catalytic steam reforming: an overview. Renewable Sustainable Energy Rev. 79: 347-357.

Ngai, K.S., Tan, W.T., Zainal, Z., Zawawi, R.M. and Juan, J.C. 2016. Electrochemical sensor based on single-walled carbon nanotube/ZnO photocatalyst nanocomposite modified electrode for the determination of paracetamol. Sci. Adv. Mater. 8: 788-796.

Nizovskii, A.I., Belkova, S.V., Novikov, A.A. and Trenikhin, M.V. 2015. Hydrogen production for fuel cells in reaction of activated aluminum with water. Procedia Eng. 113: 8-12.

Pan, Y., Liu, X., Zhang, W., Liu, Z., Zeng, G., Shao, B., Liang, Q., He, Q., Yuan, X. and Huang, D. 2020. Advances in photocatalysis based on fullerene C60 and its derivatives: Properties, mechanism, synthesis, and applications. Appl. Catal., B: Environmental 265: 118579.

Pant, B., Ojha, G.P., Kuk, Y.-S., Kwon, O.H., Park, Y.W. and Park, M. 2020. Synthesis and characterization of ZnO-TiO$_2$/carbon fiber composite with enhanced photocatalytic properties. Nanomaterials 10: 1960.

Pant, B., Park, M., Kim, H.-Y. and Park, S.-J. 2016. Ag-ZnO photocatalyst anchored on carbon nanofibers: synthesis, characterization, and photocatalytic activities. Synth. Met. 220: 533-537.

Payan, A., Fattahi, M., Jorfi, S., Roozbehani, B. and Payan, S. 2018. Synthesis and characterization of titanate nanotube/single-walled carbon nanotube (TNT/SWCNT) porous nanocomposite and its photocatalytic activity on 4-chlorophenol degradation under UV and solar irradiation. Appl. Surf. Sci. 434: 336-350.

Pu, S., Zhu, R., Ma, H., Deng, D., Pei, X., Qi, F. and Chu, W. 2017. Facile in-situ design strategy to disperse TiO_2 nanoparticles on graphene for the enhanced photocatalytic degradation of rhodamine 6G. Appl. Catal., B: Environmental 218: 208-219.

Qi, K., Selvaraj, R., Al Fahdi, T., Al-Kindy, S., Kim, Y., Wang, G.-C., Tai, C.-W. and Sillanpää, M. 2016. Enhanced photocatalytic activity of anatase-TiO_2 nanoparticles by fullerene modification: a theoretical and experimental study. Appl. Surf. Sci. 387: 750-758.

Qian, J., Shen, C., Yan, J., Xi, F., Dong, X. and Liu, J. 2018. Tailoring the electronic properties of graphene quantum dots by P doping and their enhanced performance in metal-free composite photocatalyst. J. Phys. Chem. C 122: 349-358.

Raizada, P., Sudhaik, A. and Singh, P. 2019. Photocatalytic water decontamination using graphene and ZnO coupled photocatalysts: a review. Mater. Sci. Energy Technol. 2: 509-525.

Ramesh Reddy, N., Mamatha Kumari, M., Shankar, M.V., Raghava Reddy, K., Woo Joo, S. and Aminabhavi, T.M. 2021. Photocatalytic hydrogen production from dye contaminated water and electrochemical supercapacitors using carbon nanohorns and TiO_2 nanoflower heterogeneous catalysts. J. Environ. Manage. 277: 111433.

Ran, J., Zhang, J., Yu, J., Jaronic, M. and Qiao, S.Z. 2014. Earth-abundant cocatalysts for semiconductor-based photocatalytic water splitting. Chem. Soc. Rev. 43: 7787-7812.

Reddy, K.R., Jyothi, M.S., Raghu, A.V., Sadhu, V., Naveen, S. and Aminabhavi, T.M. 2020. Nanocarbons-supported and polymers-supported titanium dioxide nanostructures as efficient photocatalysts for remediation of contaminated wastewater and hydrogen production. Nanophotocatalysis and Environmental Applications, Springer: 139-169.

Regulska, E., Rivera-Nazario, D.M., Karpinska, J., Plonska-Brzezinska, M.E. and Echegoyen, L. 2019. Zinc porphyrin-functionalized fullerenes for the sensitization of titania as a visible-light active photocatalyst: river waters and wastewaters remediation. Molecules 24.

Ren, Z., Liu, X., Chu, H. Yu, H., Xu, Y., Zheng, W., Lei, W., Chen, P., Li, J. and Li, C. 2017. Carbon quantum dots decorated $MoSe_2$ photocatalyst for Cr (VI) reduction in the UV–vis-NIR photon energy range. J. Colloid Interface Sci. 488: 190-195.

Safajou, H., Khojasteh, H., Salavati-Niasari, M. and Mortazavi-Derazkola, S. 2017. Enhanced photocatalytic degradation of dyes over graphene/Pd/TiO_2 nanocomposites: TiO_2 nanowires versus TiO_2 nanoparticles. J. Colloid Interface Sci. 498: 423-432.

Sahaym, U. and Norton, M.G. 2008. Advances in the application of nanotechnology in enabling a 'hydrogen economy'. J. Mater. Sci. 43: 5395-5429.

Saini, D., Aggarwal, R., Anand, S.R. and Sonkar, S.K. 2019. Sunlight induced photodegradation of toxic azo dye by self-doped iron oxide nano-carbon from waste printer ink. Sol. Energy 193: 65-73.

Sapkota, K.P., Lee, I., Hanif, M., Islam, M. and Hahn, J.R. 2019. Solar-light-driven efficient ZnO–single-walled carbon nanotube photocatalyst for the degradation of a persistent water pollutant organic dye. Catalysts 9: 498.

Shaban, Y.A. and Khan, S.U.M. 2008. Visible light active carbon modified n-TiO_2 for efficient hydrogen production by photoelectrochemical splitting of water. Int. J. Hydrogen Energy 33: 1118-1126.

Shahzad, K., Tahir, M.B. and Sagir, M. 2019. Engineering the performance of heterogeneous

WO_3/fullerene@Ni_3B/Ni(OH)$_2$ photocatalysts for hydrogen generation. Int. J. Hydrogen Energy 44: 21738-21745.

Shi, L., Ma, J., Yao, L., Cui, L. and Qi, W. 2018. Enhanced photocatalytic activity of $Bi_{12}O_{17}Cl_2$ nano-sheets via surface modification of carbon nanotubes as electron carriers. J. Colloid Interface Sci. 519: 1-10.

Singhal, S., Dixit, S. and Shukla, A.K. 2018. Self-assembly of the Ag deposited ZnO/carbon nanospheres: a resourceful photocatalyst for efficient photocatalytic degradation of methylene blue dye in water. Adv. Powder Technol. 29: 3483-3492.

Steplin Paul Selvin, S., Ganesh Kumar, A., Sarala, L., Rajaram, R., Sathiyan, A., Princy Merlin, J. and Sharmila Lydia, I. 2018. Photocatalytic degradation of rhodamine B using zinc oxide activated charcoal polyaniline nanocomposite and its survival assessment using aquatic animal model. ACS Sustainable Chem. Eng. 6: 258-267.

Sun, Z., Guo, J., Zhu, S., Ma, J., Liao, Y. and Zhang, D. 2014a. High photocatalytic performance by engineering Bi_2WO_6 nanoneedles onto graphene sheets. RSC Adv. 4: 27963-27970.

Sun, Z., Guo, J., Zhu, S., Mao, L., Ma, J. and Zhang, D. 2014b. A high-performance Bi_2WO_6–graphene photocatalyst for visible light-induced H_2 and O_2 generation. Nanoscale 6: 2186-2193.

Tahir, M.B., Ashraf, M., Rafique, M., Ijaz, M., Firman, S. and Mubeen, I. 2020. Activated carbon doped WO_3 for photocatalytic degradation of rhodamine-B. Appl. Nanosci. 10: 869-877.

Tan, L.L., Chai, S.P. and Mohamed, A.R. 2012. Synthesis and applications of graphene-based TiO_2 photocatalysts. ChemSusChem 5: 1868-1882.

Tang, J., Durrant, J.R. and Klug, D.R. 2008. Mechanism of photocatalytic water splitting in TiO_2. Reaction of water with photoholes, importance of charge carrier dynamics, and evidence for four-hole chemistry. J. Am. Chem. Soc. 130: 13885-13891.

Tian, H., Shen, K., Hu, X., Qiao, L. and Zheng, W. 2017. N, S co-doped graphene quantum dots-graphene-TiO_2 nanotubes composite with enhanced photocatalytic activity. J. Alloys Compd. 691: 369-377.

Tian, J., Zhang, L., Fan, X., Zhou, Y., Wang, M., Cheng, R., Li, M., Kan, X., Jin, X. and Liu, Z. 2016. A post-grafting strategy to modify gC_3N_4 with aromatic heterocycles for enhanced photocatalytic activity. J. Mater. Chem. A 4: 13814-13821.

Tonda, S., Kumar, S., Gawli, Y., Bhardwaj, M. and Ogale, S. 2017. g-C_3N_4 (2D)/CdS (1D)/rGO (2D) dual-interface nano-composite for excellent and stable visible light photocatalytic hydrogen generation. Int. J. Hydrogen Energy 42: 5971-5984.

Ulmer, U., Dingle, T., Duchesne, P.N., Morris, R.H., Tavasoli, A., Wood, T. and Ozin, G.A. 2019. Fundamentals and applications of photocatalytic CO_2 methanation. Nat. Commun. 10: 1-12.

Velu, M., Balasubramanian, B., Velmurugan, P., Kamyab, H., Ravi, A.V., Chelliapan, S., Lee, C.T. and Palaniyappan, J. 2020. Fabrication of nanocomposites mediated from aluminium nanoparticles/Moringa oleifera gum activated carbon for effective photocatalytic removal of nitrate and phosphate in aqueous solution. J. Cleaner Prod. 124553.

Wang, B., Liu, B., Ji, X.-X. and Ma, M.-G. 2018a. Synthesis, characterization, and photocatalytic properties of bamboo charcoal/TiO_2 composites using four sizes powder. Materials 11.

Wang, F., Wang, Y., Feng, Y., Zeng, Y., Xie, Z., Zhang, Q., Su, Y., Chen, P., Liu, Y. and Yao, K. 2018b. Novel ternary photocatalyst of single atom-dispersed silver and carbon quantum dots co-loaded with ultrathin g-C_3N_4 for broad spectrum photocatalytic degradation of naproxen. Appl. Catal., B: Environmental 221: 510-520.

Wang, P., Dimitrijevic, N.M., Chang, A.Y., Schaller, R.D., Liu, Y., Rajh, T. and Rozhkova, E.A. 2014. Photoinduced electron transfer pathways in hydrogen-evolving reduced graphene oxide-boosted hybrid nano-bio catalyst. ACS Nano 8: 7995-8002.

Wang, R., Lu, K.-Q., Tang, Z.-R. and Xu, Y.-J. 2017. Recent progress in carbon quantum dots: synthesis, properties and applications in photocatalysis. J. Mater. Chem. A 5: 3717-3734.

Wang, W., Zeng, Z., Zeng, G., Zhang, C., Xiao, R., Zhou, C., Xiong, W., Yang, Y., Lei, L., Liu, Y., Huang, D., Cheng, M., Yang, Y., Fu, Y., Luo, H. and Zhou, Y. 2019a. Sulfur doped carbon quantum dots loaded hollow tubular g- C_3N_4 as novel photocatalyst for destruction of Escherichia coli and tetracycline degradation under visible light. Chem. Eng. J. 378: 122132.

Wang, Z., Li, C. and Domen, K. 2019b. Recent developments in heterogeneous photocatalysts for solar-driven overall water splitting. Chem. Soc. Rev. 48: 2109-2125.

Wen, J., Xie, J., Yang, Z., Shen, R., Li, H., Luo, X., Chen, X. and Li, X. 2017. Fabricating the robust g-C_3N_4 nanosheets/carbons/NiS multiple heterojunctions for enhanced photocatalytic H_2 generation: an insight into the trifunctional roles of nanocarbons. ACS Sustainable Chem. Eng. 5: 2224-2236.

Xia, Y., Cheng, B., Fan, J., Yu, J. and Liu, G. 2019. Unraveling photoexcited charge transfer pathway and process of CdS/graphene nanoribbon composites toward visible-light photocatalytic hydrogen evolution. Small 15: 1902459.

Xie, G., Zhang, K., Guo, B., Liu, Q., Fang, L. and Gong, J.R. 2013. Graphene-based materials for hydrogen generation from light-driven water splitting. Adv. Mater. 25: 3820-3839.

Xie, Y., Yu, S., Zhong, Y., Zhang, Q. and Zhou, Y. 2018. SnO_2/graphene quantum dots composited photocatalyst for efficient nitric oxide oxidation under visible light. Appl. Surf. Sci. 448: 655-661.

Yang, J., Wang, D., Han, H. and Li, C.A.N. 2013. Roles of cocatalysts in photocatalysis and photoelectrocatalysis. Acc. Chem. Res. 46: 1900-1909.

Yang, X. and Wang, D. 2017. Photophysics and photochemistry at the semiconductor/electrolyte interface for solar water splitting. Semicond. Semimetals, Elsevier 97: 47-80.

Ye, S., Zhou, X., Xu, Y., Lai, W., Yan, K., Huang, L., Ling, J. and Zheng, L. 2019. Photocatalytic performance of multi-walled carbon nanotube/$BiVO_4$ synthesized by electro-spinning process and its degradation mechanisms on oxytetracycline. Chem. Eng. J. 373: 880-890.

Yeh, T.-F., Cihlář, J., Chang, C.-Y., Cheng, C. and Teng, H. 2013. Roles of graphene oxide in photocatalytic water splitting. Mater. Today 16: 78-84.

Yuan, A., Lei, H., Xi, F., Liu, J., Qin, L., Chen, Z. and Dong, X. 2019. Graphene quantum dots decorated graphitic carbon nitride nanorods for photocatalytic removal of antibiotics. J. Colloid Interface Sci. 548: 56-65.

Yuan, Y., Gong, X. and Wang, H. 2015. The synergistic mechanism of graphene and MoS_2 for hydrogen generation: insights from density functional theory. Phys. Chem. Chem. Phys. 17: 11375-11381.

Yue, Z., Chu, D., Huang, H., Huang, J., Yang, P., Du, Y., Zhu, M. and Lu, C. 2015. A novel heterogeneous hybrid by incorporation of Nb_2O_5 microspheres and reduced graphene oxide for photocatalytic H_2 evolution under visible light irradiation. RSC Adv. 5: 47117-47124.

Zhang, J., Fu, J., Chen, S., Lv, J. and Dai, K. 2018. 1D carbon nanofibers@ TiO_2 core-shell nanocomposites with enhanced photocatalytic activity toward CO_2 reduction. J. Alloys Compd. 746: 168-176.

Zhang, J., Yu, J., Zhang, Y., Li, Q. and Gong, J.R. 2011. Visible light photocatalytic H_2-production activity of CuS/ZnS porous nanosheets based on photoinduced interfacial charge transfer. Nano Lett. 11: 4774-4779.

Zhang, P., Song, T., Wang, T. and Zeng, H. 2017. In-situ synthesis of Cu nanoparticles hybridized with carbon quantum dots as a broad spectrum photocatalyst for improvement of photocatalytic H_2 evolution. Appl. Catal., B: Environmental 206: 328-335.

Zhang, X.-Y., Li, H.-P., Cui, X.-L. and Lin, Y. 2010. Graphene/TiO_2 nanocomposites: synthesis, characterization and application in hydrogen evolution from water photocatalytic splitting. J. Mater. Chem. 20: 2801-2806.

Zhu, Z., Chan, Y.-C., Chen, Z., Gan, C.-L. and Wu, F. 2018. Effect of the size of carbon nanotubes (CNTs) on the microstructure and mechanical strength of CNTs-doped composite SnO. 3AgO. 7Cu-CNTs solder. Mater. Sci. Eng. A 727: 160-169.

Zou, J.-P., Wang, L.-C., Luo, J., Nie, Y.-C., Xing, Q.-J., Luo, X.-B., Du, H.-M., Luo, S.-L. and Suib, S.L. 2016. Synthesis and efficient visible light photocatalytic H_2 evolution of a metal-free g-C_3N_4/graphene quantum dots hybrid photocatalyst. Appl. Catal., B: Environmental 193: 103-109.

Zouzelka, R., Kusumawati, Y., Remzova, M., Rathousky, J. and Pauporté, T. 2016. Photocatalytic activity of porous multiwalled carbon nanotube-TiO_2 composite layers for pollutant degradation. J. Hazard. Mater. 317: 52-59.

Carbon-Based Materials for Microsupercapacitors

Alisha Nanwani and Abhay D. Deshmukh*

Energy Materials and Devices Laboratory, Department of Physics, RTM Nagpur University, Nagpur, Maharashtra, India

1. Introduction

In recent decades the application of electronic products is increasing at an extremely high speed with the development of modern science and technology. To meet the living standards of the people, new electronic devices such as foldable smartphones, smart robots, health trackers, microsensors and other microelectronic devices are emerging in the industry. To meet the energy demands of these miniaturized electronics and smart autonomous devices, there is an urgent need for the development of miniaturized energy storage devices (González et al. 2016, Chen et al. 2018, Wang and Xia 2013).

Currently, microbatteries and microsupercapacitors (MSCs) are the most reliable power sources for portable devices. Generally, they provide miniaturized electronic devices with the required energy and power for some time. Particularly, MSCs have gained more interest in these fields as they provide better cyclability and charge/discharge performance. For example, in biomedical and sensor fields, microbatteries require frequent replacement due to their short lifespan. Comparatively, MSCs do not need frequent replacements due to their ultra-long lifetime with very small capacitance deterioration (Wang et al. 2020, Liu and Gao 2017). In addition, for high power applications, microbatteries are incompetent but they can provide better power performance when connected in series/parallel, which will naturally increase the device volume. Instead, MSCs can provide high power density in small volumes without any additional integration requirements and therefore can be suitable candidates for flexible integrated systems requiring high power density. Also, MSCs can be integrated on flexible substrates that prove their better mechanical properties.

Therefore, MSCs are more practical than microbatteries in these areas (Liu et al. 2017, Hu et al. 2015). Presently, there are two types of MSCs: conventional sandwich-

Corresponding author: abhay.d07@gmail.com

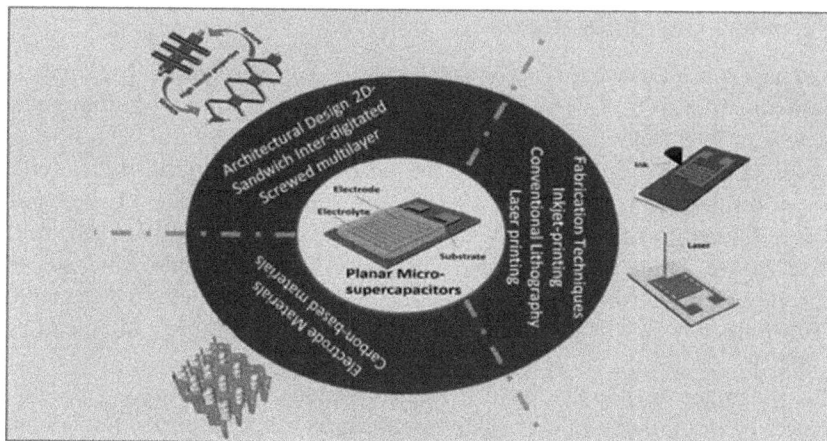

Figure 1: Schematics of contents covered in this chapter.

like structure and interdigitated structure (Qi et al. 2017). The planar interdigitated structure offers the advantage of a shorter ion diffusion distance of electrolyte and thus gives excellent power density. Therefore, it is necessary to summarize the effect of design on the performance of MSCs. Recently, much progress can be seen in the field of MSCs, but the low energy density is the major obstacle that limits their practical applications. One of the major factors that can improve the performance of MSCs is electrode material design. Recently, many new materials are investigated to improve the electrochemical performance of the device. Different types of materials have different characteristics but good electrical conductivity, better ion diffusion channels and specific surface area are the major requirements that need to be acquired in order to achieve excellent electrochemical performance. In fact, we can improve the device performance by designing all the components of MSCs (electrode, electrolyte, substrate, and current collector) (Qi et al. 2017).

A proper device fabrication technology should be chosen. Therefore, in this chapter, we summarize the structural design and fabrication methods for carbon-based materials in detail. On the whole, this chapter starts with the design concepts for MSCs and then introduces different fabrication techniques used so far in MSCs. Finally, different carbon materials and their optimal design are summarized.

2. Design Considerations and Performance Metrics for MSCs

The electrochemical performance of the MSCs strongly depends on the combination of components, such as electrodes, electrolytes, substrates, and current collectors. Additionally, the final device performance depends on the design of the device, compatibility within the components, and device assembly. Therefore, simultaneously working on the advancement of each of the components is not a better option to optimize the properties of supercapacitors (Liang et al. 2019). The major focus of this chapter is architectural approaches and parameters for standard MSC performance.

2.1 Design Considerations

A lot of new procedures and electrode materials have been reported that are feasible for designing of MSCs. At first, MSCs were fabricated similar to thin-film batteries or capacitors in a sandwiched configuration as shown in Figure 2a. But, the 2D sandwich-like configuration suffers from several disadvantages, which include unwanted displacement of electrodes under different environments of applications. Also, it is difficult to simultaneously handle both the energy and power performance of these MSCs for their superior performance. Although by increasing the mass loading the energy density can be improved, the power density deteriorates in thicker electrodes due to longer diffusion paths and increased electronic resistance (Liang et al. 2019).

As illustrated in Figure 2b, in the case of planar interdigitated architecture, each electrode of the device consists of a number of microelectrode fingers on the substrate. The fabrication process of the electrode is similar to the thin-film fabrication technique, which involves the patterning of the electrode. The interdigitated electrode architecture has many advantages over the 2D sandwich-like configuration. For example, a large number of open edges due to the interdigitated electrode fingers improves the diffusion of electrolyte ions (Liang et al. 2019). The other advantage is that the narrow interspaces between the electrodes can be controlled by using advanced patterning techniques or by applying microelectronic fabrication processes. This results in enhanced power density due to reduced ionic diffusion distance between the microelectrodes and reduced ionic resistance. Additionally, without any separators and binders, planar interdigitated MSCs offer excellent electrical and mechanical properties (Liang et al. 2019). In addition, the planar architecture of the MSC electrodes offers the ease of fabrication and integration of MSCs with other microdevices, which is advantageous for designing full microelectronic systems.

According to Liu et al. (2017), in order to obtain a high-performance MSC, some key parameters are shown in Figure 2b. As the ionic conductivity of the electrolyte is constant, the power density of the device deteriorates due to the increase in ion diffusion path and ESR of the device that results due to an increase in the gap between the adjacent electrodes. In order to improve the energy/power density performance, it is important to increase the ratio between the width of the gap to the width of the electrode. Apart from this, the energy density of the device can be enhanced either by increasing the thickness of the electrode (t) forming at the 3D structure or by increasing the length of the electrode (I). The former allows high areal mass loading of the active material that improves the device performance per unit area (Liang et al. 2019). Additionally, other than the planar interdigitated design, there is a helical structure shown in Figure 2c (Gao et al. 2018), which enhances the strength of the device.

On the other hand, the power density of the device can be improved by employing porous thin films. As electrode materials, they offer more ion diffusion channels that help shorten the ion diffusion path and increase the power density instead of using layered 2D materials since ions can diffuse only into the interlayer through edges, limiting the power performance even though it is under an optimized design. Recently, Yun et al. (2017) fabricated a porous graphene thin film as electrode material for MSC. This combination offers superior power density as well capacitance compared to other carbon-based all-solid-state MSCs as shown in Figure 2d. Recently, thin, flexible, and integrated high-performance MSCs are employed in flexible, portable/wearable electronic devices, such as sensors, artificial electronic skin, and roll-up displays. The key role in the flexibility of MSCs is played by the mechanical properties of electrode materials, electrolytes, substrates, and their assembly. Generally, flexible substrates, such as polymer films and paper, satisfy the requirements for MSCs. Meanwhile, gel and solid-state electrolytes can be used for flexible, rigid, and interdigitated planar MSCs (Liang et al. 2019).

2.2 Performance Metrics for MSCs

The amount of energy stored and the amount of power delivered per unit volume or weight are the two figures of merit of a supercapacitor. Note that it is difficult to compare different MSCs on the basis of gravimetric capacitance, as gravimetric performance depends upon total mass, thickness, and density of electrodes as well as the weights of other constituents. For planar MSCs, using gravimetric metrics is inappropriate as the weight of the electrode material is negligible, and the volume and substrate area of the device is limited. So far, reports suggest that volumetric capacitance is dependent on the properties of materials and as for thicker electrodes, the value of volumetric capacitance is essential. But the volumetric capacitance varies with the thickness of the electrodes. Specifically, electrodes for EDLCs, the volumetric capacitance decreases if the electrode consists of a complex internal pore structure and is thicker.

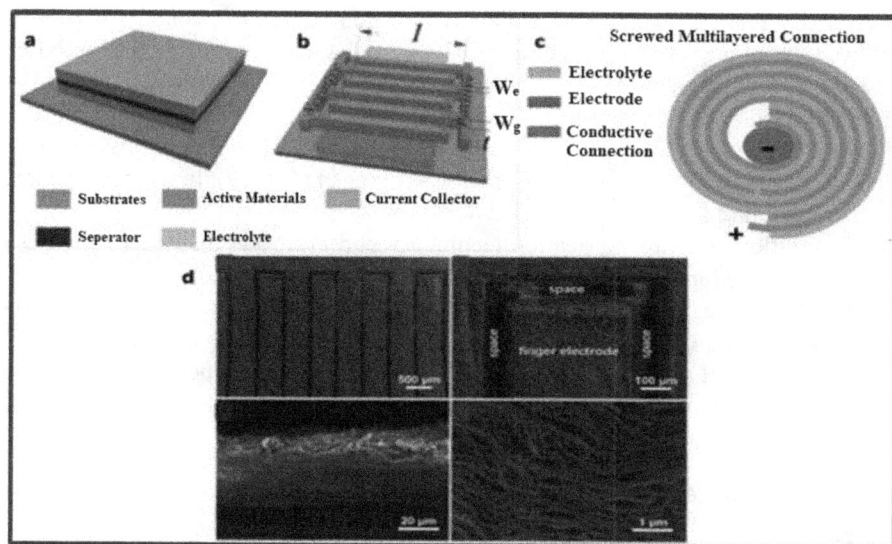

Figure 2: (a) Schematic diagram of 2D-sandwich supercapacitor and (b) in-plane interdigitated MSC structure. Reprinted with permission of Wiley-VCH, Copyright 2017 (Liu and Gao 2017). (c) Multilayer screwed connection, reprinted with permission of Wiley-VCH Copyright 2018 (Gao et al. 2018). (d) SEM image of laser carved RGO interdigitated MSC electrode. Reprinted with permission of Elsevier, copyright 2017 (Yun et al. 2017).

Also, the volumetric capacitance decreases for pseudocapacitive electrodes due to the inaccessible redox sites. The performance evaluation of the footprint area of the MSCs is the key as the integration of devices is supposed to be with the miniaturized electronic devices with limited areas (Liang et al. 2019). Therefore, areal capacitance, power, and energy density are the basis for MSCs. Areal capacitance C (F cm^{-2}), power P (W cm^{-2}), and energy densities E (Wh cm^{-2}) is calculated by the following equations:

$$C = \frac{Q}{A\Delta V} \tag{1}$$

$$E = \frac{0.5\, C\Delta V^2}{A\, 3600} \tag{2}$$

$$P = \frac{\Delta V^2}{A\, 4ESR} \tag{3}$$

where A is the total area of the microelectrode array and V is the voltage window. The capacitance and voltage window can be measured from circular cyclic voltammetry (*CV*) and galvanic charge-discharge (*GCD*) directly. Measuring total area accurately is the key to evaluating the areal performance of the electrode.

2.2.1   Microsupercapacitor Components and Architectures

Recently, many interesting materials with extraordinary properties are studied as electrode materials for MSCs. Among these materials, carbon-based materials have gained much research interest due to their good electronic conductivity, large surface area, and cyclability. Among these carbon-based materials, graphene-based MSCs can be integrated with microdevices due to their analogy with thin-film materials used in microelectronics. In addition, the compatibility of the electrolyte with as-designed electrode materials plays an important role in the improvement of electrochemical performance of planar MSCs (Liang et al. 2019). In the next section, we have summarized the carbon-based MSCs.

In addition, a lot of fabrication techniques and electrode materials have been reported. But before designing a device, it is essential to know the application purpose. In an integrated self-powering system, there are many required parameters such as power and energy densities, integrability, and cyclability for the application of MSCs (Liang et al. 2019). The above parameters strongly depend on fabrication techniques and compatibility between all the components of the device. In Section 4, we will discuss the recent fabrication techniques used for planar MSCs.

2.2.2   Electrolytes

The electrolyte is another important component of planar MSCs. Ideally, electrolyte should be electrochemically stable in a large voltage window, electronically insulating, and ionically conductive with additional requirements, such as low volatility, low toxicity, low viscosity, low cost, and availability at high purity. The operating voltage window of the device depends upon the decomposition potential of the electrolyte, which is the key factor for high specific energy values. On the other hand, the conductivity of the electrolyte influences the equivalent series resistance (ESR) which in turn contributes to the power density of supercapacitors.

Electrolytes for the SCs are classified into two types: liquid electrolyte and solid electrolyte. The liquid electrolytes are further classified as ionic electrolytes, aqueous electrolytes, and organic electrolytes. Although liquid electrolytes are developed for commercial supercapacitors, difficult packaging and electrolyte leakage hinder their use in electronic applications. In the case of MSCs, the use of solid-state electrolytes is recommended as they not only reduce external electrolyte leakage and internal shorting issues but also reduce the thickness of the device by altering the need for separators and extra encapsulation layers. The field of solid-state electrolytes is under dramatic progress.

Aqueous Electrolytes: Aqueous electrolytes (such as H_2SO_4, KOH, and Na_2SO_4) offer high ionic conductivity (1 S cm^{-1}) and low resistance but suffer from a low operating voltage window. For instance, Nagar et al. created a redox electrolyte by immersing KI into H_2SO_4. The as-fabricated graphene-based paper MSC reported enhanced performance with an outstanding volumetric capacitance of 29 mFcm^{-3} at a current density of 6.5 mAcm^{-3} (Nagar et al. 2018). But aqueous electrolytes offer a limited working voltage range of 1 V due to the low decomposition voltage of water (1.3 V). Also, these aqueous electrolytes are common as they can be prepared by simple processes.

Organic Electrolytes: The energy performance of the MSCs can be improved significantly by using organic electrolytes, a kind of mixture of salt and solvent; for instance, acetonitrile and propylene carbonate that offers a wide working voltage of almost 3 V. Carbide-derived carbon-based MSC offered an excellent voltage window of 2 V with an areal capacitance of 1.5 Fcm^{-2}, resulting in an energy density of 3 Jcm^{-2} by using 1M tetraethylammonium tetrafluoroborate (TEABF$_4$) in propylene carbonate electrolyte (Huang et al. 2013). Huang et al. designed a freestanding elastic carbon film based MSC that showed a wide working voltage range of 3 V employing 2M EMI, BF$_4$ in CH_3CN (acetonitrile) electrolyte and demonstrated a volumetric capacitance of 150 Fcm^{-3} at a scan rate of 20 mV/s, resulting in an improved energy density (Huang et al. 2016). Due to the poor ionic conductivity of organic electrolytes, their use is limited in high-power performance MSC.

Ionic Liquids: Ionic liquids are molten salts at room temperature, are non-flammable, and have a low vapor pressure that reduces the risk of explosion, as an electrolyte without solvent ionic liquids has gained tremendous attention in the field of MSCs. Also, in comparison to aqueous electrolytes, ionic liquids offer a wide voltage window but suffer from low ionic conductivity compared to the former (Liang et al. 2019). Furthermore, for next-generation integrated MSCs, ionic liquids can be developed and improved by mixing them with gel electrolytes.

Solid-state Electrolytes: The electrolytes discussed above are liquid in nature and require reliable encapsulation which limits their use in miniaturized and flexible devices. Instead, solid-state electrolytes are of particular interest for MSCs and offer the additional advantage of miniaturization and integration by combining separator and electrolyte in a single layer (Liang et al. 2019). Solid electrolytes are mainly produced by mixing a polymer matrix with an additive such as acid, salt, and ionic liquid in which ionic conduction takes place through voids and defects in the solid solution. Therefore, nowadays their performance compares very well with liquid electrolytes. Recently, different polymer matrixes have been developed, such as poly(vinylidene fluoride) (PVDF), poly(vinyl alcohol) (PVA), polyacrylonitrile (PAN), and poly(vinylpyrrolidone). Among these polymers, PVA is most extensively used. PVA/H_2SO_4, PVA/lithium chloride, and PVA/H_3PO_4 are most commonly used as solid-state electrolytes for planar MSCs. These solid-state electrolytes offer excellent cyclic stability, good mechanical properties, and low leakage current. The ionic conductivity of PVA/H_2SO_4 is reported to be 7×10^{-3} Scm^{-1}and for PVA/H_3PO_4 it is 10^{-5} to 10^{-3} Scm^{-1} at room temperature. Ionic gels or ionic liquid-based solid electrolytes offer a wide voltage window of above 2 V which in turn improves the energy density of MSCs. 1-ethyl-3-methylimidazolium bis(trifluoromethylsulfonyl) imide/fumed silica and PVA/1- butyl-3 methylimidazolium tetrafluoroborate ($BMIBF_4$) are also used as non-aqueous gel electrolytes for MSCs with a wide voltage range of 2.5 V and demonstrating excellent power performance (Liang et al. 2019).

Compared to liquid electrolytes, solid electrolytes possess low ionic conductivity. Also, to meet the requirements such as wide potential range and cyclability, attaining adequate chemical stability is critical. Additionally, solid-state electrolytes should guarantee mechanical flexibility and transparency as a requirement for their use in flexible and stretchable MSCs. Recently, Zhang et al. has reported a 3D graphene-based MSC using PVA/H_2SO_4 as a solid-state electrolyte that exhibits a high power density of 14.4 mWcm^{-2} and an energy density of 0.29 µWcm^{-2} and a potential window of 1 V (Zhang et al. 2017).

3. Carbon-Based MSCs

Carbon materials are promising candidates for supercapacitors due to their excellent properties, such as non-toxicity, high specific surface area, low cost, easy processing, good electrical conductivity, high chemical stability, and wide operating temperature range. Based on the dimensionality carbon materials are classified into three types: zero-dimensional (0D) includes carbon/graphene quantum dots, one dimensional (1D) includes carbon nanofibers, and carbon nanotubes (CNTs), and two dimensional (2D) includes graphene. Carbon-based electrochemical capacitors operate similar to electrochemical double-layer capacitors, which rely on the specific surface area for charge accumulation at the interface between electrode and electrolyte. This storage mechanism by adsorption limits the energy density performance. Therefore, based on the dimensionality of the carbon materials, there are certain remedies to improve the performance. The common strategy for 0D and 1D carbon materials is to design the nano/microporous structure to increase the active sites. A variety of strategies are used to improve the performance of graphene. The main factor that restricts the performance of graphene is restacking. In order to prevent this re-stacking, certain guest elements are introduced which act as spacers. Additionally, creating in-plane pores facilitates the electrode ion diffusion and provides sufficient reactive sites. In addition to the above two strategies, doping of heteroatom not only provides sufficient reactive sites but also improves the conductivity. A table summarizing the electrochemical performance along with fabrication techniques used for the above materials is shown in this section.

The application of carbonaceous powder in microelectrodes is hindered as it is difficult to form uniform and qualified thin films from them. With an aim to design carbon-based MSCs, Chmiola et al. accepted to tackle the problems in the technological hurdles (Chmiola et al. 2010). They found that carbide-derived carbon can be a promising candidate for MSCs with exceptionally high volumetric capacitance. The method of fabrication is shown in Figure 3. The entire synthesis and fabrication processes such as chemical vapor deposition and physical vapor deposition of precursor carbides and Au current collectors, the chlorination and plasma etching of the photolithography are compatible with the existing semiconductor industry and thus CDC is the promising candidate for MSC devices.

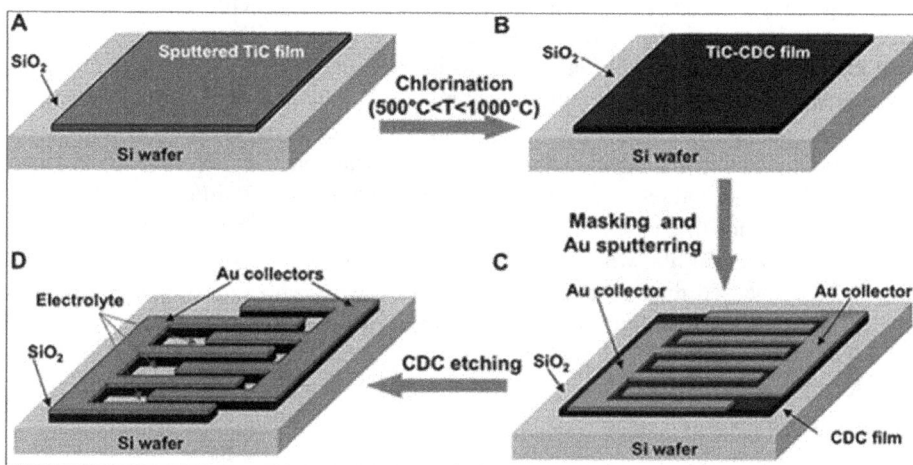

Figure 3: (A-D) Schematics of fabrication process of on-chip CDC-based MSC. Reprinted with permission from American Association for the Advancement of Science Copyright (2010) (Huang et al. 2013).

To obtain high capacitance, it is necessary to obtain high surface area within the limited volume for MSCs and activated carbon is still a better choice. But the existing fabrication routes for carbon powders are not applicable for activated carbon thin films (ACFs) which are essential for MSCs. The challenges include the formation of cracks due to polymer shrinkage at high temperatures during heating and carbonization, the weak interface between polymer and substrate due to large stresses developed during harsh synthesis and cooling conditions, and the possible damage to the brittle film due to photolithographical process. Wei et al. reported a method to fabricate the ACFs with reduced interface stresses, namely catalyst-assisted low-temperature carbonization of an organic compound solution (Wei et al. 2013). The sucrose and H_2SO_4 (as a catalyst) were spin-coated on SiO_2/Si substrate by spin coating, was dried at room temperature, annealed, and carbonized at 700°C in a vacuum in order to remove the decomposition products of carbohydrate and catalyst residues. It was further activated at 900°C in a CO_2 atmosphere to obtain a porous structure. Also, to sustain during lithographical patterning, strong substrate adhesion was achieved by further annealing at 1,100°C in the Ar atmosphere. The as-fabricated device exhibited a high volumetric capacitance of 390 F cm^{-3} at a scan rate of 1 mV/s. The volumetric performance can be further improved by increasing the activation time, as the film becomes more porous and rate capability can be further improved due to easy access to the electrolyte ions. But it is still noteworthy that the fabrication process is still harsh as high-temperature annealing is required several times which is not compatible with manufacturing the device on the chip.

The other important direction in MSCs is the utilization of its fast response time by designing open structured devices to replace electrolytic capacitors (Pech et al. 2010, Yoo et al. 2011). Pech et al. fabricated MSC by employing electrophoretic deposition for several micrometer thick layers of nanostructured carbon onions (OLC) onto inter-digitally patterned Au current collector on silicon wafers. The OLC particles were prepared by annealing nano-diamonds at 1,800°C. A device showed a stable capacitive performance within a potential window of about 3 V in 1M solution of tetraethylammonium tetrafluoroborate in PC. The micro-device demonstrated an areal capacitance of about 0.9 mF cm^{-2} at a scan rate of 100 mV/s, which is comparable to the values obtained at lower scan rates (0.4-2 mFcm^{-2}). The most interesting feature of this OLC-based MSC is its extremely small relaxation time constant of about 26 ms, which is much smaller than AC-carbon-based MSC (700 ms) and OLC-based macroscopic SC (>1 s). This research suggests that electronic conductivity should be improved in order to achieve a fast response time. Also, the micropores should be reduced within the electrodes, but the open area in contact with the electrolyte should be enlarged as the outermost surface is most easily accessible to the electrolyte ions for charge-discharge processes. But such a design hinders the volumetric capacitance and also the film thickness should be thin as well, thus limiting the areal capacitance of the device.

4. Fabrication Techniques for MSCs

The various fabrication methods reported for planar MSCs (Figure 4) are conventional photolithography fabrication method, inkjet printing method, screen printing method, electrophoretic method, laser scribing, electrode conversion, and electrolytic deposition method (Bu et al. 2020). Different active materials have different properties, hence there has not been a fabrication method suitable for all active materials. Therefore, while fabricating MSCs, it is important to take into consideration the used active materials, the type of electrolytes, and the interface between electrodes and electrolytes.

4.1 Conventional Lithography Method

Photolithography is the most convenient method in the field of microelectronics for defining patterns on the semiconductor surface. The process involves three steps: coating of photoresist, exposure, and development of photoresist. In these methods,

Table 1: Summary of recent advancements in carbon-based MSCs

Electrode Materials	Substrate	Electrolyte	Specific Capacitance (mF cm^{-2})	Cycle stability (Retention)	Energy Density	Power Density	Fabrication Technique	Reference
Activated carbon	Silicon	1 M Et$_4$NBF$_4$ in PC	11.6	-	10 mWh cm^{-3}	~40 mW cm^{-3}	Electrophoretic deposition	(Pech et al. 2010)
Activated carbon	Silicon	1 M Et$_4$NBF$_4$ in PC	2.1	-	1.8 μWh cm^{-2}	44.9 mW cm^{-2}	Ink-Jet printing	(Pech et al. 2010)
Activated carbon	Silicon	1 M NaNO$_3$	90.7	-	-	51.5 mW cm^{-2}	Photolithography	(Shen et al. 2011)
Activated carbon	Silicon	1 M Et$_4$NBF$_4$ in PC	81	-	71.4 μWh cm^{-2}	34.4 mW cm^{-2}	Si$_3$N$_4$ hard mask	(Durou et al. 2012)
Photo-resist derived carbon	Silicon	0.5 M H$_2$SO$_4$	75	87% after 1,000 cycles			Photolithography	(Beidaghi and Chen 2011)
CNT arrays	Mo/Al	Ionic liquid BMIM-PF6	0.428	No loss after 10 cycles			Photolithography	(Jiang and Zhou 2009)
CNT array	Silicon	0.1 M Na$_2$SO$_4$	36.5	-	0.4 Wh kg^{-1}	1 kW kg^{-1}	Photolithography	(Liu et al. 2011)
Carbide derived carbon	Silicon/TiC	1 M H$_2$SO$_4$/1 M TEABF4	450/300				Photolithography	(Chmiola et al. 2010)
Meso-carbon micro-bead	Silicon	Solid-state [BMIM][BF$_4$]	100	53% after 8,000 cycles	10 μWh cm^{-2}	0.575 mW kg^{-1}	Dispenser printing	(Miller et al. 2009)
Super P carbon black	Silicon	1.5 M H$_2$SO$_4$	0.8	-	-	-	Origami method	(In et al. 2006)
Porous RGO film	PET	PVA/H$_2$SO$_4$	37.95	95% after 10,000 cycles	1.45 mWh cm^{-3}		Ice drying, laser carve	(Yun et al. 2017)

(Contd.)

Table 1: (*Contd.*)

Electrode Materials	Substrate	Electrolyte	Specific Capacitance (mF cm⁻²)	Cycle stability (Retention)	Energy Density	Power Density	Fabrication Technique	Reference
RGO/MWCNTS	PET	PVA/H_3PO_4	46.6	88.65 after 10,000 cycles	6.47 mWh cm⁻³	10 mW cm⁻³	Direct Laser Writing	(Mao et al. 2018)
Graphene/CNT/cross-linked PH 1000 film	Rubber substrate	PVA/H_3PO_4	107.5	93.2% after 8,000 cycles	0.54 µWh cm⁻	1.22 mW cm⁻²	Mask-assisted filtration	(Xiao et al. 2018)
Holey polypyrrole/RGO poly oxometalate	Exfoliated graphing-nylon	PVA/H_2SO_4	115	80% after 2,000 cycles	4.8 mWh cm⁻³	645.1 mW cm⁻³	Hydrothermal reduction	(Qin et al. 2018)
RGO/MXene	PET	PVA/H_2SO_4	2.4	97% after 10,000 cycles	8.6 mWh cm⁻³		Mask-assisted printing	(Couly et al. 2018)
Exfoliated graphene/MnO_2/PEDOT:PSS	PET	PVA/LiCl	9.6	92% after 5,000 cycles	8.6 mWh cm⁻³	4.2 Wcm⁻³	Mask-assisted printing	(Zheng et al. 2018)
RGO/cellulose nanocrystals/RGO/MoS_2	Kapton	PVA/H_3PO_4	121	97% retained after 10,000 cycles	100 µWh cm⁻²	≈102 mW cm⁻²	Spinning	(Pan et al. 2018)
RGO/PANI	Polystyrene	PVA/H_2SO_4	1,329	75% retained after 1,000 cycles			3D printing	(Wang et al. 2018)

MnO$_2$@RGO	FTO	PVA/LiCl	114.3	77% after 6,000 cycles	13.9 mWh cm^{-3}	34.7 mW cm^{-3}	Laser engraving	(Du et al. 2019)
RGO/PEDOT	PET	PVA/H$_3$PO$_4$	35.12	90% after 6,000 cycles	4.876 mWh cm^{-3}		Laser scribing	(Mao et al. 2019)
G-CNT ink	Polyimide sheet	PVA/H$_3$PO$_4$	9.81	95.5% after 10,000 cycles	1.36 µWh cm^{-2}	0.25 mW cm^{-2}	Direct ink writing	(Wang et al. 2020)
Sulfur-doped graphene	Silicon	PVA/H$_2$SO$_4$	553	95% retained after 10,000 cycles	3.1 mWh cm^{-3}	1191 W cm^{-3}	Photolithography	(Wu et al. 2017)
Nanoporous carbon	Silicon	1 M Et$_4$NBF$_4$/ PC	-	95% after 10,000 cycles	-	232.6 W cm^{-3}	Photolithography	(Yang et al. 2019)

the desired pattern is transferred on the thin resist by means of interaction between the resist and particles, like photons or electrons (Bu et al. 2020).

Once the pattern is structured on resist, the devices can be obtained either by metal lift-off or by etching. Depending on the writing procedure, the lithography method is divided into two categories: series and parallel. For series writing (maskless), the pattern is transferred pixel-by-pixel on resist to obtain a more precise pattern but unfortunately, it is a slow process. For parallel writing, such as optical lithography using different wavelengths, the pattern is transferred to resist through the mask containing the pattern. Although this method has the advantage of being fast, it is limited by diffraction effects and the mask fabrication steps. On the industrial front, for mass production, optical lithography is preferred but series writing is used for mask fabrication (Bu et al. 2020). Conventional lithography needs process optimization for extremely small feature sizes, but this method is still a good choice for a feature size of more than 1 micron. Thus, in order to achieve small feature size, certain parameters need optimization, such as PR thickness, spin time, spin speed, developer concentration, developing time, and PEB time. Wang et al. (2017) employed photomask assisted photo-reduction strategy to produce a GO-TiO$_2$ hybrid for high-performance MSC. The device showed remarkable specific capacitance of about 233 F cm^{-3} in PVA/H$_2$SO$_4$ with extraordinary flexibility. The MSC showed a power density of about 7.7 mW hcm^{-3} with a power density of 312 Wcm^{-3} and also the device showed an excellent cyclability of about 10,000 cycles with no loss of capacitance (Wang et al. 2017).

4.2   Screen Printing Method

The screen printing technique is used for patterning on the substrates like cloth and paper. The technique uses a woven mesh to support an ink-blocking stencil to obtain a desired pattern. The process involves the following steps: ink is pressed through the openings of the mesh, adheres to the substrate, and lastly as the ink dries the desired pattern is obtained on the substrate. As the screen printing technique does not employ sophisticated instrumentation and processes, it is a low-cost and simple printing method and is used for electronic devices, such as field-effect emission devices and transparent electrodes (Bu et al. 2020). As the whole fabrication process can be performed in ambient conditions and does not make use of an expensive metal deposition process and the ink containing the active material can be directly printed on the desired substrate through the opening of the mesh. The process is versatile and can be used for materials that are incompatible with the existing fabrication techniques (Wang et al. 2017). The printable colloidal suspension and viscous paint can be formed by formulating electrochemical active materials with inactive additives, such as a binder, rheological agents, and surfactants. At first, functional inks to print MSCs were formulated from pseudocapacitive materials, such as metal oxides and conducting polymers possessing high capacitance. But due to their low electrical conductivity, these pseudocapacitive microdevices suffered from low power density and cyclic stability. Recently, reduced graphene oxide (rGO) and liquid exfoliated graphene are used for preparing printable inks for designing fully-printed flexible MSCs (Bu et al. 2020). Additionally, electrode geometries play a crucial role in device performance with narrower interspaces between the microelectrodes helping improve the device capacitance and rate capability. To date, the achievable feature size is 30-50 micro-meter, which is sufficient for wearable electronics. But due to the optimization challenges of the rheological behavior of the major constituents, most screen-printed MSCs are restricted to millimeter-scale resolution, which hampered their on-chip integration. Therefore, for efficient production of flexible screen-printed MSCs, the electrode materials should satisfy the following conditions: (i) high electrical conductivity without further chemical treatment for rapid charge transfer, (ii) dual charge storage mechanism (electric double layer and pseudocapacitive) to achieve high performance, (iii) excellent printability to achieve high-resolution printing and good reliability, (iv) free from additive mass to improve the performance, (v) should be printed in porous structure to promote ion diffusion, and (vi) good mechanical integrity to sustain deformation (Bu et al. 2020)

4.3 Inkjet Printing Method

Inkjet printing is a simple and high uptake method, which can simultaneously perform deposition and patterning, reducing the usage of material and process complexity. Inkjet printing can achieve computer printing, which prints a pattern by propelling ink droplets on paper, plastic, and other substrates. As the printing process is efficient and versatile to integrate planar MSCs to random connections and any scale. Large-scale fabrication of MSC arrays of more than 100 devices on a single Si wafer and flexible substrate (Kapton) are manufactured. Inkjet printing is one of the most reliable techniques due to its advantages, such as high resolution (down 50 micro-meter), scalability, cost-efficient, direct printing, low material consumption, and control on thickness. Also, the printing process is compatible with most of the materials due to which all printed MSC can be realized, which improves the scalability and resolution of the device as no manual assembly step is required for further device processing. For example, Pech et al. prepared an ink of activated carbon with PTFE as a binder in ethylene glycol with a stabilizing surfactant. This material was deposited on patterned gold as a current collector with a substrate temperature of about 140°C to maintain homogeneity. A variety of supercapacitors were designed with the width of microelectrodes varying from 40-100 μm (Yun et al. 2017).

4.4 Electrophoretic Deposition Method

The method where charged particles in a colloidal suspension are moved through the liquid due to electric field and deposited on the electrode with opposite polarity, forming a desired material or device is known as electrophoretic deposition (EPD). A wide range of structures can be deposited by EPD, from traditional to advanced, porous materials to compact coatings and thin films (nanometer scale) to thick films (mm scale) (Bu et al. 2020). These structures include different shapes and compositions and can be formed within a short experimentation span and with a single instrument.

EPD possess highly versatile applications, simple instrumentation, short processing time, cost-efficient, facile modification, dense packaging of particles

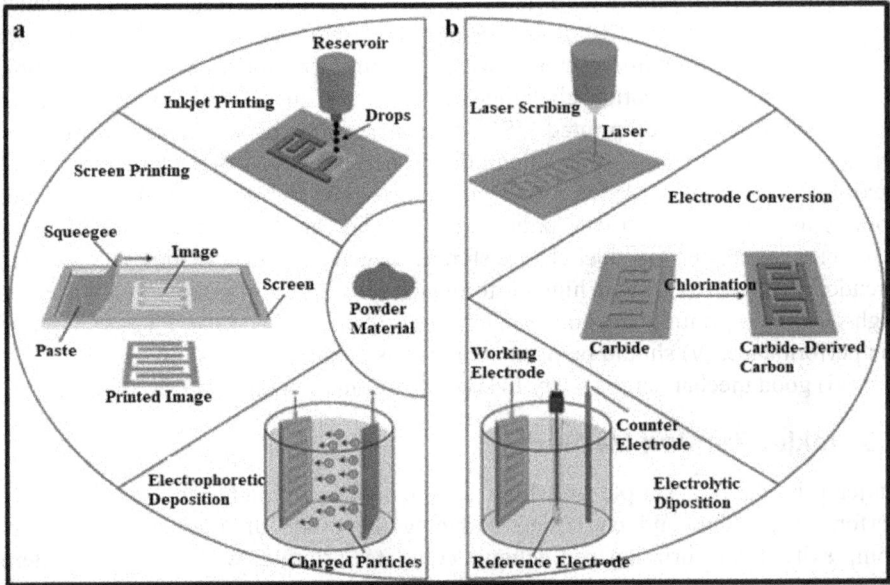

Figure 4: Different techniques for fabrication of MSCs.

in the final product, high quality of the microstructures produced, easy production of geometrically complicated shapes, and simple control of the thickness and morphology have made EPD an interesting technique both in academia and industry. In EPD, the particle state and their evolution in the suspension can be manipulated and controlled. Moreover, the production of dense, homogenous, or porous microstructures can be enabled by accurate and appropriate processing conditions (Wang et al. 2017).

4.5 Laser Scribing

A number of terminologies are included in laser scribings such as thin-film scribing and micro scribing and are used to create fine-scale groves that are precisely controlled or write lines into films and coating layers. Laser energy can be used in two ways either by focusing as a spot or scanning on the entire surface or by imaging through the mask and projecting the pattern on the surface. Laser scribing can be used for accurate writing of very fine features on the surface of the material or by removing one or more layers in a multi-layered substrate (Bu et al. 2020)

It is possible to selectively target individual layers in thin-film devices by making use of precision laser processing, such as in capacitors and batteries variable patterning process can be created. One deposited layer can be removed from the other or from the underlying substrate that is critical for the processing of isolation scribes in thin-film capacitor production by tuning the laser process. Additionally, any material can be scribed including those that are reflecting and transparent by careful selection of laser type and process parameters. For instance, Peng et. al successfully prepared boron-doped porous graphene from polyimide sheets immersed in boric acid simply by laser induction. A flexible solid-state MSC was fabricated by assembling solid electrolyte and the resulted device showed an excellent areal capacitance of 16.5 mF cm^{-2}, three times higher than the undoped electrode material (Peng et al. 2015). Selective laser machining permits the removal of metallic layers and coating with almost negligible machining of the underlying substrate. Highly dense circuitry and functional or decorative patterns can be created by writing very fine lines. Conductive and non-conductive areas can be determined by laser scribing thin oxide layers grown on flexible or rigid substrates (Bu et al. 2020).

Laser scribing creates much narrower scribe lines compared to mechanical scribing which helps improve the overall yield. The process being noncontact reduces the micro-cracking and substrate damage risk. The excellent beam quality of Spectra-Physics lasers and high peak power results in cleaner scribe lines and higher throughput. The following are the advantages of laser scribing (Bu et al 2020):

- Clean scribing of hard and brittle materials.
- Low-cost operation with non-contact processing.
- Low parts per wafer due to narrow cut widths.
- Low cost, robust and reliable manufacturing.

4.6 Electrode Conversion Method

Carbon materials produced by selectively etching metals from metal carbides using chlorine at elevated temperatures are known as carbide-derived carbon (CDC). CDC has shown excellent electrochemical performance as active material in traditional supercapacitors because its microstructure can be tuned precisely by tailoring the synthesis conditions for a particular electrolyte (Huang et al. 2013). CDC is attractive in the domain of microfabricated capacitors due to two reasons. The first reason is the conductive nature of the precursor carbide that can be deposited by well-known physical and chemical deposition techniques (CVD and PVD) as uniform thin and thick films. Next, the chlorination process can be performed at a temperature of about 200°C, and the resulting coatings are well-adhered with a clean and atomically precise interface, which improves the device impedance (Huang et al. 2013).

5. Conclusion

In this chapter, we have reviewed the recent advances in the fabrication, design, and approaches to exploit the carbon materials for planar interdigitated MSCs. The standard requirements for planar MSCs are their power and energy performance per unit area, their integrability with microdevices, and their cyclability. The above properties depend upon the combined effect of electrode and electrolyte properties, the effect of fabrication techniques, and the geometries adopted. Therefore, a direct comparison of the literature is constrained. The recent literature indicates that carbon materials, such as CNT, quantum dots, and graphene show the versatility and compatibility for the integration on both flexible and rigid planar substrates. In spite of such a development in the field of carbon-based MSCs, still there is a scope for further development from understanding the charge transport mechanism and fundamental electrochemistry to the integration and fabrication aspects.

References

Beidaghi, M., Chen, W. and Wang, C.L. 2011. Electrochemically activated carbon micro-electrode arrays for electrochemical micro-capacitors. J. Power Sources, 196(4): 2403–2409.

Bu, F., Zhou, W., Xu, Y., Du, Y., Guan, C. and Huang, W. 2020. Recent Developments of Advanced Micro-Supercapacitors: Design, Fabrication and Applications. Npj. Flex Electron. 4: 31.

Chen, D., Lou, Z., Jiang, K. and Shen, G. 2018. Device configurations and future prospects of flexible/stretchable lithium-ion batteries. Adv. Funct. Mater. 28(51): 1805596.

Chmiola, J., Largeot, C., Taberna, P.L., Simon, P. and Gogotsi, Y. 2010. Monolithic carbide-derived carbon films for micro-supercapacitors. Science 328: 480–483.

Couly, C., Alhabeb, M., Van Aken, K.L., Kurra, N., Gomes, L., Navarro-Suárez, A.M., Anasori, B., Alshareef, H.N. and Gogotsi, Y. 2018. Asymmetric flexible mxene-reduced graphene oxide micro-supercapacitor. Adv. Electron. Mater. 4(1): 1700339.

Durou, H., Pech, D., Colin, D., Simon, P., Taberna, P.L. and Brunet, M. 2012. Wafer level fabrication process for fully encapsulated micro-supercapacitors with high specific energy. Microsyst. Technol. 18: 467–473.

Gao, Y., Wan, Y., Wei, B. and Xia, Z. 2018. Capacitive enhancement mechanisms and design principles of high performance graphene oxide-based all-solid-state supercapacitors. Adv. Funct. Mater. 28(17): 1706721.

González, A., Goikolea, E., Barrena, J.A. and Mysyk, R. 2016. Review on supercapacitors: technologies and materials. Renew. Sust. Energ. Rev. 58: 1189–1206. https://doi.org/DOI: 10.1016/j.rser.2015.12.249.

Hu, H., Pei, Z. and Ye, C. 2015. Recent advances in designing and fabrication of planar micro-supercapacitors for on-chip energy storage. Energy Storage Mater. 1: 82–102.

Huang, P., Heon, M., Pech, D., Brunet, M., Taberna, P.-L., Gogotsi, Y., Lofland, S., Hettinger, J.D. and Simon, P. 2013. Micro-supercapacitors from carbide-derived carbon (CDC) films on silicon chip. J. Power Sources 225: 240–244.

Huang, P., Lethien, C., Pinaud, S., Brousse, K., Laloo, R., Turq, V., Respaud, M., Demortiere, A., Daffos, B., Taberna, P.L., Chaudret, B., Gogotsi, Y. and Simon, P. 2016. On-chip and freestanding elastic carbon films for micro-supercapacitors. Science 351(6274): 691–695.

In, H.J., Kumar, S., Shao-Horn, Y. and Barbastathis, G. 2006. Origami fabrication of nanostructured, three dimensional devices: electrochemical capacitors with carbon electrodes. Appl. Phys. Lett. 88: 083104.

Jiang, Y.Q., Zhou, Q. and Lin, L. 2009. Planar MEMS supercapacitor using carbon nanotube forest. IEEE 22nd Int. Conf. on MEMS, 587.

Liang, J., Mondal, A.K., Wang, D.W. and Iacopi, F. 2019. Graphene-based planar microsupercapacitors: recent advances and future challenges. Adv. Mater. Technol. 4: 1800200.

Liu, C.C., Tsai, D.S., Chung, W.H., Li, K.W., Lee, K.Y. and Huang, Y.S. 2011. Electrochemical micro-supercapacitors of patterned electrodes loaded with manganese oxide and carbon nanotube. J. Power Sources 196(13): 5761–5768.

Liu, L., Niu, Z. and Chen, J. 2017. Design and integration of flexible planar micro-supercapacitors. Nano Res. 10: 1524–1544.

Liu, N. and Gao, Y. 2017. Recent progress in micro-supercapacitors with in-plane interdigital electrode architecture. Small 13: 1701989.

Mao, X., Xu, J., He, X., Yang, W., Yang, Y., Xu, L., Zhao, Y. and Zhou, Y. 2018. All-solid-state flexible microsupercapacitors based on reduced graphene oxide/multi-walled carbon nanotube composite electrodes. Appl. Surf. Sci. 435: 1228–1236.

Mao, X., He, X., Xu, J., Yang, W., Liu, H., Yang, Y. and Zhou, Y. 2019. Three dimensional reduced graphene oxide/poly(3,4-ethylenedioxythiophene) composite open network architectures for microsupercapacitors. Nanoscale Res. Lett. 14: 267.

Miller, L.M., Ho, C.C., Shafer, P.C., Wright, P.K., Evans, J.W. and Ramesh, R. 2009. Integration of a low frequency, tunable MEMS piezoelectric energy harvester and a thick film micro capacitor as a power supply system for wireless sensor nodes. Proc. IEEE ECCE 30: 2627–2634.

Mu, J., Zhao, X., Zhang, Y., Zhang, Y., Huang, S., Sheng, B., Xie, Y., Zhang, Y., Xie, Z. and Du, E. 2019. Layered coating of ultraflexible graphene-based electrodes for high-performance in-plane quasi-solid-state micro-supercapacitors. Nanoscale, 11: 14392–14399.

Nagar, B., Dubal, D.P., Pires, L., Merkoci, A. and Romero, P.G. 2018. Design and fabrication of printed paper-based hybrid micro-supercapacitor by using graphene and redox active electrolyte. Chem Sus Chem 11(11): 1849–1856.

Pan, H., Wang, D., Peng, Q., Ma, J., Meng, X., Zhang, Y., Ma, Y., Zhang, S. and Zhu, D. 2018. High performance micro-supercapacitors based on bioinspired microfibers. ACS Appl. Mater. Interfaces 10(12): 10157–10164.

Pech, D., Brunet, M., Durou, H., Huang, P., Mochalin, V., Gogotsi, Y., Taberna, P. and Simon, P. 2010. Ultrahigh-power micrometer-sized supercapacitors based on onion-like carbon. Nat Nano 5: 651–654.

Pech, D., Brunet, M., Taberna, P.L., Simon, P., Fabre, N., Mesnilgrente, F., Conedera, V., Durou, H. 2010. Elaboration of a microstructured inkjet-printed carbon electrochemical capacitor. J. Power Sources, 195(4): 1266–1269.

Peng, Z., Ye, R., Mann, J., Zakhidov, D., Li, Y., Smalley, P.R., Lin, J. and Tour, J.M. 2015. Flexible boron-doped laser-induced graphene microsupercapacitors. ACS Nano 9: 5868–5875.

Qi, D., Liu, Y., Liu, Z., Zhang, L. and Chen, X. 2017. Design of architectures and materials in in-plane micro-supercapacitors: current status and future challenges. Adv. Mater. 29: 1602802.

Qin, J., Zhou, F., Xiao, H., Ren, R. and Wu, Z.-S. 2018. Mesoporous polypyrrole-based graphene nanosheets anchoring redox polyoxometalate for all-solid-state micro-supercapacitors with enhanced volumetric capacitance. Sci. China Mater, 61(2): 233–242.

Shen, C.W., Wang, X.H., Zhang, W.F. and Kang, F.Y. 2011. A high-performance three-dimensional micro supercapacitor based on self-supporting composite materials,. J. Power Sources 196: 10465–10471.

Wang, S., Wu, Z.-S., Zheng, S., Zhou, F., Sun, C., Cheng, H.-M. and Bao, X. 2017. Scalable fabrication of photochemically reduced graphene-based monolithic micro-supercapacitors with superior energy and power densities. ACS Nano 11(4): 4283–4291.

Wang, Y. and Xia, Y. 2013. Recent progress in supercapacitors: from materials design to system construction. Adv. Mater. 25: 5336–5342.

Wang, Y., Cao, Q., Guan, C. and Cheng, C. 2020. Recent advances on self-supported arrayed bifunctional oxygen electrocatalysts for flexible solid-state Zn-air batteries. Small 16: 2002902.

Wang, Y., Zhang, Y., Wang, G., Shi, X., Qiao, Y., Liu, J., Liu, H., Ganesh, A. and Li, L. 2020. Direct graphene-carbon-nanotube composite ink writing all-solid-state flexible microsupercapacitors with high areal energy density. Adv. Funct. Mater. 30(16): 1907284.

Wang, Z., Zhang, Q.E., Long, S., Luo, Y., Yu, P., Tan, Z., Bai, J., Qu, B., Yang, Y., Shi, J., Zhou, H., Xiao, Z.Y., Hong, W. and Bai, H. 2018. Three-dimensional printing of polyaniline/reduced graphene oxide composite for high performance planar supercapacitor. ACS Appl. Mater. Interfaces 10(12): 10437–10444.

Wei, L., Nitta, N. and Yushin, G. 2013. Lithographically patterned thin activated carbon films as a new technology platform for on-chip devices. ACS Nano 7: 6498–6506.

Wu, Z.-S., Tan, Y.-Z., Zheng, S., Wang, S., Parvez, K., Qin, J., Shi, X., Sun, C., Bao, X., Feng, X. and Müllen, K. 2017. Bottom-up fabrication of sulphur-doped graphene films derived from sulphur-annulated nanographene for ultrahigh volumetric capacitance micro-supercapacitors. J. Am. Chem. Soc. 139: 4506–4512.

Xiao, H., Wu, Z.-S., Zhou, F., Zheng, S., Sui, D., Chen, Y. and Bao, X. 2018. Stretchable tandem micro-supercapacitors with high voltage output and exceptional mechanical robustness. Energy Storage Mater. 13: 233–240.

Yang, Li, Huaqing Xie, Jing Li, Yoshio Bando, Yusuke Yamauchi and Joel Henzie. 2019. High performance nanoporous carbon microsupercapacitors generated by a solvent-free MOF-CVD method. Carbon 152: 688–696.

Yoo, J.J., Balakrishnan, K., Huang, J., Meunier, V., Sumpter, B.G., Srivastava, A., Conway, M., Mohana Reddy, A.L., Yu, J. and Vajtai, R. 2011. Ultrathin planar graphene supercapacitors. Nano Lett. 11: 1423–1427.

Yun, X.W., Xiong, Z.Y., Tu, L., Bai, Q. and Wang, X.G. 2017. Hierarchical porous graphene film: an ideal material for laser-carving fabrication of flexible micro-supercapacitors with high specific capacitance. Carbon 125: 308–317.

Zhang, L., DeArmond, D., Alvarez, N.T., Malik, R., Oslin, N., McConnell, C., Adusei, P., Hsieh, K.Y.Y. and Shanov, V. 2017. Flexible micro-supercapacitor based on graphene with 3D structure. Small 13(10): 1603114.

Zheng, S., Lei, W., Qin, J., Wu, Z.-S., Zhou, F., Wang, S., Shi, X., Sun, C., Chen, Y. and Bao, X. 2018. All-solid-state high energy planar asymmetric supercapacitors based on all-in-one monolithic film using boron nitride nanosheets as separator. Energy Storage Mater. 10: 24–31.

Carbon Nanocomposite for Energy Storage Applications

Pooja Zingre and Abhay D. Deshmukh*

Energy Materials and Devices Laboratory, Dept. of Physics, RTM Nagpur University, Nagpur - 440033, India

1. Introduction

With the outburst of consumption of traditional energy, currently available energy resources such as fossil fuels are just insufficient to fulfil the demand for vital energy. This ultimately leads to an urgent need of harnessing new, more efficient and environmentally friendly energy resources. Meanwhile, the increasing trend of portable electronic devices motivates for development of durable, lightweight, cost-effective and environmentally friendly energy storage devices. Fuel cells, lithium-ion batteries, sodium-ion batteries, lithium sulphur batteries, and supercapacitors (SC) are different classes of electrochemical energy storage (EES) devices and are expected to deliver comprehensive electrochemical performances (Hoang et al. 2019, Borenstein et al. 2017). Amongst these energy storage devices, supercapacitor attracted more attention under its high power generation, fast charge discharge and long cycle life, whereas electrochemical behaviour of supercapacitor strongly depends on electrode material. Amid various available electrode materials, carbon is widely used owing to its natural abundance and low cost and it can be easily commercialised.

Carbon has a combination of numerous exceptional properties like mechanical strength, good electrical conductivity, high electron mobility, high chemical and thermal stability, large surface area, etc., and fulfils the maximum conditions required for supercapacitor electrode material. The carbon family includes a large number of materials ranging from 0D to 3D consisting of quantum dots, CNT's, graphene, graphene oxide and activated carbons (Tang et al. 2014). They all serve as effective and excellent electrode materials for electrochemical energy storage. However, aside from their diverse sustainable properties, carbon electrodes suffer from some limitations in supercapacitors including low energy density, aggregation during synthesis and processing results in inferior device performance. Thus, it is recommended to recombine carbon electrode materials with pseudocapacitive or

Corresponding author: abhay.d07@gmail.com

redox-active materials. This increases the overall specific capacity and electroactive behaviour of electrode materials. Generally, the carbon-based materials execute high performances in commercially available supercapacitors because of their high surface area, which significantly enhances the specific capacitance. However, their large pores and randomly interconnected architectures restrict their practical applications. Considering the combinations of carbon nanomaterials with pseudocapacitive materials will increase the overall cell performance. Since, charge storage mechanism of carbon-based materials follows intercalation/deintercalation at electrode and electrolyte interfaces, whereas pseudocapacitive materials give added advantages of the faradic process along with intercalation/deintercalation process.

Abundantly, most researchers are interested in this type of hybrid material as it provides the synergistic effect of both EDLC and pseudocapacitive electrode materials and helps to enhance the specific capacitance, energy density, power density and cyclic stability of supercapacitors. Emerging carbon-based materials, such as carbon nanotubes and graphene, are most considerably used for nanocomposites purposes. It provides sp^2 hybridised carbon and surface dangling bonds, and they provide high accessible surface areas and porosity to the materials. More research has been done to enhance the performances of carbon-based nanocomposites. There has been growth in the more complicated sp^2 carbon nanomaterials because of the ability to use three-dimensional (3D) architecture designs with a variety of current collectors (Kumar et al. 2021). In general, pseudocapacitive materials, such as metal oxides, conducting polymers, metal-organic frameworks and transition metal carbides and nitrides, are frequently studied and are fundamental candidates. Faradic process in pseudocapacitive materials provides an open range of working potentials and increases the specific capacitance. During the faradic process, the charge passes through the surface of electrode materials which is similar to the battery mechanism (Kumar et al. 2021). Therefore, pseudocapacitive materials possess high electrochemical performances. Every composite has many added advantages due to the synergistic effect of core materials. In the last few years, intensive research efforts have been devoted to rationalising fundamental properties of nanocomposites for developing beyond traditional supercapacitors.

A major goal of the current chapter is to develop a comprehensive understanding of carbon-based nanocomposites with materials, like metal oxides, doping heteroatoms, metal-organic frameworks, transition metal carbides, nitrides, etc. Overview of the advantages of carbon-based nanocomposites irrespective of their structural and synthesis parameters. Highlighting the impact of carbon nanocomposites to overcome the complications regarding electrochemical properties of supercapacitors, the chapter also emphasizes the importance of composite nanostructures for designing highly efficient supercapacitors.

2. All Carbon Composites

2.1. Carbon Quantum Dots, Graphene Quantum Dots and Activated Carbons Composites

Carbon quantum dots (CQD), graphene quantum dots (GQD), onion-like carbons and activated carbon (AC) are included in carbon nanomaterials having very high specific surface area. These nanomaterials show an enormous combination of favourable properties, like low toxicity, high density, environmental stability, photostability, diminished photobleaching, electrocatalysis, multiphoton excitation and biocompatibility. Additionally, CQD/GCQD are enriched with enormous heteroatom contents, active sites and facilitate fast electron transfer; all these factors have drawn attention towards 0-D carbon nanomaterials in energy storage applications (Hoang et al. 2019). Lv et al. (2014) synthesised a highly porous, low density and light-weight, 3-D CQD aerogel with a large specific surface area. The preparation method involves *in-situ* assembling of CQDs in sol-gel polymerisation of resorcinol and formaldehyde solutions and subsequent polymerisation to form CQD aerogel. The obtained specific capacitance for CQD aerogel at 0.5 Ag^{-1} was 294.7 Fg^{-1} with only a 6% decrease in capacitance after 1,000 cycles. Also, CV curves reveal pure EDLC contribution to the specific capacitance of CQD aerogel. Besides CQDs activated carbon (ACs) is widely used as supercapacitor electrode material, but low capacitance, narrow potential window, amorphous microporous structure and low conductivity limit their electrochemical kinetics. Thus, to improve the electrochemical performance of ACs, highly crystallised graphene quantum dots were embedded into the activated carbon. The prepared graphene quantum dot embedded activated carbon (GEAC) shows excellent charge transfer and ion transfer kinetics along with improved electrical conductivity, facilitating the energy storage in deep and branched micropores. From N_2 adsorption-desorption isotherms, the specific area of GEAC was found to be 2,829 m^2g^{-1} and the specific capacitance calculated at 1 Ag^{-1} from the charge-discharge curve was 388 Fg^{-1} with 60% capacitance retention at a very high current density of 100 Ag^{-1} (Qing et al. 2019).

Recently, carbon quantum dot/nickel sulphide nanocomposite with superior specific capacitance was prepared. Synthesis of CQDs/NiS nanocomposite involves hydrothermal technique using lemon juice as a source of carbon quantum dots. The FESEM images of CQD/NiS nanocomposite reveal flower-like morphology with each petal length ~0.8 micrometres. The electrochemical performance of the composite was studied using CV and GCD with a 2M KOH electrolyte. CV curves of the composite materials confirm the pure capacitive behaviour where more active sites were available at low scan rates for electrostatic charging. Furthermore, specific capacitance and cyclic stability were determined from GCD curves. The nanocomposite shows specific capacitance 880 Fg^{-1} and stability up to 2,000 charge-discharge cycles at a current density of 2 Ag^{-1}. The carbon dots provides the advantage of incorporation of extra positive charges into the matrix, enhancing electrical conductivity and decreasing ion diffusion length as well as the synergetic effect of enhanced charge transfer and high specific area attributed to boosted supercapacitive behaviour of the composite material (Sahoo et al. 2018).

2.2 Carbon Nanotubes (CNT) Composites

CNTs are one-dimensional nanomaterial formed by rolling one or more graphene sheets together to form a hollow tube-like structure. Depending upon the number of rolled graphene sheets, they are categorised as single-walled CNTs (SWCNT) or multi-walled CNTs (MWCNT). CNTs are cache of numerous excellent properties, like high electrical and thermal conductivity, high tensile strength, mechanical strength, optimum chemical stability, low mass per unit volume, unique pore structure and high aspect ratio by these CNTs are a promising candidate for electroactive supercapacitors (Yang 2012). However, CNTs have the drawback of comparatively low specific surface area and low gravimetric and volumetric capacitance. Thus, the composite of CNTs with other 0-D, 1-D and 2-D carbon nanomaterials improves their electroactive behaviour. A graphene oxide (rGO)/ CNT nanocomposite was synthesised and an as-synthesised electrode exhibits specific capacitance of 279.4 Fg^{-1} along with good cycle rate stability above 90% after 6,000 cycles. The embedded CNTs prevent rGO sheets from agglomeration and wrinkle. The entangled structure is able to offer a higher surface area, which ultimately contributes to higher electrochemical capacitance (Yang et al. 2018). Furthermore, the incorporation of zero-dimensional carbon nanomaterials into the CNT boosts the EDLC by providing more electroactive sites and uniformly distribute mesopores. A recently reported, CNT/GQD nanocomposite shows a 200% boosted performance than pristine CNT. The GQDs were deposited over CNT by electrodeposition method later on SEM images, showing film-like morphology with GQDs embedded into the CNTs. The symmetric assembly of CNT/GQD composite electrode shows aerial capacitance of 44 mF cm^{-2} in a two-electrode system probably due to the formation of a continuous film-like conducting network of GQD (Hu et al. 2013). Noked et al. (2012) synthesised CNT/AC nanocomposite by dispersing CNT into polymer solution followed by carbonisation under N_2 atmosphere. Activation enhances the surface area and electrical conductivity, modifying porosity and CNTs provide excellent mechanical strength to the composite electrode material. The galvanic charge discharge over 50,000 cycles shows negligible fading in specific capacitance in an aqueous electrolyte. The prolonged cyclic stability was attributed to the presence of carbon nanotubes in the activated carbon matrix, which provides high mechanical strength and also the architecture mesoporous structure further improves the electrosorption performance.

2.3 Graphene Nanocomposites

Graphene is one atom thick layer of sp^2 hybridised, two-dimensional honeycomb lattice sheet of carbon atoms. Graphene is suitable material for electrocapacitive purposes due to physicochemical properties, extraordinary mechanical properties, larger surface area (2,630 m^2g^{-1}), good flexibility, good electrical conductivity, reliable thermal and chemical stabilities, wider potential window, low production cost and availability of abundant surface functional groups. Instead, graphene sheet undergoes restacking and agglomeration due to strong π-π interaction between carbon atoms. This affects the pore structure of the graphene sheet and damages ion transfer channels, which hinders electrochemical performance. Thus, to overcome these drawbacks activation of graphene sheet is done by various methods, like insertion of other nanomaterials within graphene sheets, exfoliation of graphene at high-temperature results in the formation of reduced graphene oxide, templating technique, etc. (Shao et al. 2015, Mohammad Khalid et al. 2019). Meanwhile, a graphene-activated carbon (GAC) nanocomposite was synthesised by facile one-step KOH activation annealing using stable graphene colloids as a precursor. The pore structure analysis obtained from BJH revealed that the GAC composite had a narrower pore distribution with an interconnected pore structure and short pore length. Morphological study of GAC using TEM showed aggregated and crumpled graphene sheets with activated hollow carbon spheres. Furthermore, the GAC nanocomposite had a specific capacitance of 122 Fg^{-1} and an energy density of 6.1 Whkg^{-1} in an aqueous KOH electrolyte. While the electrochemical studies were also performed using EMIBF$_4$, an ionic electrolyte with a wide potential window of 4 V. The maximum energy density achieved was 52.2 Whkg^{-1} at room temperature, and it was gassed up to 99.2 Whkg^{-1} at 80°C. Such high energy density at high temperature was probably due to a decrease in viscosity, which enhances the electrical conductivity and promotes fast ion transfer (Chen et al. 2012).

One of the approaches to build up the electrochemical performance of graphene sheets is to introduce CNT into the graphene matrix. The CNT present within the graphene network prevents stacking of graphene layers and enhances fast electron transfer and short electrolyte ion diffusion channels along with high surface area and high electrical conductivity. In recent years, a graphene/MWCNT nanocomposite was synthesised via *in-situ* reduction of exfoliated graphite oxide in the presence of cationic PEI followed by sequential self-assembly of resultant water-soluble PEI-modified graphene sheets with acid-oxidized MWNTs forming interconnected hybrid carbon films. Furthermore, SEM images revealed the dense and uniform porous network of carbon nanomaterials within the composite. Also, an extra bridging network between the graphene sheet and CNT due to sp^2 hybridized carbon atom was established. Cyclic voltammetry was used to determine the electrochemical behaviour of graphene/CNT nanocomposite film deposited on the ITO (indium tin oxide) glass electrode in a 1M H$_2$SO$_4$ electrolyte. The CV curves maintained rectangular shapes even at high scan rates and exhibit specific capacitance of 120 Fg^{-1}. This indicates the rapid charge/discharge and low equivalent series resistance and promotes graphene/CNT nanocomposite for electrochemical energy storage studies (Yu and Dai 2010).

2.4 Metal Oxide Nanocomposites

Transition metal oxides, such as RuO$_2$, MnO$_2$, V$_2$O$_5$, Fe$_3$O$_4$, Co$_3$O$_4$, NiO and TiO$_2$, are crucial in electrochemical energy storage, such as supercapacitors. They usually act as pseudocapacitive materials having a high value of theoretical specific capacitance. Energy storage in metal oxide electrodes is due to fast reversible redox reactions at the electrode surface as well as during charging some of the electrolyte ions that can percolate into the lattice, thus not only surface but also intercalation affects the charge storage. Instead of their excellent pseudocapacitive properties, metal oxides suffer from some limitations, such as poor electrical conductivity, destitute stability, rate capability over a large number of cycles and also low durability during the performance. This ultimately leads to low power density and restricts its application for supercapacitor purposes. While carbon nanomaterials, like CNT, graphene, mesoporous activated carbons, etc., are excellent regarding all these aspects but pure EDLC nature may suffer relatively low value of specific capacitance. In order to compensate for the drawbacks of both materials' incorporation of metal oxides into the carbon, nanomaterial leads to the composite material with the superior properties having the benefit of both faradic and nonfaradic reversible redox reactions. The synergetic effect of carbon nanomaterials with metal oxides provides high electrical conductivity, high stability, large surface area, increased mechanical strength, rate capability and durability to the electrode during charge/discharge (Figure 1). Thus, carbon nanomaterials and metal oxide binary nanocomposite provide promising electrode material for supercapacitor application. In recent years, a large number of combinations of these composites were studied by scientists worldwide for high-performance supercapacitor application some of these are discussed below (Seok et al. 2019).

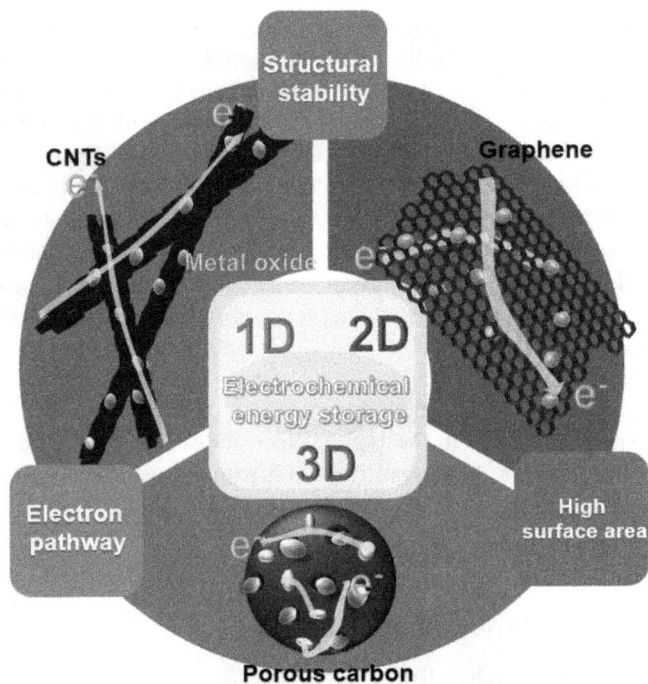

Figure 1: Schematic illustration of carbon/metal-oxide composites of various dimensions for electrochemical energy storage [reprinted with permission from *Sustainability* (Switzerland) (Seok et al. 2019)].

Amongst transition metal oxides, MnO_2 is mostly preferred as electrode material for battery-type capacitors because of its high specific capacitance, cost-effectivity and low toxicity but poor electrical conductivity and poor cyclic stability suppresses the electrochemical behaviour of MnO_2. The integration of carbon nanomaterials with the MnO_2 provides high electrical conductivity and stability to the electrode material as well as carbon nanomaterials backbone provides channels for continuous ion transport and prevents agglomeration of MnO_2 nanoparticles. For instance, Xiaomiao Feng et al. (2013) synthesised a novel sandwich-like structure of MnO_2

petal nanosheets grown on both sides of graphene by using a facile one-step hydrothermal approach. The synergetic effect of eminent electrical conductivity of graphene and the high pseudocapacitance of MnO_2 nanosheet greatly enhance the capacitance of the electrode. The electrochemical measurement of the composite electrode was done in 1M Na_2SO_4 electrolyte, which shows a high capacitance value of 516.8 Fg^{-1} at a scan rate of 1 mVs^{-1} with excellent capacity retention of 91.1% after 1,000 cycles. In an another study, a graphene sheet decorated with MnO_2 nanoflower was synthesised by simple anodic electrochemical deposition technique using two different solutions [0.1 M Na_2SO_4 and 0.1 $Mn(CH_3COO)_2$]. The 3-D electrochemical growth of MnO_2 flowers increases the separation between each graphene sheet layer, which enhances the contact between the active material and electrolyte ions. Also, the increased space between graphene nanosheets provides more active sites for electron transfer. The MnO_2 nanoflower coated graphene electrode shows specific capacitance 328 Fg^{-1} at a scan rate of 10 mVs^{-1} having an energy density of 11.4 $Whkg^{-1}$ with good cyclic stability, i.e. 99% over 1,300 cycles (Cheng et al. 2011). In addition to high specific capacitance, composite electrode also has the benefit of large surface area, high porosity, high electrical conductivity and stability. Graphene nanosheet with integrated MnO_2 nanoparticles provides more reactive sites to the composite electrode material. This results in decreasing overall equivalent series resistance and the composite electrode material become more suitable for energy storage.

CNT-MnO_2 architecture nanocomposite also gained vast interest due to the combination of high mechanical strength and electrical conductivity of CNT with the high pseudocapacitance of MnO_2. Qi et al. (2017) synthesized MnO_2 nanosheets accumulated on CNT's buckypaper by a self-controlled reduction in potassium permanganate solution. The CNT-MnO_2 nanocomposite was directly deposited as electrode material for supercapacitor without using binders or conductive agents. The *in-situ* growth of MnO_2 on buckypaper was fabricated as a flexible supercapacitor and measurement of specific capacitance at a current density of 4 Ag^{-1} provided a value of 432 Fg^{-1} and excellent retentions of 98% after 3,000 cycles. Also, ultra-high flexible and high strength CNT@MnO_2 composite was synthesised using hydrothermal synthesis which can be folded into different shapes, like nanofilms, nanoflowers, nanostars and yarns (Figure 2). The electrochemical performance of CNT@ MnO_2 composite was studied with a three-electrode system using 5 M LiCl electrolyte at 1.9 Ag^{-1} found to have a specific capacitance of 323.9 Fg^{-1}, which decreased by only 1% even at high current density 15 Ag^{-1} (Wang et al. 2019). CNT@MnO_2 nanocomposite was synthesised by microwave-assisted irradiation of potassium permanganate. The specific capacitance of the composite was 950 Fg^{-1} in 1 M Na_2SO_4 aqueous solution and had 95% retention after 500 cycles (Yan et al. 2009). This is the highest specific capacitance of the CNT@MnO_2 nanocomposite till recorded. Hence, CNT@MnO_2 nanocomposite is highly recommended as promising electrode material for electrochemical energy storage.

In the midst of transition metal oxides, RuO_2 is found to have the highest theoretical value of specific capacitance 2,200 Fg^{-1} due to a wide potential window, highly reversible redox reactions and good thermal stability, but its high cost ultimately increases the cost of fabrication of supercapacitor device. Thus, RuO_2 composite

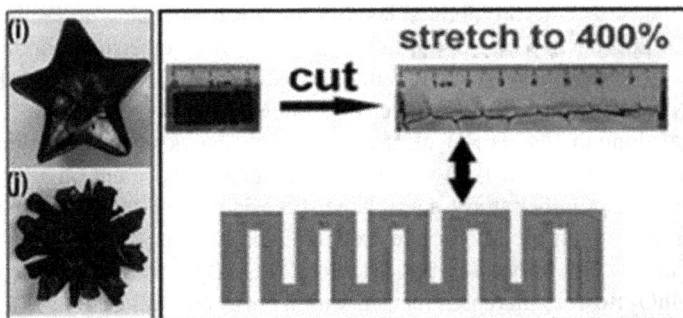

Figure 2: Different shapes folded from $CNT@MnO_2$ film [reprinted with permission from *Chemical Engineering Journal* (Q. Wang et al. 2019)].

with carbon nanomaterials is the best way to improve the overall performance of the supercapacitor electrode. Wang et al. (2016) synthesised carbon/RuO_2, CNT/RuO_2 and rGO/RuO_2 nanocomposite with the simplest and environment-friendly aqueous route. The electrochemical characteristics of the composites tested using 1 M H_2SO_4 at current density 1 Ag^{-1} have specific capacitances of 879.1 Fg^{-1} for carbon/RuO_2 with 98.4% retention, 966.8 Fg^{-1} for CNT/RuO_2 with 98% retention and 1,099.6 Fg^{-1} for RuO_2/rGO with 98.4% retention. A graphene nanosheet anchored with RuO_2 nanoparticles was recently reported for elitist supercapacitor application using a simple hydrothermal approach. The specific capacitance of the graphene/RuO_2 nanocomposite was calculated with 1 M Na_2SO_4 using galvanic charge discharge curves having a value 441.17 Fg^{-1} at current density 0.1 Ag^{-1}. Also, the specific energy density of the nanocomposite electrode has a very high value of 61.2 Whkg^{-1} with 100% retention up to 1,000 cycles. Thus, the graphene/RuO_2 nanocomposite shows an excellent combination of high specific capacitance, stability, energy density and rate capability, which is beneficial for high-performance electrochemical capacitors (Thangappan et al. 2018). Recently, a novel activated carbon/RuO_2 nanocomposite was synthesised having high specific capacitance 1,460 Fg^{-1} at 10 Ag^{-1} in an aqueous electrolyte of 0.5 M H_2SO_4. In addition to high specific capacitance, the composite materials also show very high stability and rate capability over 10,000 cycles (Hossain et al. 2018). In recent years, many other carbon/RuO_2 nanocomposites were reported and these composites are superior in electrochemical performance having high specific capacitance, energy density and excellent rate capability.

Moreover, NiO (nickel oxide) carbon nanomaterials composite also emerged as a supercapacitor electrode material because of a fantastic combination of significant electrical conductivity, electrical stability and rate capability. The carbon network within the composite material provides channels for diffusion and ion transfer, which is necessary for electrochemical energy storage. Abioye et al. (2017) synthesised oil palm-based activated carbon nickel oxide nanocomposite for electrochemical capacitor application. There were three steps involved in the preparation of OPSAC/NiO nanocomposite. Firstly, OPS activated carbon was prepared by physical activation of oil palm shell using the microwave at 800°C, followed by activation at 900°C and secondly electroless deposition of NiO on activated carbon using an alkaline bath. The electrochemical performance of the as-prepared electrode material was studied using the symmetric two-electrode system with 1M H_2SO_4. The specific capacitance of the OPSAC/NiO nanocomposite calculated from GCD curve was 411 Fg^{-1}, and it was much higher than the specific capacitance of OPSAC (134 Fg^{-1}). Shahrokhian et al. (2016) synthesised 3-D graphene/NiO nanocomposite electrode in an aqueous solution containing nickel nitrate and graphene oxide using nickel foam as substrate by a facile electrochemical co-deposition method. The specific capacitance of nanocomposite at 2 Ag^{-1} was 1,715.5 Fg^{-1}, which reduces to 1,066.7 Fg^{-1} even at a high current density 40 Ag^{-1} (Figure 3), this proves the stability of the composite electrode material. Such high specific capacitance and stability is due to the synergetic effect of the large porous surface area of graphene and the incorporation of nickel oxide enhances the electroactive sites, which contributed to the superior electrochemical behaviour. Furthermore, CNT@NiO nanocomposite was studied as electrode material for supercapacitors. The CNT@NiO nanocomposite was prepared by a simple hydrothermal synthesis method having specific capacitance 1,329 Fg^{-1} at a high current density of 84 Ag^{-1}. Such gassed-up electrochemical behaviour of CNT@NiO nanocomposite was the resultant of a superior combination of high electrical conductivity and uniformly distributed nanoporous structure of CNT with the zero-dimensional mesoporous structure of nickel oxide. This ultimately results in easier access of electrolyte ions to the electroactive sites of the composite electrode material (P. Lin et al. 2010).

Moreover, magnetite (Fe_3O_4)-carbon nanomaterials composite also provides promising electrode material for supercapacitors having relatively low cost and abundant availability. Three dimensional (3-D) graphene oxide nanosheet incorporated with iron oxide nanoparticles was synthesised. The synthesis method

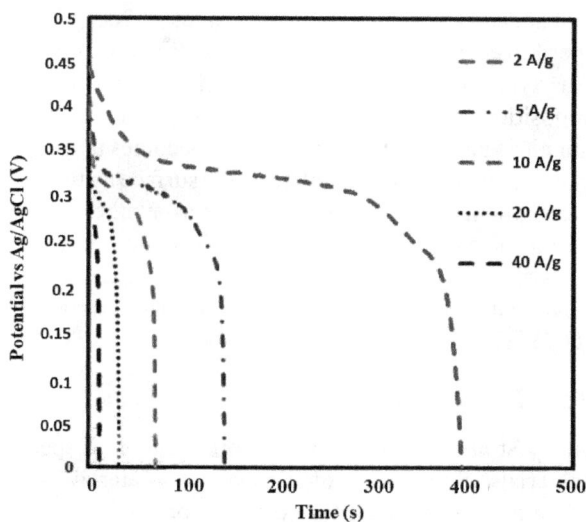

Figure 3: GCD curves of 3D graphene/NiO nanocomposite [reprinted with permission from *International Journal of Hydrogen Energy* (Shahrokhian et al. 2016)].

was very simple. Initially, rGO was prepared using modified Staudenmaier's method. After that self-assembled 3-D, rGO/Fe$_3$O$_4$ nanocomposite was prepared by facile one-pot microwave heating using FeCl$_3$·6H$_2$O as a precursor. The morphological studies revealed that the 3-D rGO nanosheet has an interconnected cotton-like fluffy structure decorated with a large number of electroactive Fe$_3$O$_4$ nanoparticles. This 3-D rGO/Fe$_3$O$_4$ exhibits a high specific capacitance of 455 Fg^{-1} at current density 8 Ag^{-1} with an exquisite rate capability of 91% over 9,500 cycles at a current density of 3.8 Ag^{-1}. Also, from EIS studies, the Rct of the prepared composite electrode was only 4 Ω, and the less inclined line shows excellent capacitive behaviour. Hence, the outstanding electrochemical capacitive properties of 3-D rGO/Fe$_3$O$_4$ nanocomposite was resultant of the sanctified effect of incorporated Fe$_3$O$_4$ in the rGO nanosheet. The uniformly distributed Fe$_3$O$_4$ nanoparticles suppress restacking of the rGO nanosheets and prevent them from agglomeration. In addition to this, iron-oxide nanoparticles within the rGO sheets provide a path for electrolyte ion diffusion during the charge/ discharge process. This makes 3D rGO/Fe$_3$O$_4$ nanocomposite a promising material for supercapacitor application (Kumar et al. 2017). A CNT@Fe$_3$O$_4$ nanocomposite was synthesised by facile hydrothermal method. SEM images of CNT@Fe$_3$O$_4$ nanocomposite shows that the spacing between CNT increases and cubic Fe$_3$O$_4$ nanoparticles were anchored on the surface of CNT as well as within the empty spaces in the CNT network. Such structural morphology of nanocomposite is highly beneficial for superior electrochemical performance. The specific capacitance of the electrode was 117 Fg^{-1} with an energy density of 16 Whkg^{-1}. CNT backbone provides iron oxide nanoparticles with a large number of electroactive sites as well as excellent stability to the overall electrode. Also, CNT@Fe$_3$O$_4$ nanocomposite provides channels for electrons to move freely between iron oxide and CNT interface, which causes a decrease in the charge transfer resistance and this ultimately leads to the enhancement of supercapacitor behaviour (Guan et al. 2013).

Apart from the above-discussed carbon-metal oxide nanocomposite, CNT/Co$_3$O$_4$ nanocomposite was synthesised by the combination of acid-treated multiwall CNT and *in-situ* decomposition of Co(NO$_3$)$_2$ in *n*-hexanol solution using ultrasonication. Due to itching with nitric acid, the MWCNT surface becomes negatively charged and thus the positively charged Co ions accumulate on the surface by the electrostatic force of attraction. The electrochemical capacitance of the as-synthesised CNT/Co$_3$O$_4$ was found to be 200.98 Fg^{-1} in a 1 M KOH electrolyte. Thus, the easier, cost-effective and eco-friendly synthesis method with high-value specific capacitance ensures CNT/ Co$_3$O$_4$ nanocomposite suitable electrode material for an electrochemical capacitor (Shan and Gao 2007).

2.5 Heteroatom Doped Nanocomposites

In recent years many strategies were adopted to enhance the specific capacitance of carbon nanomaterials. Summation of carbon nanomaterials with heteroatoms boosts chemical stability, surface polarity, electrical conductivity and electron-donor properties of composite electrode material. You et al. (2013) used facile hydrothermal treatment followed by freeze-drying and then graphene oxide-dispersed pristine CNTs were carbonized in the presence of pyrrole for the synthesis of 3-D N-doped graphene-CNT networks (NGCs). The morphological study of nanocomposite shows uniformly distributed mesoporous structure of nitrogen-doped graphene interlinked with CNT due to sufficient polymerisation of pyrrole during hydrothermal treatment. Also, the presence of CNT enhances the cycling stability and mechanical strength of the NGC composite electrode material, i.e. 96% after 3,000 cycles.

One of the strategies to improve the specific capacitance of carbon nanomaterials is co-doping of multiple heteroatoms, which enhances the surface functional groups. One such nitrogen and sulphur co-doped carbon nanocomposite was prepared by adopting a simple milling procedure by mixing the halogenated polymer, dehalogenation agent and dopants together. The N and S doped carbon was simply prepared by dechlorination reaction between PVDC and KOH using dimethylformamide solvent as a source of nitrogen, while dimethyl sulfoxide as a source of sulphur. The electrochemical characterisation of N, S doped carbon nanocomposite shows specific capacitance 427 Fg^{-1}, and rate capability was knocked off in H_2SO_4 solution compared to NaCl or KOH solution. This is probably due to pseudocapacitance provided by sulphur and nitrogen while absorbing H^+ ions (Chang et al. 2017).

Also, there is a new trend of using biomass as a carbon precursor and doping it with heteroatoms for achieving a high-performance supercapacitor. Lin et al. (2019) synthesised oxygen and nitrogen co-doped carbon nanocomposite using peach gum as a carbon source. The composite was prepared by simple hydrothermal treatment and 1,6 hexanediamine was used as a precursor for nitrogen doping. Further, the prepared sample was treated with KOH for adding an oxygen functional group. The prepared sample shows a sheet-like structure with uniform distribution of pores trapped with nitrogen and oxygen atom. The nanocomposite exhibits a large specific surface area 1,535 m^2g^{-1} and 402 Fg^{-1} of specific capacitance with a holding of 91% over 7,000 cycles. Such superior performance is attributed to the heteroatom content, which enhances the wettability and provides additional pseudocapacitance to the composite material. In addition to this, the carbon network provides a continuous conductive path for electrolyte ions and minimises the ion diffusion route and contributes to the double-layer capacitance.

The charge delocalisation of carbon atoms was successfully enhanced by phosphorous doping. Secondly, phosphorous doping enhances the surface reactive sites and thus provides ion transport/diffusion paths and contributes to the pseudocapacitance. Meanwhile, phosphor-doped carbon nanobowls (PCNB) were synthesised by simple evaporation technique using the effect of the pressure difference on a curved surface of a liquid to induce the formation of phosphorous doped carbon nanobowls. The mixture of hollow carbon spheres and P_2O_5 (1:3) was heated under an argon atmosphere at 450°C for 1 hour and later on, the temperature was raised to 900°C and held for 3 hours to obtain PCNB. The morphological studies using SEM showed that the hollow spheres were converted to nanobowls after evaporation. The gravimetric capacitance measured using three-electrode system with 6 M KOH at a current density of 1 Ag^{-1} was 273.58 Fg^{-1}. The symmetric supercapacitor device was fabricated using PCNB and tested for practical applications, which shows superb capacitance retention and cyclic stability, i.e. 94.9% over 40,000 cycles. Such excellent electrochemical performance is mostly attributed to the bowl morphology that can achieve the high-density stack and the tight interconnection between bowls, thus enhancing the electrical conductivities and improving the space utilization that can facilitate ion/electron transport. Also, phosphorous doping increases the surface wettability and surface polarity avails more sites for ion adsorption and contributes to the considerable pseudocapacitance that ultimately enhances the overall specific capacitance (Zhou et al. 2020). Moreover, many other studies have been developed regarding heteroatom-doped carbon nanocomposites, which are discussed in references (Chen et al. 2012, Ji et al. 2016, Li et al. 2016, Zhang et al. 2017, Liu et al. 2017).

2.6 Conducting Polymer-based Nanocomposite

Conducting polymers, including polyaniline (PANI), polypyrrol (Ppy), polyacetylene (PA), polythiophene (PTh) and poly 3,4-ethylenedioxythiophene (PEDOT), have distinctive properties like high conductivity, high intrinsic flexibility, ease of synthesis as well as charge storage by fast reversible redox reactions which make them a promising candidate for battery type electrode material. Conducting polymers have basically very low conductivity in the undoped state but they are able to tune their electrical conductivity upon doping from semiconductor to metallic regime. They achieve electrical conductivity through π-electron delocalisation along their polymer backbone. Besides this conducting polymers have some pitfalls including low rate capability, poor stability as well as during charging/discharging conducting polymers undergo expansion and contraction, which causes the breakdown of the electronic structure all leads to restricting the application of conducting polymers as a supercapacitor electrode. To compensate for this, a composite of conducting polymers with carbon nanomaterials is very effective. Carbon nanomaterials provide a large surface area and high mechanical strength along with an excellent electrical conductivity, which improves the electrochemical performance of the composite material (Abdelhamid and Snook 2018).

A new strategy was applied to prepare skeleton/skin-like freestanding three-dimensional carbon aerogel (CA) and polyaniline (PANI) nanocomposites. The carbon aerogels were synthesised by freeze-drying and carbonisation treatment of mixture containing graphene sheets/carbon nanotubes (CNT)/poly(amic acid) (PAA) chains. Later on, thin films of polyaniline were deposited on the surface of carbon aerogel. SEM images of the CA/PANI nanocomposite revealed that polyaniline nanorods were anchored on the surface of carbon aerogel with a uniformly aligned mesoporous structure. The electrochemical studies of the composite electrode were assessed from cyclic voltammetry and galvanic charge discharge. The cyclic voltammetry shows rectangular shapes with three redox peaks that show the significant contribution of PANI to the pseudocapacitance. Moreover, the GCD study divulges that the specific capacitance of CA/PANI nanocomposite was 966 Fg^{-1} at current density 1 Ag^{-1}, which still remains 917 Fg^{-1} at a larger current density of 100 Ag^{-1}. The CA/PANI shows great cycling stability of 93% over 5,000 cycles with constantly maintained 100% columbic efficiency. Such excellent performance of CA/PANI nanocomposite was due to the synergetic effect of carbon aerogel matrix embedded with the polyaniline nanorods interconnected with the π-π stacking. The carbon aerogel provides a larger surface area as well as a highly conducting carbon backbone and the uniformly aligned porous PANI nanorods minimises the ion diffusion paths and expose all electroactive sites by allowing rapid penetration of electrolyte ions, which contributes to the high specific capacitance (Liu et al. 2017).

Another novel nanocomposite of reduced graphene oxide and polyaniline having high electrochemical performance was synthesised using simple *in-situ* polymerisation. Firstly, rGO nanofoam was prepared by template-assisted synthesis and followed by *in-situ* polymerisation of aniline on rGO foam. The rGO/PANI nanocomposite shows unique evenly distributed porous morphology of rGO surface deposited with densely packed PANI flakes. The specific capacitance of synthesised nanocomposite showed 701 Fg^{-1} at 1 Ag^{-1}. The CV curves of rGO/PANI nanocomposite were nearly rectangular having a high peak current and large integrated area with two redox peaks. The area under the curve is directly proportional to the high specific capacitance, while the redox peaks are assigned to the transition of PANI between leucoemeraldine-emeraldine and emeraldine-pernigraniline transformation. This immense specific capacitance is a combination of EDLC and pseudocapacitance of rGO/PANI nanocomposite. The stacking between rGO and PANI improves the charge transfer rate and access electrolyte ions to utilise the full active material, which ultimately gooses up the specific capacitance. Also, rGO provides high mechanical strength support and a highly flexible backbone and protects the electrode material from damage during charge discharge, which contributes to the cycling stability of the electrode material (Sun et al. 2015).

Apart from polyaniline, polypyrrole is also used as pseudocapacitive material in electrochemical capacitor electrodes owing to its high mass density, high degree of flexibility, environmental stability and suitable electrical conductivity, but poor cycling stability limits its application. Moreover, composite polypyrrole with carbon materials overcome the lack of polypyrrole by providing high cycling stability over a large number of cycles and a suitable electrode material for supercapacitors. One such CNT/polypyrrole nanocomposite was synthesised by a simple procedure. It includes initial synthesis of SWCNT by arc discharge method and later on SWCNT/polypyrrole nanocomposite was prepared using *in-situ* polymerisation of polypyrrole monomer using ferric chloride ($FeCl_3$) and sodium *p*-toluene sulphonate as oxidants for pyrrole polymerisation. The as-prepared CNT/polypyrrole nanocomposite shows a specific capacitance of 274 Fg^{-1} with 95% capacitance retention over a large number of cycles. This sustainable performance of composite electrode material was due to a combination of large surface area and electrical conductivity of SWCNT with the pseudocapacitance of polypyrrole. Also, the polypyrrole was uniformly coated over SWCNT, thus getting the benefit of a large number of electron diffusion paths provided by SWCNT that leads to the increase of specific capacitance (Kay Hyeok An et al. 2002).

Moreover, recently there is a trend of using ternary composite as an electrode material for supercapacitor application. A facile one-step green and environmentally friendly synthesis method was employed to prepare V_2O_5/Ppy/GO nanocomposite via electrodeposition approach. The SEM image of the nanocomposite shows wrinkled morphology attributed to Ppy/GO with spherical V_2O_5 particles uniformly distributed throughout the matrix. The electrochemical performance was studied using three-electrode system with 0.5 M Na_2SO_4 electrolyte. The observed specific capacitance from charge discharge was found to be 705 Fg^{-1} with energy and power density 27.6 $Whkg^{-1}$ and 3,600 Wkg^{-1}, respectively. Such extraordinary performance of ternary nanocomposite is due to the presence of three different phases in which graphene oxide provides high electrical conductivity along with the good interfacial contact between V_2O_5 nanoparticles and Ppy with GO matrix. V_2O_5 improves charge storage capacity, while Ppy provides good electrical conductivity and contributes to the significant pseudocapacitance. In this way, nanocomposites have the benefit of both EDLC and pseudocapacitance and become suitable materials for supercapacitor electrodes (Asen et al. 2017).

2.7   Metal-Organic Framework-Based Carbon Nanocomposites

Metal-organic framework (MOF) consists of metal centres associated with organic ligands. MOF have interconnected uniformly distributed pore structures with the controlled metal ion and other elemental compositions. Additionally, metal ions in MOF's network serve as pseudocapacitive moieties contributing to the fast reversible redox reactions inclusive of a very large surface area of about ~7,000 m^2g^{-1}, all these attract attention towards MOFs as a promising electrode material for electrochemical energy storage. But the non-conductivity and chemical instability limit their application. Combining these materials with the highly conducting, high mechanically and electrochemically stable carbon nanomaterials provide electrode material agreeing effect and ensures high specific capacity, good electrical conductivity and durable stability (Du et al. 2019). A high-performance cobalt-based MOF/graphene nanocomposite was prepared through a simple one-step precipitation method (Figure 4). The synthesis involved $Co(NO_3)_2.6H_2O$ was dissolved in methanol and graphene added to the mixture and ultrasonicated for 30 minutes (solution 1). Then, 2-MeIm was dissolved in methanol (solution 2) and later on solution 2 was added to solution 1 and obtained precipitate was washed to gain the final product CoMG. The FESEM image of CoMG nanocomposite shows CoMOF nanoparticles were randomly distributed over graphene sheet and crystalline structure of CoMOF remains maintained. From Raman spectra, I_D/I_G ratio for CoMG was 0.62 indicating the presence of more defects within the composite which facilitates more electron transfer and encourages specific capacity. Further electrochemical behaviour was studied using three-electrode system with 6M KOH. CV curve shows hysteresis loop type structure with redox peaks which correspond to the reversible transition of Co^{3+} to Co^{2+} and contribute to the pseudocapacitance. The specific capacitance of CoMG was 549.96 Fg^{-1}, and it was much higher than the specific capacitance of pristine graphene and CoMOF. Thus, CoMG exhibits excellent storage capacity because of the synergetic effect of CoMOF and graphene. Graphene provides high electrical conductivity and mechanical strength, while CoM improves the capacitance and reduces the agglomeration of graphene sheets. Hence, the CoMG nanocomposite was optimal for high-energy storage applications (Azadfalah et al. 2020).

A Ni-MOF/CNT nanocomposite having enormously high specific capacitance was synthesised using a simple hydrothermal approach. TEM images showed that CNTs were enwrapped with a Ni-MOF sheet. EDS confirms the presence of C, O and

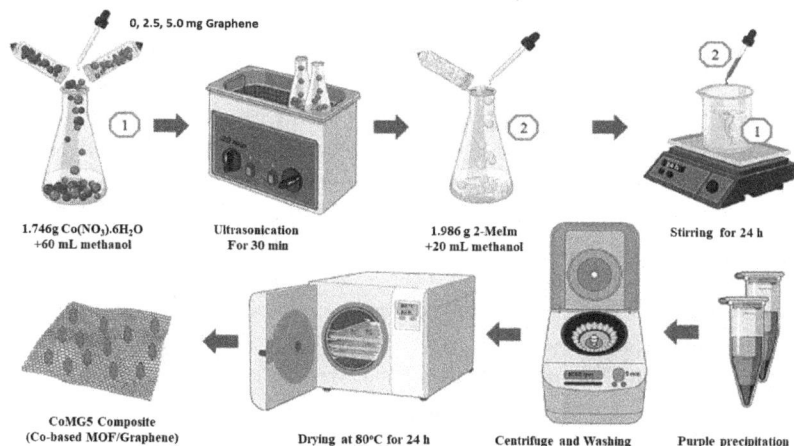

Figure 4: Schematic of synthesis of CoMG nanocomposite [reprinted with permission from *Journal of Energy Storage* 33 (Azadfalah et al. 2020)].

Ni, and Ni element was uniformly distributed throughout the outer shell and improves capacitive performance. The electrochemical behaviour was studied using CV and GCD. CV curve shows highly redox peaks due to reversible redox reaction of Ni^{2+} and Ni^{3+}. GCD curves reveal excellent specific capacitance of 1,735 Fg^{-1} at 0.5 Ag^{-1} with good retention events at high scan rates. Thus, the high specific capacitance was attributed to the well-defined pore structure of Ni-MOF, which access electrolyte ions diffusion and uncovers active sites. The CNT wrapped by Ni-MOF acts as current collectors and charge transport paths as well as the electrical conductivity of CNT improves charge transfer and ultimately decreases the equivalent series resistance. Therefore, Ni-MOF/CNT composite was implemented as promising electrode material for supercapacitor devices (Wen et al. 2015).

Moreover, Ni-MOF/rGO nanocomposite was synthesised using a facile hydrothermal route. The morphological study of Ni-MOF/rGO nanocomposite revealed uniform distribution of Ni-MOF in the rGO sheet and improves ion diffusion or movements between the active electrode material and electrolyte. The rGO in the nanocomposite act as substrate as well as active material and promotes charge supply. To measure the electrochemical activity composite was loaded over Ni-foam substrate and 6 M KOH solution was used as electrolyte. The specific capacitance was calculated from GCD curves found to be 1,154.4 Fg^{-1} at 0.66 Ag^{-1}. The composite electrode material was benefited from both EDLC and pseudocapacitance additionally high electrical conductivity and stability of rGO lift up the energy storage behaviour (Kim et al. 2020).

2.8 Transition Metal Nitride and Carbide (Mxenes) Based Nanocomposites

In recent years, Mxenes (M stands for transition metals like vanadium (V), titanium (Ti), molybdenium (Mo), etc., and X stands for C and/or N) are developed as a supercapacitor electrode material because of their metallic electrical conductivity, high melting point, a large number of surface functional groups, wear and corrosion resistance, high reaction selectivity, energy storage by pseudocapacitive mechanism and ease of forming composites (Yuan et al. 2020). The nanocomposite of Mxene with carbon provides combinational advancement properties to the electrode material. Amongst Mxenes vanadium nitride/carbide is mostly used for energy storage applications because of its high electrical conductivity and large negative potential working window. Recently, a facile solution combustion synthesis method was employed for the preparation of two-dimensional (2D) vanadium nitride/carbon nanosheet (VN/C) nanocomposite. A solution combustion method involves fast exothermal redox reactions in a homogeneous solution, this promotes *in-situ* synthesis of homogeneous composite. The XRD pattern of VN/C displays gentle and decreased intensities peaks which ensures incorporation of VN into carbon nanosheet. Morphological studies using SEM reveal sheet-like structures with uniform pore size distribution and increased glucose additives. It confirms the compliance of interconnected vanadium nitride particles on uniformly porous carbon nanosheets. Furthermore, the electrochemical behaviour of VN/C nanocomposite as electrode material was investigated with a three-electrodes system using a 1M KOH electrolyte. The CV curve shows a quasi rectangular shape with two redox peaks; this confirms the contribution of both EDLC and pseudocapacitance to overall specific capacitance (Figures 5a and 5b). Also, from galvanic charge discharge curves, the calculated specific capacitance was 249 Fg^{-1} at current density 0.1 Ag^{-1} and cycling stability was excellent having 75% retention at 1 Ag^{-1} after 5,000 cycles. The admirable electrochemical outcomes of VN/C nanocomposite are attributed to the unique electrode nanostructure. The carbon nanosheet within the composite act as an electrolyte ion migration channel as well as the VN/C nanocomposite possess a large surface area and enhances fast surface sorption reaction. Additionally, VN/C nanocomposite electrode provided the high energy storage capacity for supercapacitor by interfacial contact between VN and carbon nanosheet. All these factors reveal excellent electrochemical properties and ensure VN/C nanocomposite as anode material for supercapacitors (Wu et al. 2019).

Recently, another novel nanocomposite of Mxene and graphene sheets was synthesised with extraordinary toughness and high volumetric specific capacitance. Initially, GO was synthesised using Hummer's method and Ti$_3$AlC$_2$ was used as a precursor for obtaining Mxene monolayer nanosheet. Later on, Mxene functionalised graphene sheets (MrGO) were prepared using vacuum-assisted filtration followed by the formation of long-chain AD molecules to obtain MrGO-AD sheets that were obtained by dissolving MrGO in AD solution (mixture of aminopyrene (AP) and disuccinimidyl suberate (DSS) in molar ratio 1:1). The MrGO nanocomposite shows uniquely architecture Mxene functionalised GO nanosheets were interlinked through Ti-O-C covalent. HRTEM images confess a few layers stacked Mxene layers were dispersed within rGO platelets and MrGO-AD sheet shows uniform distribution of Mxene sheets and long-chain AD providing π-π interactions between rGO nanosheets. Wide-angle X-ray scattering (WAXS) results proved MrGO-AD had a higher degree of orientation, i.e. 85.5% and improved alignment of rGO nanosheets; along with this XPS reveal the presence of Ti and N within the composite. From stress-strain curve MrGO-AD sheet has a tensile strength of 699.1 ± 30.6 MPa and an ultra high

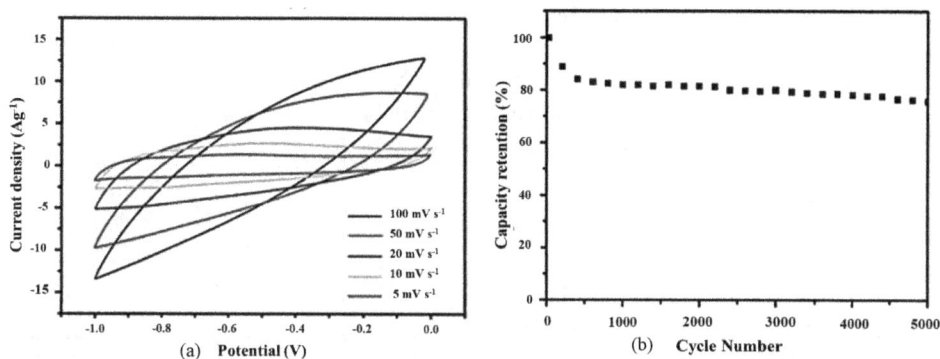

Figure 5: (a) CV curves of VN/C nanocomposite; (b) cyclic performance of VN/C at 1 Ag^{-1} [reprinted with permission from *Applied Surface Science* 466 (Wu et al. 2019)].

toughness of 42.7 ± 3.4 MJm^{-3}. Therefore, such high strength and toughness were due to covalent bonding between rGO - Mxene and π-π bonding between long-chain AD molecule - rGO sheet. Furthermore, the electrochemical studies affirm the high volumetric capacitance of 645 Fcm^{-3} at the current density 1.0 Acm^{-3} with excellent capacitance retention ~100% over 20,000 cycles. Also, the solid-state symmetric supercapacitor device with MrGO-AD electrode exhibits excellent capacitance retention during cycles of 180° bending with only a 2% decrease in capacitance. The high strength and toughness of the nanosheet contributed to superior electrochemical performance during bending. Thus, the synergetic effect of Mxene and AD within the rGO nanosheet provides Ti-O-C covalent bonding and π-π bridging interaction that enhances the mechanical strength, toughness, electrical conductivity and increases the charge transport of the MrGO-AD sheet and exploited as highly flexible auspicious electrode material for superior volumetric energy storage and power generation devices (Zhou et al. 2020).

3. Summary

Supercapacitor gained attention in the field of modern energy storage devices by its instantaneous power delivery, long cycle life, fast charge discharge rates and high specific capacitance. However, it is still a challenge to improve the energy density of the supercapacitors. In this chapter, recent studies and developments regarding the nanocomposite electrode material, containing at least a carbon nanomaterial is discussed. The nanocomposite electrode materials scrutinised here usually exhibit high specific capacitance, high conductivity, a large number of electrolyte ions diffusion paths, high charge transfer rate, rich electroactive sites, large surface area as well as long cycle life. Herein, we explained the composite of carbon nanomaterial with pseudocapacitive materials prepared by a simple synthesis route. The individual component of the composite contributes to the electrochemical performance, like carbon backbone within the nanocomposite enhances the mechanical strength and provides ion diffusion channels that ensure the cyclic stability of the electrode. Additionally, the pseudocapacitive material contributes to the high specific capacitance through fast redox reactions and confirms the utilisation of the whole mass of electrode material in charge storage thus, increasing the specific capacitance and energy density of the electrode material. It concludes that the composite material is much more beneficial than the individual and the presence of carbon nanomaterial uplifts the overall electrochemical performance.

References

Abdelhamid, Muhammad E. and Graeme A. Snook. 2018. Conducting polymers and their application in supercapacitor devices. Enc. Polymer Sci. Techno. 1–20. https://doi.org/10.1002/0471440264.pst666.

Abioye, Adekunle Moshood, Zulkarnain Ahmad Noorden and Farid Nasir Ani. 2017. Synthesis and characterizations of electroless oil palm shell based-activated carbon/nickel oxide nanocomposite electrodes for supercapacitor applications. Electrochim. Acta 225: 493–502. https://doi.org/10.1016/j.electacta.2016.12.101.

Asen, Parvin, Saeed Shahrokhian and Azam Iraji Zad. 2017. One step electrodeposition of V_2O_5/polypyrrole/graphene oxide ternary nanocomposite for preparation of a high performance supercapacitor. Int. J. Hydrog. Energy. 42(33): 21073–21085. https://doi.org/10.1016/j.ijhydene.2017.07.008.

Azadfalah, Marziyeh, Arman Sedghi, Hadi Hosseini and Hamideh Kashani. 2021. Cobalt based metal organic framework/graphene nanocomposite as high performance battery-type electrode materials for asymmetric supercapacitors. J. Energy Storage 33(September): 101925. https://doi.org/10.1016/j.est.2020.101925.

Borenstein, Arie, Ortal Hanna, Ran Attias, Shalom Luski, Thierry Brousse and Doron Aurbach. 2017. Carbon-based composite materials for supercapacitor electrodes: a review. J. Mater. Chem. A 5(25): 12653–12672. https://doi.org/10.1039/c7ta00863e.

Chang, Yingna, Guoxin Zhang, Biao Han, Haoyuan Li, Cejun Hu, Yingchun Pang, Zheng Chang and Xiaoming Sun. 2017. Polymer dehalogenation-enabled fast fabrication of N,S-codoped carbon materials for superior supercapacitor and deionization applications. ACS Appl. Mater. Interfaces 9(35): 29753–29759. https://doi.org/10.1021/acsami.7b08181.

Chen Li-Feng, Xu-Dong Zhang, Hai-Wei Liang, Mingguang Kong, Qing-Fang Guan, Ping Chen, Zhen-Yu Wu and Shu-Hong Yu. 2012. Synthesis of nitrogen-doped porous carbon nanofibers as an efficient electrode material for supercapacitors. ACS Nano 6(8): 7092–7102. https://doi.org/10.1021/nn302147s.

Chen, Yao, Xiong Zhang, Haitao Zhang, Xianzhong Sun, Dacheng Zhang and Yanwei Ma. 2012. High-performance supercapacitors based on a graphene-activated carbon composite prepared by chemical activation. RSC Advances 2(20): 7747–7753. https://doi.org/10.1039/c2ra20667f.

Cheng, Qian, Jie Tang, Jun Ma, Han Zhang, Norio Shinya and Lu Chang Qin. 2011. Graphene and nanostructured MnO_2 composite electrodes for supercapacitors. Carbon 49(9): 2917–2925. https://doi.org/10.1016/j.carbon.2011.02.068.

Du, Lei, Lixin Xing, Gaixia Zhang and Shuhui Sun. 2020. Metal-organic framework derived carbon materials for electrocatalytic oxygen reactions: recent progress and future perspectives. Carbon 156: 77–92. https://doi.org/10.1016/j.carbon.2019.09.029.

Feng, X., Yan, Z., Chen, N., Zhang, Y., Ma, Y., Liu, X., Fan, Q., Wanga, L. and Huang, W. 2013. The synthesis of shape-controlled MnO_2/graphene composites via a facile one-step hydrothermal method and their application in supercapacitors. J. Mater. Chem. A 1: 12818–12825. https://doi.org/10.1039/C3TA12780J.

Guan, Dahui, Zan Gao, Wanlu Yang, Jun Wang, Yao Yuan, Bin Wang, Milin Zhang and Lianhe Liu. 2013. Hydrothermal synthesis of carbon nanotube/cubic Fe_3O_4 nanocomposite for enhanced performance supercapacitor electrode material. Mater Sci. Eng. B Solid State Mater Adv Technol 178(10): 736–743. https://doi.org/10.1016/j.mseb.2013.03.010.

Hoang, Van Chinh, Khyati Dave and Vincent G. Gomes. 2019. Carbon quantum dot-based composites for energy storage and electrocatalysis: mechanism, applications and future prospects. Nano Energy 66: 2211–2855. https://doi.org/10.1016/j.nanoen.2019.104093.

Hossain, M. Nur, Shuai Chen and Aicheng Chen. 2018. Fabrication and electrochemical study of ruthenium-ruthenium oxide/activated carbon nanocomposites for enhanced energy storage. J. Alloys Compd. 751: 138–147. https://doi.org/10.1016/j.jallcom.2018.04.104.

Hu, Yue, Yang Zhao, Gewu Lu, Nan Chen, Zhipan Zhang, Hui Li, Huibo Shao and Liangti Qu. 2013. Graphene quantum dots-carbon nanotube hybrid arrays for supercapacitors. Nanotechnology 24(19). https://doi.org/10.1088/0957-4484/24/19/195401.

Ji, Hongmei, Ting Wang, Yang Liu, Chunliang Lu, Gang Yang, Weiping Ding and Wenhua Hou. 2016. A novel approach for sulfur-doped hierarchically porous carbon with excellent capacitance for electrochemical energy storage. Chem. Commun. 52(86): 12725–12728. https://doi.org/10.1039/c6cc05921j.

Kay Hyeok An, Kwan Ku Jeon, Jeong Ku Heo, Seong Chu Lim, Dong Jae Bae and Young Hee Lee. 2002. High-capacitance supercapacitor using a nanocomposite electrode of single-walled carbon nanotube and polypyrrole. J. Electrochem. Soc. 149. https://doi.org/10.1149/1.1491235.

Kim, Jeonghyun, Soo-Jin Park, Sungwook Chung and Seok Kim. 2019. Preparation and capacitance of Ni metal organic framework/reduced graphene oxide composites for supercapacitors as nanoarchitectonics. J. Nanosci. Nanotechnol. 20(5): 2750–2754. https://doi.org/10.1166/jnn.2020.17469.

Kumar, Rajesh, Rajesh K. Singh, Alfredo R. Vaz, Raluca Savu and Stanislav A. Moshkalev. 2017. Self-assembled and one-step synthesis of interconnected 3D network of Fe_3O_4/reduced graphene oxide nanosheets hybrid for high-performance supercapacitor electrode. ACS Appl. Mater. Interfaces 9(10): 8880–8890. https://doi.org/10.1021/acsami.6b14704.

Kumar, Sachin, Ghuzanfar Saeed, Ling Zhu, Kwun Nam Hui, Nam Hoon Kim and Joong Hee Lee. 2021. 0D to 3D carbon-based networks combined with pseudocapacitive electrode material for high energy density supercapacitor: a review. Chem. Eng. J. Elsevier B.V. https://doi.org/10.1016/j.cej.2020.126352.

Li, Bing, Fang Dai, Qiangfeng Xiao, Li Yang, Jingmei Shen, Cunman Zhang and Mei Cai. 2016. Nitrogen-doped activated carbon for a high energy hybrid supercapacitor. Energy Environ. Sci. 9(1): 102–106. https://doi.org/10.1039/c5ee03149d.

Lin, Pei, Qiujie She, Binling Hong, Xiaojing Liu, Yining Shi, Zhan Shi, Mingsen Zheng and Quanfeng Dong. 2010. The nickel oxide/CNT composites with high capacitance for supercapacitor. J. Electrochem. Soc. 157(7): A818. https://doi.org/10.1149/1.3425624.

Lin, Yi, Zeyu Chen, Chuying Yu and Wenbin Zhong. 2019. Heteroatom-doped sheet-like and hierarchical porous carbon based on natural biomass small molecule peach gum for high-performance supercapacitors. ACS Sustain. Chem. Eng. 7(3): 3389–3403. https://doi.org/10.1021/acssuschemeng.8b05593.

Liu, Mingkai, Bomin Li, Hang Zhou, Cong Chen, Yuqing Liu and Tianxi Liu. 2017. Extraordinary rate capability achieved by a 3D 'Skeleton/Skin' carbon aerogel-polyaniline hybrid with vertically aligned pores. Chem. Commun. 53(19): 2810–2813. https://doi.org/10.1039/c7cc00121e.

Liu, Simin, Yijin Cai, Xiao Zhao, Yeru Liang, Mingtao Zheng, Hang Hu, Hanwu Dong, Sanping Jiang, Yingliang Liu and Yong Xiao. 2017. Sulfur-doped nanoporous carbon spheres with ultrahigh specific surface area and high electrochemical activity for supercapacitor. J. Power Sources 360: 373–382. https://doi.org/10.1016/j.jpowsour.2017.06.029.

Lv, Lingxiao, Yueqiong Fan, Qing Chen, Yang Zhao, Yue Hu, Zhipan Zhang, Nan Chen and Liangti Qu. 2014. Three-dimensional multichannel aerogel of carbon quantum dots for high-performance supercapacitors. Nanotechnology 25(23). https://doi.org/10.1088/0957-4484/25/23/235401.

Mohmmad Khalid, Bhardwaj, Prerna, Varela, Hamilton. 2019. Carbon-based composites for supercapacitor. Sci. Techno. Adv. Appl. Supercapacitors 1–23. https://doi.org/10.5772/intechopen.80393.

Noked, Malachi, Sivan Okashy, Tomer Zimrin and Doron Aurbach. 2012. Composite carbon nanotube/carbon electrodes for electrical double-layer super capacitors. Angew. Chem. Int. Ed. 51(7): 1568–1571. https://doi.org/10.1002/anie.201104334.

Qi, Wen, Xuan Li, Ying Wu, Hong Zeng, Chunjiang Kuang, Shaoxiong Zhou, Shengming Huang and Zhengchun Yang. 2017. Flexible electrodes of MnO_2/CNTs composite for enhanced performance on supercapacitors. Surf. Coat. Technol. 320: 624–629. https://doi.org/10.1016/j.surfcoat.2016.10.038.

Qing Yan, Yuting Jiang, He Lin, Luxiang Wang, Anjie Liu, Yali Cao, Rui Sheng, Yong Guo, Chengwei Fan, Su Zhang, Dianzeng Ji and Zhuangjun Fan 2019. Boosting the Supercapacitor Performance of Activated Carbon by Constructing Overall Conductive Networks Using Graphene Quantum Dots. J. Mater. Chem. A 7(11): 6021–6027. https://doi.org/10.1039/c8ta11620b.

Sahoo, Srikant, Ashis Kumar Satpati, Prasanta Kumar Sahoo and Prakash Dattatray Naik. 2018. Incorporation of carbon quantum dots for improvement of supercapacitor performance of nickel sulfide. Research-article. ACS Omega 3(12): 17936–17946. https://doi.org/10.1021/acsomega.8b01238.

Seok, Dohyeong, Yohan Jeong, Kyoungho Han, Do Young Yoon and Hiesang Sohn. 2019. Recent progress of electrochemical energy devices: metal oxide-carbon nanocomposites as materials for next-generation chemical storage for renewable energy. Sustainability (Switzerland) 11(13). https://doi.org/10.3390/su11133694.

Shahrokhian, Saeed, Rahim Mohammadi and Elham Asadian. 2016. One-step fabrication of electrochemically reduced graphene oxide/nickel oxide composite for binder-free supercapacitors. Int. J. Hydrog. Energy 41(39): 17496–17505. https://doi.org/10.1016/j.ijhydene.2016.07.087.

Shan, Yan and Lian Gao. 2007. Formation and characterization of multi-walled carbon nanotubes/Co_3O_4 nanocomposites for supercapacitors. Mater. Chem. Phys. 103 (2–3): 206–210. https://doi.org/10.1016/j.matchemphys.2007.02.038.

Shao, Yuanlong, Maher F. El-Kady, Lisa J. Wang, Qinghong Zhang, Yaogang Li, Hongzhi Wang, Mir F. Mousavi and Richard B. Kaner. 2015. Graphene-based materials for flexible supercapacitors. Chem. Soc. Rev. 44(11): 3639–3665. https://doi.org/10.1039/c4cs00316k.

Sun, Hang, Ping She, Kongliang Xu, Yinxing Shang, Shengyan Yin and Zhenning Liu. 2015. A self-standing nanocomposite foam of polyaniline@reduced graphene oxide for flexible super-capacitors. Synth. Met. 209: 68–73. https://doi.org/10.1016/j.synthmet.2015.07.001.

Tang, Cheng, Qiang Zhang, Meng Qiang Zhao, Gui Li Tian and Fei Wei. 2014. Resilient aligned carbon nanotube/graphene sandwiches for robust mechanical energy storage. Nano Energy 7: 161–169. https://doi.org/10.1016/j.nanoen.2014.05.005.

Thangappan, R., Arivanandhan, M., Dhinesh Kumar, R. and Jayavel, R. 2018. Facile synthesis of RuO_2 nanoparticles anchored on graphene nanosheets for high performance composite electrode for supercapacitor applications. J. Phys. Chem. Solids 121: 339–349. https://doi.org/10.1016/j.jpcs.2018.05.049.

Wang, Pengfei, Hui Liu, Yuxing Xu, Yunfa Chen, Jun Yang and Qiangqiang Tan. 2016. Supported ultrafine ruthenium oxides with specific capacitance up to 1099 F G-1 for a supercapacitor. Electrochim. Acta 194: 211–218. https://doi.org/10.1016/j.electacta.2016.02.089.

Wang, Qiufan, Yun Ma, Xiao Liang, Daohong Zhang and Menghe Miao. 2019. Flexible supercapacitors based on carbon nanotube-MnO_2 nanocomposite film electrode. Chem. Eng. J. 371(January): 145–153. https://doi.org/10.1016/j.cej.2019.04.021.

Wen, Ping, Peiwei Gong, Jinfeng Sun, Jinqing Wang and Shengrong Yang. 2015. Design and synthesis of Ni-MOF/CNT composites and RGO/carbon nitride composites for an asymmetric supercapacitor with high energy and power density. J. Mater. Chem. A 3(26): 13874–13883. https://doi.org/10.1039/c5ta02461g.

Wu, Haoyang, Mingli Qin, Zhiqin Cao, Xiaoli Li, Baorui Jia and Xuanhui Qu. 2019. Highly efficient synthesis of 2D VN nanoparticles/carbon sheet nanocomposites and their application as supercapacitor electrodes. Appl. Surf. Sci. 466(February): 982–988. https://doi.org/10.1016/j.apsusc.2018.10.102.

Yan, Jun, Zhuangjun Fan, Tong Wei, Jie Cheng, Bo Shao, Kai Wang, Liping Song and Milin Zhang. 2009. Carbon nanotube/MnO_2 composites synthesized by microwave-assisted method for supercapacitors with high power and energy densities. J. Power Sources 194(2): 1202–1207. https://doi.org/10.1016/j.jpowsour.2009.06.006.

Yang, Dongfang. 2012. Application of nanocomposites for supercapacitors: characteristics and properties. Nanocomposites-New Trends Dev. 299–328. https://doi.org/10.5772/50409.

Yang, Wenyao, Yan Chen, Jingfeng Wang, Tianjun Peng, Jianhua Xu, Bangchao Yang and Ke Tang. 2018. Reduced graphene oxide/carbon nanotube composites as electrochemical energy storage electrode applications. Nanoscale Res. Lett. 13. https://doi.org/10.1186/s11671-018-2582-6.

You, Bo, Lili Wang, Li Yao and Jun Yang. 2013. Three dimensional N-doped graphene-CNT networks for supercapacitor. Chem. Commun. 49(44): 5016–5018. https://doi.org/10.1039/c3cc41949e.

Yu, Dingshan and Liming Dai. 2010. Self-assembled graphene/carbon nanotube hybrid films for supercapacitors. J. Phys. Chem. Lett. 1(2): 467–470. https://doi.org/10.1021/jz9003137.

Yuan, Shuoguo, Sin Yi Pang and Jianhua Hao. 2020. 2D transition metal dichalcogenides, carbides, nitrides, and their applications in supercapacitors and electrocatalytic hydrogen evolution reaction. Appl. Phys. Rev. 7(2). https://doi.org/10.1063/5.0005141.

Zhang Weili, Chuan Xu, Chaoqun Ma, Guoxian Li, Yuzuo Wang, Kaiyu Zhang, Feng Li, Chang Liu, Hui-Ming Cheng, Youwei Du, Nujiang Tang and Wencai Ren. 2017. Nitrogen-superdoped 3D graphene networks for high-performance supercapacitors. Adv. Mater. 29(36): 1–9. https://doi.org/10.1002/adma.201701677.

Zhou, Tianzhu, Chao Wu, Yanlei Wang, Antoni P. Tomsia, Mingzhu Li, Eduardo Saiz, Shaoli Fang, Ray H. Baughman, Lei Jiang and Qunfeng Cheng. 2020. Super-tough MXene-functionalized graphene sheets. Nat. Commun. 11(1): 1–11. https://doi.org/10.1038/s41467-020-15991-6.

Zhou Yan, Zixin Jia, Lingling Shi, Zhen Wu, Binyong Jie, Siyuan Zhao, Liyuan Wei, Aiguo Zhou, Junwu Zhu, Xin Wang and Yongsheng Fu. 2020. Pressure difference-induced synthesis of P-doped carbon nanobowls for high-performance supercapacitors. Chem. Eng. J. 385(September 2019): 123858. https://doi.org/10.1016/j.cej.2019.123858.

Multidimensional Graphene-Based Advanced Materials for Electrochemically Stable Supercapacitors

Deepa B. Bailmare[1], Pranjali Khajanji[2] and Abhay D. Deshmukh[1]*

[1] Energy Materials and Devices Laboratory, Department of Physics, RTM Nagpur University, Nagpur - 440033, India
[2] Metal Power Analytical, Mumbai - 400047, India

1. Introduction

Graphene is a thin layer of carbon that stacks together to form a honeycomb lattice known as a wonder material and has the potential to endow its application with astonishing traits. It is the best conductor with amazing flexibility and mechanical strength, hence emerged as an ecologically good carbon material to be used in a variety of applications. Apart from this, graphene possesses high charge mobility, a specific surface area (2,650 m^2/g) with high thermal conductivity and low bandgap (Zhu et al. 2010). Due to its highly abundant properties and flexible structure, it is widely explored in energy storage devices. Graphene is an extremely diverse material that can fit with other elements (including gases and metals) and produce superior materials with extremely different properties. Hence, graphene has attracted researchers' attention for exploring its various properties with wider applications. In a most trendy way, graphene is applied in various applications like sensors, composites materials, nanoelectronics and energy storage devices. The schematic illustration in Figure 1 shows the properties of graphene material and its utilization in various applications. In the case of energy storage systems, most favorably, the materials should be rich in properties, like high specific surface area, high mechanical and chemical stability without sacrificing structural ability. In energy storage systems, graphene is often suggested to replace the activated carbon-based materials in part due to its high specific surface area and better pore size distribution than a carbon-based material.

Corresponding author: abhay.d07@gmail.com

The requirement of high surface area and porosity still limit the capacitance and energy density of energy storage devices. Graphene plays a major role in enhancing the performances of energy storage systems, like supercapacitors (Bai et al. 2020). In the case of supercapacitors, maintaining the electrodes with high specific surface area, flexible nature and high mechanical strength are very clumsy. Although the graphene supercapacitors are said to store high energy than aluminum ion batteries, it charges in a second and maintains the discharging cycles up to tens to thousand cycles. However, producing graphene on large scale is indeed a major challenge. Dozens of companies are struggling to make graphene-based materials with high quality with cost-effectiveness. The large scale production of graphene-based materials is done by the liquid-phase exfoliation method, whereas the graphene-based films are synthesized by chemical vapor deposition methods. Apart from this method, chemical excision, electrochemical synthesis and flash joule heating methods are also in consideration. Although graphene has high theoretical capacitance, the theoretical capacitance of graphene has not been accompanied so far due to the self-restacking of graphene sheets, which leads to reduced surface area and unfortunately reduces the capacitance as well and hence the dimensionality comes into the picture. The three-dimensional structures with open pores will provide high electrical conductivity to the graphene and thus exceed the terminology of stacking in multidimensional frames. Consequently, the effect of structures and architectures has been studied in many studies till now. Multidimensionality in graphene with the incorporation of one-, two- and three-dimensional stacking provides a highly flexible structure, which generates high ion absorption and less resistance to the electrode materials. Based on the dimensions of graphene varieties of structures and morphologies were created and ranged from one-dimension, two-dimensions and three-dimension architectures.

Although the production cost of graphene is much higher due to its highly critical synthesis methods, still graphene-based materials rule the energy storage market. According to Zhang et al. (2020) versions of batteries and supercapacitors have many problems related to their energy-storing capability and limited flexibility. Hence, they have provided the island bridge design of connecting cells in supercapacitors, which allows the stretching of the device up to 100%. They have made graphene foam decorated ZnP electrode; all in all, the electrode gives high stretchability with extra high sensitivity to the sensor-based wearable device. Yu et al. (2020) reported the 3-D graphene with extra thickness and rich ion transport path with significant practical application. This newly developed structure provides a balanced contradiction between the thickness of electrode and ion transport which show specific capacitances as high as 2,412.2 $mFcm^{-2}$ at 0.5 $mAcm^{-2}$. Accordingly, flexible solid-state micro-supercapacitors were constructed with a great energy density of 134.4 $\mu Whcm^{-2}$ at a power density of 325 μWcm^{-2}. These findings provide a potential application of graphene in the energy storage system. In particular, this chapter discusses the overview on multidimensional graphene-based electrode materials with the fundamental of synthesis of graphene, structural and morphological studies and finally the flexible graphene-based multidimensional structures for supercapacitors.

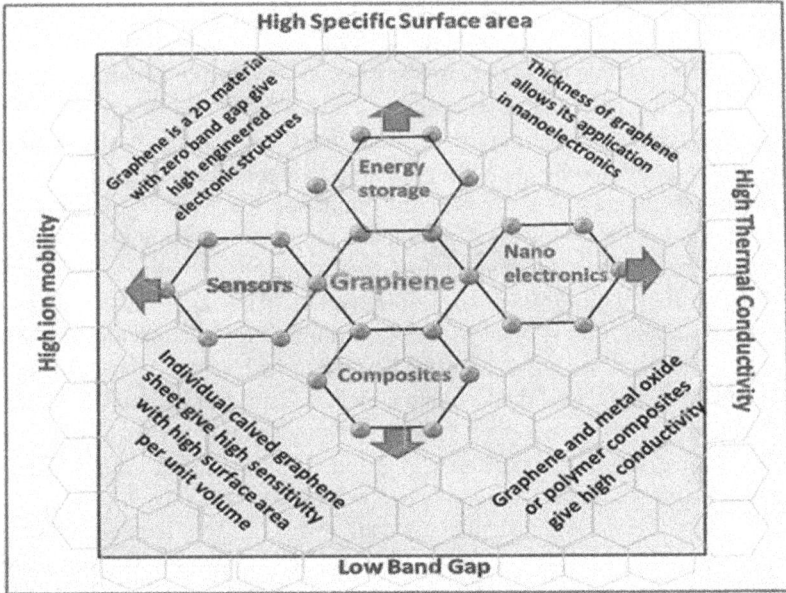

Figure 1: An illustrative version of graphene as a wonder material with varying features, advantages and various practical applications.

2. Nomenclature and Terminology of Graphene

The graphene is a two-dimensional monolayer of sp2 hybridized carbon atoms, which is freely suspended into a variety of substrates (Bianco et al. 2013). Graphene sheets are formed by the layered arguments of carbon-based graphite with a thickness of atomic scales. Basically, graphene is a two-dimensional graphitic layer formed by structural elements in graphene and 3-D carbon atoms. Carbon state of art, mostly graphene layer term, is considered for the three-dimensional sp^2 bonded carbon atoms, i.e. the carbon atoms lose their planarity and hence convert themselves to a layered structure. Such compounds are mostly said to be more electronegative and irrelevance to the π-compounds, which drastically decreases their electrical conductivity. Whereas in the case of the charge transfer mechanism, a carbon sheet in its planarity keeps their bonded structures apart from the defects and the reactive materials forms expanded intervals in the host structure; such compounds are named as graphene intercalation compounds. The term graphite is originated due to the modifications in chemical elements of carbon. In graphite, each carbon atom forms a layered stacking in regular three-dimensional ways. Boehm et al. (1986) suggested the various nomenclatures related to graphene to understand its terminology. They said carbon is a hypothetical member of infinite sized polycyclic aromatic hydrocarbons, which involves naphthalene, coronene, ovalene, anthracene, phenanthrene, tetracene, etc. (Boehm et al. 1986). These series of compounds have 'ene' in their common ends with carbon-carbon double bonds, and its last member contains 'graph' from the graphite. They also said that graphene is not isolated with

3-D carbon atoms, but the layers of graphene from the same carbon sources that can be utilized for many applications (Bianco et al. 2013, Boehm et al. 1986).

The lateral dimensions and width of the graphene sheets range from tens of nanometers to micrometers or in the macroscopic dimensions. These ranges may affect the percolation, threshold, bandgap and interacting behavior of nanosheets (Bianco et al. 2013). Moreover, it is assumed that the density and height of the graphene nanosheets are dependent on the lateral dimensions. For novel nanosheets, the lateral dimensions are more likely to be determined, especially for single-layer graphene or 2D materials. Hence, the nomenclature is necessary to determine the exact 2-D material in its way to capture the exact thickness and layers of the graphene nanosheets or microsheets (Walter et al. 2014). Furthermore, when talking about crystallography or morphologies of the materials, it is very important to define a material whether based on its materialistic features or properties of the materials. The high-quality graphene crystallites and size of nanosheets is the most important focus of the researchers to be applied for various applications. Unfortunately, only materials features and properties are not enough to understand their practical applications. Hence, we believe in the crystallographic approach because atomic arrangements of materials are more likely to determine the phases of matters in a most fundamental way. Whereas, the classification of nanoscale of the materials is also limited to the shape, size and thickness of materials in their nano state. As supercapacitors electrode, graphene enormously gives various structures and morphologies depending upon its synthesis techniques. The morphology is more conveniently depicted as the number of layers depending upon the thickness of graphene, few layers graphene structures, in a plane shape ribbon-like structures, roughly isotropic or polygonal-shaped ragged structures. Moreover, apart from the armchair and zig-zag graphene ribbons, the most focus has been established for thickness and layers of graphene nanosheets in a plane. Since, the thickness and layers of graphene nanosheets are basically needed for various applications, like nanoelectronics, energy storage, crystallographic filler phases, etc. (Geim 2007). Depending upon various applications, graphene has been synthesized through a number of synthesis methods.

3. Fundamentals of Synthesis of Graphene

The graphene can be assembled in varieties of structures, like free-standing graphene, one dimensional (1-D), two dimensional (2-D), three dimensional (3-D) and recent focus of graphene-based electrode materials in its multidimensional structures according to their macro-structural complexity. There are many methods that have been implanted to synthesize graphene nanosheets; most likely used methods are the synthesis of graphene oxides through modified Hummer's method and graphene nanosheets through optimized chemical vapor deposition, chemical excision by hydrothermal, solvothermal and chemical reduction of graphene oxide to form a variety of multidimensional structures. The solvent exfoliation method is also implanted for the synthesis of graphene-based electrodes to achieve great conductivity with additional functional groups. Figure 2 shows a schematic of various synthetic routes of multidimensional graphene to be used in a variety of applications. Two-dimensional graphene structures were mostly synthesized via modified Hummer's method. Layers of graphene sheets are formed through the conversion of carbonaceous graphite material to graphene oxide. Furthermore, it is speculated that the functionalization of graphene with the introduction of heteroatom leads to improvement in electrical and structural properties. Graphene oxides (GO) serve as a conducting porous framework to anchor different metal nanoparticles. GO is a carbon material that shows high chemical, structural and electrical properties that are similar to graphene and graphene-based framework. In 1958, Hummer's and Offeman developed the method to synthesize graphene oxides. In this proposed method, H_2SO_4 is used as an exfoliating agent and $NaNO_3$ and $KMnO_4$ are used to add oxygen functionality to the graphene (Hummers and Offeman 1958). Graphene oxide (GO) contains high-density oxygen functionality with hydroxyl, epoxy and carboxyl groups in its basal plane and its edges, respectively. Hence, they give high wettability to the material with ease of functionalization. This method firmly used graphitic carbon to get oxidized graphene. Zahed et al. (2018) reported that the graphene oxide synthesis with modified Hummer's method. According to this method, graphite and $NaNO_3$ were mixed with H_2SO_4 and stirred in an ice bath at a temperature of about 0 to 5°C. Then, $KMnO_4$ was slowly added to the mixture at a temperature of less than 15°C. The ice bath was then removed and the mixture was stirred for 2-D at 35°C temperature. Furthermore, 100 ml water was slowly

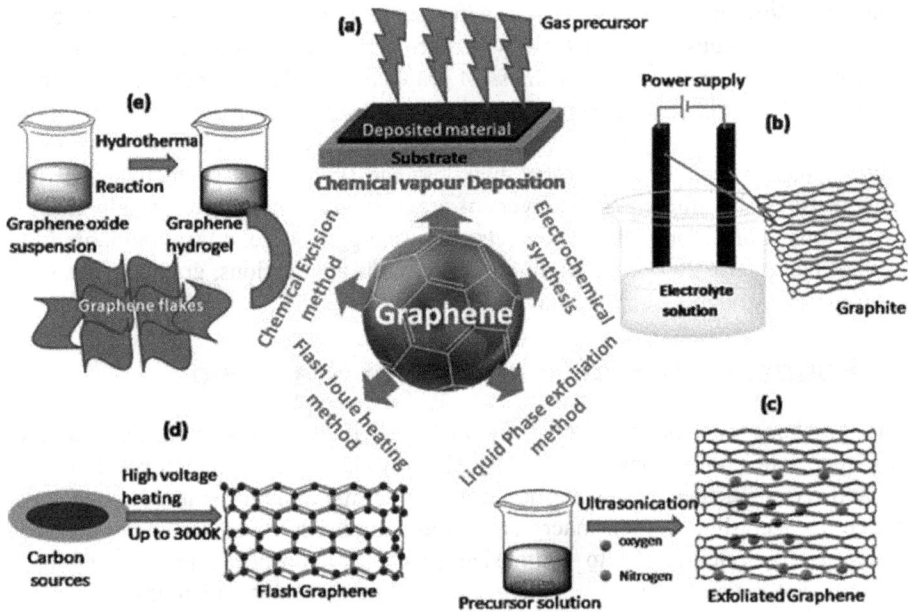

Figure 2: Schematic synthetic routes of graphene nanomaterials followed by various synthesis techniques: (a) graphene film synthesis by a chemical vapor deposition method, (b) graphene electrode synthesis through electrochemical deposition method, (c) the exfoliation of graphene done by liquid-phase exfoliation under ultrasonication, (d) illustration of flash joule heating method for graphene formation and (e) illustrative showing the hydrothermal route for synthesizing graphene aerogels.

added and the temperature of the sample was raised to 98°C rapidly. The color of the mixture changed to brown and the reaction completes with the addition of H_2O_2 to react with additional $KMnO_4$. The product obtained was thoroughly washed with HCl and deionized water and then dried in a vacuum at room temperature, and the resultant graphene oxide was obtained. Yu et al. (2016) reported the highly efficient graphene oxide synthesis through modified Hummer's method. This method uses improvization of $NaNO_3$ free modified Hummers's method with the replacement of $KMnO_4$ with K_2FeO_4 as an oxidizing agent so as to reduce the amount of concentrated sulfuric acid. Most importantly, the Hummers method distinguishes itself from others by eliminating $NaNO_3$ since it causes toxic gases, like NO^{2-}, N_2O_2, NO^{3-}, etc., in order to avoid the contents of $NaNO_3$. Excessive amounts of $KMnO_4$ and acids were added, whereas sulfuric acids used to increase the interlayer spacing during oxidation of graphite. Additionally, H_3PO_4 may be used to exfoliate graphite along with sulfuric acid (Samuel et al. 2019). This method is specially made for synthesis of graphene oxide that is further utilized as starting material for synthesizing multidimensional graphene nanostructures.

3.1 Optimized Chemical Vapor Deposition Technique

The synthesis of graphene through chemical vapor deposition is a new approach to develop a few layers of graphene sheets over substrates. In this method, the route involves hydrocarbons on the surface of metals, like nickel and copper with the substantial transfer of conducting substrate. This yield is a highly-conducing graphene sheet; the CVD method enables the substitution of the carbon atom within the defected area of the substrates and allows the growth of a few layers of graphene over conducting substrates (López et al. 2009). CVD technique generally involves one-step synthesis based on breaking of molecules of precursors (gas, liquids and solids) in their gaseous state and reforms it over a conducting substrate. In 2009, the first graphene sheet developed over a Cu foil conducting substrate with CVD method utilizes large area and high-quality graphene film (X. Li et al. 2009). Basically, the nucleation and growth dynamics of graphene can serve as an excellent example for the study of the physics of two-dimensional graphene materials. More specifically, in the CVD method the grain size, structures and mode of operation of graphene are conducted by the optimum ratio of CH_4/H_2, temperature, pressure and pretreatment of substrates, which finally results in the tuned structural, electrical and chemical properties of graphene films. The layer-by-layer or bilayer graphene manipulation is achieved by the CVD method by nucleation and growth dynamics of graphene. Under these conditions, the bilayer growth of graphene can be done by high temperature and high hydrogen to methane ratio in CVD chamber. Moreover, the formation of each first layer grain involves the addition of multiple adlayer grains with wide size distribution. Notably, each twisted angle of bilayer grain is different from that of the single crystal of the first layer grain. Hence, it is difficult for us to understand what role the residual carbon material plays during the formation of the adlayer and what will happen if the adlayer grains merge and interact. So, in that sense of synthesis, Chu and Woon et al. (2020) reported the unique second-growth approach to clarify the role of residual carbon atom during the growth of TBLG. Here, they had grown multilayer graphene with the help of the CVD method. In brief, copper foil is placed on a silicon boat for thermal processing to allow forming of multilayer graphene. The growth is completed at various gas ratios H_2/CH_4. Initially, the copper substrate was partially oxidized with a flow of Ar (500 sccm, 900 mtorr) with low residues of oxygen. Afterwards, multilayer graphene was grown CH_4/H_2 ratio of 30:1 for 30 minutes. The multilayered graphene covers the copper substrate with the front and backside. In the overall process, the chamber pressure is maintained at 1.4 torr by accompanying Ar flow. After the growth of graphene film, the cooling process takes place (cooled down below 800°C within 1 minute) (Chu and Woon 2020). Different research groups analyzed different morphologies of CVD synthesized graphene, which was further utilized for many potential applications like energy storage, sensors, gas storage, etc.

3.2 Chemical Excision

This technique includes hydrothermal and solvothermal methods to form a graphene-based material for supercapacitors. Although, hydrothermal and solvothermal approaches are more likely to form graphene hydrogels without further use of a solvent or purifying agents. Hence, most research has followed the one-step synthesis of graphene gels through hydrothermal and solvothermal approaches. Xu et al. (2010) reported the self-assembly of graphene hydrogels through one-step hydrothermal approach wherein different concentrations of GO was dispersed in 20 ml Teflon lined autoclave at 180°C for 1 to 12 hours. The prepared hydrogel was taken out and used for the experiments. The prepared graphene hydrogel was highly electrically conductive, provided high specific capacitance with great stability to the electrode materials. Luo et al. (2019) reported the one-step hydrothermal synthesis method for three-dimensional graphene and graphene hydrogel for supercapacitors. Briefly, the technique usage 14 ml graphene oxide (GO) dispersed in 22 ml-Teflon sealed stainless steel autoclave and treated at 180°C for 4 hours. Accordingly, different concentration of solutions was also taken to prepare three-dimensional graphene materials. Furthermore, the graphene quantum dots were also synthesized through two-step hydrothermal techniques in which as prepared 3DG was utilized. The prepared 3DG and GQD composite material delivers high specific capacitance of 242 Fg^{-1} which in accordance helped composite material to achieve high specific capacitance and better electrical conductivity. The graphene in its multidimensional state gives a high possibility to the electrode materials to achieve high electrical conductivity and better ionic mobility during the percolation reaction.

3.3 Electrochemical Synthesis

The electrochemical method is the most reliable and green synthesis method to form graphene nanosheets or graphene quantum dots. For most energy storage devices, the flexibility of electrode material is the most general requirement. Moreover, by selecting appropriate electrolytes, the graphene nanosheets and graphene quantum dots can be synthesized in a multidimensional way. Most exclusively, Hilder et al. (2011) reported the simplest and least expensive method of direct electrodeposition of graphene through its suspension. The graphene oxide (GO) suspension was mixed with (1:1) ratio of NaCl electrolyte (Hilder et al. 2011). The constant potential deposition was carried out in 1 to 1.4 V of potential. The pH value of the solution was adjusted through a diluted solution of sodium hydroxide or hydrochloric acid. The deposition time was taken to be 15 minutes and black spongy material was coated over a substrate. The idea of direct deposition of graphene nanosheet agitates desirable conductivity with lower oxygen levels during the reduction in the electrolyte suspension.

In the case of electrochemically deposited graphene films, the whole substrate gets covered and serves good electrical conductivity with low barrier resistances between electrode and electrolyte interfaces. By this method, 3-D interconnected surface fully exposed to electrolyte and hence escalated ion percolation during the electrochemical process. Li et al. (2013) reported the fiber shaped solid electrochemical supercapacitors through electrochemical synthesis of graphene oxide (GO) wherein 3, 6, 9 and 12 mgmL^{-1} graphene oxide suspended in 0.1 M lithium per chlorate (LiClO$_4$) and deposited at -1.2 V of potential for 2 to 40 seconds. After electrodeposition, the prepared electrode materials were thoroughly washed with de-ionized water to remove excess salt and residues. It was measured that the prepared electrode material delivers a specific capacitance of 7.5 to 10 µFcm^{-1} at 50 mV/s to 10 V/s. Flexible electrochemically reduced graphene oxide electrode material gives high specific capacitance, great rate capability and chemical stability (Li et al. 2013). Furthermore, Pham et al. (2015) diluted graphene oxide in a 2 mgmL^{-1} concentration and pH was adjusted to 1.35 with the help of 1 M H$_2$SO$_4$ solution. SS mesh was dipped into the solution of active material. DC constant voltage is applied between the carbon paper (current collector) and SS mesh working electrode. A 5 V of potential was applied for 60 minutes to deposit graphene oxide over the SS mesh substrate. The directly deposited graphene oxide-oriented SS mesh electrode material delivers high specific performances. Such a system in 1 M H$_2$SO$_4$ and 1M H$_2$SO$_4$ + hydroquinone redox electrolyte exhibited high capacitances of 147 Fg^{-1} and 223 Fg^{-1}. Direct growth of ERGO hydrogel as electrode material gives high electrochemical performances with extreme stability (Pham et al. 2015). Purkait et al. (2018) also reported the electrochemical deposition technique as the least expensive and easy to scale up industrially applicable method to form reduced graphene oxide. Here, a porous reduced graphene oxide network was formed over a Cuf/Cu wire electrode through the electrolysis of graphene oxide solution. The electrolysis was carried out over different concentrations (1, 3, 5 and 7 mgmL^{-1}) of graphene oxide mixed with 1 M phosphoric acid at -1 V of potential. The deposition process carries over 200 s. The maximum specific capacitance has been achieved by 5 mgmL^{-1} solution concentration. The flexible wire-shaped supercapacitors device delivers a high specific capacitance of 81 Fg^{-1}/40.5 mFcm^{-1} in PVA/H$_3$PO$_3$ electrolyte with 1 V of the potential window. Furthermore, the device was most stable up to 5,000 cycles with 94.5% capacitance retention (Purkait et al. 2018). Likewise, the *in-situ* deposition of graphene oxide over substrate will open high chances of achieving a specific surface area exposed to the electrolyte solution. The electrodeposition is directly proportional to the processed and controllable formation of thin films over conducting substrates.

3.4 Solvent Exfoliation Method

The key factor for producing ambient graphene is low oxygen contents, layer-by-layer insertion and large lateral size. Hence, during the solvent exfoliation of graphite in an ambient condition, commonly the oxygen, is introduced in the lattice of carbon and breaks the graphene into a small piece. There were various solvents used to disperse the graphene to form pristine material for various applications. Most commonly, organic solvents like benzene, toluene and nitrobenzene are used to disperse graphene. Apart from this, polar solvents like N, methyl pyrrolidone (NMP) and N,N-dimethyl formamide (DMF) are also used to create homogeneous graphene dispersion but most reports are showing orthodichloro benzene for the formation of homogeneous graphene dispersion (Niyogi et al. 2003). However, the synthesis of graphene faces a critical trade-off in high yield without sacrificing its properties, hence attracting a crucial research focus. The exfoliation of graphene without the functionalization results in very poor product yield, which is not suitable for commercialization purposes. Whereas the personification of graphene through the conversion of graphite oxide to graphene results in high yield and helps to glorify the commercialization techniques. The major effect of solvent exfoliation with functionalization may give less carbon to oxygen ratio with slightly different properties than graphene. However, many synthesis approaches were applied to do the synthesis of graphene with the solvent exfoliation method. But synthesizing graphene with pristine properties, cost-effectiveness and an environmentally friendly approach is still a challenge. In the early years, Sumantha Sahoo et al. (2013) reported the synthesis of graphene with solvent exfoliation method using ortho dichlorobenzene as a solvent. ODCB has a high boiling point and is used as a preferred solvent for single and multi-walled CNTs dispersion. Here, they used a simple one-step synthesis approach. The specific proportion of graphite is dispersed in ODCB solvent homogenized for a different interval of time in an ultrasonic bath. All the solutions were heated to 185°C for solvent evaporation and sonicated for a 60 to 300 minutes longer duration and after each sonication the vials were took out for characterization. This method used to radiate the graphite in a solvent and forms bubbles. These large numbers of tiny bubbles grow up in the negative pressure in ultrasonic wave propagation. Furthermore, in usual cases, synthesizing 3-D graphene with complex structures is considerably difficult. Emilie Bordes et al. (2018) reported the dispersion and exfoliation of graphite in ionic liquid solution; in a typical synthesis the suspended graphite solution was prepared by the graphite flakes in a 2 ml concentration of ionic liquid to form a highly concentrated solution of monolayer graphene. Furthermore, the suspension was sonicated for several durations at a maximum temperature of 328K. The temperature-controlled ultrasonication generates thermal energy, which expanded the graphite layer and makes the exfoliation more precise. The advantage of taking ionic liquids as a solvent is that with increasing temperature the solvent does not degrade. For maintaining the long term stability of graphene suspension, centrifugation at 10,000 rpm for 1 hour was carried out (Emilie Bordes et al. 2018). Likewise, the solvent exfoliation in different solvents provides quality yield of graphene and hence used for many applications.

3.5 Flash Joule Heating Method

The flash graphene (FG) comprises conversion of amorphous carbon into flash graphene. This technique has the advantages of quick and cheap transformation of coal, food or waste materials into flash graphene in a fraction. A tremendous study has been carried out for the synthesis of graphene, whereas the deconvolutions of waste materials like coal, plastics and, many hazardous materials are a genuine concern worldwide. The flash graphene required milliseconds to convert the waste materials into graphene. This method uses 3,000 Kelvin temperature to heat carbon-containing materials. The starting materials can be anything that contains carbons such as food waste, coal, biochar, wood waste, etc. The yield is totally dependent on the sources of carbon. The 2-D materials' synthesis with such a cost-effective method is quite a difficult task compared to other synthesis methods discussed earlier. The flash graphene provides scalable production of 2-D graphene in a much shorter duration. Herein, 2D materials are extracted through a high pulse voltage directly applied to the respective precursor. Most recently, Luong et al. (2020) reported the gram-scale synthesis of flash graphene through flash joule heating. In this method, amorphous carbon is situated in lightly compressed quartz or ceramic tubes between two electrodes, which can be copper or graphite. The whole system can be run at atmospheric pressure or under mild vacuum conditions. High pulse voltage is bombarded between the two electrodes wherein the carbon-containing material inside the chamber experiences around 3,000 K temperature within 100 ms, which effectively converts the amorphous carbon into flash graphene (Luong et al. 2020). Furthermore, Wala et al. (2020) reported the synthesis of graphene through plastic waste. The work relates to the conversion of plastic waste products into flash graphene. Here, they have used plastic wastes like milk jugs, plumbing pipes, food packaging items and disposable coffee cups. The plastic waste was cut and powdered by using a cutter with a grain size up to 1 to 2 nm. Furthermore, waste mixed with 5wt% of carbon black for making a conducting mixture. Later on, the powder was packed in between two copper electrodes and placed in a quartz tube. The sample was compressed to obtain the resistivity of 120 ohms and an alternating voltage of 120 V was applied in electrodes for 8 s. They have performed two variations of AC and DC voltages i.e., for DC flash joule heating process was performed in the material prepared by AC flash joule heating. A 450 V and 60 mF of the capacitor was charged up to 110 V and applied over electrodes for 500 ms discharge time, which results in high-quality flash graphene (Wala et al. 2020). The method involves the scalable and cost-effective synthesis of graphene with very abundant methods of preparation.

4. Structural and Morphological Studies of Graphene

The structural and morphological diversity is defined by the synthesis approaches. Fundamental properties, like electronic properties, conducting properties and structural properties, are interred related to one another. The graphene-based materials show excellent flexibility with respect to the surface morphologies and provide high impact in various applications. Graphene is a diverse material that fascinates excellent crystallographic properties, morphological flexibility and disordered structures. These studies can be done with different characterization techniques depending upon the graphene orientation.

4.1 Crystallographic Studies

Depending on the synthesis methods, accordingly structural and crystallographic feature varies. This section provides a short overview of the crystallinity and interplanar spacing of graphene with XRD measurements. The X-ray diffraction technique is a well-known technique for knowing the structural parameters of any material. This technique provides dependable approaches to scale up the formation of materials. The XRD technique is the most valuable method for determining structural transformations in carbon-based materials, like activated carbon, carbon nanotubes, carbon aerogels, carbon composites and multidimensional graphene. There were different nomenclatures of carbon as graphite, graphene and graphene oxides. The revelation of XRD data assesses the effectiveness of oxidation in graphene, expansion of graphene and reduction of graphene through various synthesis techniques. Typically, the XRD pattern of graphite reveals two peaks at positions $\Theta = 26°$ and $\Theta = 43°$, whereas the XRD pattern of graphene oxide as synthesized by Hummer's method is located at the lowest angle with (0 0 1) crystal plane with an interlayer spacing of $d = 8.33$ Å. It is worth noting that the interlayer spacing of graphite is 3.37 Å. It was suggested that the interlayer spacing increases with the introduction of oxygen in graphite. After conversion into graphene, the interlayer spacing decreases and peak broadening was observed that suggested the lower ordered stacking of graphene (Johra and Lee 2014) whereas after the exfoliation of graphite, natural graphite is converted into graphene. The natural graphite shows the characteristics of XRD pattern with a peak at $\Theta = 26.4°$ and after exfoliation, the peak intensity dramatically decreases for a similar $\Theta = 26.4°$ peak position. This decrease in intensity occurs due to the decrease in thickness of graphite because of the breaking of inter-carbon bonds in graphite structures and there will be no other crystal plane observed after exfoliation indicating non-functionalized graphene sheets (Badri et al. 2017). Apart from this, after the oxidation of graphite flakes (0 0 2) plane shifts to a lower angle ($\Theta = 9.9°$), which indicates the formation of graphite oxide. Usually, the graphite oxide indicates peak position at a lower angle of incidence, whereas the exfoliation of graphite oxide makes the peak broader and increases the d spacing (Cai et al. 2017) [Figure 3(c)].

4.2 Morphological Studies of Graphene

The dimensional features of graphene depend upon the formation and newly optimized trends of preparing graphene-based materials. It is important to explore the material properties with HR-TEM or SEM analysis for knowing the exact surface analogies and morphological characteristics of materials. The SEM and TEM are well-known characterization methods to examine surface morphologies of materials. Graphene is formed by the structural transformation of graphite. Most predictably the graphite materials have irregular grains, acicular grains or natural flakes like morphologies (Wu et al. 2009). Vertically grown graphene shows varieties of

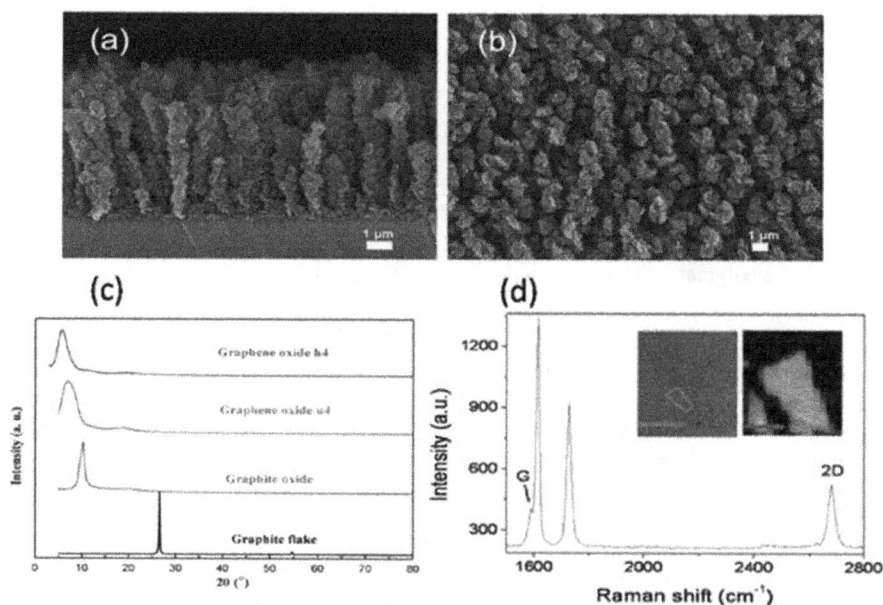

Figure 3: Morphology of graphene forest: (a) SEM image of cross-section and (b) detailed graphene structure over quartz substrate (reprinted with permission of Royal society of chemistry (Yifei Ma et al. 2015). (c) XRD measurement of graphite flakes, graphite oxide, graphene oxide (reprinted with permission of Journal of environmental nanoscience [Cai et al. 2017)]. (d) Raman analysis of graphene on PET. The insets show optical image (left) and integrated Raman intensity image (right) of the 2-D band (reprinted with permission American chemical society) (Yu et al. 2008), copyright (@ 2008).

morphological structures, which make them a higher surface area for most of the applications. The microscale graphene structure seems like a resemblance of a tree and packed graphene film look like a forest [Figures 3(a) and (b)]. The graphene tree has 6.4 micrometers of height and inter-tree spacing was hundreds of nanometers. Each graphene tree comprises thousands of graphene sheets stacked through random arrangement (Yifei Ma et al. 2015). While in the case of exfoliation of graphene from graphene oxide (GO), SEM images show graphene oxides flakes distribute over a substrate. This is further proven by FESEM images, an average size of graphene flakes is estimated to be 500 nm. The high-temperature ultrasonication reduces the graphene flakes size (Yifei Ma et al. 2015). This confirms that graphene in its various forms shows varieties of morphological analogies to the exhibitory performances. K.K et al. (2019) give the approach of free-standing graphene conductive matrix for $Ni(OH)_2$. Since graphene properties are tailored by atomic doping, and hence they try heteroatom doping regardless of getting high electrochemical performances. The as-prepared N doped graphene material gives wrinkled paper-like tiny structures compared to typical pristine graphene materials. The structures and morphologies of graphene give abundant electrochemically active electrode materials, which provide great stability to the materials. According to the method of synthesis, a varying morphological feature of graphene is summarized in Table 1.

Table 1: Various morphological studies of graphene-based material

Materials	Synthesis Methods	Morphology	Reference
Fiber shaped r-GO	Electrochemical synthesis	Three-dimensional interpenetrating networks	(Pham et al. 2015)
r-GO hydrogel	Electrochemical synthesis	3D interconnected framework	(Purkait et al. 2018)
3D graphene/polymer	Hydrothermal	3D interconnected loosely stacked graphene	(Li et al. 2019)
Graphene nanosheet	Exfoliation of graphite	Layered structures with ripple surface	(Siburian et al. 2018)
3-D Graphene	Chemical vapor deposition	Interconnected 3D scaffold structure of the nickel foam	(Chen et al. 2011)
3-D Graphene	Employing polymerization	Layer by layer disordered structures	(Matsuyama et al. 2016)
Chemically-converted graphene (CCG)	Hydrothermal	Densely packed and randomly oriented, forming a porous network structure	(Yu et al. 2016)
Polypyrrole graphene	Hummers method/ electrochemical polymerization	Knitted structure on substrate	(Bhargava et al. 2020)
CC@rGO/ polypyrrole	Ultrasonication/ modified chemical oxidative polymerization method	Sheet-like and wrinkled morphology	(Chen et al. 2017)

4.3 Disorders in Graphene

A typical analogy of defects and crystal disorders in graphene is studied by Raman spectroscopy. Raman spectrum of carbon-based materials, including carbon nanotubes, carbon nanofibres, carbon composites, activated carbon, graphite, diamonds and graphene nanosheets, are usually characterized at 800 to 2,000 cm^{-1}. Among these modes, the D bands are located at 1,330-1,360 cm^{-1} and the G band located near 1,580 cm^{-1} [Figure 3(d)] (Yu et al. 2008). Raman spectra of a few layers graphene, multilayer graphene and graphite powder will determine the aim of investigating D band and G band intensities, which determine the degree of defects created in the graphene structures. If the defect ratio is larger, the quantity of defects also increases in the graphene structure. The occurrence of the D band is located at the signal of ~1,342 cm^{-1}, which justifies the defects created in the graphitic structure before exfoliation into graphene with G band ~1,571 cm^{-1} and 2-D intense band at ~2,693 cm^{-1}. Structural variations and degrees of disorders are largely depending on the raw graphite material. Larger the number of defects created in the graphite, the larger the degree of disorders of graphene. First-order G bands are permitted in graphene because of the hexagonal symmetry of graphene lattice. Two Raman modes in band level symmetry and in-plane vibrations are responsible for the G band occurrence. D band is also another important feature of Raman spectra. This band discusses the degree of disorders in graphene. D band is usually more intense in amorphous carbon material, whereas defeated in graphitic materials. It is evidenced that the ratio of I_d/I_g increases with the defect density with the low defect density because of elastic scattering and decreases with the high defect density attenuated Raman peaks (Ferrari 2007). The intense D band is considered to be having high Sp2 and low Sp3 hybridized carbon atoms. Likewise, when the carbon is more amorphous, then the D band shows strong intensity whereas it disappears for a honeycomb-like structure. Hence, the low-intensity D band together with the 2-D band shows a high-quality graphene layer and indicates more precision to the formation of graphene sheets (Lan et al. 2018).

5. Application of Multidimensional Graphene in Supercapacitors

The terminology of supercapacitors owes high performances in terms of its energy density, power density, response time and unlimited cycle process. Apart from this, supercapacitors categorically uses various new types of materials according to their structures and features in which a hybrid type material is most likely to be used. Whereas, supercapacitors applications may be constrained due to the compromised energy density concerning new practical applications which further need to deal with the challenges, like durability in devices, reliability, cost-effectiveness, long term cyclic performances and life cycle cost. The pervasive reliability of supercapacitors is a major area of interest nowadays. The specific capacitance of supercapacitors is proportional to high specific surface area and ion separable dielectric, which maximizes the specific capacitance of supercapacitors than conventional capacitors and hence offers high energy storing capability than capacitors. On the other hand, supercapacitors give 100 times more power delivering performances compared to rechargeable batteries, which makes it more promising to future energy storage applications (Muzaffar et al. 2019). In addition to this, there are some more advantages of supercapacitors, like high-rate capability, long term cycle life, resistance effectiveness, operational temperature, lightweight, unified packaging, flexible outlet, etc. However, unlike batteries (Li-ion Batteries) that provide energy density in the ranges of 100-1,000 Whkg^{-1}) (Qiao et al. 2019), supercapacitors suffer energy density which is a widespread shortcoming of supercapacitors for applicability. The comparative Ragone plot of energy density and power density is shown in Figure 4. Considerable efforts have been made by researchers to enlighten a way to enhance the energy density of supercapacitors to make them cost-friendly comparable to batteries. The operation of supercapacitors is dependent upon type electrode, electrolyte and operating potential, resistances and potential ranges that need to be used. However, the parametric performances, such as high specific surface area, specific capacitance,

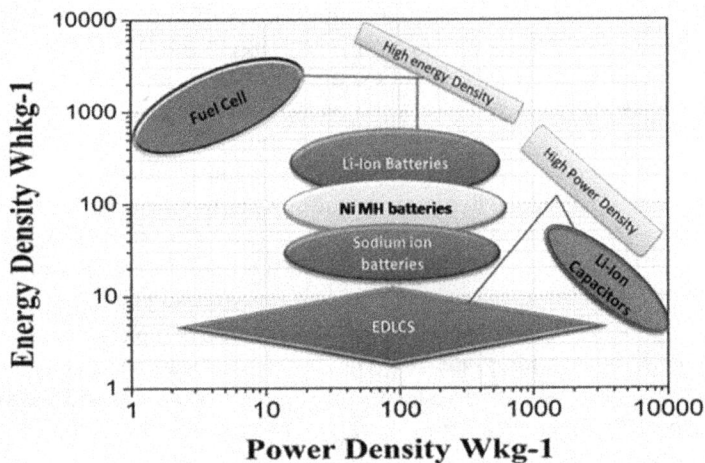

Figure 4: Ragone plot of comparison of energy density (Whkg[-1]) and power density (Wkg[-1]) of different energy storage devices.

resistances and cycle life of supercapacitors are determined by the structure and type of electrode materials. Hence, electrode materials play a major role in enhancing the overall performances of supercapacitors. Basically, it performs according to the categorically distinguished charge storage mechanisms of electrode materials, which involves EDLC and pseudocapacitive materials. In the case of EDLC (electric double layer capacitors), the mechanism is faradic, i.e. absorption/desorption process takes place at electrode and electrolyte interface; on the other hand, pseudocapacitors are termed to have a non-faradic mechanism, i.e. the charge storage involves typical redox reaction along with the intercalation/deintercalation processes.

Pseudocapacitors show battery type terminology and hence suffer from various limitations such as low power delivering performances, less stability, structural distortion and less electrical conductivity. Graphene has been generally reported as flexible material for supercapacitors. Graphene, a monolayer of carbon atoms stacked in a two-dimensional lattice, which gives good electrical conductivity and a high specific surface area of 2,657 m^2/g with the theoretical capacitance of ~21 μ Fcm^{-2} and a corresponding specific capacitance of 550 Fg^{-1} (Xia et al. 2009, Ke Q. and Wang J. 2016). Hence, it is believed to be a wonder material for energy storage applications. Most specifically, as a supercapacitors electrode, graphene has attracted considerable attention. Various nanomaterials are synthesized with the help of graphene fundamentals, which are further utilized for flexible or wearable devices.

5.1 Multidimensional Graphene Nano-materials for Supercapacitors

With the persistent development of nanotechnologies, such as nanoelectronics and nanoengineering, energy storage demands new requirements for the materials for wearable and temperature resistant applications. The scenario is quite obvious because the development of energy storage devices with high electrical, mechanical and electrochemical performance is a monotonous task. The flexible supercapacitors can achieve high mechanical performances without sacrificing electrochemical outcomes. At the same time, the design and structures of flexible supercapacitors vary with the type of electrode materials and substrates used. In most cases, the flexible devices are manufactured from one-dimensional fiber structures, two-dimensional plane structures and three-dimensional laminated structures (Yang and Mai 2014). The schematic approach of fabricating flexible devices is shown in Figure 5.

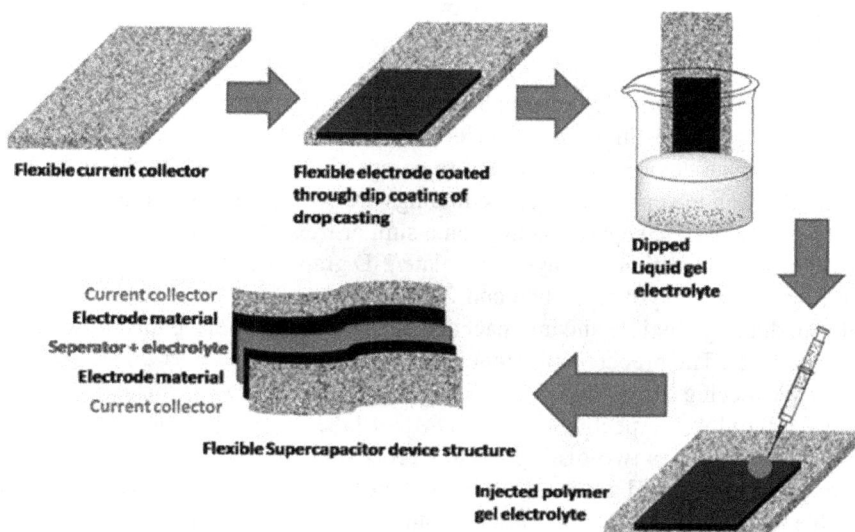

Figure 5: Schematic illustration of the effective ways for configuration of flexible asymmetric/ symmetric supercapacitors devices.

Furthermore, the two-dimensional graphene structures give high mechanical strength and excellent performances, which makes them flexible materials with high mechanical robustness. These endured features of 2-D graphene materials ensure the high applicability of material as freestanding graphene films and flexible supercapacitors electrode. The multidimensional structure of graphene interestingly provides quality interest for the manufacturing of graphene-based flexible devices. The porous and loose features of the 3-D building block give highly access to the growth of supercapacitors. While in the case of 3-D nanostructures when a compressive force is enacted, the 3D nanostructures get destroyed which creates major disadvantages for flexible supercapacitor devices. The involvement of compressive force distorted the ion absorption mechanism of electrode and electrolyte which tends to decrease the electrochemical performances of supercapacitor electrode materials. These issues generalize the idea of improving the structural stability of the electrode materials for enhancing the electrochemical performances of flexible supercapacitors. The two-dimensional graphene-based materials have the potential to meet the requirement of flexible supercapacitors in terms of high electrical conductivity and mechanical strength. Sahoo et al. (2019) reported the two-dimensional graphene electrodes sandwiched with the porous PVDF ionic liquid electrolyte. The symmetric graphene-based device was configured in a coin cell using two ideal graphene electrodes separated by a TEABF4-porous PVDF separator. At first, the operating potential window of the device was evaluated which interestingly shows 3.0 V of operating potential window without any evolution. Furthermore, the two-dimensional graphene-based device at a scan rate of 5 mV/s possesses a high specific capacitance of 36.62 Fg^{-1} (40.69 $mFcm^{-2}$); excluding this, the charge discharge profile recorded using applied current and device potential window showing a high gravimetric capacitance of 28.46 Fg^{-1} with the areal capacitance of 31.63 $mFcm^{-2}$. The device also provides a high energy density of 35.58 $Whkg^{-1}$ at a power density of 750 Wkg^{-1}. The self-discharging performance of the graphene device was evaluated with the help of piezoelectric PVDF and $TEABF_4$ separation applied compressive forces of 5 to 20 N, respectively. Interestingly, the device was able to self-charge up to 70, 77 and 112 mV when the device was subjected to compressive forces of 10, 15 and 20 N, respectively (Sahoo et al. 2019). These findings are very generous to the application of graphene in flexible supercapacitors on a similar trend. Shi et al. (2019) reported the design and fabrication of polyoxometalate/2-D graphene nanosheets through the nanocomposite of polyoxometalate and 2-D graphene. This approach provides the highly conducting, flexible and compact device used for monitoring the pulse rate of the human body. The prepared electrode material was deposited over a Ti foil, which delivers the specific capacitance of 64.73 Fg^{-1} and 295 Fg^{-1} for graphene nanosheets and POM/2D GNs, respectively. The POM/2D GNs especially show high cyclic stability because it has two distinguishable processes, i.e. a rapid decrease process before 15,000 cycles and a stable process with gentle decrease. The results are more favorable with the incorporation of polyoxometalate in 2-D graphene nanosheets for the increasing cyclic stability of two-dimensional graphene nanosheets. The POM/2D GNs were used as a flexible device to run a sensor for monitoring human pulse rate. The as-designed sensor clearly recorded the pulse within 15 seconds, and it basically coincides with the results of the commercial optical detection heart-beat monitoring device, such as photoplethysmography (PPG) (Shi et al. 2019). The various three-dimensional graphene-based materials were also used for the supercapacitor applications as it provides high structural diversity and abundant network to build a 3-D porous architecture. Zhou et al. (2019) reported the one-step self-assembly method for 3D CNTS/RGO aerogels using CNTF and graphene oxide solution as a starting precursor. CNTF acts as a cross-linking agent to connect the adjacent graphene and prevents restacking of graphene to occupy 3-D porous framework. The resultant 3D porous architecture achieves high specific surface area with low density and abundantly porous network. The prepared 3D CNTS/RGO delivers a high specific capacitance of 412.8 Fg^{-1} at 1 Ag^{-1} current density. The electrode further runs for 10,000 cycles with 89.8% cyclic stability. On another approach of three-dimensional graphene-based supercapacitors, Du et al. (2020) constructed a functionalized 3-D graphene network that was constructed by decorating with p-phenylenediamine (PPD) regulating through the hydrothermal method. The well-defined drGO/PPD/CDs exhibits a three-dimensional network and it favors high ion percolation and hence exhibits high specific performances of 468 Fg^{-1} at 0.5 Ag^{-1} current density. The flexible solid-state device further delivers high electrochemical performances such that it gives high gravimetric capacitance of 322 Fg^{-1} at 0.5 Ag^{-1} current density with good cyclic stability of 5,000 cycles with 76% capacitance retention. The various materials are fabricated for the state-of-the-art graphene-based flexible devices for reaching the requirements of highly stable supercapacitors (Table 2).

Table 2: The supercapacitors based on multidimensional graphene electrodes and their derivatives

Electrode Materials	Current Collector	Electrolyte	Specific Capacitance	Energy Density	Power Density	Reference
Graphene/ PANI	Ni foam	PVA/ H_3PO_3	261 Fg^{-1} at 0.38 Ag^{-1}	23.2 $Whkg^{-1}$	399 Wkg^{-1}	(Xie et al. 2014)
Graphene paper pillered	Stainless steel cell	1M $LiPF_6$	83.2 Fg^{-1} at 10 mV/s	26 $Whkg^{-1}$	5.1 $kWkg^{-1}$	(Wang et al. 2011)
3-D graphene/ Fe_2O_3	-	2M LiCl	76.8 Fg^{-1} at 16 Ag^{-1}	41.7 $Whkg^{-1}$	13.5 $kWkg^{-1}$	(Yang et al. 2015)
Multilayer rGO	Au sputtered	PVA/ H_3PO_3	247.3 Fg^{-1} at 0.176 Ag^{-1}	-	-	(Yoo et al. 2011)
CNT/rGO hybrid		PVA/ H_2SO_4	207 Fg^{-1} /158 $mFcm^{-1}$	20 mWh/cm^{-2}	15 $mWcm^{-2}$	(Gao et al. 2009)
Ag/GF/ OMC electrode		6M KOH	213 Fg^{-1}	4.5 $Whkg^{-1}$	5,040 Wkg^{-1}	(Zhi et al. 2014)
3-D graphene hydrogel	Au coated PI	PVA/ H_2SO_4	186 Fg^{-1}	0.61 $Whkg^{-1}$	0.67 $kWkg^{-1}$	(Xu et al. 2013)
MnO_2/ GRP	Graphene paper	Na_2SO_4	76.8 $mFcm^{-2}$	6140 $\mu Whcm^{-2}$	36 $mWcm^{-2}$	(Sadak Omer et al. 2019)
PANI/3D graphene	SS mesh	1M H_2SO_4	165 Fg^{-1} at 1 Ag^{-1}	25.3 $Whkg^{-1}$	553.4 Wkg^{-1}	(Zhang et al. 2019)
$Ni(OH)_2$/ graphene	Flexible graphite foil	PVA/KOH	61.7 mF/cm^{-2}	17 $\mu Whcm^{-2}$	1,200 μWcm^{-2}	(Li et al. 2018)
CNT/G hybrid	PET coated with Au	1M Na_2SO_4	200 Fg^{-1} at 0.1 Vs^{-1}	-	-	(Cheng et al. 2013)

5.2 Substrates for Device Structures

The device structures are based on the substrate materials used for the fabrication of flexible supercapacitors. Substrates play a crucial role in delivering mechanical flexibility and often provide strength to the devices. The most commonly used substrate materials for the fabrication of graphene-based flexible devices are flexible plastic substrate (polyethylene terephthalate PET), regular conventional paper, carbon cloth and carbon paper, etc. (Yang and Mai 2014). The highly conducting and porous substrate provides high conductivity and strength to the electrode materials. The presence of mechanical strength along with the conductive frameworks serves as an excellent building block to specify a high specific surface area and porosity of the materials. Most recently, the carbon-based substrates are also attracting interest because it provides high conductivity, mechanical strength and flexibility to the electrode materials. The carbon nanofibers (CNF) are one-dimensional substrate materials mostly synthesized by electro-spinning technique with the help of polymer precursors. They are a promising electrode material that provides high specific surface area and porosity to the material which generates high energy storing power of the flexible devices. Further graphene fibers are also used as a macroscopic one-dimensional ensemble for regulating the graphene nanosheets over building blocks in continuous uniaxial directions. The GFs are enormous excellent properties of graphene nanosheets like mechanical strengths and hence are emerging as a unique advantageous material than carbon nanofibers (Xu 2011). Similarly, more efforts have been devoted to the synthesis of three-dimensional structures over a macroscopic level. Mostly, the three-dimensional substrates are used to grow the three-dimensional architecture of graphene flakes and their nanocomposites. The 3-D substrates especially provide porous structures to build the high specific surface area which reimbursed the high specific capacity and cyclic stability to the electrode materials. Likewise, the multidimensional structures based on graphene provide considerable tuning of substrates with morphologies. Likewise, for improving the flexible device performances, the choice of substrate for electrode materials also plays an important role. Various papers have reported the use of multidimensional substrates for the flexible device formations (Xie et al. 2014, Wang et al. 2011, Cheng et al. 2013).

6. Summary and Perspective

To summarize, two-dimensional graphene has been explored through numerous synthetic approaches for obtaining different graphene structures. However, most of them were commercialized depending upon their production capacity and technical maturity. Mass production of graphene for different applications was synthesized through liquid-phase exfoliation methods. The large surface area graphene films or electrodes were also being synthesized by chemical vapor deposition methods, whereas the large-scale production of graphene materials are indispensable when expanding the production in the future. This is because the controlled morphology, size and structural parameter of graphene on large scale are clumsy. Hence, it is peremptory to optimize the controlled and tunable morphology, flake size, structures, purity and surface area of graphene. Moreover, the production cost and scalability of graphene should not get neglected. On the other hand, developing highly reliable and cost-effective synthesis methods for the production of graphene should be under consideration to reduce the production cost of the materials. On the contrary various methods have been implanted to fabricate graphene-based electrodes with cost-effectiveness and richer properties. These fabricated materials collectively embraced the fascinating properties and performances of graphene electrodes. All in all, we anticipated that the environmentally friendly and cost-effective methods provide flexible graphene structures used in enormous applications. Specifically, in the case of energy storage devices like supercapacitors, the basic requirement of the electrode materials is their high specific surface area and porosity. There are various successful fabrication methods for developing flexible devices with the use of graphene-based materials. (Gao et al. 2013, Zhi et al. 2014 and Sadak Omer et al. 2019). Additionally, the mechanical strength of graphene-based devices is significant to provide a strong flexible structure and optimistic fabricated devices. On the whole, the current chapter discussed the origin of graphene as two-dimensional material with excellent structural diversity. Furthermore, it shows the fascinating synthetic approaches of multidimensional graphene-based materials and their structural and morphological studies. Lastly, the chapter summarizes the significance and uniqueness of graphene nanomaterials in the fabrication of electrochemically stable supercapacitors.

References

Bai, L., Zhang, Y., Tong, W., Sun, Li., Huang, H., An, Q., Tian, N. and Chu, P.K. 2020. Graphene for energy storage and conversion: synthesis and interdisciplinary applications. Electrochem. Energ. Rev. 3: 395–430.

Bianco, A., Cheng, H.-M., Enoki, T., Gogotsi, Y., Hurt, R.H., Koratkar, N. and Zhang, J. 2013. All in the graphene family – a recommended nomenclature for two-dimensional carbon materials. Carbon, 65: 1–6.

Boehm, H.P., Setton, R. and Stumpp, E. 1986. Nomenclature and terminology of graphite intercalation compounds. Carbon, 24(2): 241–245.

Cai, C., Sang, N., Shen, Z. and Zhao, X. 2017. Facile and size-controllable preparation of graphene oxide nanosheets using high shear method and ultrasonic method. J. Exp. Nanosci. 12(1): 247–262.

Chen, Z., Liao, W. and Ni, X. 2017. Spherical polypyrrole nanoparticles growing on the reduced graphene oxide-coated carbon cloth for high performance and flexible all-solid-state supercapacitors. Chem. Eng. J. 327: 1198–1207.

Chen, Z., Ren, W., Gao, L., Liu, B., Pei, S. and Cheng, H.-M. 2011. Three-dimensional flexible and conductive interconnected graphene networks grown by chemical vapour deposition. Nature Materials, 10(6): 424–428.

Cheng, H., Dong, Z., Hu, C., Zhao, Y., Hu, Y., Qu, L., Chen, N. and Dai, L. 2013. Textile electrodes woven by carbon nanotube-graphene hybrid fibers for flexible electrochemical capacitors. Nanoscale, 5: 3428–3434.

Chu, C.-M. and Woon, W.-Y. 2020. Growth of twisted bilayer graphene through two-stage chemical vapor deposition. Nanotechnology, 31: 43.

Du, P., Kang, H. and Dong, Y. 2020. Carbon dots regulate crosslinking of functionalized three-dimensional graphene networks decorated with p-phenylenediamine for superior performance flexible solid-state supercapacitors. J. Energy Storage, 30: 101586.

Emilie Bordes, Bishoy Morcos, David Bourgogne, J.-M.A., Pierre-Olivier Bussière, Catherine C. Santini, Anass Benayad, M.C. and Gomes, A.A.H.P. 2018. Dispersion and stabilisation of exfoliated graphene in ionic liquids. Front. Chem., 7: 223.

Fatima Tuz Johra, Jee-Wook Lee, W.-G.J. 2014. Facile and safe graphene preparation on solution based platform. J. Ind. Eng. Chem. 20, 2883–2887.

Ferrari, A.C. 2007. Raman spectroscopy of graphene and graphite: disorder, electron–phonon coupling, doping and nonadiabatic effects. Solid State Commun. 143: 47–57.

Gao, K., Shao, Z., Li, J., Wang, X., Peng, X., Wang, W., Wang, F. (n.d.). Cellulose nanofiber-graphene all solid-state flexible supercapacitors. J. Mater. Chem. A, 2013(1), 63–67.

Geim, A.K.N.K. 2007. The rise of graphene. Nat Mater, 6: 183–191.

Hilder, M., Winther-Jensen, B., Li, D., Forsyth, M. and MacFarlane, D.R. 2011. Direct electro-deposition of graphene from aqueous suspensions. Physical Chemistry Chemical Physics, 13(20): 9187.

Ke, Q. and Wang, J. 2016. Graphene-based materials for supercapacitor electrodes – a review. J. Mater, 2(1): 37–54.

Lan, Y., Zondode, M., Deng, H., Yan, J.-A., Ndaw, M., Lisfi, A. and Pan, Y.-L. 2018. Basic concepts and recent advances of crystallographic orientation determination of graphene by Raman spectroscopy. Crystals, 8(10): 375.

Li, C., Yang, Z., Tang, Z., Guo, B., Tian, M. and Zhang, L. 2019. A scalable strategy for constructing three-dimensional segregated graphene network in polymer via hydrothermal self-assembly. Chem. Eng. J. 363: 300–328.

Li, X., Cai, W., An, J., Kim, S., Nah, J., Yang, D., Piner, R., Jung, I., Tutuc, E., Banerjee, S.K., Colombo, L. and Ruoff, R.S. 2009. Large-area synthesis of high-quality and uniform graphene films on copper foils. Science 324: 1312.

Li, Y., Sheng, K., Yuan, W. and Shi, G. 2013. A high-performance flexible fibre-shaped electrochemical capacitor based on electrochemically reduced graphene oxide. Chem. Commun. 49: 291–293.

Li, Y., Ye, H., Chen, J., Wang, N., Sun, R. and Wong, C.-P. 2018. Flexible β-Ni(OH)$_2$/graphene electrode with high areal capacitance enhanced by conductive interconnection. J. Alloy. Compd. 737: 731–739.

López, V., Sundaram, R.S., Gómez-Navarro, C., Olea, D., Burghard, M., Gómez-Herrero, J. and Kern, K. 2009. Chemical vapor deposition repair of graphene oxide: a route to highly-conductive graphene monolayers. Adv. Mater. 21(46): 4683–4686.

Luo, P., Guan, X., Yu, Y., Li, X. and Yan, F. 2019. Hydrothermal synthesis of graphene quantum dots supported on three-dimensional graphene for supercapacitors. Nanomaterials, 9(2): 201.

Luong, D.X., Bets, K.V., Algozeeb, W.A., Stanford, M.G., Kittrell, C., Chen, W., Salvatierra, R.V., Ren, M., McHugh, E.A., Advincula, P.A., Wang, Z., Bhatt, M., Guo, H., Mancevski, V., Shahsavari, R., Yakobson, B.I. and Tour, J.M. 2020. Gram-scale bottom-up flash graphene synthesis. Nature 577: 647–651.

Matsuyama, S., Sugiyama, T., Ikoma, T. and Cross, J.S. 2016. Fabrication of 3D graphene and 3D graphene oxide devices for sensing VOCs. MRS Advances 1(19): 1359–1364.

Muhammad Ashraf Saiful Badri, M.M.S., Noor, N.F.M. and Mohd Yusri Abd Rahman, A.A.U. 2017. Green synthesis of few-layered graphene from aqueous processed graphite exfoliation for graphene thin film preparation. Mater. Chem. Phys. 193: 212–219.

Muzaffar, A., Ahamed, M.B., Deshmukh, K. and Thirumalai, J. 2019. A review on recent advances in hybrid supercapacitors: design, fabrication and applications. Renew. Sust. Energ. Rev. 101: 123–145.

Niyogi, S., Hamon, M.A., Perea, D.E., Kang, C.B., Zhao, S.K. Pal, A.E. Wyant, M.E.I. and Haddon, R.C. 2003. Ultrasonic dispersions of single-walled carbon nanotubes. J. Phys. Chem. B 107: 8799–8804.

Panel Priya Bhargava, Wenwen Liu, Michael Pope Ting, and Tsui Aiping Yu. 2020. Substrate comparison for polypyrrole-graphene based high-performance flexible supercapacitors. Electrochimica Acta 358: 136846.

Pham, V.H., Gebre, T. and Dickerson, J.H. 2015. Facile electrodeposition of reduced graphene oxide hydrogels for high-performance supercapacitors. Nanoscale 7(14): 5947–5950.

Purkait, T., Singh, G., Kumar, D., Singh, M. and Dey, R.S. 2018. High-performance flexible supercapacitors based on electrochemically tailored three-dimensional reduced graphene oxide networks. Scientific Reports 8(1): 640.

Qiao, Y., Jiang, K., Deng, H. and Zhou, H. 2019. A high-energy-density and long-life lithium-ion battery via reversible oxide-peroxide conversion. Nature Catalysis 2(11): 1035–1044.

Sadak, Omer, Wang, W., Guan, J., Sundramoorthy, A.K. and Gunasekaran, S. 2019. MnO_2 nanoflowers deposited on graphene paper as electrode materials for supercapacitors. ACS Appl. Nano Mater. 2, 7: 4386–4394.

Sahoo, S., Krishnamoorthy, K., Pazhamalai, P., Mariappan, V.K., Manoharan, S. and Kim, S.-J. 2019. High performance self-charging supercapacitors using a porous PVDF-ionic liquid electrolyte sandwiched between two-dimensional graphene electrodes. J. Mater. Chem. A 7: 21693–21703.

Samuel, M., Bhattacharya, J., Raj, S., Santhanamb, N. and Singh, H,S.P. 2019. Efficient removal of chromium (VI) from aqueous solution using chitosan grafted graphene oxide (CS-GO) nanocomposite. Int. J. Biol Macromol. 121: 285–292.

Shi, Y., Wang, X., Luo, J. and Xie, Q. 2019. Fabrication and characterization of polyoxometalate/2D graphene-based flexible supercapacitors for wearable electronic pulse-beat application. J. Mater. Sci. Mater. Electron. 30: 3692–3700.

Siburian, R., Sihotang, H., Lumban Raja, S., Supeno, M. and Simanjuntak, C. 2018. New route to synthesize of graphene nano sheets. Orient. J. Chem. 34(1): 182–187.

Sumanta Sahoo, Goutam Hatui, Pallab Bhattacharya, Saptarshi Dhibar and Das, C.K. 2013. One pot synthesis of graphene by exfoliation of graphite in ODCB. Graphene 2: 42–48.

Upadhyay, K.K., Bundaleska, N., Abrashev, M., Bundaleski, N., Teodoro, O.M.N.D., Fonseca, I. and Montemor, M.F. 2019. Free-standing N-Graphene as conductive matrix for $Ni(OH)_2$ based supercapacitive electrodes. Electrochimica Acta 334: 135592.

Wala, A. Algozeeb, Paul E. Savas, Duy Xuan Luong, Weiyin Chen, Carter Kittrell, Mahesh Bhat, Rouzbeh Shahsavari and Tour, J.M. 2020. Flash graphene from plastic waste. ACS Nano 14(11): 15595–15604.

Walter, J., Nacken, T.J., Damm, C., Thajudeen, T., Eigler, S. and Peukert, W. 2014. Determination of the lateral dimension of graphene oxide nanosheets using analytical ultracentrifugation. Small 11(7): 814–825.

Wang, G., Sun, X., Lu, F., Sun, H., Yu, M., Jiang, W., ... Lian, J. 2011. Flexible pillared graphene-paper electrodes for high-performance electrochemical supercapacitors. Small 8(3): 452–459.

Wu, Z.-S., Ren, W., Gao, L., Liu, B., Jiang, C. and Cheng, H.-M. 2009. Synthesis of high-quality graphene with a pre-determined number of layers. Carbon 47: 493–499.

Xia, J., Chen, F., Li, J. and Tao, N. 2009. Measurement of the quantum capacitance of graphene. Nat. Nanotechnol. 4: 505–509.

Xie, Y., Liu, Y., Zhao, Y., Tsang, Y.H., Lau, S.P., Huang, H. and Yang, C. 2014. Stretchable all-solid-state supercapacitor with wavy shaped polyaniline/graphene electrode. J. Mater. Chem. A 2: 9142–9149.

Xu, Y., Sheng, K., Li, C. and Shi, G. 2010. Self-assembled graphene hydrogel via a one-step hydrothermal process. ACS Nano, 4(7): 4324–4330.

Xu, Z. and Gao, C. 2011. Graphene chiral liquid crystals and macroscopic assembled fibres. Nat. Commun. 2: 571.

Xu, Y., Lin, Y., Huang, X., Liu, Y., Huang, Y. and Duan, X. 2013. Flexible solid-state supercapacitors based on three-dimensional graphene hydrogel films. ACS Nano 7: 4042–4049.

Yang, M., Lee, K.G., Lee, S.J., Lee, S.B., Han, Y.-K. and Choi, B.G. 2015. Three-dimensional expanded graphene–metal oxide film via solid-state microwave irradiation for aqueous asymmetric supercapacitors. ACS Appl. Mater. Interfaces. 7(40): 22364–22371.

Yang, P. and Mai, W. 2014. Flexible solid-state electrochemical supercapacitors. Nano Energy 8: 274–290.

Yifei Ma, Mei Wang, Namhun Kim, Jonghwan Suhrc and Heeyeop Chae. 2015. A flexible supercapacitor based on vertically oriented 'Graphene Forest' electrodes. J. Mater. Chem. A, 3: 21875–21881.

Yoo, J.J., Balakrishnan, K., Huang, J., Meunier, V., Sumpter, B.G., Srivastava, A., Conway, M., Mohana Reddy, A.L., Yu, J. and Vajtai, R. 2011. Ultrathin planar graphene supercapacitors. Nano Lett 11: 1423–1427.

Yu, H., Zhang, B., Bulin, C., Li, R. and Xing, R. 2016. High-efficient synthesis of graphene oxide based on improved Hummers method. Scientific Reports 6(1): 36143.

Yu, J., Wu, J., Wang, H., Zhou, A., Huang, C., Bai, H. and Li, L. 2016. Metallic fabrics as the current collector for high-performance graphene-based flexible solid-state supercapacitor. ACS Appl. Mater. Interfaces 8(7): 4724–4729.

Yu, T., Ni, Z., Du, C., You, Y., Wang, Y. and Shen, Z. 2008. Raman mapping investigation of graphene on transparent flexible substrate: the strain effect. J. Phys. Chem. C 112: 12602–12605.

Yu, X., Li, N., Zhang, S., Liu, C., Chen, L., Han, S. and Wang, Z. 2020. Ultra-thick 3D graphene frameworks with hierarchical pores for high-performance flexible micro-supercapacitors. J. Power Sources 478: 229075.

Zahed, M., Parsamehr, P. Sadat, Tofighy, M.A. and Mohammadi, T. 2018. Synthesis and functionalization of graphene oxide (GO) for salty water desalination as adsorbent. Chem. Eng. Res. Des. 138: 358–365.

Zhang, C., Peng, Z., Huang, C., Zhang, B., Xing, C., Chen, H., ... Tang, S. 2020. High-energy all-in-one stretchable micro-supercapacitor arrays based on 3D laser-induced graphene foams decorated with mesoporous ZnP nanosheets for self-powered stretchable systems. Nano Energy 81: 105609.

Zhang, T., Yue, H., Gao, X., Yao, F., Chen, H., Lu, X. and Guo, X. 2019. Polyaniline nanowire arrays on three-dimensional hollow graphene balls for high-performance symmetric supercapacitor. J. Electroanal. Chem. 855: 113574.

Zhi, J., Zhao, W., Liu, X., Chen, A., Liu, Z. and Huang, F. 2014. Highly conductive ordered mesoporous carbon based electrodes decorated by 3D graphene and 1D silver nanowire for flexible supercapacitor. Adv. Funct. Mater. 24: 2013–2019.

Zhou, L., Wang, J., Liu, Z., Yang, J., Chen, M., Zheng, Y. and Xiong, C. 2019. Facile self-assembling of three-dimensional graphene/solvent free carbon nanotubes fluid framework for high performance supercapacitors. J. Alloys Compd. 153157.

Zhu, Y., Murali, S., Cai, W., Li, X., Suk, J.W., Potts, J.R. and Ruoff, R.S. 2010. Graphene and graphene oxide: synthesis, properties, and applications. Adv. Mater. 22: 3906–3924.

Part II: Mechanical and Engineering Applications

Effect of Ultrasonic Treatment on Graphite in Metal Matrix Composite

Ramendra Kumar Gupta*, V. Udhayabanu and D.R. Peshwe
Department of Metallurgical and Materials Engineering,
Visvesvaraya National Institute of Technology, Nagpur - 440010, India

1. Introduction

In material development, metal matrix composite (MMC) is a great innovation (Ibrahim I. A. and F. A. Mohamed 1991). Among the various metals, aluminum (Al) and its alloys were chosen as matrix material because of their high specific strength (Yang Lan and Li 2004, Macke et al. 2012). Therefore, aluminum composites are the global expectation in the transport industries (electrical vehicle, automobile, and aerospace) (Yang Lan and Li 2004, Macke and Schultz 2012, Rana et al. 2012). The selection of a material for a specific application is a challenging task for a material engineer. To obtain a material with all required properties, a materialist has to tailor the material. To obtain desired properties, various ceramic particles are to be injected into aluminum and its alloys by researchers (Yang Lan and Li 2004, Barekar et al. 2009, Zhou et al. 2014, Pramod et al. 2015, Mathew et al. 2017). This idea of reinforcing hard particles in metals innovates the area of metal matrix composite.

1.1 Metal Matrix Composites (MMCs)

The metal matrix composite is the combination of at least one metal or one alloy, and the second is a ceramic material or an organic compound where the hybrid composite is the combination of metal and two other materials (Pramod et al. 2015). It has been seen that the property of any material is affected by the atmospheric condition or working environment. Therefore, two or more materials were combined and analyzed for different useful properties and optimized in different materials. In general, for lightweight, aluminum and magnesium alloys were selected as a matrix material, and Al_2O_3, CNT, graphite, B_4C, SiC, zircon, mica, and other hard and soft particles were chosen as reinforcement material (Macke et al. 2012). The synthesized MMCs have greater strength, less density, enhanced stiffness, advanced

Corresponding author: remendragupta@gmail.com

thermal property, enhanced wear and abrasion resistance, enhanced damping, and many more changes in mechanical properties when compared to the unreinforced matrix material. Lightweight high strength was the reason for accepting Al-MMCs in transport industries, which provided new scope to develop MMCs during the last 40 years (Polmear 2006).

1.2 Why Graphite Particles?

To tailor desired composite materials, various ceramic particles were injected into aluminum alloys. Among these particles, graphite (Gr_p) is a natural, black mineral, which is an allotrope of carbon (Krishnan et al. 1981, Barekar et al. 2009). It is a low-density material that is easily mined in many places in the world. Its flaky shape is the most common morphology found in nature. Graphite is a bunch of tightly bonded hexagonal lattice network sheets of the carbon atom. These sheets of carbon atoms are weakly bonded with Vander wall force of attraction. It can also be said that graphite is a bunch of n-number of graphene sheets. Graphite layers are attached with Vander walls force of attraction and if somehow this bond breaks, these layers can separate and nano-size flake can be obtained (Roshini et al. 2015). Moreover, there is a large difference in the coefficient of thermal expansion (CTE) between aluminum and graphite, which makes graphite more suitable for reinforcement. This is the reason for choosing graphite particles as reinforcement in aluminum alloy.

There are some factors which are very important in the selection of any reinforcement particle.

1. **Size Factor:** A large number of ceramic particles are available in micron and nanosize. The properties of alloy or matrix may differ because of reinforcement particles and their size. The reinforcement size plays a dominant role in mechanical and microstructural properties. It has been observed that nanosize reinforcement strengthens the metal matrix where coarse particles may weaken the matrix. Moreover, it has been also noticed that nano reinforcement has a large amount of agglomeration, which can destroy mechanical properties (Rana et al. 2012).
2. **Cost Factor:** The production cost of any material composite in any industry should be cost-effective. This cost depends upon the following things (Rana et al. 2012):
    (a) Availability of particles: If the particle is easily available, the cost of materials will be less or else it will be costly.
    (b) Reinforcement size: The reinforcement particles are available in micron size and nanosize. Nanosize particles are costly as compared to micron size particles.

1.3 Challenges in the Fabrication of Al-Gr$_p$ MMCs

To fabricate particle reinforced composites, three roosts, liquid processing, semi-solid processing, and powder metallurgy are available. Among these techniques, liquid processing, with stir cast arrangement is the most popular and widely acceptable technique for large-scale production in industries. In this process, a stir is used to create a vortex in molten metal as particles are added slowly. While reinforcing graphite particles in aluminum melt, the following challenges were observed (Ramani et al. 1991, Pillai et al. 1995, Sharma et al. 1996, Yang Lan and Li 2004, Leng et al. 2008, Barekar et al. 2009, Yan et al. 2011, Liu Han and Li 2011, Macke et al. 2012, Wang et al. 2014, Suresh et al. 2014, Christy et al. 2015, Pai et al. 2015).

(a) Porosity: It is a basic casting issue, generally present in cast specimens but it immediately intensifies with particle reinforcement.

(b) Agglomeration of particles and non-uniform distribution of particles: Mainly two reasons cause agglomeration. One is surface energy in which micron sizes and nanosized particles stick together to overcome their surface energy. Another reason is while reinforcing particles into the melt, these particles prefer to attach gas molecules rather than melt surface which causes agglomeration. The particles segregate at some places on the matrix, so the distribution of particles is limited to some place.

(c) Particle wettability issue: The particles which are broken into small sizes have dangling bonds. These bonds or particles cover up their surface with the gaseous layer to overcome their surface energy. These particles while reinforcing are attached to the gaseous molecule and cause agglomeration and non-wettability in the matrix.

These are the basic issues that are subjected as casting the issue in all *ex-situ* particles reinforced composite.

1.4 Overview and Need for Ultrasonic Treatment (UT)

The basic issues related to castings such as porosity, agglomeration, and non-wettability of particles degrade the mechanical properties of composite material. But recently developed technique ultrasonic treatment (UT) of the composite melt can resolve all mentioned issues.

Ultrasonic treatment is capable to resolve all issues such as porosity, agglomeration, and non-wettability of particles generated during the casting process. Moreover, UT can break reinforced particles into finer sizes.

Li et al. (2008), who worked on pure Al and Al-Si composite, reported that the UT for 5 minutes is sufficient to remove dissolved gases or porosity in the composite. Similarly, Christy et al. (2015), Eskin et al. (1995), Su et al. (2012), and Nampoothiri et al. (2018) worked with different UT times on aluminum composites and reported reduction porosity in synthesized metal ingots or composites when compared to its untreated composite. Nampoothiri et al. (2016), who worked on Al-TiB_2 composite, reported grain refinement in UT composite. Recently Christy et al. (2015) worked on UT Al-Gr_p composite and reported uniform distribution of particles by braking agglomeration, reduction in porosity, and breakage of reinforced particles. In the last two decades, many research papers were published on UT-based different composites with various reinforcement. They all have common results like reduction in porosity, breakage of agglomerations, and some newfound improvement in wettability of particles (Ramani et al. 1991, Sri Harini et al. 2015, Kumar et al. 2020). But the purpose of lightening the weight of composite with acceptable strength was still missing in their works. So, the present study is based on addressing all basic issues related to composite casting and fulfilling the aim of lightweight high strength. Thus, our research group selected micro-graphite particles as reinforcement in Al-Cu is the most common alloy, which is used for the automobile and aerospace industries.

2. Experimental Detail

A batch of pure aluminum (99.85%) and pure copper (99.75%) metals were melted in a resistance furnace. This batch is collected in 200 g ingot. This ingot again melted in the same furnace with the stir casting setup where graphite particles were preheated at 200°C for two hours while being slowly injected. After mixing graphite with zircon coated blade in Al alloy, the UT was injected with the help of Ti-6Al-4V probe for 5 minutes. After UT this melt was immediately poured in a cast-iron mold preheated at 400°C for 4 hours. A similar process was followed for the casting of non-UT composite for comparison purposes.

From the synthesized specimen, $10 \times 10 \times 10$ mm size specimens were cut with the help of electric discharge machining (EDM). These specimens were polished with emery paper up to 2,000 grade and then cloth polishing has been done with silvo reagent for getting mirror finishing. The samples were observed in JEOL-JSM-6380A, Japan Scanning Electron Microscopy (SEM), JEOL, JSM-7610F Field Emission Scanning Electron Microscopy (FEG-SEM) and JEOL, and JEM 2100 (Japan) Transmission Electron Microscopy (TEM). The hardness test was conducted on Micro Vickers Hardness (Mitutoyo Hironodai, Japan) at a 500g load for 15 seconds dwell time. A minimum of 10 values were measured and reported.

3. Results and Discussion

Figure 1 (a and b) shows the SEM-EDS images of Al-Gr_p composite without UT. The agglomeration of graphite particles can be identified in Figure 1 (a) with the help of Figure 1 (b) where the red color shows the particle distribution. Porosity also can be identified with these images. It can be observed that the porosity is associated with the agglomerated particles. The morphology of $CuAl_2$ in the untreated composite is continuous and near the grain boundary. Figure 1 (c and d) shows the SEM-EDS images of UT composite. The uniform distribution of particles can be identified in Figure 1 (d) wherein the red color mapping of carbon elements shows the distribution of graphite. The observation and comparison of both images (c and d) confirm that graphite particles are situated near the grain boundary. The morphology of $CuAl_2$ also changed from continuous to globular due to UT.

Figure 2 (a) shows the FESEM micrograph of UT Al-Gr_p composite. The particles are embedded near the grain boundary. The measured size ranges between 300-1,000 nanometers (nm). It can also be observed that particles are fully wetted in the melt. Figure 2 (b) shows the TEM micrograph of composite with UT. In a micrograph, a flake of graphite particles can be observed; the thickness of these flakes ranges between 200-700 nm. Near the flakes, a large amount of dislocation pileup can be noticed. Moreover, the interface of the flake and matrix is fully attached, which confirms particles are fully wetted in the melt. Figure 3 shows the hardness result comparison plot in composite without UT and composite with UT. With the graph, it can be said that the hardness of the UT composite is uniform all over the surface.

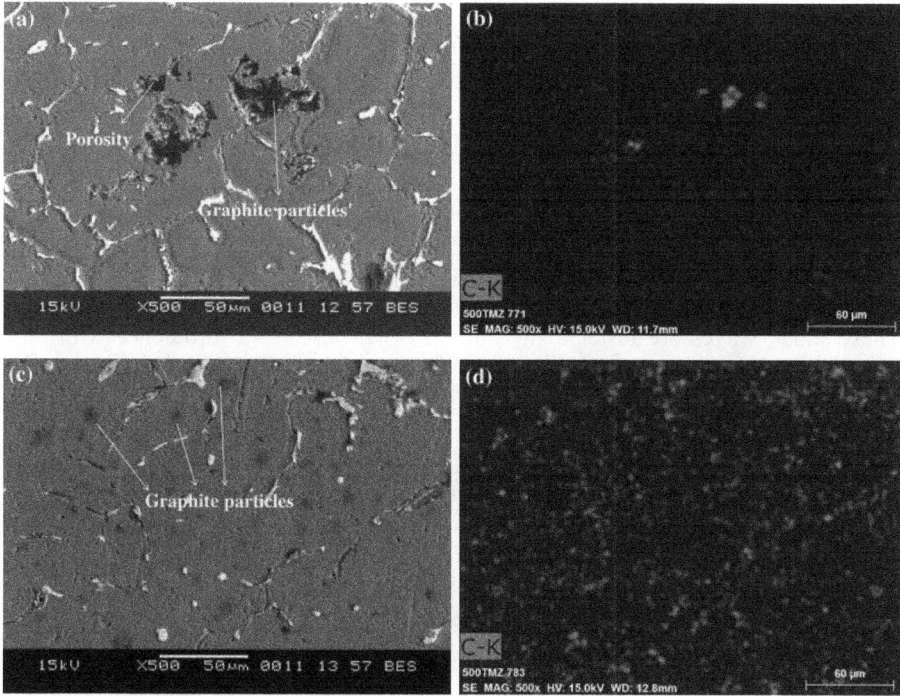

Figure 1: SEM images of (a) composites without UT, (b) EDS mapping of carbon element distribution for composites without UT, (c) composites with UT, and (d) EDS mapping of carbon element distribution for composites with UT.

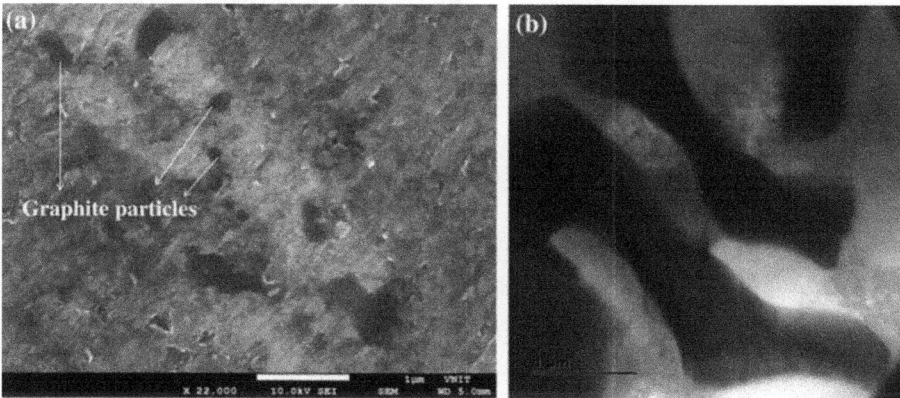

Figure 2: (a) FESEM micrograph of UT composite and (b) TEM micrograph of UT composite.

In the casting process, porosity is a major casting issue. While reinforcing particles, the atmospheric air also entraps in the melt. The result that porosity gets is an increase in the composite. To degas or remove porosity from composite in melt condition ultrasonic vibration is injected, which produces cavitation in the melt. The cavitation generates in the cavity or bubble present in the melt. During UT, two types of pressure were applied; one is the compression phase in which positive pressure is exerted on molecules which results in a pulling pressure applied on molecules; the other is the rarefaction phase in which negative pressure pushes a small amount of vapor into the cavity. When the amplitude of pressure waves becomes greater, the tensile stress melt and then these bubbles grow into an unstable state where these cavitation bubbles collapse. The whole process which includes the formation, growth and implosion of the cavitation bubble during UT is known as acoustic cavitation (Eskin et al. 1995, Roshini et al. 2015, Nampoothiri et al. 2016).

The implosion of the cavitation bubble generates shock waves into the melt. These shock waves are faster than 100 m/s, which creates a turbulence whirlpool (Suslick et al. 1999, Zhang et al. 2009). The shock wave has sufficient energy to break agglomerations and the whirlpool washes the particles. This resolve issues related to porosity, agglomerations, and non-wettability. The effect of the whirlpool can be reflected in terms of the uniform distribution of particles (Nampoothiri et al. 2015). Moreover, this effect can also be observed in the change in morphology of $CuAl_2$ from continuous to a globular form.

The implosion of cavitation bubbles takes place near weak locations, such as agglomerations and porosity (Eskin et al. 2015). The implosions have sufficient energy to break agglomerations and Vander walls force between two graphite layers. So, it can be said that their nanographite can be obtained in the aluminum alloy by reinforcing micron size particles. Uniform distribution of nanoparticles is the reason for getting uniform hardness at the surface of UT composite. Moreover, a

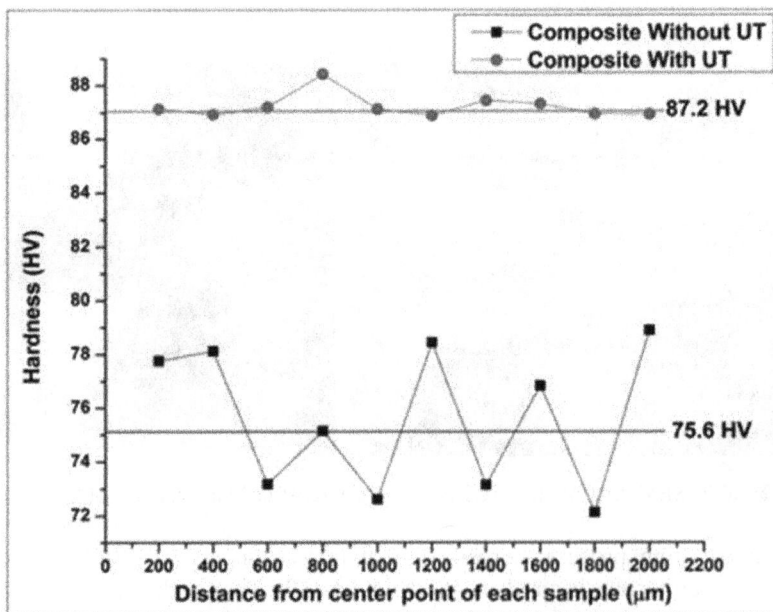

Figure 3: Micro hardness plot of composites.

large amount of CTE difference between Al and Gr_p also adds strength to the matrix. Large amounts of dislocation pile-up near the graphite in TEM support this. It has also been observed that the reinforcement of nanoparticles increases the mechanical properties of the composite. The density of graphite ranges between 1.65-2.23 g/cc and the density of aluminum alloy ranges between 2.68-2.82 g/cc. So with the rule of mixture, it can be said that the density of its composite will be less than its monolithic matrix density. That means weight reduction in any component can be gained with this composite.

3. Overall Discussion and Conclusion

The Al-Gr_p composite without UT and with UT have been fabricated successfully. Both composites were compared based on microstructural and mechanical properties. The following microstructural changes have been observed in the UT composite:

1. Agglomerations are broken and particles are uniformly distributed in UT composite wherein non-UT agglomeration was associated with porosity.
2. In UT composite, particles are broken from micron size to sub-micron size. This reduction in the size of particles is possible due to cavitation bubble formation during UT.
3. Morphology of $CuAl_2$ changed from continuous to globular after UT.
4. The particles are wetted in UT composite wherein in non-UT composite, these particles were associated with gas molecules.

The synthesized composite is a lightweight composite material that can be used for structural applications in transport industries. The composite also can be used in defense industries for lightweight applications. The current study is based on materials as-cast condition analysis. This material is heat-treatable alloys that can be explored more after processing. The material is synthesized with low-density reinforcement with excellent distribution of sub-micron particles where wettability is also good. The composite offers a new area of research for further studies.

Acknowledgment

The author would like to thank the Science and Engineering Research Board, Govt. of India for the financial support (Grant No: ECR/2017/001010).

References

Barekar, N., Tzamtzis, S., Dhindaw ,B.K., Patel, J., Hari Babu, N. and Fan, Z. 2009. Processing of aluminum-graphite particulate metal matrix composites by advanced shear technology. J. Mater. Eng. Perform. 18(9): 1230–1240. doi: 10.1007/s11665-009-9362-5.
Eskin, G.I., Makarov, G.S. and Pimenov, Y.P. 1995. Effect of ultrasonic processing of molten metal on structure formation and improvement of properties of high-strength Al-Zn-Mg-Cu-Zr alloys. Adv. Perform. Mater. 2: 43–50.

Eskin, G.I. and Eskin, D.G. 2015. *Ultrasonic Treatment of Light Alloys*. Second edition. CRC Press Taylor & Francis Group. doi: 10.15713/ins.mmj.3.

Gupta, R.K., Nampoothiri, J., Dhamodharan, S., Ravi, K.R., Udhayabanu, V. and Peshwe, D.R. 2020. Ultrasonic assisted synthesis of Al e Cu/2 vol % Gr p composite and its characterization. J. Alloys Compd. Elsevier B.V, 845: 156087. doi: 10.1016/j.jallcom.2020.156087.

Ibrahim, I.A., Mohamed, F.A. and Lavernia, E.J. 1991. Particulate reinforced metal matrix composites – a review. J. Mater. Sci. Technol. 26: 1137–1156.

Krishnan, B.P., Surappa, M.K. and Rohatgi, P.K. 1981. The UPAL process: a direct method of preparing cast aluminium alloy-graphite particle composites. J. Mater. Sci. 16(5): 1209–1216. doi: 10.1007/BF01033834.

Leng, J., Wu, G., Zhou, Q., Dou, Z. and Huang, X. 2008. Mechanical properties of SiC/Gr/Al composites fabricated by squeeze casting technology. Scr. Mater. 59(6): 619–622. doi: 10.1016/j.scriptamat.2008.05.018.

Li, J., Momono, T., Tayu, Y. and Fu, Y. 2008. Application of ultrasonic treating to degassing of metal ingots. Mater. Lett. 62(25): 4152–4154. doi: 10.1016/j.matlet.2008.06.016.

Liu, Z., Han, Q. and Li, J. 2011. Ultrasound assisted in situ technique for the synthesis of particulate reinforced aluminum matrix composites. Compos. Part B. Elsevier Ltd, 42(7): 2080–2084. doi: 10.1016/j.compositesb.2011.04.004.

Macke, A., Schultz, B.F. and Rohatgi, P. 2012. Metal matrix: composites offer the automotive industry an opportunity to reduce vehicle weight, improve performance. Adv. Mater. Process. 170(3): 19–23.

Mathew, J., Mandal, A., Kumar, S.D., Bajpai, S. and Chakraborty, M. 2017. Effect of semi-solid forging on microstructure and mechanical properties of in-situ cast Al-Cu-TiB$_2$ composites. J. Alloys Compd. Elsevier B.V, 712: 460–467. doi: 10.1016/j.jallcom.2017.04.113.

Nampoothiri, J., Raj, B. and Ravi, K.R. 2015. Effect of ultrasonic treatment on microstructure and mechanical property of in-situ Al/2TiB$_2$ particulate composites. Mater. Sci. Forum 463–466. doi: 10.4028/www.scientific.net/MSF.830-831.463.

Nampoothiri, J., Harini, R.S., Nayak, S.K., Raj, B. and Ravi, K.R. 2016. Post in-situ reaction ultrasonic treatment for generation of Al-4.4Cu/TiB$_2$ nanocomposite: a route to enhance the strength of metal matrix nanocomposites. J. Alloys Compd. Elsevier B.V, 683: 370–378. doi: 10.1016/j.jallcom.2016.05.067.

Nampoothiri, J., Balasundar, I., Raj, B., Murty, B.S. and Ravi, K.R. 2018. Porosity alleviation and mechanical property improvement of strontium modified A356 alloy by ultrasonic treatment. Mater. Sci. Eng. A. Elsevier Ltd, 724: 586–593. doi: 10.1016/j.msea.2018.03.069.

Pai, A., Shankar, S., Errol, R., Silva, D. and Nikhil, R.G. 2015. Effect of graphite and granite dust particulates as micro-fillers on tribological performance of Al 6061-T6 hybrid composites. Tribology Int. Elsevier, 92: 462–471. doi: 10.1016/j.triboint.2015.07.035.

Pillai, U.T.S., Pai, B.C., Satyanarayana, K.G. and Damodaran, A.D. 1995. Fracture behaviour of pressure die-cast aluminium-graphite composites. J. Mater. Sci. 30(6): 1455–1461. doi: 10.1007/BF00375248.

Polmear, L. 2006. Light Alloys from Traditional Alloys to Nanocrystals. Fourth Ed. Elsevier Butterworth-Heinemann Publications.

Pramod, S.L., Bakshi, S.R. and Murty, B.S. 2015. Aluminum-based cast in situ composites: a review. J. Mater. Eng. Perform. Springer US, 24(June): 2185–2207. doi: 10.1007/s11665-015-1424-2.

Ramani, V., Pillai, R.M., Pai, B.C. and Ramamohan, T.R. 1991. Factors affecting the stability of non-wetting dispersoid suspensions in metallic melts. Composites, 22(2): 143–150. doi: 10.1016/0010-4361(91)90673-5.

Rana, R., Purohit, R. and Das, S. 2012. Review of recent studies in Al matrix composites. Int. J. Sci. Eng. Res. 3(6): 1–16. Available at: http://www.ijser.org/researchpaper%5CReview-of-recent-Studies-in-Al-matrix-composites.pdf.

Roshini, P.C., Nagasivamuni, B., Raj, B. and Ravi, K.R. 2015. Ultrasonic-assisted synthesis of graphite-reinforced Al matrix nanocomposites. J. Mater. Eng. Perform. Springer US, 26: 4–9. doi: 10.1007/s11665-015-1491-4.

Sharma, S.C., Rao, G.S.K., Nagarajan, M., Girish, B.M. and Kamath, R. 1996. Mechanical property evaluation of aluminium-copper-graphite particulate composites. Mater. Sci. Forum. 217–222(Part 3): 1871–1876. doi: 10.4028/www.scientific.net/msf.217-222.1871.

Sri Harini, R., Nampoothiri, J., Nagasivamuni, B., Raj, B. and Ravi, K.R. 2015. Ultrasonic assisted grain refinement of Al-Mg alloy using in-situ MgAl$_2$O$_4$ particles. Mat. Lett. Elsevier, 145: 328–331. doi: 10.1016/j.matlet.2015.01.132.

Su, H., Gao, W., Feng, Z. and Lu, Z. 2012. Processing, microstructure and tensile properties of nano-sized Al$_2$O$_3$ particle reinforced aluminum matrix composites. Mater. Des. Elsevier Ltd, 36: 590–596. doi: 10.1016/j.matdes.2011.11.064.

Suresh, S., Shenbaga Vinayaga Moorthi, N., Vettivel, S.C., Selvakumar, N. and Jinu, G.R. 2014. Effect of graphite addition on mechanical behavior of Al6061/TiB$_2$ hybrid composite using acoustic emission. Mater. Sci. Eng. A. Elsevier, 612: 16–27. doi: 10.1016/j.msea.2014.06.024.

Suslick, K.S., Didenko, Y., Fang, M.M., Hyeon, T., Kolbeck, K.J., Namara Mc, W.B., Mdleleni, M.M. and Wong, M. 1999. Acoustic cavitation and its chemical consequences. Philos. Trans. R. Soc. A Math. Phys. Eng. Sci. 357(1751): 335–353. doi: 10.1098/rsta.1999.0330.

Wang, M., Chen, D., Chen, Z., Wu, Y., Wang, F., Ma, N. and Wang, H. 2014. Mechanical properties of in-situ TiB$_2$/A356 composites. Mater. Sci. Eng. A. 590: 246–254. doi: 10.1016/j.msea.2013.10.021.

Yan, J., Xu, Z., Shi, L., Ma, X. and Yang, S. 2011. Ultrasonic assisted fabrication of particle reinforced bonds joining aluminum metal matrix composites. Mater. Des. Elsevier Ltd, 32(1): 343–347. doi: 10.1016/j.matdes.2010.06.036.

Yang, Y., Lan, J. and Li, X. 2004. Study on bulk aluminum matrix nano-composite fabricated by ultrasonic dispersion of nano-sized SiC particles in molten aluminum alloy. Mater. Sci. Eng. A. 380(1): 378–383. doi: 10.1016/j.msea.2004.03.073.

Zhang, S., Zhao, Y., Cheng, X., Chen, G. and Dai, Q. 2009. High-energy ultrasonic field effects on the microstructure and mechanical behaviors of A356 alloy. J. Alloys Compd. 470(1–2): 168–172. doi: 10.1016/j.jallcom.2008.02.091.

Zhou, D., Qiu, F., Wang, H. and Jiang, Q. 2014. Manufacture of nano-sized particle-reinforced metal matrix composites: a review. Acta Metall. Sin. (English Letters), 27(5): 798–805. doi: 10.1007/s40195-014-0154-z.

Novel Applications of Graphene in the Aerospace Industry

Radhika Wazalwar* and Megha Sahu

Department of Materials Engineering, Indian Institute of Science, Bengaluru - 560012, India

1. Introduction

Nanomaterials are of significant interest to the research community due to their unique properties and countless applications. Graphene, a two-dimensional nanomaterial, first reported in 2004, gained tremendous attention, particularly in the field of composites. Bottom-up approaches, like chemical vapor deposition (CVD) (Muñoz and Aleixandre 2013), epitaxial growth (SiC substrate), and top-down approaches, like micromechanical exfoliation (Bhuyan et al. 2016) and liquid-phase exfoliation of graphite (Ciesielski and Samorì 2014), can be used to synthesize graphene. Chemical oxidation of graphite is a common method to obtain graphite oxide, which can further be exfoliated to obtain graphene oxide. Some of the common graphite oxidation routes are Brodie (1859), Staudenmaier (1898), and Hummer's method (Hummers and Offeman 1958). These routes require strong acids and oxidizing agents but guarantee easy and large-scale synthesis of graphene oxide. Graphene displays excellent properties such as Young's modulus of 1 TPa and tensile strength of 130 GPa (Lee et al. 2008). It has a theoretical surface area of 2,630 m^2 g^{-1}, exceptional optical properties, ultrahigh electronic mobility ($>$200,000 cm^2 V^{-1} s^{-1}) (Morozov et al. 2008), and high thermal conductivity ($>$5,000 W m^{-1} K^{-1}) (Balandin et al. 2008). These properties make graphene an attractive filler material for different types of composites.

Several review articles have summarized the status of composites reinforced with carbon-based nanomaterials, especially graphene. The mechanical, thermal, and electrical properties of such composites have been discussed at length in these reviews (Domun et al. 2015, Szeluga et al. 2020, Atif et al. 2016, Singh et al. 2019). The excellent strength of graphene and its high aspect ratio contribute immensely toward improved mechanical properties of aircraft-grade epoxy composites. This chapter will comprehensively address graphene applications in the aerospace

---

Corresponding author: radhikaw@iisc.ac.in

industry, particularly corrosion resistance, electromagnetic interference (EMI) shielding, adhesives, fire retardancy, and anti-icing (Figure 1). The coming sections will focus on the mechanism through which graphene contributes to enhancing the properties mentioned above. The advantages and limitations of graphene over conventional materials have also been discussed.

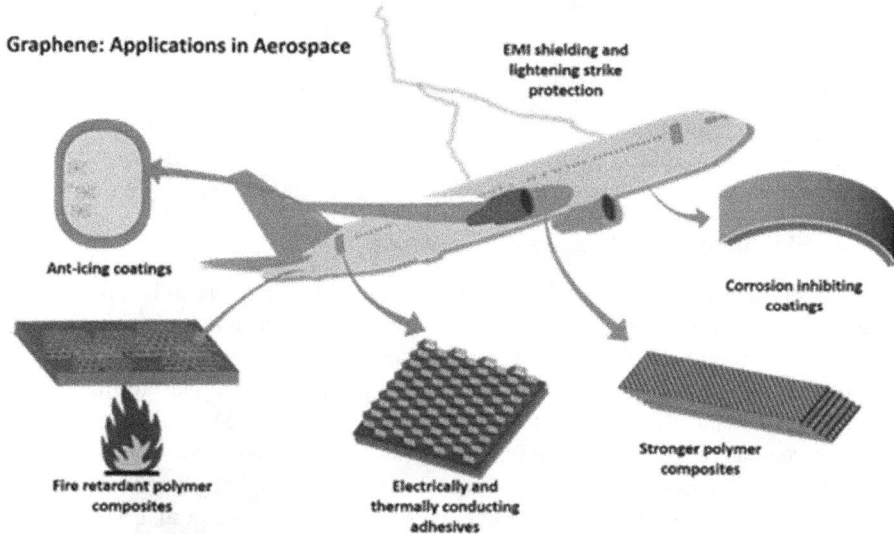

Figure 1: Applications of graphene in the aerospace industry.

2. Graphene Applications in the Aerospace Industry

2.1 EMI Shielding

Electromagnetic interference occurs when electromagnetic energy gets transmitted from one electronic device to another via induction, conduction, or electrostatic coupling. Such interference can manifest as static in radio and have more dangerous implications, like affecting the operation of an aircraft, medical devices, automobiles, etc. Particularly in the aerospace industry, EMI can lead to harmful consequences, such as loss of information, faulty communication between the control tower and pilot, interference with the radar signals, etc. EMI's common sources on an aircraft are devices, such as laptops and mobiles inside the plane, lightning, solar flare, radar, and control systems in the cockpit (Figure 2a). It is essential to shield the aircraft from disruptive electromagnetic interference to avoid failure of these controls and prevent the airplane from going off course. EMI shielding can either be done at the circuit board or by enclosing the electrical device into an enclosure to protect it from EMI. Conventionally, metal alloy enclosures were preferred due to their high electrical and thermal conductivity. Commonly used alloys are mu-metal, alloys of magnesium, nickel, aluminum, stainless steel, etc. The significant limitations with metal alloys are high density and have a tendency to rust. Plastic shields embedded with silver, copper, or aluminum foils were used to overcome these limitations.

Plastics are inert to corrosive environments and are lighter in weight. With the advent of high conductivity nanomaterials, such as graphene and carbon nanotubes, conducting polymer composites became a popular solution for EMI shielding. The choice of EMI shielding material depends on shielding thickness, flexibility, electrical conductivity, dielectric properties, permeability, corrosion resistance, EM wave absorption, absorption frequency bandwidth, thermal and mechanical properties, processability, cost, etc. These factors play a crucial role in determining the shielding effectiveness of the material.

EMI shielding effectiveness (SE) is measured in terms of decibels (dB). It is the ratio of the magnitude of the incident field to the magnitude of the transmitted field. An SE value below 10 dB is considered very low, and an SE of 40 dB represents 99% attenuation and is the minimum target value for most applications.

2.1.1 EMI Shielding Attributes of Graphene

Graphene is a low-density material capable of imparting high strength and corrosion resistance to its polymer composites. The critical property under consideration for EMI shielding applications is the electrical conductivity of the material. According to the Drude model, the electrical conductivity is proportional to the charge carrier density and mobility of the material (Liu et al. 2016). Graphene displays high charge carrier mobility ($250,000$ cm^2 V^{-1} s^{-1}) (Compton et al. 2011) but a low carrier density. To achieve high shielding effectiveness, graphene can be doped with electron acceptors or donors to improve its carrier density (Liu et al. 2016). Wan et al. (2017) compared the shielding effectiveness (SE) of large, medium, and small-sized graphene sheets doped with iodine. They reported that the iodine-doped large graphene sheets (over 10 μm in size) displayed an SE of 52.2 dB at 8.2 GHz for a thickness of 12.5 μm. The undoped sheets showed an SE of 47 dB, which was much lower than the doped sheets (Wan et al. 2017). Apart from the carrier charge density and mobility, the SE is also a strong function of complex permittivity (ε_r) and complex permeability (μ_r) responsible for the dielectric and magnetic losses, respectively (Colaneri and Schacklette 1992).

The high aspect ratio, high mechanical strength, electrical conductivity, surface modification, and graphene sheets' defects contribute toward the EMI shielding properties. The conductivity determines the conduction losses, while the surface functionalization and defects determine the polarization losses. Therefore, chemically modified graphene is often used for EMI shielding applications despite its low conductivity because it increases polarization losses (Wang et al. 2018, 2014). Figure 2 (b) shows the mechanisms through which EMI shielding occurs. Single-layer graphene displays a conductivity of 10^8 S/m and as the number of layers increases, the conductivity reduces (Pei et al. 2010, Zhao et al. 2011). Therefore, the SE increases with an increase in the state of exfoliation of graphene (Shen et al. 2014).

2.1.2 Composites for EMI Shielding

Several carbon nanostructures have been explored for their EMI shielding applications (Gupta et al. 2019, Wang et al. 2018). Recently, graphene polymer

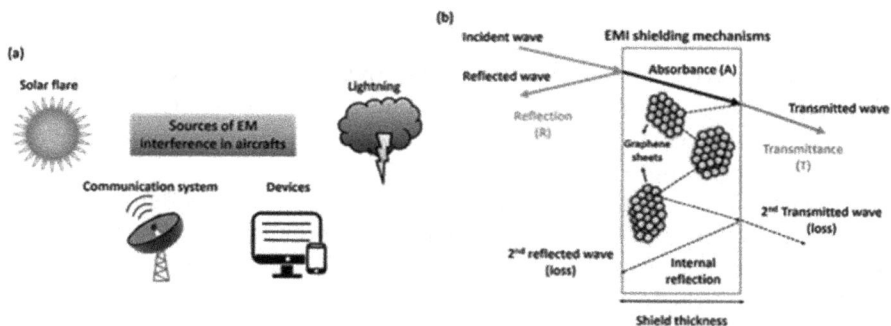

Figure 2: (a) Major sources of electromagnetic wave interference in aircraft, (b) EMI shielding mechanisms.

composite foams have grabbed the attention of scientists. Graphene reinforced foams of various polymers, such as polystyrene (Yan et al. 2012), PEI (Ling et al. 2013), PVDF (Eswaraiah et al. 2011), PDMS (Chen et al. 2013), PI (Li et al. 2015), PMMA (Zhang et al. 2011), etc., with varying graphene loads that have been synthesized. Such composites can provide EMI shielding effectiveness up to almost 30 dB for 0.8-2.5 mm shielding thickness. In the case of foams, the predominant mechanism of shielding is absorption rather than reflection because energy loss occurs through absorption induced by multiple reflections. The foams' porous structure enables the EM wave to penetrate easily and get attenuated as it passes through the interconnected walls of the foam. Porous materials typically display a lower surface conductivity than non-porous materials, in turn aiding the impedance matching at the air/surface interface. As a result of this, the reflection of the EM wave is reduced, and it can penetrate through the material and get absorbed (Gupta and Tai 2019). Shen et al. (2016) extensively worked on polyurethane (PU) graphene composites and have successfully synthesized them at low density (0.027-0.03 g/cc) PU/graphene foams. They proved that PU/graphene foams can provide a comprehensive EMI shielding performance via absorption-dominant mechanisms. The EM wave energy is dissipated as heat by either conductive dissipation or multiple reflections. It was also reported that on the application of compressive strain, the EMI shielding performance of the foam was reduced due to partial cracking of the porous composite network. Shen et al. reported that at 10 wt.% graphene loading in the PU foam, the EMI SE increased by 19.9 dB. As the sponge thickness increased, the total SE increased and finally reached 57.7 dB for a thickness of 6 cm (Shen et al. 2016).

Graphene-coated materials are also popular for EMI applications due to their lightweight, low density, high electrical conductivity, mechanical strength, and flexibility. Graphene paper was used by Zhang et al. (2015), and an EMI SE of 100 dB for a thickness of 100 μm was achieved. Some authors have reported hybrid nanoparticles such as reduced graphene oxide (rGO) coated Fe_3O_4@SiO_2@ polypyrrole for shielding applications. A film of thickness 0.27 mm displayed an EMI SE of 32 dB attributable to graphene's high electrical conductivity, magnetic losses due to Fe_3O_4 particles, and multiple reflections due to the foam-like morphology of the film (Yuan et al. 2018). Other examples of hybrid fillers used for EMI shielding were CNT/graphene (Song et al. 2017), Fe/Co graphene (Li et al. 2015), reduced GO/MoS$_2$ (Wang et al. 2015), etc. Absorption, reflection, and multiple reflection phenomena are driven by the electrical conductivity, shielding thickness, and skin effect, respectively. EMI SE of 47.7 dB was achieved between 9-12 GHz for coatings made of a multi-layered composite of graphene and wax. The shielding mechanism was dominated by absorption (Song et al. 2014).

Graphene/epoxy composites have shown a percolation threshold as low as 0.52 vol.%. EMI SE of 21 dB was reported in the X band for 15 wt. % (8.8 vol.%) loading. Therefore, such epoxy composites can be considered lightweight and practical EMI shielding materials (Liang et al. 2009). Marka et al. (2015) added layered graphene to PVA by using solution mixing and casting. The presence of defects and functional groups on graphene's surface caused polarization losses and contributed toward shielding (Marka et al. 2015).

The ability of graphene to form stable and conducting thin films opens up many opportunities in EMI shielding. Recently, Haydale (Ammanford, U.K.) has commercialized graphene used for EMI shielding and has launched graphene-modified composite for lightning strike protection. The use of graphene in real-life industrial applications is set to bring about a paradigm shift in aerospace material solutions.

2.1.3 Electrically Conducting Adhesives

The aerospace industry has demonstrated the reliability of carbon fiber reinforced plastics (CFRP) composites in structural parts. The commercial composite aircraft from the USA and Europe-based aerospace giants have completed more than a decade since their launch and have established the niche for CFRP structures in the aerospace industry. The use of adhesives in the fuselage, wings, and empennage components for EMI shielding is limited by the electrically insulating nature of most adhesives. To protect the aircraft from damage caused by lightning strikes, it is extremely important to provide an electrically conducting path for the grounding of lightning current. Electrically conducting adhesives (ECA) with silver nanoparticles, flakes, and nanowires are commercially available. However, these have high loading (>30 vol.%) of silver additives to make an electrically conducting percolating network, leading to increased cost. Therefore, there have been reports where the hybrid conductive filler complements the silver additive for improved conductivity. Graphene has superior electrical conductivity (10^6 S/cm) (Zhang et al. 2010) and has an enormous surface area (2,630 m^2 /g), which is much greater than CNT (Pumera 2009) and most other nanomaterials. Hence, very little graphene addition can make a percolating network with a very in the adhesive and is ideal for electrically conducting adhesives. Researchers have explored various forms of graphene for application in electrically conducting adhesives, such as N-doped graphene sheet (Pu et al. 2014, Ma et al. 2016), surfactant stabilized graphene (Amoli et al. 2015), silver graphene nanocomposite (Amoli et al. 2015, Zeng et al. 2016), etc. The summary of the results from various such studies is listed in Table 1. Though these studies show the potential of graphene for conducting adhesives, graphene's true potential considering its excellent electrical conductivity is yet to be harnessed. Graphene does have the potential to entirely replace the silver additives in ECA. However, to achieve that, dispersion and structural defects-related limitations have to be overcome in the coming years.

Table 1: Results from key research studies for improving the electrical conductivity of commercial electrical adhesives using graphene-based materials

Sr. No	Conducting Filler	Commercial ECA Silver Loading (wt.%)	Conducting Filler Loading (wt.%)	Resistivity Ω.cm/ Conductivity S/m	Reference
1	N doped graphene	30	1	4.4×10^{-2} Ω.cm	(Pu et al. 2014)
2	Hydrothermally synthesized graphene	69.5	0.5	5.0×10^{-5} Ω cm	(Ma et al. 2016)
3	SDS stabilized graphene	30	1.5	7.6×10^{-3} Ω.cm	(Amoli et al. 2015)
4	Silver nanoparticle decorate graphene	60		7.69×10^{-4} Ω.cm	(Amoli et al. 2015)
5	Silver graphene nanoparticle composite	69.8	0.2	8.76×10^{-5} Ω.cm	(Zeng et al. 2016)
6	Silver flakes and silver-graphene nanocomposite	65 (silver flakes:silver graphene NC = 20:80)		2.37×10^{-4} Ω.cm	(Peng et al. 2014)
7	Polypyrrole + graphene	0	Pyroll 5 Graphene 1	2.33×10^{-9} S/m	(Aradhana R. 2019)

2.2 De-Icing and Anti-Icing Systems

Ice formation on aircraft surfaces poses a big challenge for the regular operation of aircraft. The accumulation of ice during flight creates an uneven surface, resulting in altered aerodynamics. Hence, it affects the flow of air over wings and tail, which leads to a reduction in lift, aircraft controllability issues, and an increase in weight and drag. All these problems lead to increased fuel consumption. Removal of ice before the take-off of flight is also mandatory for the safe operation of aircraft. Currently used methods involve the use of a spray of de-icing and anti-icing fluids (Thomas et al. 1996, Parent and Ilinca 2011) bleed air method, ultrasound-based method (Wang 2017), microwaves-based method (Feher and Thumm 2011), and electro-thermal (Raji et al. 2016) de-icing methods. Among all the mentioned methods, the electro-thermal de-icing method is the most promising and power efficient. Commercial aircraft use the electro-thermal de-icing method (Sinnett 2007). In the electro-thermal de-icing technology, the spray metal mat is embedded in the composite structure, which acts as a conducting and electro-thermal element and enables local heating of the wing skins and other flight-critical surfaces (Figure 3).

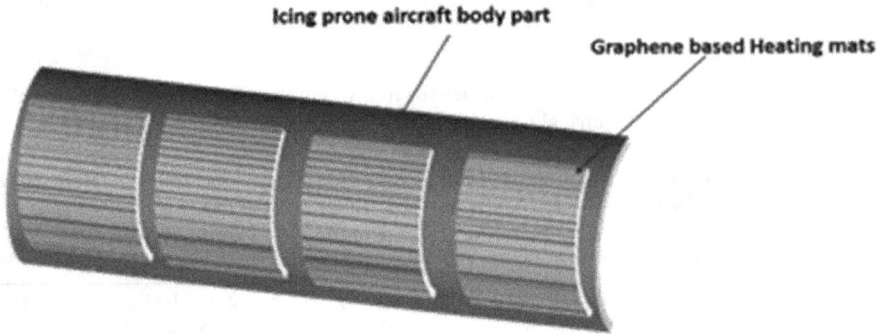

Figure 3: Heating mats containing graphene conductors for de-icing and anti-icing protections.

2.2.1 Graphene for De-Icing and Anti-Icing

Studies have reported metal, CNT, and conducting polymers as an electro-thermal element for polymer composite. However, these solutions consume a lot of energy; for example, a spray-on metal mat as an electro-thermal element for de-icing requires 140-200 kW of power supply for a commercial aircraft. The metal mat also comes with a weight penalty. Therefore, there is a need for a lightweight conductive material, which overcomes these limitations and provides exceptional electro-thermal properties without huge energy consumption.

Graphene, an sp^2 hybridized array of C-atoms in a honeycomb lattice, has a Π electron cloud over and above the 2-D plane containing the atoms, leading to excellent thermal (Geim 2009) and electrical properties (He et al. 2012). Graphene has a thermal conductivity of ~5,000 W mK^{-1}, which is significantly higher than any pure metal, metal alloy (Cu- 385.0 W mK^{-1}), CNTs (~3,000 W mK^{-1}) (Pop et al. 2006), and other 2-D materials. Researchers have reported graphene used for developing ink with extremely high conductivity, which can be used to develop the conducting materials by using spray-on technology. Graphene has been demonstrated at a lab-scale for de-icing and anti-icing applications in the form of graphene nanoribbons (Raji et al. 2016, Volman 2014), functionalized graphene nanoribbons (Volman 2014), and graphene nanoplatelets based ink (Figure 4). These graphene-based solutions involve forming a percolation network to complete the conducting path in the application region, which may require a high concentration of graphene filler (2.5-100 vol.%) (Raji et al. 2016, Wang 2016). This limitation can be overcome by using graphene foam (Bustillos 2018), where graphene is interconnected in all three dimensions forming a percolating network with embedded air pockets. This leads to a lightweight, interconnected network of graphene sheets without high filler loading required for the energy-efficient electro-thermal conductors for de-icing applications.

2.3 Fire Retardants

For aerospace applications, from a fire safety perspective, special attention is given to flight interiors. It is mandatory for the materials used for flight interiors, including composite panels on the wall, the storage bin's door, the fabric of chairs, and the

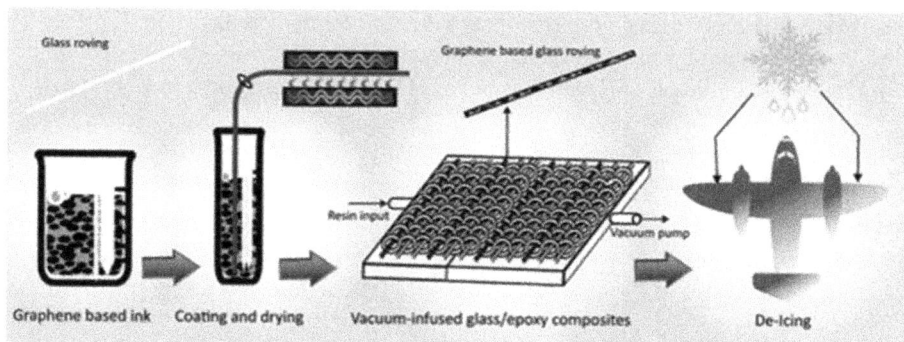

Figure 4: Graphene-based glass/epoxy composite manufacturing method for de-icing applications (Karim et al. 2018). (Reprinted with permission from ref 105, Copyright © 2018 RSC publication)

carpet on the floor to be flame retardant. The aerospace industry is moving toward polymers and polymer composites for their lighter weight, design flexibility, and low cost. There are concerns related to the fire safety and fire retardant ability of these polymer composites. The various fiber-reinforced polymers (FRPs) used in the aircraft interior, such as glass fiber reinforced phenolic composite or carbon fiber reinforced polymer (CFRPs), are organic and flammable. Toxic gases and smoke is released on the combustion of these polymers. Hence, conventionally, flame retardant (FRs) additives, such as organophosphorus compounds, metal hydroxides, nitrogen-containing FRs, and halogenated organic fire retardants, are added to improve the fire resistance. However, these additives are being reconsidered because of their environmental concerns. The fire retardants are of two types: reactive FRs or additive FRs. The reactive FRs are chemically bonded to polymer chains, whereas more widely used additive FRs are blended with polymers and not chemically bonded to polymer chains. The additive FRs pose a greater environmental challenge as the probability of the polymer leaching out during manufacturing, disposal, and recycling are high. These chemicals find their way in the air, water bodies, and affect employee health and the environment.

2.3.1 Graphene Composites as Fire-Retardant Systems

Graphene sheets have excellent barrier properties, and the same has been explored by researchers for flame retardant application for polymer composites. Graphene sheets act as insulating barriers, which inhibit mass loss during the combustion process. Another mechanism uses functionalized graphene sheet where functional group enables the formation of a continuous char layer, which separates oxygen from bulk materials and suppresses thermal degradation (Figure 5A). Graphene sheets have strong van der Walls interactions, which poses a challenge for the uniform distribution of these 2-D sheets in the polymer matrix. The uniform distribution is extremely important for it to act as a fire-retardant. To overcome this issue, researchers have demonstrated the use of functionalized graphene as a flame retardant additive for epoxy (Chen et al. 2017, Li et al. 2019) and polyurethane (Yao et al. 2020). The summary of results from key research studies for improving the fire safety of aerospace polymers using graphene-based flame retardant is listed in Table 2. To find the efficiency of a fire retardant UL94 vertical burning test is conducted, where limiting oxygen index is calculated as per ASTM D-2863 as shown in Figure 5B.

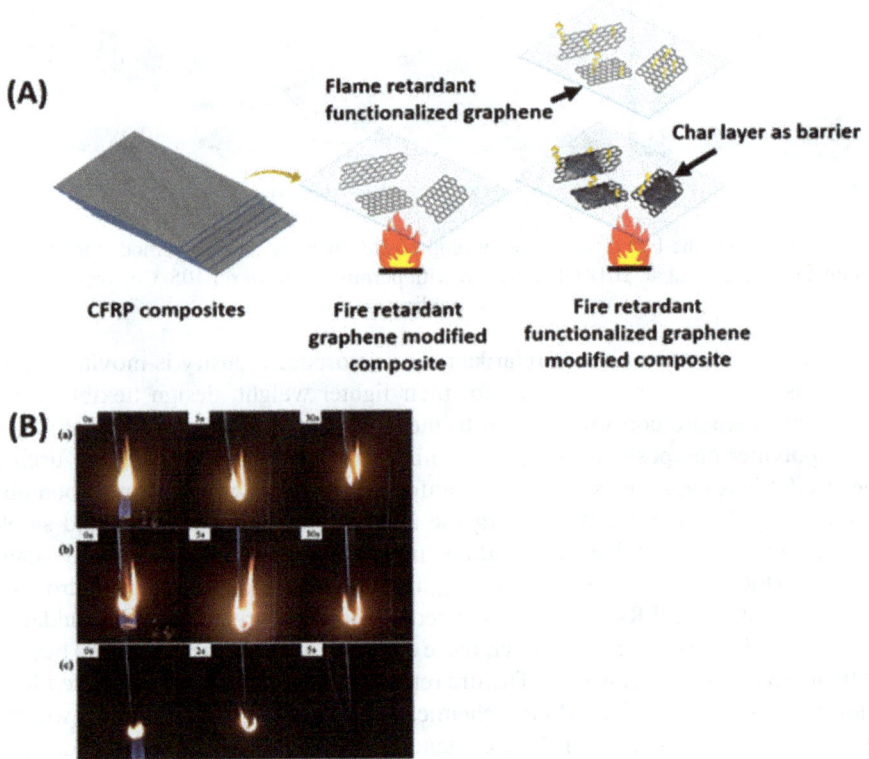

Figure 5: (A) Fire retardant mechanism of graphene and functionalized graphene derivatives and (B) Vertical flame test for long-chain phosphaphenanthrene containing graphene epoxy composite (Chen et al., 2017). (Reprinted with permission from ref 65, Copyright © 2017 Nature)

2.4 Corrosion Protection

According to IUPAC 2012, corrosion is defined as "an irreversible interfacial reaction of a material (metal, polymer or ceramic) with its environment, resulting in the consumption of the material or dissolution into the material of a component of the environment". Corrosion can occur in both wet and dry environments. Many factors cause corrosion and affect the rate of corrosion: (i) type of material and its surface and electrochemical properties, (ii) material design, (iii) the environment (acid, alkali, salts, moisture, gases, solvents, etc.) and (iv) the conditions (temperature, concentration, humidity, pollution, ventilation, etc.).

Corrosion can lead to premature in-service failures, which could be devastating, especially in aircraft. Hence, it is essential to take sufficient care at the design and manufacturing stage itself to prevent corrosion. Some of the conventional methods to

Table 2: Summary of results from key research studies for improving the fire safety of aerospace polymers using graphene-based flame retardant

Base Polymer	Filler	% of Flame Retardant	Method of Incorporation	LOI % Improvement	Reference
Epoxy	GO-f-POSS based copolymer	2% wt., 4%	Chloroform solvent dispersion	4% 7%	(Li et al. 2019)
Epoxy	Graphene-f-long-chain phosphaphenanthrene	2, 4 wt.%	Acetone solvent dispersion	3.8% 5.4%	(Chen et al. 2017)
Polyurethane foam	Graphene	0.25, 0.5, 0.75, 1 wt.%	One pot graphene-PU synthesis	5.8%	(Yao et al. 2020)
Polyurethane foam	Reduced GO	30 wt.%	Dip coating	4.6%	(Wu et al. 2021)

prevent corrosion are (i) protective coatings of inert metals and conductive polymers, (ii) alloying, (iii) design modifications and (iv) corrosion inhibitors, etc.

Out of these methods, the protective coating is a widely used method due to its easy scale-up and effectiveness. Conventional coatings are thick and tend to alter the substrate's properties and often cannot withstand high temperatures. In this regard, graphene and its various composites have played a crucial role in making lightweight corrosion-resistant coatings that do not dramatically alter the substrates' optical, thermal, and electrical properties but can halt charge transfer at the interface of metal and electrolyte (Kirkland et al. 2012). Graphene is considered inert to corrosive environments (Heer et al. 2007), provides high-temperature oxidation resistance in the absence of electrolyte (Chen et al. 2011), is impermeable to oxygen (Bunch et al. 2008), and has a high surface area and high adsorption capacity.

Corrosion is monitored using effective methods to test, such as salt spray test, humidity test, copper sulfate test, electrochemical corrosion test, weight loss methods, etc. The corrosion resistance of a material can also be measured quantitatively using electrochemical impedance spectroscopy (EIS).

2.4.1 Pure Graphene Coatings

For the first time in 2011, Chen et al. (2011) coated a pure Cu and Cu/Ni alloy with pristine graphene synthesized using CVD. Some standard methods to coat a surface with graphene are: (i) CVD, (ii) high-temperature pyrolysis of organic molecules, (iii) rapid thermal annealing, (iv) powder spray, (v) spray coating, (vi) dip coating, (vii) electrochemical methods, etc. Spray coating, dip coating, and electrochemical methods give poor quality and low coverage of the coated surface. For these reasons, CVD is a more preferred method. CVD is usually done on Cu or Nickel substrates. More recently, also on Rh, Au, Pt, Co, Pd, etc. But some commercial metals like Fe, Al, and Mg cannot withstand the high-temperature conditions during a CVD but need to be protected against corrosion. To increase the range of metals that can be coated using the CVD technique, Zhao et al. (2013) devised a method to transfer the CVD coatings to the target surface mechanically. The authors made a polymethyl methacrylate (PMMA)/graphene/Cu interlayer followed by etching out of the Cu layer. The PMMA/graphene film was deposited on Ag, the target metal, and the PMMA was dissolved using acetone. However, the residues from polymers such as PMMA tend to alter the coating properties and compromise its performance compared to a pure graphene coating (Li et al. 2013). Using a polymer-free mechanical transfer method (Lin et al. 2014, Wang et al. 2013) or annealing (Lin et al. 2012) of the coating has also been reported to overcome these difficulties. However, no one method is suitable for all coating applications. The coating's strength comes from the interaction of the coating with the substrate through physisorption or chemisorption. Stronger interactions are more efficient in inhibiting corrosion (Nine et al. 2015). Weatherup et al. (2015) studied the interaction of the graphene coating with the substrate (Figure 6). They reported that when the graphene coating interacts strongly with the substrate and the substrate can form a passivating oxide layer, the corrosion protection efficiency was maximum.

Pure graphene coatings offer enhanced protection from corrosive environments but have some limitations. Wrinkles, cracks, and defects in the pure graphene layer

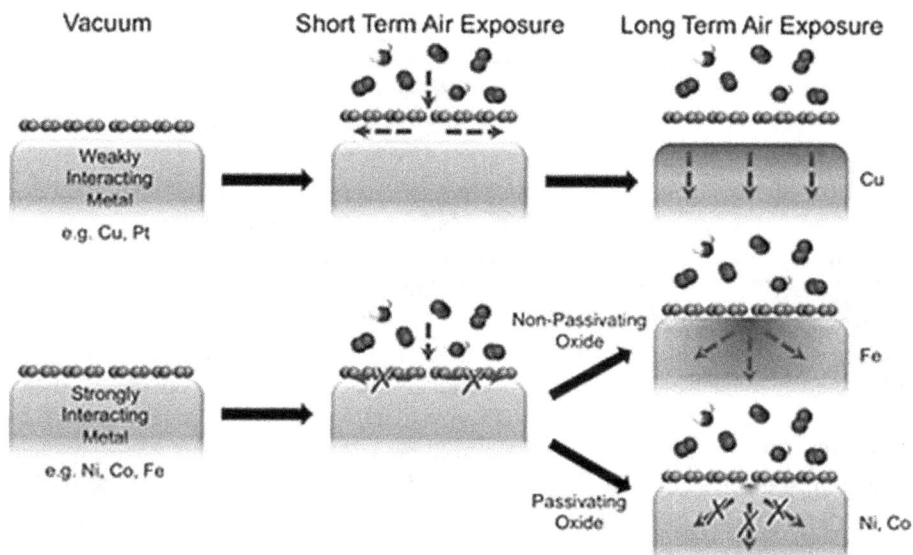

Figure 6: Passivating behaviour of different metals covered with graphene (Weatherup et al. 2015). Reproduced with permission from ref 78 (Copyright © 2015 ACS publication)

are sites at which corrosion can begin. Such defects are susceptible to oxidation and can cause micro-galvanic corrosion, leading to more rapid metal substrate degradation (Prasai et al. 2014). During the service, graphene coating can undergo mechanical damage that compromises the coating's integrity, allowing corrosive media to pass through the coating and cause damage to the metal substrate. Once the corrosion begins, the substrate's oxide layer can cause the graphene coating to peel off, further exposing the metal surface (Wu et al. 2019).

Researchers used various techniques to overcome the limitations mentioned above. The interaction of the graphene coating with the substrate was studied, CVD parameters were optimized (Anisur et al. 2018), and multilayer coatings were preferred over single-layer coatings (Yu et al. 2018). Some researchers have also used atomic layer deposition (Du et al. 2008) to passivate graphene defects selectively.

2.4.2 Graphene Composite Coatings

Pure graphene coatings are difficult to scale up to an industrial level and have various other limitations, including higher processing costs. Composite coatings offer an advantage over pure graphene coatings because they combine the anti-corrosion properties of graphene and the polymer's film-forming capacity. The conventional coating methods can be used to apply graphene composite coatings, and therefore they are easier to scale up. Some authors have reported composites containing silane-modified graphene oxide for corrosion resistance (Li et al. 2016, Parhizkar et al. 2018). Polydopamine functionalized graphene oxide was also reported in some cases as a filler material for composite coatings (Cui et al. 2018). Hybrid fillers, such as GO/CNT (Hu et al. 2019), TiO_2-GO (Liu et al. 2018), Fe_3O_4/GO (Zhan et al. 2018), silica-graphene (Sun et al. 2015), etc., improve the dispersion of the filler in the composite matrix, which contributes toward corrosion resistance. Apart from the state of dispersion, the graphene sheets' orientation to the substrate is also critical. Sheets that are oriented parallel to the substrate are more efficient in offering corrosion protection than the sheets that are oriented randomly or perpendicular to the substrate (Cui et al. 2016, Jiao et al. 2014).

Some authors have reported covalently and non-covalently modified graphene reinforced epoxy resin coatings for enhanced corrosion resistance (Ding et al. 2018, He et al. 2019). Zinc-rich epoxy coatings have been reported for corrosion inhibition through sacrificial anode reactions (Park and Shon 2015). To achieve sufficient electron flow, composites are loaded with >80% Zn, which seriously compromises the coating's toughness and strength. Researchers have established the importance of graphene to zinc-rich coatings to enhance the mechanical properties and the corrosion resistance of such coatings (Ding et al. 2018, Teng et al. 2018). Apart from epoxy, functionalized graphene oxide and hybrid graphene fillers have been added to polymers, such as polyurethane (Li et al. 2014), PVA (Li et al. 2018), polylactic acid (PLA) (Zhang et al. 2015), nylon (Jin et al. 2013), etc., to make corrosion protection coatings.

2.4.3 Mechanism of Corrosion Resistance in Graphene Coatings

The graphene coating obstructs the penetration of corrosive media. The honeycomb structure of graphene and its distribution in the polymer matrix provides a shielding effect by providing a tortuous infiltration path, thereby improving the corrosion resistance and coating life. According to the Neilsen model (Nielsen 1967), the permeation path's tortuosity is governed by the aspect ratio and volume fraction of the filler in the composite coating. Another way graphene contributes toward corrosion resistance is by improving the composite coating's mechanical strength and toughness. Cracks on the coating surface are an easy entry point for the corrosive media to contact the substrate. As these cracks grow, the corrosion protection will be compromised. Graphene's presence prevents the coating from developing such cracks under mechanical loads and prevents the corrosive media from coming in contact with the substrate. In the case of zinc-rich coatings, graphene's presence acts as the cathode for the anodic zinc powder. Hence, when the electrolyte encounters graphene and zinc powder together, it does not permeate further in search of a cathode and thereby protecting the substrate.

3. Limitations of Graphene

The previous sections highlighted the superiority of graphene over other carbon-based nanomaterials owing to its high strength, thermal and electrical conductivity, high aspect ratio, and ease of functionalization. These attributes undoubtedly make graphene a popular filler material for the fabrication of aircraft composites. However, some limitations, such as dispersion issues, lack of standardization, lack of consensus on health and environmental hazards, industrial scale-up issues, etc., still need to be addressed.

4. Conclusion

Research in the field of aircraft materials has provided sufficient evidence that graphene can significantly improve the performance of polymer composites. Graphene-loaded electrically conducting polymer composites are capable of serving the dual purpose of EMI shielding and lightning protection. Other crucial properties, such as corrosion resistance and flame retardant behavior of polymer composites, can also be enhanced by using graphene-based nano-fillers. The use of graphene in adhesives can decrease aircraft weight, reduce the number of joints and lower the threat of corrosion and fatigue. The aviation industry dictates stringent quality control guidelines and regulations because passenger safety is at stake. The coming decade is crucial in terms of the advancement in the commercialization of graphene-based aircraft materials. Researchers need to find innovative solutions to overcome graphene's limitations and demonstrate reliability to deem it holistically fit for aerospace industry use.

Note

Further permissions related to the material excerpted (Figures 4, 5B, and 6) should be directed to the authors and publication house of the original article.

References

Amoli, B.M., Trinidad, J., Rivers, G., Sy, S., Russo, P., Yu, A., Zhou, N.Y. and Zhao, B. 2015. SDS-stabilized graphene nanosheets for highly electrically conductive adhesives. Carbon 91: 188-199.

Amoli, B.M., Trinidad, J., Hu, A., Zhou, Y.N. and Zhao, B. 2015. Highly electrically conductive adhesives using silver nanoparticle (Ag NP)-decorated graphene: the effect of NPs sintering on the electrical conductivity improvement. J. Mater Sci-Mater El. 26(1): 590-600.

Anisur, M.R., Banerjee, P.C., Easton, C.D. and Singh, R.K. 2018. Controlling hydrogen environment and cooling during CVD graphene growth on nickel for improved corrosion resistance. Carbon 127: 131–140.

Aradhana, R., Mohanty, S. and Nayak, S.K. 2019. Synergistic effect of polypyrrole and reduced graphene oxide on mechanical, electrical and thermal properties of epoxy adhesives. Polymer 166: 215-228.

Atif, R., Shyha, I. and Inam, F. 2016. Mechanical, thermal, and electrical properties of graphene-epoxy nanocomposites—a review. Polymers 8: 281.

Balandin, A.A, Ghosh, S., Bao, W., Calizo, I., Teweldebrhan, D., Miao, F. and Lau, C.N. 2008. Superior thermal conductivity of single-layer graphene. Nano Lett. 8(3): 902-907.

Bhuyan, Md. S.A., Md. N., Uddin, Islam, Md. M., Bipasha, F.A. and Hossain, S.S. 2016. Synthesis of graphene. Int Nano Lett 6: 65-83.

Brodie, B.C. 1859. On the atomic weight of graphite. Philos. Trans. R. Soc. London 14: 249-259.

Bunch, J.S., Verbridge, S.S., Alden, J.S., van der Zande, A.M., Parpia, J.M., Craighead, H.G. and McEuen, P.L. 2008. Impermeable atomic membranes from graphene sheets. Nano Lett. 8(8): 2458.

Bustillos, J.Z. 2018. Three-dimensional graphene foam–polymer composite with superior deicing efficiency and strength. ACS Appl. Mater. & Interfaces 10(5): 5022-5029.

Chen, S., Brown, L., Levendorf, M., Cai, W., Ju, S.Y., Edgeworth, J., Li, X., Magnuson, C.W., Velamakanni, A., Piner, R.D. and Kang, J. 2011. Oxidation resistance of graphene-coated Cu and Cu/Ni alloy. ACS Nano. 5(2): 1321.

Chen, W., Liu, Y., Liu, P., Xu, C., Liu, Y. and Wang, Q. 2017. The preparation and application of a graphene-based hybrid flame retardant containing a long-chain phosphaphenanthrene. Sci. Rep. 7(1): 8759.

Chen, Z., Xu, C., Ma, C., Ren, W. and Cheng, H.M. 2013. Lightweight and flexible graphene foam composites for high-performance electromagnetic interference shielding. Adv. Mater. 25(9): 1296-1300.

Ciesielski, A. and Samorì, P. 2014. Graphene via sonication assisted liquid-phase exfoliation. Chem. Soc. Rev. 43: 381-398. DOI: 10.1039/C3CS60217F.

Colaneri, N.F. and Schacklette, L.W. 1992. EMI shielding measurements of conductive polymer blends. IEEE Transactions on Instrumentation and Measurement 41(2): 291-297.

Compton, O.C. 2011. Chemically active reduced graphene oxide with tunable C/O ratios. ACS Nano. Jun 28; 5(6): 4380-4391.

Cui, M., Ren, S., Zhao, H., Xue, Q. and Wang, L. 2018. Polydopamine coated graphene oxide for anticorrosive reinforcement of water-borne epoxy coating. Chem. Eng. J. 335: 255-266.

Cui, Y., Kundalwal, S.I. and Kumar, S. 2016. Gas barrier performance of graphene/polymer nanocomposites. Carbon 98: 313-333.

De Heer, W.A., Berger, C., Wu, X., First, P.N., Conrad, E.H., Li, X., Li, T., Sprinkle, M., Hass, J., Sadowski, M.L. and Potemski, M. 2007. Epitaxial graphene. Solid State Commun. 143(1–2): 92.

Ding, J., Rahman, O., Peng, W., Dou, H. and Yu, H. 2018. A novel hydroxyl epoxy phosphate monomer enhancing the anticorrosive performance of waterborne graphene/epoxy coatings. Appl. Surf. Sci. 427: 981-991.

Ding, R., Zheng, Y., Yu, H., Li, W., Wang, X. and Gui, T. 2018. Study of water permeation dynamics and anti-corrosion mechanism of graphene/zinc coatings. J. Alloy. Compd. 748: 481-495.

Domun, N., Hadavinia, H., Zhang, T., Sainsbury, T., Liaghat, G.H. and Vahid, S. 2015. Improving the fracture toughness and the strength of epoxy using nanomaterials – a review of the current status. Nanoscale 7: 10294. DOI: 10.1039/c5nr01354b.

Du, X., Skachko, I., Barker, A. and Andrei, E.Y. 2008. Approaching ballistic transport in suspended graphene. Nat Nanotechnol. 3(8): 491-495.

Eswaraiah, V., Sankaranarayanan, V. and Ramaprabhu, S. 2011. Functionalized graphene–PVDF foam composites for EMI shielding. Macromol. Mater. Eng. 296(10): 894-898.

Feher, L. and Thumm, M. 2011. High-frequency microwave anti-/de-icing system for carbon-reinforced airfoil structures. Intense Microwave Pulses VIII: International Society for Optics and Photonics 99-110.

Geim, A.K. 2009. Graphene: status and prospects. Science 324(5934): 1530-1534.

Gupta, S. and Tai, N.H. 2019. Carbon materials and their composites for electromagnetic interference shielding effectiveness in X-band. Carbon 152: 159-187.

He, Y., Chen, C., Xiao, G., Zhong, F., Wu, Y. and He, Z. 2019. Improved corrosion protection of waterborne epoxy/graphene coating by combining non-covalent and covalent bonds. React. Funct. Polym. 137: 104-115.

He, Q., Wu, S., Yin, Z. and Zhang, H. 2012. Graphene-based electronic sensors. Chem. Sci. 3(6): 1764-1772.

Hsieh, Y.P., Hofmann, M., Chang, K.W., Jhu, J.G., Li, Y.Y., Chen, K.Y., Yang, C.C., Chang,

W.S. and Chen, L.C. 2014. Complete corrosion inhibition through graphene defect passivation. ACS Nano 28; 8(1): 443-448. doi: 10.1021/nn404756q.

Hu, H., He, Y., Long, Z. and Zhan, Y. 2017. Synergistic effect of functional carbon nanotubes and graphene oxide on the anti-corrosion performance of epoxy coating. Polym. Adv. Technol. 28: 754-762.

Hummers, W.S. and Offeman, Jr. R.E. 1958. Preparation of graphitic oxide. J. Am. Chem. Soc. 80(6): 1339-1339 https://doi.org/10.1021/ja01539a017.

Jiao, W., Shioya, M., Wang, R., Yang, F., Hao, L., Niu, Y., Liu, W., Zheng, L., Yuan, F., Wan, L. and He, X. 2014. Improving the gas barrier properties of Fe_3O_4/graphite nanoplatelet reinforced nanocomposites by a low magnetic field induced alignment. Compos. Sci. Technol. 99: 124-130.

Jin, J., Rafiq, R., Gill, Y.Q. and Song, M. 2013. Preparation and characterization of high performance of graphene/nylon nanocomposites. Eur. Polym. J. 49: 2617-2626.

Karatas, E., Gul, O., Karsli, N.G. and Yilmaz, T. 2019. Synergetic effect of graphene nanoplatelet, carbon fiber and coupling agent addition on the tribological, mechanical and thermal properties of polyamide 6,6 composites. Composites Part B 163: 730-739.

Karim, N., Zhang, M., Afroj, S., Koncherry, V., Potluri, P. and Novoselov, K.S. 2018. Graphene-based surface heater for de-icing applications. RSC Advances 8(30): 16815-16823.

Kirkland, N.T., Schiller, T., Medhekar, N. and Birbilis, N. 2012. Exploring graphene as a corrosion protection barrier. Corros. Sci. 56: 1-4.

Lee, C., Wei, X., Kysar, J.W. and Hone, J. 2008. Measurement of the elastic properties and intrinsic strength of monolayer graphene. Science 321(5887): 385-388 doi: 10.1126/science.1157996.

Li, J., Cui, J., Yang, J., Ma, Y., Qiu, H. and Yang, J. 2016. Silanized graphene oxide reinforced organofunctional silane composite coatings for corrosion protection. Prog. Org. Coat. 99: 443-451.

Li, M., Zhang, H., Wu, W., Li, M., Xu, Y., Chen, G. and Dai, L. 2019. A novel POSS-based copolymer functionalized graphene: an effective flame retardant for reducing the flammability of epoxy resin. Polymers 11(2): 241.

Li, X., Feng, J., Du, Y., Bai, J., Fan, H., Zhang, H., Peng, Y. and Li, F. 2015. One-pot synthesis of $CoFe_2O_4$/graphene oxide hybrids and their conversion into FeCo/graphene hybrids for lightweight and highly efficient microwave absorber. J. Mater. Chem. A 3(10): 5535-5546.

Li, Y., Yang, Z., Qiu, H., Dai, Y., Zheng, Q., Li, J. and Yang, J. 2014. Self-aligned graphene as anticorrosive barrier in waterborne polyurethane composite coatings. J. Mater.Chem. A 2: 14139-14145.

Li, Z., Wang, Y., Kozbial, A., Shenoy, G., Zhou, F., McGinley, R., Ireland, P., Morganstein, B., Kunkel, A., Surwade, S.P. and Li, L. 2013. Effect of airborne contaminants on the wettability of supported graphene and graphite. Nat Mater. 12(10): 925-931.

Li, X., Bandyopadhyay, P., Guo, M., Kim, N.H. and Lee, J.H. 2018. Enhanced gas barrier and anticorrosion performance of boric acid induced cross-linked poly(vinyl alcohol co-ethylene)/graphene oxide film. Carbon 133: 150-161.

Li, Y., Pei, X., Shen, B., Zhai, W., Zhang, L. and Zheng, W. 2015. Polyimide/graphene composite foam sheets with ultrahigh thermostability for electromagnetic interference shielding. RSC Adv. 5: 24342-24351.

Liang, J., Wang, Y., Huang, Y., Ma, Y., Liu, Z., Cai, J., Zhang, C., Gao, H. and Chen, Y. 2009. Electromagnetic interference shielding of graphene/epoxy composites. Carbon 47(3): 922-925.

Lin, W.H., Chen, T.H., Chang, J.K., Taur, J.I., Lo, Y.Y., Lee, W.L., Chang, C.S., Su, W.B. and Wu, C.I. 2014. A direct and polymer-free method for transferring graphene grown by chemical vapor deposition to any substrate. ACS Nano 25; 8(2): 1784-1791.

Lin, Y.C., Lu, C.C., Yeh, C.H., Jin, C., Suenaga, K. and Chiu, P.W. 2012. Graphene annealing: how clean can it be? Nano Lett. 11; 12(1): 414-419. doi: 10.1021/nl203733r.

Ling, J., Zhai, W., Feng, W., Shen, B., Zhang, J. and Zheng, W. 2013. Facile preparation of lightweight microcellular polyetherimide/graphene composite foams for electromagnetic interference shielding. ACS Appl Mater Interfaces. Apr 10; 5(7): 2677-2684.

Liu, J., Yu, Q., Yu, M., Li, S., Zhao, K., Xue, B. and Zu, H. 2018. Silane modification of titanium dioxide-decorated graphene oxide nanocomposite for enhancing anticorrosion performance of epoxy coatings on AA-2024. J. Alloy. Compd. 744: 728-739.

Liu, Y., Xu, Z., Zhan, J., Li, P. and Gao, C. 2016. Superb electrically conductive graphene fibers via doping strategy. Adv Mater. 28(36): 7941-7947.

Ma, H., Ma, M., Zeng, J., Guo, X. and Ma, Y. 2016. Hydrothermal synthesis of graphene nanosheets and its application in electrically conductive adhesives. Materials Lett. 178: 181-184.

Marka, S.K., Sindam, B., James, Raju K.C. and Srikanth, V.V.S.S. 2015. Flexible few-layered graphene/poly vinyl alcohol composite sheets: synthesis, characterization and EMI shielding in X-band through the absorption mechanism. RSC Adv. 5(46): 36498-36506.

Morozov, S.V., Novoselov, K.S., Katsnelson, M.I., Schedin, F., Elias, D.C., Jaszczak, J.A. and Geim, A.K. 2008. Giant intrinsic carrier mobilities in graphene and its bilayer. Phys. Rev. Lett. 100: 016602.

Muñoz, R. and Aleixandre, C.G. 2013. Review of CVD synthesis of graphene. Chemical Vapor Deposition 191: 297-322 https://doi.org/10.1002/cvde.201300051.

Nielsen, L.E. 1967. Models for the permeability of filled polymer systems. J. Macromol. Sci.: Part A -Chem. 1: 929-942.

Nine, Md J., Cole, M.A., Tran, D.N.H. and Losic, D. 2015. Graphene: a multipurpose material for protective coatings. J. Mater. Chem. A. 3: 12580.

Parent, O. and Ilinca, A. 2011. Anti-icing and de-icing techniques for wind turbines: critical review. 6. Cold. Reg. Sci. Technol. 65(1): 88-89.

Parhizkar, N., Ramezanzadeh, B. and Shahrabi, T. 2018. Corrosion protection and adhesion properties of the epoxy coating applied on the steel substrate pre-treated by a sol-gel based silane coating filled with amino and isocyanate silane functionalized graphene oxide nanosheets. Appl. Surf. Sci. 439: 45-59.

Park, S.M. and Shon, M.Y. 2015. Effects of multi-walled carbon nano tubes on corrosion protection of zinc rich epoxy resin coating. J. Ind. Eng. Chem. 21: 1258-1264.

Pei, S., Zhao, J., Du, J., Ren, W. and Cheng, H. 2010. Direct reduction of graphene oxide films into highly conductive and flexible graphene films by hydrohalic acids. Carbon, 48: 4466-4474.

Peng, X., Tan, F., Wang, W., Qiu, X., Sun, F., Qiao, X. and Chen, J. 2014. Conductivity improvement of silver flakes filled electrical conductive adhesives via introducing silver–graphene nanocomposites. J. Mater. Sci.: Mater. Electron.25(3): 1149-1155.

Pop, E., Mann, D., Wang, Q., Goodson, K. and Dai, H. 2006. Thermal conductance of an individual single-wall carbon nanotube above room temperature. Nano Lett. 6(1): 96-100.

Prasai, D., Tuberquia, J.C., Harl, R.R., Jennings, G.K. and Bolotin, K.I. 2012. Graphene: corrosion-inhibiting coating. ACS Nano 28; 6(2): 1102-1108. doi: 10.1021/nn203507y.

Pu, N.W., Peng, Y.Y., Wang, P.C., Chen, C.Y., Shi, J.N., Liu, Y.M., Ger, M.D. and Chang, C.L. 2014. Application of nitrogen-doped graphene nanosheets in electrically conductive adhesives. Carbon 67: 449-456.

Pumera, M. 2009. Electrochemistry of graphene: new horizons for sensing and energy storage. Chem. Rec. 9(4): 211-223.

Raji, A.R.O., Varadhachary, T., Nan, K., Wang, T., Lin, J., Ji, Y., Genorio, B., Zhu, Y., Kittrell, C. and Tour, J.M. 2016. Composites of graphene nanoribbon stacks and epoxy for joule heating and deicing of surfaces. ACS Appl. Mater. & Interfaces 8(5): 3551-3556.

Shen, B., Zhai, W. and Zheng, W. 2014. Ultrathin flexible graphene film: an excellent thermal conducting material with efficient EMI shielding. Adv. Funct. Mater. 24(28): 4542-4548.

Shen, B., Li, Y., Zhai, W. and Zheng, W. 2016. Compressible graphene-coated polymer foams with ultralow density for adjustable electromagnetic interference (EMI) shielding. ACS Appl Mater Inter. 8(12): 8050-8057.

Singh, N.P., Gupta, V.K. and Singh, A.P. 2019. Graphene and carbon nanotube reinforced epoxy nanocomposites: a review. Polymer 180: 121724 .

Sinnett, M. 2007. Electro-thermal ice protection system for the B-787. Aircr. Eng. Aerosp. Technol. 79(6).

Song, Q., Ye, F., Yin, X., Li, W., Li, H., Liu, Y., Li, K., Xie, K., Li, X., Fu, Q. and Cheng, L. 2017. Carbon nanotube-multilayered graphene edge plane core shell hybrid foams for ultrahigh-performance electromagnetic-interference shielding. Adv. Mater. 29(31): 1701583.

Song, W.L., Cao, M.S., Lu, M.M., Bi, S., Wang, C.Y., Liu, J., Yuan, J., Fan and L.Z. 2014. Flexible graphene/polymer composite films in sandwich structures for effective electromagnetic interference shielding. Carbon 66: 67-76.

Staudenmaier, L. 1898. Verfahren zur darstellung der graphitsäure. Ber. Deut. Chem. Ges. 31: 1481.

Sun, W., Wang, L., Wu, T., Pan, Y. and Liu, G. 2015. Inhibited corrosion-promotion activity of graphene encapsulated in nanosized silicon oxide. J. Mater. Chem. A 3: 16843-16848.

Szeluga, U., Pusz, S., Kumanek, B., Olszowska, K., Kobyliukh, A. and Trzebicka, B. 2020. Effect of graphene filler structure on electrical, thermal, mechanical, and fire retardant properties of epoxy-graphene nanocomposites – a review. Crit. Rev. Solid State 46: 152-187.

Teng, S., Gao, Y., Cao, F., Kong, D., Zheng, X., Ma, X. and Zhi, L. 2018. Zinc-reduced graphene oxide for enhanced corrosion protection of zinc-rich epoxy coatings. Prog. Org. Coat. 123: 185-189.

Thomas, S.K., Cassoni, R.P. and MacArthur, C.D. 1996. Aircraft anti-icing and de-icing techniques and modeling. J. Aircraft 33(5): 841-854.

Volman, V.T.-R. 2014. Conductive graphene nanoribbon (GNR) thin film as anti-icing/de-icing heater. 2014 20th International Conference on Microwaves, Radar and Wireless Communications (MIKON): IEEE, p. 1-6.

Wan, Y.J., Zhu, P.L., Yu, S.H., Sun, R., Wong, C.P. and Liao, W.H. 2017. Graphene paper for exceptional EMI shielding performance using large-sized graphene oxide sheets and doping strategy. Carbon 122: 74-81.

Wang, T., Zheng, Y., Raji, A.R.O., Li, Y., Sikkema, W.K. and Tour, J.M. 2016. Passive anti-icing and active deicing films. ACS Appl. Mater. Interfaces 8(22): 14169-14173.

Wang, Y., Chen, D., Yin, X., Xu, P., Wu, F. and He, M. 2015. Hybrid of MoS$_2$ and reduced graphene oxide: a lightweight and broadband electromagnetic wave absorber. ACS Appl. Mater. Interfaces 7(47): 26226-26234.

Wang, C., Murugadoss, V., Kong, J., He, Z., Mai, X., Shao, Q., Chen, Y., Guo, L., Liu, C., Angaiah, S. and Guo, Z. 2018. Overview of carbon nanostructures and nanocomposites for electromagnetic wave shielding. Carbon 140: 696-733.

Wang, D.Y., Huang, I.S., Ho, P.H., Li, S.S., Yeh, Y.C., Wang, D.W., Chen, W.L., Lee, Y.Y., Chang, Y.M., Chen, C.C. and Liang, C.T. 2013. Clean-lifting transfer of large-area residual-free graphene films. Adv Mater. 27; 25(32): 4521-4526. doi: 10.1002/adma.201301152.

Wang, Z. 2017. Recent progress on ultrasonic de-icing technique used for wind power generation, high-voltage transmission line and aircraft. Energy Build. 140: 42-49.

Weatherup, R.S., D'Arsié, L., Cabrero-Vilatela, A., Caneva, S., Blume, R., Robertson, J., Schloegl, R. and Hofmann, S. 2015. Long-term passivation of strongly interacting metals with single-layer graphene. J. Am. Chem. Soc. 18; 137(45): 14358-14366.

Wen, B., Wang, X.X., Cao, W.Q., Shi, H.L., Lu, M.M., Wang, G., Jin, H.B., Wang, W.Z., Yuan, J. and Cao, M.S. 2014. Reduced graphene oxides: the thinnest and most lightweight materials with highly efficient microwave attenuation performances of the carbon world. Nanoscale 6(11): 5754-5761.

Wu, Q., Liu, C., Tang, L., Yan, Y., Qiu, H., Pei, Y., Sailor, M.J. and Wu, L. 2021. Stable electrically conductive, highly flame-retardant foam composites generated from reduced graphene oxide and silicone resin coatings. Soft Matter. 1: 68-82.

Wu, Y., Zhu, X., Zhao, W., Wang, Y., Wang, C. and Xue, Q. 2019. Corrosion mechanism of graphene coating with different defect levels. J. Alloy Comp. 777: 135-144.

Yan, D.X., Ren, P.G., Pang, H., Fu, Q., Yang, M.B. and Li, Z.M. 2012. Efficient electromagnetic interference shielding of lightweight graphene/polystyrene composite. J. Mater. Chem. 22: 18772-18774.

Yao, Y., Jin, S., Ma, X., Yu, R., Zou, H., Wang, H., Lv, X. and Shu, Q. 2020. Graphene-containing flexible polyurethane porous composites with improved electromagnetic shielding and flame retardancy. Compo. Sci. Technol. 200: 108457.

Yu, F., Camilli, L., Wang, T., Mackenzie, D.M., Curioni, M., Akid, R. and Bøggild, P. 2018. Complete long-term corrosion protection with chemical vapor deposited graphene. Carbon 132: 78-84.

Yuan, Y., Yin, W. and Yang, M. 2018. Lightweight, flexible and strong core-shell non-woven fabrics covered by reduced graphene oxide for high-performance electromagnetic interference shielding. Carbon 130: 59-68.

Zeng, J., Tian, X., Song, J., Wei, Z., Harrington, S., Yao, Y., Ma, L. and Ma, Y. 2016. Green synthesis of AgNPs/reduced graphene oxide nanocomposites and effect on the electrical performance of electrically conductive adhesives. J. Mater. Sci.: Mater. Electron. 27(4): 3540-3548.

Zhan, Y., Zhang, J., Wan, X., Long, Z., He, S. and He, Y. 2018. Epoxy composites coating with Fe_3O_4 decorated graphene oxide: modified bio-inspired surface chemistry, synergistic effect and improved anti-corrosion performance. Appl. Surf. Sci. 436: 756-767.

Zhang, L.L., Zhou, R. and Zhao, X. 2010. Graphene-based materials as supercapacitor electrodes. J. Mater. Chem. 20(29): 5983-5992.

Zhang, L., Alvarez, N.T. and Zhang, M. 2015. Preparation and characterization of graphene paper for electromagnetic interference shielding. Carbon 82: 353-359.

Zhang, H.B., Yan, Q., Zheng, W.G., He, Z. and Yu, Z.Z. 2011. Tough graphene-polymer microcellular foams for electromagnetic interference shielding. ACS Appl. Mater. Interfaces. 3(3): 918-924.

Zhang, L., Li, Y., Wang, H., Qiao, Y., Chen, J. and Cao, S. 2015. Strong and ductile poly (lactic acid) nanocomposite films reinforced with alkylated graphene nanosheets. Chem. Eng. J. 264: 538-546.

Zhao, Y., Xie, Y., Hui, Y.Y., Tang, L., Jie, W., Jiang Y., Xu, L., Lau, S.P. and Chai, Y. 2013. Highly impermeable and transparent graphene as an ultra-thin protection barrier for Ag thin films. J. Mater. Chem. C. 1: 4956-4961.

Zhao, X., Hayner, C.M., Kung, M.C. and Kung, H.H. 2011. Flexible holey graphene paper electrodes with enhanced rate capability for energy storage applications. ACS Nano 22; 5(11): 8739-8749.

A Review of the Tribological and Thermal Effectiveness of Graphene-Based Nano-Lubricants

T.P. Kulkarni[1], B.G. Toksha[2]*, Prashant Gupta[2], S.E. Shirsath[3] and A.T. Autee[1]

[1] Department of Mechanical Engineering, Maharashtra Institute of Technology, Aurangabad, India
[2] Centre for Advanced Materials Research and Technology (M-CAMRT), Maharashtra Institute of Technology, Aurangabad, India
[3] University of New South Wales, Sydney, Australia

1. Introduction

The twenty-first century economic, trade, productivity and development growth has led to the greater need for advanced machining from manufacturing to the transportation sector. There are huge demands and expectations of precision manufacturing and miniaturization in terms of higher productivity at lower energy consumption. The remedies to address this situation for improving energy efficiency and mechanical durability are the incorporation of newer, more energy-effective technologies and the installation of upgraded machinery. However, the replacement of existing plants and machinery with newer ones is a constraint due to involved economic implications. One of the main considerations in this regard is friction and wear, which induce enormous power consumption and curtailed operating life. Though wear and heat losses are inevitable, they can be reduced to acceptable levels. Therefore, the most essential part is to reduce friction and make the mechanical parts wear-resistant. The machinery with low friction and longer life will lead to energy savings, reduction in hazardous emissions and will address the environmental concerns (Guo and Zhang 2016). Lubrication is the process employed to diminish the friction and wear between two sliding surfaces, which are machined close and moved relative to one another by introducing a substance in between the sliding surfaces. There are multifold roles of lubrication, like providing a sliding film between moving parts, cooling by heat transfer, sealing of uneven surfaces by fillers, cleaning, dampening, cushioning of machine parts under high stress, protection from oxidation and corrosion, vibration

Corresponding author: mittoksha@gmail.com

and noise reduction, dirt and contaminant removal and leakage prevention. The key properties that the lubricant must possess are high boiling point, adequate viscosity, appropriate pour points, lower freezing point, demulsibility, higher thermal stability and oxidation and corrosion resistance (Shafi and Charoo 2020).

Though the unprecedented situation of Covid-19 has severely affected the consumption of lubricants in the industrial and automotive sector, it is projected that the lubricants market is going to grow at a CAGR of 3.6% and reach USD 182.6 billion by 2027 (2020). The nano-lubricants are lightweight lubricants with more durability in operating conditions. These nano-lubricants offer excellent thermal transport properties to address the environmental problems in improved ways as compared to conventional lubricants along with better load-carrying capabilities. There are numerous applications wherein the nano-lubricants are being used extensively, like the extreme tribological environments involved in military combat devices (Ramsden 2012), aerospace (Sayuti et al. 2013), automotive (Hemmat Esfe et al. 2017), thermal power plants (Chandrabhan et al. 2017), space components (Fan and Wang 2015) and industrial applications (Kalita et al. 2012).

Graphene is a one-layered carbon atom structured in a honeycomb lattice with sp2 hybridization. The graphene material exhibits the properties, such as high chemical inertness, liquids and gases impermeability, ultra-thin stable state, higher aspect ratio, enormously high mechanical strength, tensile stress and high thermal conductivity. These properties, as compared to other additives like WS_2, MoS_2, copper, boron, zirconium and platinum, manifest a few of the important features required in an additive of a nano-lubricant. These properties lead to wear protection, excess heat removal and prevention from the corrosion and oxidation tendencies of a substrate. The properties of graphene that could be employed in the field of coolants and lubricants with enhanced heat transfer and tribological applications are summarized in Table 1. There is a huge scope for researchers to create understanding about the inclusion of graphene in view of suspension stability, performances and toxicity concerns. It is sincerely believed that the insights contained in the present book chapter will be of value for scholars in the field, policymakers and manufacturing leaders about the effectiveness of graphene additives-based nano-lubricants.

2. Thermal Behavior of Graphene-Based Nano-Lubricants

The role of graphene as a nano-filler is to improve the thermal characteristics of lubricants with modifications in viscosity, viscosity index and pour point of the base liquids. These characteristics affect the working conditions under the influence of ambient temperature leading to undesirable oxidation and an increase in wear. The sedimentation and movement of nanoparticles are controlled by graphene concentration, which also affects the thermal conductivity. The occurrence of sedimentation and agglomeration in nano-lubricant can reduce the thermal conductivity and have adverse effects on the friction coefficient. The hydrophobic nature of graphene refrains it from dispersion in polar solvents and forms agglomerates in water-based suspensions while improving the viscosity of ionic

Table 1: Important thermal and mechanical properties of graphene for the tribological and thermal nano-lubricants applications

Thermal/ Mechanical	Properties	Values
Thermal	Thermal conductivity	~ 5,000 W/m-K (Balandin et al. 2008)
	Melting point	~ 4,125 K (Josphat Phiri et al. 2017)
	Thermal resistance (interface)	4,108 Km^2/W (rGr–SiO$_2$) (Josphat Phiri et al. 2017)
Mechanical	Young's modulus	~ 1 TPa (Josphat Phiri et al. 2017)
	Tensile strength	~ 130 GPa (Josphat Phiri et al. 2017)
	Fracture toughness	~ 4 MPam$^{0.5}$ (Josphat Phiri et al. 2017)
	C-C bond length	0.142 nm (Josphat Phiri et al. 2017)
	Graphene sheets Inter-planar spacing	0.335 nm (Josphat Phiri et al. 2017)
	Sheet resistance	1.310^4-5.110^4 Ω^2 (Nair et al. 2008)
	Optical transmittance	97.7% (Nair et al. 2008)
	Bulk density	0.2-0.4 g/cm^3 (Safira et al. 2020)
	Relative gravity	2.0-2.25 g/cm^3 (Safira et al. 2020)
	Specific surface area	300 m^2/g (Chae et al. 2004)
	Thickness	Less than 2 nm (Safira et al. 2020)
	Density	2.2 g/cm^2 (Safira et al. 2020)

liquids (Abdul Khaliq Rasheed et al. 2016, Sanes et al. 2017). It is reported that dispersion stability is improved and result in high thermal stability with graphene oxide as an additive in Polyalphaolefin 4 (PAO4) oil (Bao et al. 2019). The thermal conductivity and stability in polyester-based graphene nano-lubricants were reported by Aws S. Al-Janabi et al. (2020) wherein in this study, a stable nano-lubricant along with the considerable increment in thermal conductivity of 26% was observed. It was also reported that the addition of nanoparticles increased the viscosity of the lubricant by 7% for a sample of 0.3 vol.% along with sedimentation observed after 14 days. Mbambo et al. reported thermal conductivity enhancements in ethylene glycol-based nano-fluids by dispersing Ag-decorated 2-D graphene and reported the highest thermal conductivity of 32% (Mbambo et al. 2020). The study of thermal conductivity of nanofluids synthesized by ultrasound irradiation with graphene nano-platelets by Lee and Rhee (2014) claimed higher thermal conductivities over theoretical values and attributed this observation with two-dimensional structure and high specific area offered by graphene. The graphene nanoparticles spread over metal to metal contact because of their sheet-like structure, which increases surface area and effectively transmits more heat (Zhang et al. 2017).

The improvement in thermal conductivity and viscosity with the addition of graphene nano-flakes as compared to lubricant fluid is reviewed by Rasheed et al. (2016). Literature reports in this direction indicate an uneven effect on the stability of graphene-based suspensions with the surfactants, including hexadecyltrimethyl ammonium bromide (CTAB), octadecylamine, dicyclohexyl carbodiimide gum Arabic and sodium dodecyl sulfate (SDS) have (Uddin et al. 2013, Wu et al. 2021). The water-based nanofluid was found to be suitable in three criteria of highest stability, lower viscosity and enhanced thermal conductivity (Sarsam et al. 2016). The lab-scale procedures of dispersing graphene in base lubricants involve the addition of surfactants and centrifugation to get the well-dispersed samples that are difficult to reproduce at a large scale because of the wastage of graphene involved (Texter 2014). The use of surfactants might not be a feasible solution before addressing the concerns of manufacturers and professional bodies regarding viscosity standards.

The role of the concentration of dispersed graphene and the temperature response is a critical parameter to study nano-lubricant usability. There is ample literature present to conclude that the greater fraction of graphene and graphene oxide results in increased thermal conductivity (Yu et al. 2009, Baby and Ramaprabhu 2010, Yu et al. 2010, Pawan K Singh and V Manoj Siva 2011). Esfahani et al. reported that the particle size and viscosity lead to significant improvement in thermal conductivity at an optimal concentration (Esfahani et al. 2016). Sajid and Ali (2018) reviewed and reported conclusively the positive effect of nanoparticles concentration on the thermal response of nano-fluids (Sajid and Ali 2018). The significant role of the parameters such as graphene nano-sheet size and thermal resistance at the sheet interface has a significant impact on the improvement of thermal conductivity in graphene-based nano-lubricants. The reason for this kind of behavior is understood on the basis of various approaches. A micro-convection model accentuating the role of increased specific surface area and random movement depicting Brownian motion was discussed by Patel et al. (2005). Another report on similar lines has claimed Brownian random movement of nanoparticles as a key mechanism along with the concentration and size of the nanoparticles as key parameters, controlling the thermal behavior of nano-fluids (Jang and Choi 2004).

The transfer of heat generated via a fluid medium is a significant concern in numerous engineering applications, including refrigerators, heat exchangers, automobiles and power plants (Suneetha and Reddy 2016). The optimization of energy consumption and reduction in the processing time of a system is the demands of cooling/heating systems with ultra-high performance. It is an essential feature in the heat transfer sector directly affecting the refrigeration industry. The modified lubricants with nano-graphene dispersed in base oil with a wide range of doping concentrations from 0.001% to 60% have been patented for their better waste heat removal, stability and viscosity (Zhamu et al. 2011). The role of R600a being an eco-friendly refrigerant in a vapor compression system was studied for improvement in the pull downtime, lesser power consumption and better cooling capacities. It was concluded that the graphene-based nano-lubricants could be used in the vapor compression system as a promising replacement for pure mineral oil (Babarinde et al. 2020). There are studies about domestic refrigerators using graphene-based nano-lubricants with improvement in working parameters, such as lower cabinet temperature, compressor power consumption, power per ton of refrigeration, pull-down time, shell compressor temperature eliminating safety concerns and significant improvement in the performance (Adelekan et al. 2019, Lou et al. 2015). In all, for the state-of-the-art commercial use of graphene-based nano-lubricants for the improved thermal conductivity, better heat transfer and optimal viscosity in various applications need more studies to be conducted considering the stability of graphene material in various base oils with varying concentrations over a longer period and under varying operating temperatures.

3. Tribological Behavior of Graphene-Based Nano-Lubricants

The various factors involved in the tribological behavior of graphene as a lubricant additive in the base oil for improving the friction and wear properties are discussed in this section. Table 2 represents the selected studies in chronological order about the use of graphene as an additive in base lubricant. The parameters such as coefficient of friction (CoF) and the wear scar diameter (WSD) are of the utmost importance for analyzing the tribological behavior of lubrication (Sahoo 2005). The laboratory test employed, size of nanoparticles along with their concentration in base oil with prominent findings provided in the comment section is presented in Table 2.

Paul et al. (2019) explored the tribological properties of nano-lubricant (dodecylamine functionalized graphene–DFG/5w-30). At a fixed applied load as the concentration of the dispersed graphene in the base oil increased, CoF decreased which was attributed to the formation of the lubricating film. However, with increased applied load and graphene concentration the margin of reduction of CoF decreased. The mechanism involved here may be with an increase in load, the tribo-film is disturbed as nano-sheets are being pushed out of the wear track. The optical microscope images of the wear track indicated the increase in the width of the scars with the increase in the applied load (Paul et al. 2019). Another reason for increased CoF is the sliding of nanoparticles instead of rolling. At low contact, pressure rolling of nanoparticles is the main action mechanism for friction and wear reduction, whereas at higher pressure sliding of nanoparticles on the contact surface is expected to take place instead of rolling (Tevet et al. 2011).

Zin et al. (2016) studied Poly-Alkylene Glycol (PAG) lubricant used in compressors operating with CO_2 refrigerant. Friction and wear properties of nano-lubricant (carbon nano-horns-CNHs/ PAG) were investigated in the regime of boundary lubrication and under severe plastic dominance wear conditions, a significant reduction in CoF by 18% at 25°C and 7% at 70°C for lubricant containing 0.1 wt.% of CNHs. The cone-shaped morphology of carbon nano-horns (CNH) is suitable for rolling/sliding mechanisms, thereby reducing the CoF. The decreased CoF in the working of compressor could correspond to increased efficiency and reduced power requirement, and increased anti-wear properties possibly will ensure extended durability of devices (Zin et al. 2016). Kumar et al. (2020) studied the used engine oil (UEO) with graphene nanoparticles for their tribological properties (Graphene/SAE 15W40). The UEO was collected from heavy earth-moving machines after the usage of 250 hours employed in the coal mine. It has been reported that 0.5 wt.% sample has minimum average CoF and wear as compared to other nano-lubricant concentrations and drain oil. The mechanism for improvement in the lubricating properties of drain oil is the presence of nanoparticles between the layers of lubricant, which makes the relative motion between the layers easier and results in decreased friction and wear.

Table 2: Summary of experimental results on tribology of graphene

Sr. No.	Reference	Base Fluid	Material	Size of Nanoparticles (nm)	Concentration	Unit of Concentration	Type of Study	Remarks
1	Eswaraiah et al. (2011)	Engine oil	Ultrathin graphene (UG)	Thickness - 2 nm	0.0125, 0.025	mg/mL	Four-ball test	With graphene 0.025 mg/mL in engine oil, decrease in CoF and WSD by 80% and 33%, respectively.
2	Zhang et al. (2011)	PAO 9	Graphene sheets	-	0.02-0.06	wt.%	Four-ball test	Reduction in friction coefficient and WSD by 17% and 14%, respectively.
3	Wang et al. (2011)	SN 350	Modified graphene platelets (MGP)	Average diameter-1.2 μm, thickness-10-15 nm	0.075	wt.%	Four-ball test	At a concentration of 0.075 wt.% improvement in the wear resistance and load-carrying capacity of the lubricating oil.
4	Senatore et al. (2013)	SN 150	Graphene oxide nanosheets	-	0.1	wt.%	Ball-on-disc	The average CoF reduced by 20% in comparison with the base lubricant at the set parameters value 1. Average contact pressure of 1.17 GPa. 2. Temperature in the range 25-80°C.

5	Kinoshita et al. (2014)	Water-based lubricant	Graphene oxide (GO) monolayer sheets	-	1	wt.%	Reciprocating sliding test	The CoF of the GO dispersion = 0.05. WC ball and flat plate exhibited no wear for 60,000 friction cycles.
6	Zhang et al. (2014)	IL	Graphene oxide sheets	-	0.075	mg/mL	Ball-on-disk	Minimum friction coefficient of 0.057 with addition of GO sheets as compared to 0.14 for pure IL at load of 30N.
7	Fan et al. (2014)	PAO 40	Multilayer graphene (MLG)	-	0.1	wt.%	Reciprocation friction test	Bentone grease with 0.1 wt.% MLG with significant reduction of approximately 10.4 % in CoF.
8	Azman et al. (2016)	PAO 10 + TMP	Graphene nanoplatelets (GNPs)	Thickness - 5 nm	0.05	wt.%	Four-ball test	Reduction in friction = 5% Reduction in wear = 15% for 0.05 wt.% GNP in the base lubricant.
9	Dou et al. (2016)	PAO 4	Crumpled graphene balls	Diameter - 500 µm	0.01- 0.1	wt.%	Pin-on-disk test	For crumpled graphene balls 1. Reduction in CoF = 20% 2. Reduction in wear = 85%.

(Contd.)

Table 2: *(Contd.)*

Sr. No.	Reference	Base Fluid	Material	Size of Nanoparticles (nm)	Concentration	Unit of Concentration	Type of Study	Remarks
10	Zin et al. (2016)	PAG	Carbon nanohorns (CNHs)	Length - 30-50 nm Diameter - 2-5 nm	0.04, 0.1, 0.2, 0.5 and 1	wt.%	Ball-on-disk	Significant reduction in CoF = 18% at 25°C CoF = 7% at 70°C for lubricant containing 0.1% wt of CNHs
11	Rashmi et al. (2017)	TMP	Nano graphene platelets (NGPs)	Average diameter - 12 nm	0.01 to 0.1	wt.%	Four-ball test	Maximum reduction of 7% in the CoF for NGP-Palm oil TMP Ester, and WSD reduction of 16.2% in the at 0.05 wt.% NGP at 80 kg load.
12	Ali et al. (2018)	A5 (5W-30)	Graphene (Gr)	Thickness - 3–10 nm	0.03, 0.2, 0.4 and 0.6	wt.%	Reciprocating tribometer test	Improvement in anti-friction and anti-wear properties in the range of 29-35% and 22-29%, respectively.
13	Río et al. (2018)	TMPTO	Graphene nanoplatelets (GnP).	Thickness - 11-15 nm	0.05, 0.10, 0.25 and 0.5	wt.%	Ball on plate reciprocating test	The CoF was lower for all the nanolubricants compared to pure TMPTO. For the nanolubricant with a 0.25 wt.% in GnP = minimum wear width.

	Author	Base fluid	Nanomaterial	Size/Thickness	Concentration	Unit	Test	Remarks
14	Avilés et al. (2019)	Water + Dsu	Graphene (G)	Thickness 0.6–3.8 nm	0.1, 0.05	wt.%	Pin-on-disk test	Addition of a lower graphene concentration of 0.05 wt.% achieves the lowest initial CoF.
15	Shit et al. (2019)	5w-30	Dodecylamine -f-graphene (DAG)	-	0.01, 0.05, 0.1	wt.%	Ball-on-plate test	The reduction in COF was ~19, ~26 and ~40% for 0.01, 0.05 and 0.1 wt.% concentration of DAG dispersed in base oil.
16	Avilés et al. (2020)	IL	Graphene (G)	Thickness - 0.55-3.74 nm	0.5	wt.%	Pin-on-disk test	Ionic liquid with 0.5 wt.% of graphene could reduce the CoF by 40% for thin film lubrication
17	Kumar et al. (2020)	SAE 15W40	Graphene (G)	Thickness – 2-5 nm	0.2, 0.3, 0.4, 0.5 and 0.9	wt.%	Pin-on-disk test	The WSD and CoF reduced by 48.47% and 73.77%, respectively, with 0.5 wt.% graphene based used oil.

However, when the concentration of graphene increased above 0.5 wt. %, it resulted in the increase of CoF and wear. The wear scar on the pin post tribological test with EDX confirmed that there was an increase in carbon and oxygen content. This increase was related to the agglomeration of graphene on the pin surface (Kumar et al. 2020). As the graphene concentration increases, lubricant could not absorb graphene physically or chemically and thus reducing suspension stability (Liñeira del Río et al. 2018).

Extreme pressure properties of lubricants are important for the mating surfaces exposed to very high temperatures and pressures specifically present in automotive vehicles as gear elements and industrial machines. The characterization of lubricant's 'extreme pressure' properties is carried out by a four-ball test in which the maximum non-seizure load reflects the strength of lubrication film between metal to metal surface contacts (Yu et al. 2017). Extreme pressure properties of nano-lubricant (ultrathin graphene/engine oil) were investigated by Eswaraiah et al. (2011). It was reported that the non-seizure load for pure engine oil was 550N. As the graphene concentration in the engine oil increased, load-carrying capacity increased initially and then decreased with an increase in graphene concentration. The maximum load-carrying capacity nearly doubled to a value of 935 N for an optimal concentration of 0.025 mg/mL of graphene. Elastic deformation of the graphene at higher load and nano-bearing behavior of graphene layers between moving surfaces can be attributed to this increased load-carrying capacity of the nano-lubricants. It is interesting to note that the excessive addition of graphene (more than 0.025 mg/mL) in base engine oil decreased the load-carrying capacity, which may be due to the agglomeration of graphene on the surface (Eswaraiah et al. 2011). In another study, Lin et al. reported for nano-lubricant (Modified graphene platelets – MGP/SN 350) with the addition of 0.075 wt. % MGP to the base oil and load-carrying capacity was increased from 418.5 N to 627.2 N wherein a similar decline in the load-carrying capacity with an increase in MGP higher than 0.075 wt. % was observed (Lin et al. 2011).

Agglomeration of graphene nanoparticles is a major concern as the concentration of these nanoparticles deviates from its optimum value leads to agglomeration, which reduces anti-friction and anti-wear properties. It is very difficult to maintain uniformity for such a precise optimum nanoparticles concentration in working conditions. One solution could be the use of surfactants; however, there is a scarcity of systematic studies on the effectiveness of surfactants to improve the wear properties. Dou et al. (2016) reported the use of crumpled, paper-ball-like graphene, which exhibits the self-dispersing behavior. This study reported better anti-friction and anti-wear performance of crumpled graphene balls as compared to other instances of carbon, i.e. graphite, reduced graphene oxide and carbon black, and additionally these properties are not much dependent on their concentration in oil (Dou et al. 2016).

4. Toxicity Concerns

Even though the use of nanomaterials has revolutionized the world of composites with lubricants being one of the applications benefiting from their use, the environmental impact of the same should also be considered. They have the potential to inadvertently pollute the environment along with the people involved in handling, production, logistics, consumption and the discarding of lubricants have made the developed countries join hands with international organizations, such as OECDWPMN, ISO TC229, REACH, RoHS, etc. (Park and Yeo 2016). As the research on these aspects in graphene-based nano-lubricants is limited, environmental and health hazards concerning this nanomaterial are crucial in decision-making toward the absorption of such disruptive technologies. There have been frameworks, such as EU-FP7 future nano needs (Stone 2018) which respond to the regulatory needs of future markets pertaining to the use of nanomaterials. The crucial overall exposure situations across the value chain and service life of currently researched nanomaterials have been interpreted to know the *in-situ* effects of the same. It is, therefore, of paramount importance to consider the environmental influx of nanomaterials due to their release from the lubricants along with tangible situations like its physical contact.

The graphene nanomaterial has been tested to understand the hazards it poses to living cells. For instance, nano-graphene platelets (GNP) have been tested for their cytotoxicity via the MTT assay method using mouse macrophage cell line J774A.1. Cytotoxicity, in this case, is calculated as the cell viability (%) subtracted by one cent. The test reported that the sample containing only graphene was 79% cytotoxic as the cell viability was only 21%. The cytotoxicity tests conducted by fluorescence microscopy analysis exhibited a higher mortality rate of the cells with increasing GNP content in the polymer as the cells contained corroded or ruptured cell walls probably due to the slicing effect of sharp corners/edges in GNP (Manikandan et al. 2020).

There have been reports of toxicology studies of graphene family materials on various cellular models, such as human neuronal cells (Chang et al. 2011),

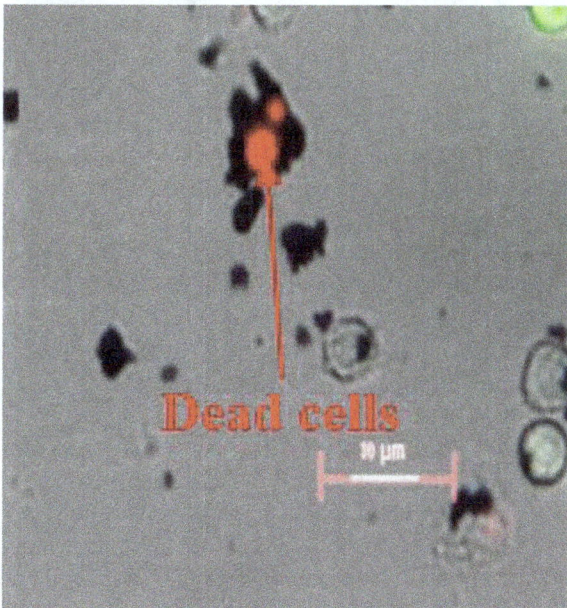

Figure 1: Fluorescent microscopy analysis of cells from graphene sample well. Reprinted with permission from Elsevier (Manikandan et al. 2020).

macrophages (Mendes et al. 2015), lungs (Zhang et al. 2010, Hu et al. 2011), liver and colorectal epithelial cells (Lammel et al. 2013, Kucki et al. 2017). A specific report involving toxicity analysis for reduced graphene oxide (rGO) in lubricants is discussed for human lung epithelial cell line no A549 and murine macrophages no. 264.7 (Esquivel-Gaon et al. 2018). The *in-vitro* analysis of nano-graphene lubricant composed of pristine graphene (rGO) dispersed in base oil (72%) on eukaryotic cells was carried out using the high throughput screening method. The parameters measured for toxicity analysis are cell count, mitochondrial membrane potential, cell membrane integrity and lysosomal acidification in the lung exposure model cell line. Also, various media were used to disperse rGO nanosheets for quantifying the toxicology effect with the help of its growth, DNA fragmentation and morphology on P. Putida, a bacterium commonly found in the soil. The exposure of epithelial cells to increasing rGO content does not report any significant decrease in cell viability till 100 µgmL^{-1} upon 24 hours of exposure. Furthermore, there was an increase in the lysosomal activity indicating the endocytic pathway activation. The exposure of rGO on P. Putida at concentrations of 100 µgmL^{-1} did not exhibit any notable effect as well. To summarize, any case of acute toxicity of mass-produced rGO was ruled out after the observations of this study. However, as this is just a short-term approach, better understanding in regard to chronic effects at non-harmful concentrations may be sought with long-term exposure and subsequent testing.

An overall generality on graphene should be evaded as its hazards are dependent on the use and the methodology of synthesis/manufacture (Bianco 2013). The hydrophobic behavior of graphene induces instability and thereby becomes prone to trapping by aggregating in liquid organ (physiological) environments. These aggregations result in an increased level of reactive oxygen species, which are responsible for cell toxicity (Liao et al. 2011, Sanchez et al. 2012). Poor cell proliferation and tissue regeneration result from restriction in the activity with the surrounding tissues due to the hydrophobic nature of the graphene surface. However, due to oxygen groups, graphene oxide (GO) displays good dispersion in a variety of solvents and hydrophilic and adsorption of protein, adhesion to cells and cell growth and exponential division can be improved with GO, which makes it a material for choice in tissue engineering and implants that are biomedical applications. Thus, GO can be employed in place of neat graphene in applications of cytotoxic safety (Dong and Qi 2015).

A comparison has been reported between graphene and carbon nanotubes providing information over physicochemical attributes, such as surface and structure for graphene and carbon nanotubes along with the health effect at cellular, tissue and body levels. The application of single and tiny graphene sheets that can be removed from the deposition spot due to efficient internalization by macrophages present in the living body is reported. Furthermore, the utilization of colloidal dispersions of graphene sheets that are stable and hydrophilic should be carried out to curtail the *in-vivo* formation of aggregates. Lastly, the utilization of chemically modified graphene which can be efficiently broken down or excretable graphene should be used so that it does not harm the living body. This kind of information could be strategically employed to tone the toxic effects of graphene (Bussy et al. 2013).

5. Applications

A lubricant is intended to form a layer between two components, thereby getting rid of their contact with each other in skidding or rotary-type of movement. These movements are sort of omnipresent in engineering applications that include IC engines, gearboxes, hydraulic systems, power systems (such as turbines and compressors), mechanical components (such as gears and electrical components), etc. (Mang and Dresel 2007, Ramón-Raygoza et al. 2016). The advent of lubrication technology is en-route toward the use of nano-lubricants, and graphene family nanomaterials have been a beneficiary as they are used in various industries, such as machining, power plants, automotive, space components, micro/nano-electromechanical systems, etc. (Paul et al. 2019), the account of which will be given in this section.

5.1 Lubricants for Machining-Related Operations

The terminology of cutting fluids used in machining have bi-function use of reduction in friction between the cutting tool; for example, an end mill cutter and work job which is being worked over to make the product piece and reduce the temperature of the contact parts. However, there are operations such as drilling and turning that have a trifunctional effect of cutting fluid, i.e. cooling, waste chips removal and lubrication (Sharma et al. 2016). This gives us a perspective of what a salient part the cutting fluid plays in growing the output and efficiency of the machining operation industry. Generally, the cutting fluids which have been traditionally used by the machining industry in the past have raised environmental concerns and health hazards when used in excess quantities (Singh et al. 2017).

The operations like turning and drilling have been recently explored by researchers for use of cutting fluids with nanoparticles dispersed in the liquid medium. Sharma et al. has published a comprehensive review that exhibited the enhancements in the life of the cutting tool and decreases the machining region temperature with improvements in the finishing of the machined surface due to improved thermal characteristics of the nano-lubricants (Sharma et al. 2016). A significant improvement in the thermal conductivity that resulted in the reduction of coefficient of friction and subsequent cutting forces has been reported with the use of alumina-graphene hybrid nano-lubricant (Singh et al. 2017). The concept of minimum quality lubrication has been explored with graphite-based lubricant (Amrita et al. 2014). The dispersion of graphene oxide in a conventional cutting fluid for drilling of titanium (Ti-6Al-4V) alloy with a tungsten carbide drill bit yielded a substantial decrease in the wearing of the tool flank and cutting force during the drilling operation (Yi et al. 2016). Also, a 15.21% improvement of surface finishing of the bore due to lesser thermal cracks was observed to go along with a 17.21% reduction in cutting force (Yi et al. 2017).

5.2 Lubricants for Thermal Power Plants

In thermal power plants, parts such as turbines, hydraulic governor systems, valves such as a wicket, radial and butterfly ones, gears, wire rope, etc. (Bureau of Reclamation, HRTSG, Denver, Colorado 2004) require lubrication with different demands for each of them. An oil dispersed with nitrogen-doped rGO has been reportedly used for the lubrication of induced draft fans, which is responsible for developing a negative pressure environment in the steam boiler for thermal power plant systems (Chandrabhan et al. 2017). They exhibit a substantial amount of decrease in the coefficient of friction (25%) and wear scar diameter upon compared with the base oil. A batch worth 700 liters was made to test it commercially for an NTPC thermal power plant because of which lower power consumption is reported for all five months of tested data. The power savings were found in the range of 5-13% with 9% being the average value of total power savings.

5.3 Lubrication of Space Components

Space missions are very cost-intensive, and it is pertinent to have accurate and reliable technological preparations; an example of which is the various drive mechanisms for its success. The functioning of drives depends on low friction, mechanical noise, wear and durability and lubrication is an important factor to have everything working as per the requirements. A synthetic hydrocarbon oil (multialkylated cyclopentane) was tested along with graphene as an additive for the lubrication of space components by simulating the space-like environment in terms of vacuum, temperature and radiation for its performance (Fan and Wang 2015). The addition of graphene into multialkylated cyclopentane (MCP) has decreased wear volume and its scar depth profile. Also, irradiation tests suggest the self-assembly of exited hexagonal structures into pentanes and heptane's due to the exposure of high-energy particles as shown in Figure 2. These, however, are atomic-scale flaws and do not show any damage in the graphene structure exhibiting its excellent adaptability in a space environment. This feature adds to the superior lubrication capacity in space

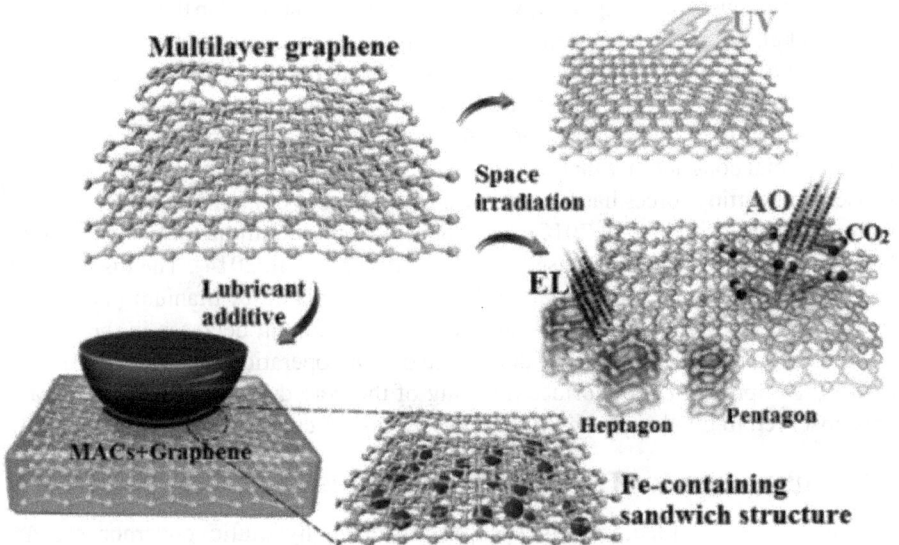

Figure 2: Irradiation effect on graphene sheets and the mechanism of friction as an additive for improving lubrication characteristics. Reprinted with permission from Royal Society of Chemistry (Paul Hirani et al. 2019)

conditions by saving the lubricant from irradiation attack, thereby allowing it to function in the best way possible.

5.4 Lubrication in Automotive Industries

Due to mobility, the performance of automobiles has a dependence on losses due to friction and has the potential to ruin it as well. The use of nano-lubricant has the potential to exhibit assuring results in terms of decreasing the wear and friction in automotive components. The important components that are expected to have high wear are internal combustion (IC) engines and gearbox along with some others. The use of multilayer graphene impregnated with copper/polyaniline recipes has been reported for automotive engine applications (Ramón-Raygoza et al. 2016). The coefficient of friction and wear decreased by 43% and 63%, respectively, with copper-based multilayer graphene dispersions in motor oil. The performance of graphene flakes-based engine oil lubricant was investigated on a 4-stroke IC engine test rig (A.K. Rasheed et al. 2016). The size and length of flakes were reported to affect the thermal performance of nano-lubricants. At 80°C, the enhancement in thermal conductivity was a 21% decrease in coefficient of friction and a 70% increase in heat transfer rate. Also, the microscopic images of piston rings that were taken away after 100 hours of service indicated lesser wear in case the lubricant contained graphene additive.

5.5 Lubrication in Refrigeration Compressors

The environmental footprint of society has a big role to play in the coming future in regards to nature. The depletion of natural resources has put a lot of pressure on finding alternative means of energy. Alternatively, the reduction in energy consumption by increasing the efficiency by technological innovation can be a solution to the increasing energy needs arising from both the industrial and residential sectors. Around 14% of the total energy consumed in the residential sector is by refrigerators. Among the important components in a refrigeration system, a compressor (rolling piston/vane type) is very significant in terms of the consumption of electricity. The rotary compressor is more popular out of the two due to its simplicity and small size along with high reliability and good equilibrium in operation (Kim 2005). The compressor parts have to be more wear-resistant and require efficient lubrication for the reliable operation of moving parts. The high wear on moving components in both rotary and vane type compressors make it imminent to have proper lubrication as it ensures lesser power consumption along with good durability of the machine part. The testing of graphene-based nanostructures in polyalkylene glycol as a base oil for a refrigeration system with CO_2 refrigerant is reported for vane-on-roller systems at 25°C and 70°C (Zin et al. 2016). The coefficient of friction was reduced by 18% along with a 75% decrease in wear scar volume at 0.1% of graphene filler. There was no change in viscosity and a slight increase in thermal conductivity was observed, thus indicating improved heat transfer characteristics. The resulting changes could improve the durability and efficiency of the compressor as a result of which subsequent reduction in energy consumption for a similar number of cycles.

5.6 Micro/Nanoelectromechanical Systems (MEMS/NEMS)

These systems are often referred to as micromachines and have attracted massive attention in today's world of miniaturization. As the systems that we are talking about are at the micro-level, it speaks for itself in terms of the lubrication systems as well for uninterrupted and effective operations. There have been attempts to investigate the tribological characteristics at the micron level by applying lubricant films dispersed with graphene oxide on micro/nano components (Pu et al. 2014, Hu et al. 2015). An innovative method comprising micro-measurement of forces to friction with the help of a microsphere connected with a colloidal probe is developed, which is capable to measure micro- and nano-coefficient of friction, respectively. When analyzed, they are directly proportional to the force of adhesion that is inferior in graphene layered films. The film over the substrate has shown a reduction in friction, capacity to carry the load, anti-wearability due to the effect of low surface energy and skidding effect due to graphene nano-sheets (Hu et al. 2015). However, the preparation methods of graphene, i.e. substrate-assisted chemical vapor deposition that is used in such a case. Even though the quality of the product made through the above method is high, the problem is with the removal of the substrate due to the generation of toxic liquid along with the economic feasibility of the process (Sun and Du 2019).

6. Characterization Tools and Techniques for Tribological and Thermal Behavior

The friction and wear behavior of contact surfaces under lubrication requires qualitative and quantitative measurements. The aim of the standard tests and/or small-scale tests is to extrapolate the laboratory results to real-life performance. The words tribometer and tribotester are used invariably for broad terms intended for equipment used to perform experiments to study and simulate the actual working conditions of wear, friction and lubrication. The tribometers are specific in their function and desired applications (Kaleli 2016). They are fabricated to study the long-term performance of the end products (Marjanovic et al. 2006). The parameters involved in tribological problems often exhibit complex behavior and their understanding and analysis are mainly based on experimental findings of laboratory tests. There are numerous test methods available for this purpose, and the experimental findings depend on the choice of the test method and test conditions. The standardization of tribology testing is set to meet the global requirements and to ensure the best performance and effectiveness. There are various agencies setting the standards, regulating and defining a particular tribological wear test rig to meet real-world scenarios for various applications. The most common standardizing agencies operating worldwide are International Organization for Standardization (ISO), ASTM International (formerly American Society for Testing and Materials-ASTM), American Society of Mechanical Engineers (ASME), Deutsches Institut für Normung (DIN), Japanese Industrial Standards (JIS), Society of Automotive Engineers (SAE), American Foundry Society (AFS), British Standard European Norm (BSEN) and China National Standard (GB). The testing parameters for which the standards are defined are applied load, contact surfaces, sliding velocity/

conditions, repeated cycles (distance), working temperature/wet/dry environment, electrification, shape and sizes of the specimens. The continuous monitoring of sliding surfaces tribological parameters, such as friction coefficient, normal load, and temperature with a number of usage cycles on the time scale is of utmost importance to predict the wear accurately. The current tribology in the direction of automation faces challenges, like sensing, accuracy, control of testing, improving the test methods and laboratory techniques of the involved instruments. Different tribometers at the laboratory scale are used to mimic the real-world environments and test the performance of materials, surfaces and lubricants before applying them to real-world conditions (Paul et al. 2019). The various requirements for which the tribometers are developed and used are:

- A specific machining situation tribo-contact simulation
- Established criteria-based selection of lubricants
- Assess bearing materials candidate for an application with high friction
- Evaluation of lubricants for a specific application
- Surface contamination monitoring
- Generate experimental data for comparing, analyzing and creating new materials, coatings or lubricants
- Investigate the basic phenomenological behavior of surfaces under friction

Thus, for studying the suitability of nano-lubricants, it is important to note the frictional response and performance under the specific of the tribometers system and the various test conditions (Gulzar et al. 2016). There are a variety of tribological test instruments with different types of configurations available as standard testing equipment. The tribological testing could be carried out effectively with these techniques available in slide testing mode, rolling mode and abrasion test mode. The various configurations available are pin-on-disc (Song et al. 2011), pin-on-ring (Vencl et al. 2009), pin-on-flat (Strahin and Doll 2018), ball-on-disc (Choudhary et al. 2012), ball-on-plate (Das et al. 2012), block-on-ring, ball-on-three-plates (He et al. 2017), four-ball (Abdullah et al. 2016) and roller-on-roller (Brizmer et al. 2017). The tribological testing is also available for specific requirements such as vacuum tribometers, low/high-temperature tribometers along with commercial and universal tribometers wherein a wide variety of tests may be carried out on the same equipment. The various tribotesting configurations for the interacting surfaces are shown in Figure 3.

The ball-on-disc is designed on standard ASTM G133-05 and ISO 7148, ISO 20808 standard as shown in Figure 3a and pin-on-disc as shown in Figure 3b is designed based on ASTM G99 standard. These two configurations are most popular owing to their relative simplicity and abundance of tribological contacts that can be easily realized. The two configurations have the same working principle, which differs in the type of contact established by the pin or the ball. In both cases, out of the two surfaces in contact, one is the upper ball or pin and the other is the lower circular flat disc. In these configurations, the ball/pin is usually held stationary while the flat surface is rotated about its central axis. The ball/pin is fixed to the holder and the load is applied to it. The generated frictional force between the rotating disc and the fixed ball/pin is measured with the help of measuring devices provided with sensors.

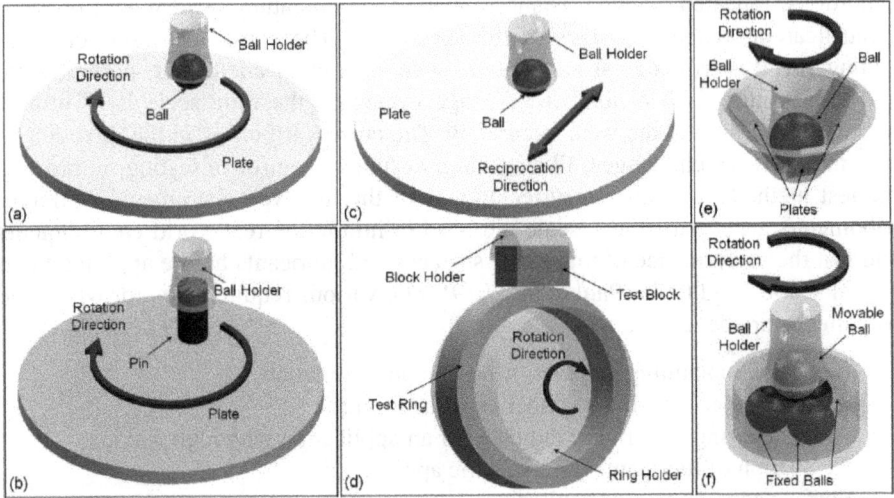

Figure 3: Tribotesting configurations for the interacting surface. Reprinted with permission from Royal Society of Chemistry (Paul Hirani et al. 2019).

P. Jimbert et al. reported the comparison between these two configurations with three different tool steel used to produce metal cutting shears (Jimbert et al. 2015). The materials samples under consideration were subjected to dry sliding tests for metal surfaces in order to reproduce the tool working conditions. The cryogenically treated samples were compared against untreated ones for their wear resistance due to cryogenic treatment. The wear rates obtained for both configurations were compared with wear data for real metal cutting tools. It was concluded with the experimental results that the ball-on-disc configuration proved to be a more realistic prediction of real tool life improvement under the cryogenic treatment as compared to the pin-on-disc configuration. Aviles et al. reported the use of pin-on-disc tribological performance of nano-lubricant with ionic base liquid dispersed with graphene material, while José M. Liñeira del Río used ball-on-disc configuration to study the tribological behavior of Trimethylolpropane Trioleate based nano-lubricants under the coating of magnetic nanoparticles (Avilés et al. 2020b, Liñeira del Río et al. 2020).

A ball-on-plate configuration is based on ASTM G194 - 08 as shown in Figure 3c. The only difference between the ball/pin-on-disc configurations is that in this configuration the ball has linear sliding velocity, and it keeps reciprocating against the plate in a back and forth motion while the movement is constrained in a unidirectional path in ball/pin-on-disc configurations. Four-ball configurations are also known as shell four-ball tester which is as shown in Figure 3d and is based on ASTM D2783 (extreme pressure test for lubricating fluids), DIN 51350-02 (extreme pressure properties test for liquid lubricants), ASTM D4172 (WP tests for lubricating fluids), ASTM D5183 (coefficient of friction of lubricants) and DIN 51350-03 (wear test for liquid lubricants). This configuration is used to evaluate the wear prevention (WP), extreme pressure (EP) and frictional behavior of lubricant. The tester consists of four balls in the configuration of an equilateral tetrahedron, wherein the upper ball rotates and is in contact with the fixed lower three balls. B. Suresha et al. used Four-Ball Tester for studying tribological behavior of Neem Oil under the influence of graphene nano-platelet (Suresha et al. 2020). Alghani et al. also reported the use of a four-ball test rig for testing the enhancement in the tribological behavior with the addition of graphene nanoparticles (Alghani et al. 2019).

Block-on-three plates testing configuration is based on ASTM G133 and is as shown in Figure 3e. This configuration is sometimes also called the ball-on-pyramid configuration. The configuration is designed by having a spherical ball inserted in three plates, which is allowed to move freely without any direction constraints. The bottom plate is made flexible for normal force to be evenly distributed on all the three contact points of the upper ball as uneven distribution of the normal load may cause flawed results. The ball and the plates are inter-exchanged so that the system can be adapted to material combinations. Block-on-Ring testing configuration is based on ASTM G77 and is as shown in Figure 3f. This testing configuration is a widely used technique for evaluating the sliding wear behavior of materials in different laboratory conditions, allowing reliable testing of the specimen under consideration. This test rig is employed in specific tribological applications where the complex wear mechanisms in sliding wear mode are taking place at the contact surface. The wear behavior under the influence of normal loading, speed, corrosion and lubrication can be studied with this versatile tribometer that can simulate the different realistic work conditions, which will be useful for wear evaluation. The role of porous graphene-based polyamide composite in the block-on-ring mode was reported by Wang et al., and it was concluded that graphene-based composite is a promising material in severe tribological conditions (Wang et al. 2020).

Temperature, pressure, thermal conductivity, heat transferability and power consumption are important points for studying thermal behavior as wear and friction in the case of tribological studies. Though not directly, an employed parameter, availability of combustion air in sufficient amounts and a safe heat exchanger are also critical in appliances. The devices used for the thermal testing ranges from sophisticated calorimeters to in-house designed heat exchangers (Mehrali et al. 2014, Yarmand et al. 2016, Ali and Xianjun 2020). Researchers integrate the required experimental setups for thermal behavior using a domestic refrigerator, capillary tube, evaporator and air-cooled condenser, etc. (D.S Adelekan et al. 2019). F. Agresti et al devised a thermal diffusivity measuring instrument based on the photoacoustic effect (Agresti et al. 2015). A laser modulated at the frequency of 2 Hz impinged on a Si slab immersed in the liquid and the generated thermal waves traveled through the sample were detected by using a photoacoustic chamber. The Si slab-photoacoustic chamber distance was varied and thermal diffusivity was estimated. Such an instrument was found useful for studying tribological and thermal properties of graphene-based nano-lubricants (Zin et al. 2016). Kim et al. reported the use of the thermal bridge method for thermal conductivity measurement of individual multiwalled carbon nanotubes (Kim et al. 2008). This thermal bridge method is claimed to be more effective as compared to other techniques and is also used to study thermal conductivity and thermal power of graphene single-layer material. This method is superior to other thermal conductivity and thermopower characterization involving nanotubes and nanowires.

The selection of appropriate measurement configurations and analytical techniques for tribological and thermal behavior is particularly challenging with consideration to collecting data points, data storage and data analysis. The accuracy and applicability of testing techniques in complex operating conditions involve many parameters and theoretical considerations. The advances in thermal engineering have enabled the incorporation of nano-lubricants because of the features, like reduction in energy losses during heat transfer along with the huge mechanical strength brought in due to nanosize. The testing instruments are available from manual, semi-automated and automated to ready-to-use sophisticated mode. From the literature available, it is reasonable to conclude that all the tribological and thermal testing discussed herein are effective and could be used for testing graphene-based nano-lubricants.

7. Conclusion

The latest developments in the field of graphene-based nano-lubricants along with their applications are reviewed which indicate that graphene nano-lubricant can be successfully employed to decrease friction, wear and improve thermal characteristics in various mechanical applications. There is adequate support available from characterization tools and techniques at the lab scale for the assessment of tribological anti-friction and anti-wear mechanism, lubrication mechanism and thermal behavior. The improvement in the friction behavior, reduced wear losses and improved thermal characteristics of mechanical systems due to the various properties of graphene material are discussed. Graphene-based nano-lubricants provide superior mechanical properties and thermal stability, which are supplemented by features like easier dispersion, increased stability, enhanced adsorption, improved interlayer slip and increase in film formation strength in comparison with conventional lubricants. However, the environmental aspects and health hazards, which are mainly due to the potentially toxic nature of graphene via its incorporation, liquid effluents generated during its synthesis and its removal from lubricants after use are yet to be conclusively understood. An account of health hazards and toxicity concerns given in the present chapter culminates that future research should be conducted on these aspects in the form of mature theories to use graphene in lubricants for high-performance functionalities and tribological purposes beyond lab-scale results to real-world applications. The cost of graphene material is high due to its smaller scale of production, which also leads to limitations in the reproducibility of uniform performance by nanomaterial. As the tribological and thermal properties of graphene nano-lubricants are dependent on the performance of nano-graphene, the commerciality of graphene-based nano-lubricants remains to be seen.

References

Abdullah, M.I.H.C., Abdollah, M.F.B., Amiruddin, H., Tamaldin, N. and Nuri, N.R.M. 2016. The potential of hBN nanoparticles as friction modifier and antiwear additive in engine oil. Mech. Ind. 17: 104-110.

Adelekan, D.S., Ohunakin, O.S., Gill, J., Okokpujie, I.P. and Atiba, O.E. 2019. Performance of an iso-butane driven domestic refrigerator infused with various concentrations of graphene based nano-lubricants. Procedia Manuf. 35: 1146–1151.

Agresti, F., Ferrario, A., Boldrini, S., Miozzo, A., Montagner, F., Barison, S., Pagura, C. and Fabrizio, M. 2015. Temperature controlled photoacoustic device for thermal diffusivity measurements of liquids and nanofluids. Thermochim. Acta 619: 48-52.

Alghani, W., Karim, M.S.A., Bagheri, S., Amran, N.A.M. and Gulzar, M. 2019. Enhancing the tribological behavior of lubricating oil by adding TiO_2, graphene, and TiO_2/graphene nanoparticles. Tribol. Trans. 62: 452-463.

Ali, M.K.A. and Xianjun, H. 2020. Improving the heat transfer capability and thermal stability of vehicle engine oils using Al_2O_3/TiO_2 nanomaterials. Powder Tech. 363: 48-58.

Ali, M.K.A., Xianjun, H., Abdelkareem, M.A.A., Gulzar, M. and Elsheikh, A.H. 2018. Novel approach of the graphene nano-lubricant for energy saving via anti-friction/wear in automobile engines. Tribol. Int. 124: 209-229.

Al-Janabi, A.S. and Hussin, M. 2020. Stability and thermal conductivity of graphene in polyester nano-lubricant. AIP Conference Proceedings 2267: 1-7.

Amrita, M., Srikant, R.R. and Sitaramaraju, A.V. 2014. Performance evaluation of nanographite-based cutting fluid in machining process. Mater. Manuf. Process. 29: 600-605.

Avilés, M.D., Carrión-Vilches, F.J., Sanes, J. and Bermúdez, M.D. 2019. Diprotic ammonium succinate ionic liquid in thin film aqueous lubrication and in graphene nano-lubricant. Tribol. Lett. 67: 2601-2610.

Avilés, M.-D., Pamies, R., Sanes, J. and Bermúdez, M.-D. 2020a. Graphene-ionic liquid thin film nano-lubricant. Nanomaterials 10: 535-553.

Azman, S.S.N., Zulkifli, N.W.M., Masjuki, H., Gulzar, M. and Zahid, R. 2016. Study of tribological properties of lubricating oil blend added with graphene nanoplatelets. J. Mater. Res. 31(13): 1932-1938.

Babarinde, T., Akinlabi, S., Madyira, D. and Ekundayo, F. 2020. Enhancing the energy efficiency of vapour compression refrigerator system using R600a with graphene nano-lubricant. Energy Rep. 6: 1-10.

Baby, T.T. and Ramaprabhu, S. 2010. Investigation of thermal and electrical conductivity of graphene based nanofluids. J. Appl. Phys. 108: 124308.

Balandin, A.A., Ghosh, S., Bao, W., Calizo, I., Teweldebrhan, D., Miao, F. and Lau, C.N. 2008. Superior thermal conductivity of single-layer graphene. Nano Lett. 8: 902-907.

Bao, T., Wang, Z., Zhao, Y., Wang, Y. and Yi, X. 2019. Long-term stably dispersed functionalized graphene oxide as an oil additive. RSC Adv. 9: 39230-39241.

Bianco, A. 2013. Graphene: safe or toxic? The two faces of the medal. Angew. Chem. Int. Ed. 52: 4986-4997.

Brizmer, V., Stadler, K., Drogen, M. van, Han, B., Matta, C. and Piras, E. 2017. The tribological performance of black oxide coating in rolling/sliding contacts. Tribol. Trans. 60: 557-574.

Bureau of Reclamation, HRTSG, Denver, Colorado 2004. FIST 2-4, Lubrication of Powerplant Equipment, Bureau of Reclaimation, Denver, Colorado.

Bussy, C., Ali-Boucetta, H. and Kostarelos, K. 2013. Safety considerations for graphene: lessons learnt from carbon nanotubes. Acc. Chem. Res. 46: 692-701.

Chae, H.K., Siberio-Pérez, D.Y., Kim, J., Go, Y., Eddaoudi, M., Matzger, A.J., O'Keeffe, M. and Yaghi, O.M. 2004. A route to high surface area, porosity and inclusion of large molecules in crystals. Nature 427: 523-527.

Chandrabhan, S.R., Jayan, V., Parihar, S.S. and Ramaprabhu, S. 2017a. Development of a nitrogen-doped 2D material for tribological applications in the boundary-lubrication regime. Beilstein J. Nanotechnol. 8: 1476-1483.

Chang, Y., Yang, S.-T., Liu, J.-H., Dong, E., Wang, Y., Cao, A., Liu, Y. and Wang, H. 2011. In vitro toxicity evaluation of graphene oxide on A549 cells. Toxicol. Lett. 200: 201-210.

Choudhary, S., Mungse, H.P. and Khatri, O.P. 2012. Dispersion of alkylated graphene in organic solvents and its potential for lubrication applications. J. Mater. Chem. 22: 21032-21039.

Das, L., Aggarwal, M., Rajkumar, K., Aravindan, S. and Gupta, M. 2012. Tribological properties of magnesium nano-alumina composites under nano-graphite lubrication. Tribol. Trans. 55: 334-344.

Dong, H.S. and Qi, S.J. 2015. Realising the potential of graphene-based materials for biosurfaces – A future perspective. Biosurf Biotribol. 1: 229-248.

Dou, X., Koltonow, A.R., He, X., Jang, H.D., Wang, Q., Chung, Y.-W. and Huang, J. 2016. Self-dispersed crumpled graphene balls in oil for friction and wear reduction. PNAS 113: 1528-1533.

Esfahani, M.R., Languri, E.M. and Nunna, M.R. 2016. Effect of particle size and viscosity on thermal conductivity enhancement of graphene oxide nanofluid. Int. Commun. Heat Mass Transf. 76: 308-315.

Esquivel-Gaon, M., Nguyen, N.H.A., Sgroi, M.F., Pullini, D., Gili, F., Mangherini, D., Pruna, A.I., Rosicka, P., Sevcu, A. and Castagnola, V. 2018. In vitro and environmental toxicity of reduced graphene oxide as an additive in automotive lubricants. Nanoscale 10: 6539-6548.

Eswaraiah, V., Sankaranarayanan, V. and Ramaprabhu, S. 2011. Graphene-based engine oil nanofluids for tribological applications. ACS Appl. Mater. Interfaces 3: 4221-4227.

Fan, X. and Wang, L. 2015a. Graphene with outstanding anti-irradiation capacity as multialkylated cyclopentanes additive toward space application. Sci. Rep. 5: 12734.

Fan, X., Xia, Y., Wang, L. and Li, W. 2014. Multilayer graphene as a lubricating additive in bentone grease. Tribol. Lett. 55: 455-464.

Gulzar, M., Masjuki, H.H., Kalam, M.A., Varman, M., Mohd Zulkifli, N.W., Mufti, R. and Zahid, R. 2016. Tribological performance of nanoparticles as lubricating oil additives. J. Nanopart Res. 18: 223.

Guo, Y.-B. and Zhang, S.-W. 2016. The tribological properties of multi-layered graphene as additives of PAO$_2$ oil in steel–steel contacts. Lubricants 4: 30.

He, A., Huang, S., Yun, J.-H., Wu, H., Jiang, Z., Stokes, J., Jiao, S., Wang, L. and Huang, H. 2017. Tribological performance and lubrication mechanism of alumina nanoparticle water-based suspensions in ball-on-three-plate testing. Tribol. Lett. 65: 40.

Hemmat Esfe, M., Saedodin, S., Rejvani, M. and Shahram, J. 2017. Experimental investigation, model development and sensitivity analysis of rheological behavior of ZnO/10W40 nano-lubricants for automotive applications. Physica E Low Dimens. Syst. Nanostruct. 90: 194-203.

Hu, W., Peng, C., Lv, M., Li, X., Zhang, Y., Chen, N., Fan, C. and Huang, Q. 2011. Protein corona-mediated mitigation of cytotoxicity of graphene oxide. ACS Nano 5: 3693-3700.

Hu, Y., Ma, H., Liu, W., Lin, Q. and Liu, B. 2015. Preparation and investigation of the microtribological properties of graphene oxide and graphene films via electrostatic layer-by-layer self-assembly. J. Nanomater. https://dl.acm.org/doi/pdf/10.1155/2015/282369.

Jang, S.P. and Choi, S.U.S. 2004. Role of Brownian motion in the enhanced thermal conductivity of nanofluids. Appl. Phys. Lett. 84: 4316-4318.

Jimbert, P., Iturrondobeitia, M., Ibarretxe, J. and Fernandez-Martinez, R. 2015. Pin on disk against ball on disk for the evaluation of wear improvement on cryo-treated metal cutting shears. AIP Conference Proceedings, 1653: 020048.

Josphat Phiri, Patrick Gane and Thad C. Maloney. 2017. General overview of graphene: production, properties and application in polymer composites. Mater. Sci. Eng. B 2017: 9-28.

Kaleli, H. 2016. New universal tribometer as pin or ball-on-disc and reciprocating pin-on-plate types. Tribol. Ind. 38(2): 6.

Kalita, P., Malshe, A.P. and Rajurkar, K.P. 2012. Study of tribo-chemical lubricant film formation during application of nano-lubricants in minimum quantity lubrication (MQL) grinding. CIRP Annals 61(1): 327-330.

Kim, B., Beskok, A. and Cagin, T. 2008. Thermal interactions in nanoscale fluid flow: molecular dynamics simulations with solid-liquid interfaces. Microfluid Nanofluidics 5: 551-559.

Kim, H.J. 2005. Lubrication oil pumping by utilizing vane motion in a horizontal rotary compressor. Int. J. Refrig. 28: 498-505.

Kinoshita, H., Nishina, Y., Alias, A.A. and Fujii, M. 2014. Tribological properties of monolayer graphene oxide sheets as water-based lubricant additives. Carbon 66: 720-723.

Kucki, M., Diener, L., Bohmer, N., Hirsch, C., Krug, H.F., Palermo, V. and Wick, P. 2017. Uptake of label-free graphene oxide by Caco-2 cells is dependent on the cell differentiation status. J. Nanobiotechnology 15: 1-18.

Kumar, A., Deval, P., Shrinet, E.S. and Ghosh, S.K. 2021. Investigation on tribological properties of used engine oil with graphene. P I Mech. Eng. J-J Eng. 235(7): 1420-1429.

Lammel, T., Boisseaux, P., Fernández-Cruz, M.-L. and Navas, J.M. 2013. Internalization and cytotoxicity of graphene oxide and carboxyl graphene nanoplatelets in the human hepatocellular carcinoma cell line Hep G2. Part. Fibre Toxicol. 10: 1-21.

Lee, G.-J. and Rhee, C.K. 2014. Enhanced thermal conductivity of nanofluids containing graphene nanoplatelets prepared by ultrasound irradiation. J. Mater. Sci. 49: 1506-1511.

Liao, K.-H., Lin, Y.-S., Macosko, C.W. and Haynes, C.L. 2011. Cytotoxicity of graphene oxide and graphene in human erythrocytes and skin fibroblasts. ACS Appl. Mater. Interfaces 3: 2607-2615.

Lin, J., Wang, L. and Chen, G. 2011. Modification of graphene platelets and their tribological properties as a lubricant additive. Tribol. Lett. 41: 209-215.

Liñeira del Río, J.M., Guimarey, M.J.G., Comuñas, M.J.P., López, E.R., Amigo, A. and Fernández, J. 2018. Thermophysical and tribological properties of dispersions based on graphene and a trimethylolpropane trioleate oil. J. Mol. Liq. 268: 854-866.

Liñeira del Río, J.M., López, E.R., González Gómez, M., Yáñez Vilar, S., Piñeiro, Y., Rivas, J., Gonçalves, D.E.P., Seabra, J.H.O. and Fernández, J. 2020. Tribological behavior of nano-lubricants based on coated magnetic nanoparticles and trimethylolpropane trioleate base oil. Nanomaterials 10: 683.

Lou, J., Zhang, H. and Wang, R. 2015. Experimental investigation of graphite nano-lubricant used in a domestic refrigerator. Adv. Mech. Eng. 7: 1687814015571011.

Mang, T. and Dresel, W. 2007, Lubricants and Lubrication. John Wiley and Sons.

Manikandan, N.A., Pakshirajan, K. and Pugazhenthi, G. 2020. Preparation and characterization of environmentally safe and highly biodegradable microbial polyhydroxybutyrate (PHB) based graphene nanocomposites for potential food packaging applications. Int. J. Biol. Macromol. 154: 866-877.

Marjanovic, N., Tadic, B., Ivkovic, B. and Mitrovic, S. 2006. Design of modern concept tribometer with circular and reciprocating movement. Tribol. Ind. 28: 6.

Mbambo, M.C., Khamlich, S., Khamliche, T., Moodley, M.K., Kaviyarasu, K., Madiba, I.G., Madito, M.J., Khenfouch, M., Kennedy, J., Henini, M., Manikandan, E. and Maaza, M. 2020. Remarkable thermal conductivity enhancement in Ag-decorated graphene nanocomposites based nanofluid by laser liquid solid interaction in ethylene glycol. Sci. Rep. 10: 10982.

Mehrali, Mohammad, Tahan Latibari, S., Mehrali, Mehdi, Mahlia, T.M.I., Sadeghinezhad, E. and Metselaar, H.S.C. 2014. Preparation of nitrogen-doped graphene/palmitic acid shape stabilized composite phase change material with remarkable thermal properties for thermal energy storage. Appl. Energy 135: 339-349.

Mendes, R.G., Koch, B., Bachmatiuk, A., Ma, X., Sanchez, S., Damm, C., Schmidt, O.G., Gemming, T., Eckert, J. and Rümmeli, M.H. 2015. A size dependent evaluation of the cytotoxicity and uptake of nanographene oxide. J. Mater. Chem. B 3: 2522-2529.

Nair, R.R., Blake, P., Grigorenko, A.N., Novoselov, K.S., Booth, T.J., Stauber, T., Peres, N.M.R. and Geim, A.K. 2008. Fine structure constant defines visual transparency of graphene. Science 320(5881), 1308-1308.

Park, H.-G. and Yeo, M.-K. 2016. Nanomaterial regulatory policy for human health and environment. Mol. Cell. Toxicol. 12(3): 223-236.

Patel, H.E., Sundararajan, T., Pradeep, T., Dasgupta, A., Dasgupta, N. and Das, S.K. 2005. A micro-convection model for thermal conductivity of nanofluids. J. Phys. 65: 7.

Paul, G., Hirani, H., Kuila, T. and Murmu, N.C. 2019a. Nano-lubricants dispersed with graphene and its derivatives: an assessment and review of the tribological performance. Nanoscale, 11: 3458-3483.

Paul, G., Shit, S., Hirani, H., Kuila, T. and Murmu, N.C. 2019b. Tribological behavior of dodecylamine functionalized graphene nanosheets dispersed engine oil nano-lubricants. Tribol. Int. 131: 605-619.

Pawan K. Singh, S.S.G. and V. Manoj Siva, S.K. 2011. Thermal conductivity enhancement of nanofluids containing graphene nanosheets. J. Appl. Phys. 110.

Pu, J., Mo, Y., Wan, S. and Wang, L. 2014. Fabrication of novel graphene–fullerene hybrid lubricating films based on self-assembly for MEMS applications. Chem. Comm. 50: 469-471.

Ramón-Raygoza, E.D., Rivera-Solorio, C.I., Giménez-Torres, E., Maldonado-Cortés, D., Cardenas-Alemán, E. and Cué-Sampedro, R. 2016. Development of nano-lubricant based on impregnated multilayer graphene for automotive applications: analysis of tribological properties. Powder Technol. 302: 363-371.

Ramsden, J. 2012. Nanotechnology for military applications. Nanotechnol. Percept. 8: 99-131.

Rasheed, A.K., Khalid, M., Javeed, A., Rashmi, W., Gupta, T. and Chan, A. 2016. Heat transfer and tribological performance of graphene nano-lubricant in an internal combustion engine. Tribol. Int. 103: 504-515.

Rasheed, Abdul Khaliq, Khalid, M., Rashmi, W., Gupta, T. and Chan, A. 2016. Graphene based nanofluids and nano-lubricants – review of recent developments. Renew. Sust. Energ. Rev. 63: 346-362.

Rashmi, W., Khalid, M., Xiao, Y. and Arwin, G.Z. 2017. Tribological studies on graphene/TMP based nano-lubricant. J. En. Sci. Technol. 12: 365-373.

Safira, L., Putra, N., Trisnadewi, T., Kusrini, E. and Mahlia, T.M.I. 2020. Thermal properties of sonicated graphene in coconut oil as a phase change material for energy storage in building applications. Int. J. Low Carbon Technol. 15: 629-636.

Sahoo, P. 2005. Engineering Tribology, PHI Learning Pvt. Ltd.

Sajid, M.U. and Ali, H.M. 2018. Thermal conductivity of hybrid nanofluids: A critical review. Int. J. Heat Mass Transf. 126: 211-234.

Sanchez, V.C., Jachak, A., Hurt, R.H. and Kane, A.B. 2012. Biological interactions of graphene-family nanomaterials: an interdisciplinary review. Chem. Res. Toxicol. 25: 15-34.

Sanes, J., Avilés, M.-D., Saurín, N., Espinosa, T., Carrión, F.-J. and Bermúdez, M.-D. 2017. Synergy between graphene and ionic liquid lubricant additives. Tribol. Int. 116: 371-382.

Sarsam, W.S., Amiri, A., Kazi, S.N. and Badarudin, A. 2016. Stability and thermophysical properties of non-covalently functionalized graphene nanoplatelets nanofluids. Energy Convers. Manag. 116: 101-111.

Sayuti, M., Sarhan, A.A.D. and Hamdi, M. 2013. An investigation of optimum SiO_2 nanolubrication parameters in end milling of aerospace Al6061-T6 alloy. Int. J. Adv. Manuf. Syst. 67: 833-849.

Senatore, A., D'Agostino, V., Petrone, V., Ciambelli, P. and Sarno, M. 2013. Graphene oxide nanosheets as effective friction modifier for oil lubricant: materials, methods, and tribological results. ISRN Tribology, 2013.

Shafi, W.K. and Charoo, M.S. 2020. An overall review on the tribological, thermal and rheological properties of nano-lubricants. Tribol. - Mater. Surf. Interfaces 1-35.

Sharma, A.K., Tiwari, A.K. and Dixit, A.R. 2016. Effects of Minimum Quantity Lubrication (MQL) in machining processes using conventional and nanofluid based cutting fluids: a comprehensive review. J. Clean. Prod. 127: 1-18.

Singh, R.K., Sharma, A.K., Dixit, A.R., Tiwari, A.K., Pramanik, A. and Mandal, A. 2017. Performance evaluation of alumina-graphene hybrid nano-cutting fluid in hard turning. J. Clean. Prod. 162: 830-845.

Song, H.-J., Jia, X.-H., Li, N., Yang, X.-F. and Tang, H. 2011. Synthesis of α-Fe_2O_3 nanorod/ graphene oxide composites and their tribological properties. J. Mater. Chem. 22: 895-902.

Stone, V., Führ, M., Feindt, P.H., Bouwmeester, H., Linkov, I., Sabella, S., Murphy, F., Bizer, K., Tran, L., Ågerstrand, M. and Fito, C. 2018. The essential elements of a risk governance framework for current and future nanotechnologies. Risk Anal. 38(7): 1321-1331.

Strahin, B.L. and Doll, G.L. 2018. Tribological coatings for improving cutting tool performance. Surf. Coat. Technol. 336: 117-122.

Sun, J. and Du, S. 2019. Application of graphene derivatives and their nanocomposites in tribology and lubrication: a review. RSC Advances 9: 40642-40661.

Suneetha, S. and Reddy, P.B.A. 2016. Investigation on graphene nanofluids and its applications: a brief literature review. Res. J. Pharm. Technol. 9: 655-663.

Suresha, B., Hemanth, G., Rakesh, A. and Adarsh, K.M. 2020. Tribological behaviour of neem oil with and without graphene nanoplatelets using four-ball tester. Adv. Tribol. 2020: e1984931.

Tevet, O., Von-Huth, P., Popovitz-Biro, R., Rosentsveig, R., Wagner, H.D. and Tenne, R. 2011. Friction mechanism of individual multilayered nanoparticles. PNAS. 108: 19901-19906.

Texter, J. 2014. Graphene dispersions. Curr. Opin. Colloid Interface Sci. 19: 163-174.

Uddin, Md.E., Kuila, T., Nayak, G.C., Kim, N.H., Ku, B.-C. and Lee, J.H. 2013. Effects of various surfactants on the dispersion stability and electrical conductivity of surface modified graphene. J. Alloys Compd. 562: 134-142.

Vencl, A., Mrdak, M. and Banjac, M. 2009. Correlation of microstructures and tribological properties of ferrous coatings deposited by atmospheric plasma spraying on Al-Si cast alloy substrate. Metall. Mater. Trans. A 40: 398-405.

Wang, L., Pan, B., Gao, J., Huang, S., Xie, M., Li, C., Luo, D., Liu, J. and Wang, H. 2020. Tribological behaviors of porous 3D graphene lubricant reinforced monomer casting polyamide 6 composite. Adv. Eng. Mater. 22: 1901170.

Wu, P., Chen, X., Zhang, C., Zhang, Jiping, Luo, J. and Zhang, Jiyang 2021. Modified graphene as novel lubricating additive with high dispersion stability in oil. Friction 9: 143-154.

Yarmand, H., Gharehkhani, S., Shirazi, S.F.S., Goodarzi, M., Amiri, A., Sarsam, W.S., Alehashem, M.S., Dahari, M. and Kazi, S.N. 2016. Study of synthesis, stability and thermo-physical properties of graphene nanoplatelet/platinum hybrid nanofluid. Int. Commun. Heat Mass Transf. 77: 15-21.

Yi, S., Li, G., Ding, S., Mo, J. and Rahman, M. 2016. Experimental study of graphene oxide suspension in drilling Ti-6Al-4V, The 2nd Information Technology and Mechatronics Engineering Conference (ITOEC 2016), Atlantis Press.

Yi, S., Li, G., Ding, S. and Mo, J. 2017. Performance and mechanisms of graphene oxide suspended cutting fluid in the drilling of titanium alloy Ti-6Al-4V. J. Manuf. Process. 29: 182-193.

Yu, W., Xie, H. and Bao, D. 2009. Enhanced thermal conductivities of nanofluids containing graphene oxide nanosheets. Nanotechnology 21: 055705.

Yu, W., Xie, H. and Chen, W. 2010. Experimental investigation on thermal conductivity of nanofluids containing graphene oxide nanosheets. Int. J. Appl. Phys. 107: 094317.

Yu, X., Zhan, R., Deng, J. and Huang, X. 2017. Prediction of the maximum nonseizure load of lubricant additives. J. Theor. Computa. Chem. 16: 1750014.

Zhamu, A. and Jang, B.Z. 2011. Nano graphene-modified lubricant. US Patent. 12 (583,320).

Zhang, B., Xue, Y., Qiang, L., Gao, K., Liu, Q., Yang, B., Liang, A. and Zhang, J. 2017. Assembling of carbon nanotubes film responding to significant reduction wear and friction on steel surface. Appl. Nanosci. 7: 835-842.

Zhang, L., Pu, J., Wang, L. and Xue, Q. 2014. Frictional dependence of graphene and carbon nanotube in diamond-like carbon/ionic liquids hybrid films in vacuum. Carbon 80: 734-745.

Zhang, W., Zhou, M., Zhu, H., Tian, Y., Wang, K., Wei, J., Ji, F., Li, X., Li, Z. and Zhang, P. 2011. Tribological properties of oleic acid-modified graphene as lubricant oil additives. J. Phys. D. Appl. Phys. 44: 205303.

Zhang, Y., Ali, S.F., Dervishi, E., Xu, Y., Li, Z., Casciano, D. and Biris, A.S. 2010. Cytotoxicity effects of graphene and single-wall carbon nanotubes in neural phaeochromocytoma-derived PC12 cells. ACS Nano 4: 3181-3186.

Zin, V., Barison, S., Agresti, F., Colla, L., Pagura, C. and Fabrizio, M. 2016. Improved tribological and thermal properties of lubricants by graphene based nano-additives. RSC Advances 6: 59477-59486.

2020, Lubricants Market Size, Share | Global Industry Report, 2020-2027.

Fabrication of Polycarbonate Filaments Infused with Carbon from Coconut Shell Powder for 3D Printing Applications

Deepa Kodali, Chibu O. Umerah, S. Jeelani and Vijaya K. Rangari*

Department of Materials Science and Engineering, Tuskegee University, Tuskegee, Al-36088

1. Introduction

Over the past few decades, additive manufacturing (AM) techniques have surged tremendously and have become a well-integrated area of research for industrialists as well as academicians. Among the various technologies that are available for 3D printing, fused deposition modeling (FDM) has become the most prominent one which can be attributed to its simplicity and reliability (Mazzanti et al. 2019). FDM technique utilizes thermoplastic filaments that are fed into the printer heads, which melts the filament during the printing process and later allows it to solidify to produce the desired parts (Ning et al. 2017). The melt extrusion from the nozzle follows a sequential layer deposition process which acts in accordance with the enumerated CAD or Stereo Lithography (STL) geometry file (Mazzanti et al. 2019). Despite the advancements made in FDM, the inferior mechanical property of the 3D printed parts has been one of the pertinent limitations and thus has become an exigent issue that needs to be addressed immediately (Rahim et al. 2019). The prime contribution to this existing limitation arises from the lack of suitable materials that are compliant with the prevailing techniques (Belter and Dollar 2015). Thus far, plastics such as Polylactic Acid (PLA), Acrylonitrile Butadiene Styrene (ABS), nylon and polycarbonate (PC) have been widely used for 3D printing given that their mechanical strength varies between 30-100 MPa and elastic modulus lies between 1.3 to 3.6 Gpa (Mazzanti et al. 2019, Smith and Dean 2013, Sodeifian et al. 2019).

While conventional plastics serve in a wide variety of applications ranging from food packaging to aerospace, the intended use of these plastics has also grown tremendously during the past years and is challenging the equilibrium of the

---

Corresponding author: vrangari@tuskegee.edu

environment. One of the alternatives to overcome this challenge is to enhance the use of bio-based materials that are obtained from bioresources, such as plants and animals (Yang et al. 2018). These bio-based materials have successfully served as fillers in the polymer matrices facilitating the refinement of thermomechanical properties (Ji et al. 2020). In view of the characteristics and benefits that are offered by bio-based materials, the development of bio-based materials suitable for 3D printing applications has gained a significant amount of interest in recent years. Driven by the need for the demand of multifarious proclivities of the consumers, the quest for the development of bio-based and biodegradable composites has become undeniable (Wang et al. 2017).

Especially carbon derived from the bioresources, which are known as biochar or biocarbon, has become an irrefutable source of filler material based on its extensive commercial applications (Mohanty et al. 2018). Furthermore, biochar also immensely contributes to the development of novel functional materials (Bourmaud et al. 2018). The addition of pinewood-based biochar as a filler element in the polypropylene composites has improved the flame retardancy of the composites (Das et al. 2016). Silica from bagasse ash has improved the thermomechanical properties of Bioplast GF 106/02 composites when added as a filler material (Imam et al. 2019). Biochar obtained from the packaging waste is successfully used to develop 3D printed parts when blended with poly(ethylene terephthalate) exhibiting superior thermomechanical and dynamic properties (Idrees et al. 2018). Biochar synthesized from bamboo charcoal has improved the electrical conductivity in addition to the thermal and mechanical properties (Li et al. 2016, Ho et al. 2015). Carbon from coconut shell powder is used as a filler to develop 3D printable bioplast filaments (Umerah et al. 2020). This work focuses on developing the polycarbonate filaments with carbon from the coconut shell powder (CCSP) as a filler material. Carbon from the coconut shell powder is blended with PC to develop biobased 3D printable nanocomposite filaments. The carbon obtained from the pyrolysis of coconut shell powder is subjected to a ball milling process to obtain carbon nanopowder. This powder is infused into the polycarbonate by the solution blending process. The precipitate thus obtained is dried and stored for characterization and extrusion of filaments. The extruded filaments are used for 3D printing to produce thin films, which are further characterized and analyzed.

2. Materials and Methods

2.1 Materials

Polycarbonate pellets were purchased from Makrolon® LED 2245 produced by Plastics Covestro, Germany, with a melt flow rate at $300^\circ C/1.2$ kg of 34 $cm^3/10$ minutes. The coconut shell powder with a particle size of about 150 microns was obtained from Essentium Materials LLC, Texas.

Polycarbonate with carbon nano-powder was dissolved using chloroform ($CHCl_3$, $\geq$ 99%) in order to create well-dispersed composites. Polymer blend was precipitated from chloroform by methanolysis for which methanol (CH_3OH, $\geq$ 99.9%) of HPLC grade was used. Both chloroform and methanol were purchased from Sigma-Aldrich Inc.

2.2 Synthesis of Carbon Nanopowder

In order to synthesize carbon from the coconut shell powder, an autogenic pyrolysis process was used. Approximately 30 gms of coconut shell powder was taken into the MTI Inc autogenic pressure reactor. The sample was heated at a rate of 5°C/min until it reaches 800°C and was then held for 2 hours isothermally with no external pressure. The carbon was then collected and subjected to ball milling for 10 hours for particle-size reduction.

Ball milling is a top-down approach that is used for the synthesis of nanoparticles which uses mechanical attrition to reduce the particle size. Mechanical attrition is a process that involves working on the particle by impact and collision to reduce the size. In a ball mill, the energy generated by the ball-to-ball and ball-to-wall collision breaks the lumps of the materials into smaller pieces. In this study, a vibratory ball mill (8,000D mixer/mill) from SPEX SamplePrep was used for ball milling. It uses two zirconia vials (6.35 cm diameter and 68 cm long) with two zirconia balls each (12.7 mm diameter). The carbon powder was loaded into the vials and then was ball milled for 10 hours after which the powder was collected, dried and stored for further characterization. The as-obtained carbon after pyrolysis is labeled as CCSP, whereas the carbon powder after ball milling is labeled as CCSP 10.

2.3 Characterization of Carbon Nanopowder

2.3.1 X-ray Diffraction

A Rigaku DMAX 2200 diffractometer with monochromatic Cu Kγ radiation (γ = 0.154056 nm) was used for X-ray diffraction (XRD) analysis. The diffraction analysis was carried out at 40 kV, 30 mA and 1.2 kW within 2θ range of 10° to 80° at a scan rate of 2°/minute.

2.3.2 Raman Analysis

Thermo-Scientific DXR Raman spectroscopy with a laser excitation wavelength of 785 nm was used for Raman spectrometry. The graphitization degree of CCSP was analyzed within the spectrum range of 0 to 2,500 cm^{-1} with laser power of 5.0 mW.

2.3.3 Microscopy

The surface morphology and the microstructure of the synthesized CCSP were examined using scanning electron microscopy (SEM) and transmission electron microscopy (TEM). A JSM-7200F field emission scanning electron microscope (JEOL USA, Peabody, MA) was used for SEM.

A JOEL 2010 TEM was used for transmission electron microscopy at an operating voltage of 200 kV. The carbon nanopowder was dispersed in ethanol and then administered on a Cu grid prior to characterization.

2.4 Extrusion of Composite Filaments

The polycarbonate pellets were dissolved in chloroform. The carbon nanopowder was added to this polymer solution at different loadings of 0.3, 0.7, 1 and 3 wt% of carbon. The carbon-added polymer solution was magnetically stirred for 12 hours and is further subjected to mechanical mixing to obtain a homogenous solution. Methanol was then added to the solution to precipitate the polymer composite. The precipitate was then filtered and dried overnight in a vacuum. The obtained precipitate is crushed into powder using mortar and pestle for extrusion.

The polymer blend thus obtained is fed into the EX2 Fil-a-Bot extruder (VT, USA) continuously to extrude the polycarbonate bio composite filaments. The extrusion temperature was maintained at 260°C for all the blends with different loadings. The diameter of the extruded filaments lies within the range of 1.5 to 1.75 mm, which is suitable for 3D printing application.

2.5 3D Printing of Thin Films

The polycarbonate filaments were 3D printed using Hyrel Systems 3D printer with a bed temperature set to 70°C and a printing temperature of 260°C. FreeCAD and Slic3r were used to design the samples for the printing process. The thickness and dimension of the single-layer was $120 \times 20 \times 0.11$ mm^3. The layer printed was in a 0° motion with a printing speed of 30 mm/s. The rectilinear infill was 100%. The retraction length was 1 mm, while the retraction speed was 30 mm/s and the nozzle diameter was 0.6 mm. A detailed description of the printing technique was discussed in a similar study by Umerah et al. (2020).

2.6 Characterization of Nanocomposite Filaments and Films

2.6.1 Thermogravimetric Analysis

A TA Q500 TGA was used to carry out thermogravimetric analysis (TGA) to obtain decomposition temperatures and the weight of the residue left from the heat flow of the specimens. The sample of approximately 15 mg was placed in a platinum pan and is heated at a rate of 10°C/minute from 30°C to 500°C in the presence of nitrogen gas.

2.6.2 Differential Scanning Calorimetry

A TA Q20000 DSC was used to analyze the thermal behavior of the composite filaments. The sample of approximately 10 to 12 mg was sealed in an aluminum pan and is heated at a rate of 5°C/min from 30°C to 380°C and then again cooled down to 30°C in the presence of nitrogen gas.

2.6.3 Tensile Test

The tensile test was performed using the Zwick/Roell Z2.5 Universal Mechanical Testing Machine of 2.5 kN load cell was used to determine the mechanical properties of the filaments as well as the thin films. The composite filaments were tested following ASTM D3379 standard. The gage of the filament was 50 mm with the specimen length being 100 mm, the pre-load tension of 0.1 N, the pre-load speed of 0.5 mm and the test speed of 0.5 mm/min. The gage length for the thin films was considered as 20 mm with the pre-load tension being 0.1 N and with a pre-load speed of 0.5 mm and a test speed of 3 mm/min. Five specimens for each blend system were tested and the average of the results was considered. The data was acquired from TestXpert data acquisition and analysis software and tensile modulus and strength were evaluated from the data obtained.

3. Results and Discussion

3.1 Morphology and Characterization of Carbon Nanopowder

The morphology and the characteristic features of the carbon obtained from the coconut shell powder are analyzed by using XRD and Raman spectroscopy as shown in Figure 1. The XRD of the CCSP exhibits two peaks at 23° which corresponds to the 002 plane and at 44.5° which corresponds to the 001 plane (Figure 1a). These two peaks suggest the presence of partially crystalline graphitic planes in the as-obtained carbon powder (Idrees et al. 2018, Pechyen et al. 2007). The characteristic peaks of normal graphite 2H are present at 26.38° (002 plane) and 44.39° (101 plane) according to JCPDS (Joint Committee on Powder Diffraction Standard) file 41-1487 (Goni et al. 2019). These two peaks are observed at 23° and 44.6° for the carbon powder that is obtained after ball milling, which does not show any significant change compared to CCSP. However, it was observed that wide peaks were present for CCSP10 confirming that the size of the particles was reduced compared to CCSP.

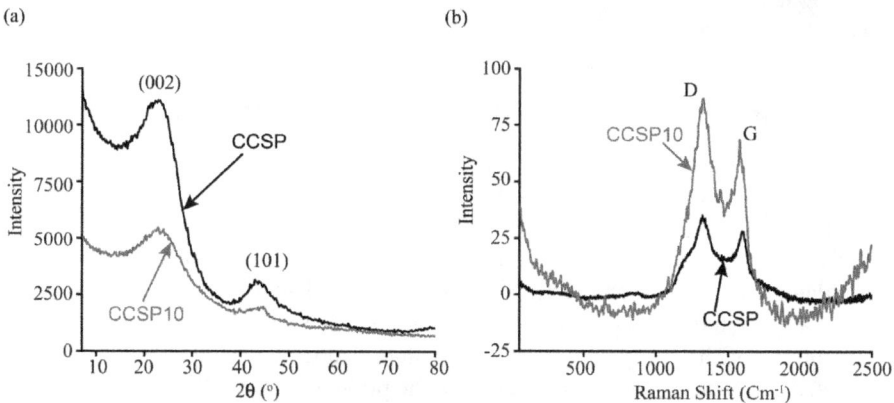

Figure 1: (a) XRD of the carbon before and after ball milling and (b) Raman spectroscopy of the carbon powder before and after ball milling.

The degree of graphitization in CCSP and CCSP10 were analyzed by Raman spectroscopy as shown in Figure 1(b). The increase in the intensity for the spectra confirms the size reduction of the particles for CCSP10. The D and G bands for CCSP were observed at 1,323.7 cm^{-1} and 1,598.5 cm^{-1}, whereas for CCSP10 D and G bands were present at 1,323.5 cm^{-1} and 1,576.2 cm^{-1}, respectively. Although significant change is not present for the D band, the wavenumber of the G band from CCSP10 decreases compared to CCSP. This may be due to the reduction in particle size due to ball milling, which might have disintegrated the graphitic structure while inducing defects (Umerah et al. 2020). This is also confirmed by the increase in the

ID/IG ratio of CCSP10, which is 1.28 when compared to the ID/IG ratio of CCSP that is 1.26 where ID and IG represent the intensities D and G band, respectively.

The surface morphology and the microstructure of the carbon powder were analyzed using SEM and TEM as shown in Figure 2. The SEM micrographs of as-obtained carbon (CCSP) and its magnified view are shown in Figure 2(a). The micrograph shows that the carbon particles are irregular in shape and size and have sheet-like structures. Figure 2(b) shows that the size of the carbon particles was reduced to nano-size after ball milling. The size of the particles was observed between 45 to 87 nm. The sheet structures were also observed to be deformed in Figure 2(b). The TEM micrograph confirms that the size of the particles is in the nano-range and the presence of sheet-like structures, irrespective of the irregular shape of the particles.

Figure 2: Scanning electron microscopy of (a) CCSP (b) CCSP10 and (b) Transmission electron microscopy of CCSP10.

3.2 Thermomechanical Analysis of Polymer Nanocomposites

The thermal properties of the PC carbonate filaments infused with carbon from coconut shell powder were analyzed by TGA and DSC as shown in Figure 3. The percentage weight change of the filaments is shown in Figure 3 (a). There is no significant change in the onset temperatures of composite filaments compared to neat filaments. The decomposition temperatures and residues of the composites were analyzed and are shown in Table 1. Figure 3(b) suggests that the degradation temperatures of the composites were lower than that of neat filaments which might be caused due to the disordered structure of the polymer chains due to the infusion of carbon particles (Abdelaziz 2015). The residue amount increased significantly with the increase in the carbon content (Table 1), suggesting that the carbon nanoparticles might have caused hindrance to the sensitive chemical groups present in the polymer and thus reduced the weight loss. Thus, the incorporation of carbon nanoparticles is highly favorable in case of weight loss issues. The thermal stability is further analyzed by considering the glass transition temperature (T_g) that is obtained from DSC thermographs as shown in Figure 3(c). The T_g of the neat polycarbonate filament is obtained at 148.71°C. There is no significant change in the T_g of the filaments and is slightly lowered with the addition of carbon particles as shown in Table 1. The decrease in the T_g might be due to the restricted mobility of the polymer chains, which might have been caused due to the addition of carbon nanoparticles (Abdelaziz 2015).

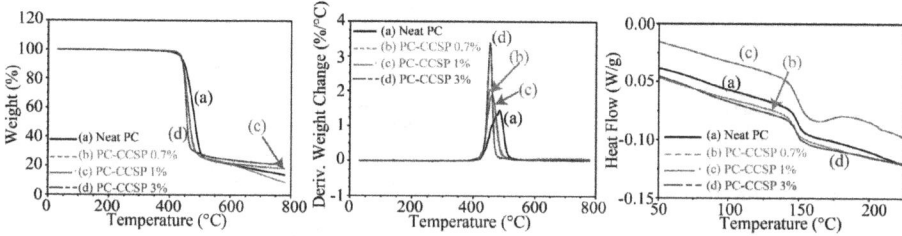

Figure 3: Thermal analysis of PC filaments infused with CCSP: (a) weight change from TGA, (b) derivative weight change from TGA and (c) DSC thermographs.

Table 1: Summary of thermal analysis

Specimen	Major Degradation (°C)	Residue Left (%)	Glass Transition Temperature (°C)
Neat PC	484.56	14.03	148.71
PC-CCSP 0.7%	457.34	8.97	146.68
PC-CCSP 1%	457.17	18.43	147.25
PC-CCSP 3%	453.64	21.28	147.8

Table 2: Summary of tensile properties

Specimen		Elastic Modulus (GPa)	Tensile Strength (MPa)
Neat PC	Filament	3.26±0.62	63.2±1.02
	3D printed film	2.17±0.28	54.61±0.71
PC-CCSP 0.7%	Filament	3.77±0.14	63.69±2.8
	3D printed film	2.68±0.84	53.71±1.7
PC-CCSP 1%	Filament	4.1±0.28	65.85±1.44
	3D printed film	3.33±0.36	56.91±4.04
PC-CCSP 3%	Filament	3.06±0.65	64±1.62
	3D printed film	2.16±0.33	53.44±1.26

Figure 4 displays the stress-strain curve of extruded and 3D printed polycarbonate-CCSP composites. CCSP as a filler increases the mechanical strength in the polycarbonate matrix. The tensile strength of polycarbonate increases with the addition of CCSP as shown in Figure 4a. Moderate loading of CCSP significantly increases the tensile strength. Similar results were observed when graphene sheets were induced in PC at low loadings (Wang et al. 2017). Elastic modulus increases with filler content; however, the modulus does decrease with the 3 wt% sample shown in Figure 4(a). This may be due to the agglomeration of particles, which have caused the strength to drop significantly at 3 wt%. Similar behavior is observed for the 3D printed films for elastic modulus as well as the tensile strength. PC composite with

(a)   (b)

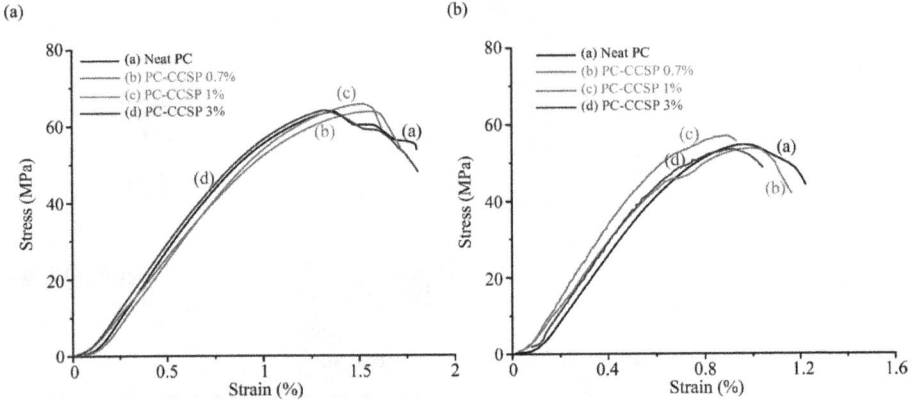

Figure 4: Tensile behavior of carbon infused polycarbonate nanocomposite:
(a) filaments and (b) 3D printed films.

1% CCSP loading exhibited superior mechanical properties compared to the other composites. When the loading reached 3%, the properties degraded and were lower than the PC. The tensile strength and the elastic modulus of the thin films were lower than the filaments. This might be due to the formation of voids between the printed polymer layers. In order to enhance the properties of the 3D printed thin films, they can be subjected to a surface treatment process, which is considered for future work.

4. Conclusion

Carbon nanopowder from waste coconut shell powder is successfully synthesized and then characterized using XRD, Raman, SEM and TEM. The SEM and TEM results confirmed that the synthesized carbon powder is in the nano-range and has sheet-like structures. The carbon nanopowder thus obtained is infused into the polycarbonate matrix to obtain polymer nanocomposite systems at various loadings. These blends were extruded into polymer nanocomposite filaments and then were 3D printed into thin films. The thermal properties of the filaments were analyzed using TGA and DSC which showed that the onset temperature from TGA as well as the glass transition temperature from DSC was slightly decreased due to the addition of carbon nanopowder. However, the residue amount was significantly increased due to the emission of volatile gases. The mechanical properties were enhanced at the lower loadings of the carbon powder. As the filler loading increases, the mechanical properties tend to decrease. The thermomechanical analysis of these polycarbonate composite systems suggests that the carbon can be used in moderate quantities as a filler to enhance the properties of the filaments and can be successfully used for 3D printing applications.

Acknowledgments

The authors would like to acknowledge the financial support of NSF-RISE #1459007 NSF-CREST# 1735971, Al-EPSCoR# 1655280, and NSF-MRI-1531934.

Conflicts of Interest

The authors do not have any conflicts of interest.

References

Abdelaziz, M. 2015. The effects of carbon nanoparticles on thermal and dielectric properties of bisphenol A polycarbonate. J. Thermoplast. Compos. Mater. 28: 1026-1046. https://doi.org/10.1177/0892705713495436

Belter, J.T. and Dollar, A.M. 2015. Strengthening of 3D printed fused deposition manufactured parts using the fill compositing technique. PLoS ONE 10: 1-19. https://doi.org/10.1371/journal.pone.0122915

Bourmaud, A., Beaugrand, J., Shah, D.U., Placet, V. and Baley, C. 2018. Towards the design of high-performance plant fibre composites. Prog. Mater. Sci. 97: 347-408. https://doi.org/10.1016/j.pmatsci.2018.05.005

Das, O., Bhattacharyya, D., Hui, D. and Lau, K.-T. 2016. Mechanical and flammability characterisations of biochar/polypropylene biocomposites. Compos. Part B Eng. 106: 120-128. https://doi.org/https://doi.org/10.1016/j.compositesb.2016.09.020

Goni, A., Rampe, M.J., Kapahang, A. and Turangan, T.M. 2019. Making and characterization of active carbon from coconut shell charcoal. Int. J. Adv. Educ. Res. 4: 40-43.

Ho, M.P., Lau, K.T., Wang, H. and Hui, D. 2015. Improvement on the properties of polylactic acid (PLA) using bamboo charcoal particles. Compos. Part B Eng. 81: 14-25. https://doi.org/10.1016/j.compositesb.2015.05.048

Idrees, M., Jeelani, S. and Rangari, V. 2018. Three-dimensional-printed sustainable biochar-recycled PET composites. ACS Sustain. Chem. Eng. 6: 13940-13948. https://doi.org/10.1021/acssuschemeng.8b02283

Imam, M.A., Jeelani, S. and Rangari, V.K. 2019. Thermal decomposition and mechanical characterization of poly (lactic acid) and potato starch blend reinforced with biowaste SiO_2. J. Compos. Mater. 53: 2315-2334. https://doi.org/10.1177/0021998319826377

Ji, A., Zhang, S., Bhagia, S., Yoo, C.G. and Ragauskas, A.J. 2020. 3D printing of biomass-derived composites: application and characterization approaches. RSC Adv. 10: 21698-21723. https://doi.org/10.1039/d0ra03620j

Li, S., Li, X., Chen, C., Wang, H., Deng, Q., Gong, M. and Li, D. 2016. Development of electrically conductive nano bamboo charcoal/ultra-high molecular weight polyethylene composites with a segregated network. Compos. Sci. Technol. 132: 31-37. https://doi.org/https://doi.org/10.1016/j.compscitech.2016.06.010

Mazzanti, V., Malagutti, L. and Mollica, F. 2019. FDM 3D printing of polymers containing natural fillers: a review of their mechanical properties. Polymers (Basel). 11. https://doi.org/10.3390/polym11071094

Mohanty, A.K., Vivekanandhan, S., Pin, J.M. and Misra, M. 2018. Composites from renewable and sustainable resources: challenges and innovations. Science 362: 536-542. https://doi.org/10.1126/science.aat9072

Ning, F., Cong, W., Hu, Y. and Wang, H. 2017. Additive manufacturing of carbon fiber-reinforced plastic composites using fused deposition modeling: effects of process parameters on tensile properties. J. Compos. Mater. 51: 451-462. https://doi.org/10.1177/0021998316646169

Pechyen, C., Atong, D., Aht-Ong, D. and Sricharoenchaikul, V. 2007. Investigation of pyrolyzed chars from physic nut waste for the preparation of activated carbon. J. Solid Mech. Mater. Eng. 1: 498-507. https://doi.org/10.1299/jmmp.1.498

Rahim, T.N.A.T., Abdullah, A.M. and Md Akil, H. 2019. Recent developments in fused deposition modeling-based 3D printing of polymers and their composites. Polym. Rev. 59: 589-624. https://doi.org/10.1080/15583724.2019.1597883

Smith, W.C. and Dean, R.W. 2013. Structural characteristics of fused deposition modeling polycarbonate material. Polym. Test. 32: 1306-1312. https://doi.org/https://doi.org/10.1016/j.polymertesting.2013.07.014

Sodeifian, G., Ghaseminejad, S. and Yousefi, A.A. 2019. Preparation of polypropylene/short glass fiber composite as Fused Deposition Modeling (FDM) filament. Results Phys. 12: 205-222. https://doi.org/10.1016/j.rinp.2018.11.065

Umerah, C.O., Kodali, D., Head, S., Jeelani, S. and Rangari, V.K. 2020. Synthesis of carbon from waste coconut shell and their application as filler in bioplast polymer filaments for 3D printing. Compos. Part B Eng. 202: 108428. https://doi.org/https://doi.org/10.1016/j.compositesb.2020.108428

Wang, J., Li, C., Zhang, X., Xia, L., Zhang, X., Wu, H. and Guo, S. 2017. Polycarbonate toughening with reduced graphene oxide: toward high toughness, strength and notch resistance. Chem. Eng. J. 325: 474-484. https://doi.org/10.1016/j.cej.2017.05.090

Wang, X., Jiang, M., Zhou, Z., Gou, J. and Hui, D. 2017. 3D printing of polymer matrix composites: a review and prospective. Compos. Part B Eng. 110: 442-458. https://doi.org/10.1016/j.compositesb.2016.11.034

Yang, E., Miao, S., Zhong, J., Zhang, Z., Mills, D.K. and Zhang, L.G. 2018. Bio-based polymers for 3D printing of bioscaffolds. Polym. Rev. 58: 668-687. https://doi.org/10.1080/15583724.2018.1484761

Section II – Biomedical, Wastewater Treatment and Environmental Applications

Part III: Bioscience and Medical Applications

Nanotechnology Against SARS-CoV-2 and Other Human Coronavirus

Shabnam Sharmin[1], Md. Mizanur Rahaman[1], Olubunmi Atolani[2] and Muhammad Torequl Islam[1]*

[1] Department of Pharmacy, Life Science Faculty, Bangabandhu Sheikh Mujibur Rahman Science and Technology University, Gopalganj 8100, Bangladesh
[2] Department of Chemistry, University of Ilorin, P.M.B. 1515, Ilorin, Nigeria

1. Introduction

Novel coronavirus (SARS-CoV-2), which is currently causing a stir in the world, was allegedly spread from China's Hubei province of Wuhan in December 2019 (Gaurav et al. 2020, Medhi et al. 2020, Zu et al. 2020). So far, 138,843,235 persons are reportedly infected with the virus as the number of fatalities increased to 2,985,676 globally (WHO COVID-19 dashboard 2021). In cases when the disease starts with mild fever, tiredness, and shortness of breath, followed by acute cough, diarrhea, failure of the immune system, its last consequence is often death in severe patients (Rabiee et al. 2020, WHO COVID-19).

While the main causative agent behind this SARS-CoV-2 associated with pneumonia is a novel coronavirus, officially termed as a severe acute respiratory syndrome-related coronavirus or SARS-CoV-2 (Du Toit 2020, Li et al. 2020, Wu et al. 2020), the morphological analysis of SARS-CoV-2 claims it is a positive-sense single-stranded ribonucleic acid [(+) ssRNA], which belongs to the family *Coronaviridae* (Udugama et al. 2020, Zhou et al. 2020). This enveloped virus is highly defended by four structural proteins namely, the spike glycoprotein (S protein) for attachment and entrance into the host cell; the membrane protein (M protein), for membrane's integrity maintenance; the envelope protein (E protein), for assembly and budding; and the nucleocapsid protein (N protein), for nucleocapsid formation and SARS-CoV RNA binding (Tooze et al. 1984, Mortola and Roy 2004, Masters 2006, Liu et al. 2014, Neuman et al. 2011). From the genome structure study, it has been confirmed that SARS-CoV-2 shares the same genome structure as the coronaviruses (CoVs), more specifically β-CoVs. SARS-CoV-2 shares higher

---

Corresponding author: dmt.islam@bsmrstu.edu.bd

similarities with SARS-CoV since they both use angiotensin-converting enzyme-2 as their main entry receptor (Cui and Shi 2019).

Structurally, the SARS-CoV-2 virus has a diameter of 60-140 nm, which falls into the same range of size for nanoparticles. For this reason, nanoparticles have been receiving increasing attention as likely therapeutic agents against the virus. Nanoparticles, having sizes similar to those of viral particles, may interact and alter the structural integrity of the viral proteins (Chan 2020, Liu et al. 2020, Udugama et al. 2020). Nanoparticles have a diameter within the range of 1-100 nm (Vert et al. 2012). Due to this unique small size range, nanoparticles are being investigated for use in tackling SARS-CoV-2 (Chan 2020). Nanoparticles exhibit potent antiviral activity when tested against different viral strains, including Herpex Simplex Virus 1 (HSV-1), human immunodeficiency virus (HIV), influenza A virus subtype H1N1 (swine flu), and hepatitis C virus (HCV) (Cavalli et al. 2009, Vijayakumar and Ganesan 2012, Mori et al. 2013, Hang et al. 2015, Kumar et al. 2018, Haggag et al. 2019). Therefore, it is plausible that nanoparticles have an inhibitory effect on the virus from the *Coronaviridae* family, such as porcine epidemic diarrhea virus (PEDV), transmissible gastroenteritis virus (TGEV), human coronavirus (HCoV-NL63, HCoV-229E) along with other viruses with lethal effects (Staroverov et al. 2011, Lv et al. 2014, Ciejka et al. 2017, Ting et al. 2018, Łoczechin et al. 2019). Abo-Zeid et al. (2020) from their recent molecular docking study claimed that FDA-approved iron oxide nanoparticles (Fe_2O_3 and Fe_3O_4) for the treatment and control of SARS-CoV-2 through a dual interaction led to reactive oxygen species (ROS) release and eventual viral inhibition. As an effort to develop a potent COVID-19 vaccine is underway, alternative treatment procedures such as the use of nanoparticles are being identified to prevent the disease. Thus, this review aims to summarize the antiviral effect of nanoparticles against SARS-CoV-2 and other CoVs that affect humans as well as their possible treatment modalities.

2. Methodology

Information was collected from the following databases: PubMed, Science Direct, MedLine and Google Scholar with the keywords 'Nanoparticles' sharing antiviral activity/effects as opposed to SARS-CoV-2 or other human CoVs. There are no restrictions imposed on language. Searched articles were evaluated for taking information about the antiviral activity of nanoparticles, possible dose or concentrations, microbial strains (test system), and their possible mechanism of action against SARS-CoV-2 or other human CoVs. The inclusion criteria were studies focused on nanoparticles and synthesized nanoparticles from different sources, studies with coronavirus strain and their mechanism of action, and studies on nano-weapons to fight against SARS-CoV-2. While the exclusion criteria were titles and/or abstract not meeting the inclusion criteria and nanoparticles with other studies obscuring the current subject of interest.

3. Findings

Among the vast shreds of evidence obtained, some randomly selected published articles found in the databases that contain screening reports on nanoparticles against viral infections of SARS-CoV-2 or other human CoVs are herein summarized.

3.1 Antiviral Activity of Nanoparticles against Different Human CoVs

From different studies, it has been demonstrated that nanomaterials (NMs) possess strong antiviral activity (Cojocaru et al. 2020). The current SARS-CoV-2 pandemic is a viral infection associated with pneumonia. Nanoparticles are reported to have antiviral efficiency when tested against SARS-CoV-2 and other models of coronavirus. Lv et al. (2014) reported the antiviral activity of silver (Ag) nano-materials, including nanoparticles and nanowires against transmissible gastroenteritis virus (TGEV), have a significant CoV due to decreased cell apoptosis and activation of p38/mitochondria-caspase-3 signaling in ST cells. Meanwhile, Ag_2S nanoclusters (3.7-5.3 nm) inhibit the synthesis of negative-strand RNA and viral budding of porcine epidemic diarrhea virus (PEDV), a model of CoV and exhibits antiviral activity (Du et al. 2018).

A number of nanoparticles also exhibit antiviral activity against SARS-CoV-2 by blocking angiotensin-converting enzyme 2 (ACE2), the main entry receptor. Silver nanoparticles and nanoparticle composite TPNT1 inhibit the viral entry step of SARS-CoV-2 via disrupting viral integrity (Chang et al. 2020, Jeremiah et al. 2020).

Carbon quantum dots are also reported for their activity against SARS-CoV-2. Cationic carbon dots (CCM-CDs) change the structure of the virus, thus inhibiting the entry and suppressing the synthesis of negative-strand RNA in PEDV and pathogenic human coronavirus HCoV-229E (Ting et al. 2018, Łoczechin et al. 2019). Quantum dots (semiconductor nanomaterials) ranging from 1-10 nm increase the drug release profile, subsequently targeting SARS-CoV-2 and thus inhibiting the activity of the SARS-CoV-2 virus (Manivannan and Ponnuchamy 2020). Some reported virucidal potencies of nanoparticles are indicated in Table 1.

3.2 Possible Nano-Weapons to Fight in COVID-19

Since nasal and oral routes are a primary route of transmission of SARS-CoV-2, the enforcement of social distancing, as well as personal hygiene maintenance, became the main mode of prevention against further spread of the virus. However, nanotechnology-based products has being recommended for use at the primary level of prevention (Nanotechnology Products Database 2020, WHO 2020b).

3.2.1 Nanotechnology-Based Facial Masks

A nanofiber-based facial mask suitable for SARS-CoV-2 has been developed by a KAIST research team, represented by professor Il-Doo Kim. This re-usable mask offers excellent filtering efficiency and maintains the alignment of nanofibers with a diameter of 100-500 nm in orthogonal or unidirectional configuration. Flextrapower Inc., a USA-based company, has developed a graphene filter mask using the antiviral property of graphene nanoparticles. Likewise, another American company, LIGC applications limited, manufactured a Guardian G-volt mask using graphene that is 99% effective against particles over 0.3 micrometers and shows 80% effectiveness against smaller particles. Integricote Inc., in collaboration with the University of

Table 1: Antiviral activity of nanoparticles on different types of Coronaviruses (CoVs)

Nanoparticle	Size (diameter)	CoV Type	Mechanism (in-vivo/in-vitro data)	Reference
Colloidal gold nanoparticles	15 nm	Enteropathogenic swine transmissible gastroenteritis (STG) coronavirus	Increased peritoneal macrophages respiratory activity and increased plasma interferon gamma (IFN-γ) level due to antigen-colloidal gold conjugation.	Staroverov et al. 2011
Ag$_2$S nanoclusters (NCs): Ag$_2$S NCs-681, Ag$_2$S NCs-722	3.7 nm 5.3 nm	Porcine epidemic diarrhea virus (PEDV) as a model of coronavirus	Inhibited synthesis of viral negative-strand RNA and viral budding. Interferon-stimulated gene (ISG) protein and pro-inflammatory cytokines activation.	Du et al. 2018
Cationic carbon dots (CCM-CDs)	-	PEDV as a model of coronavirus	Structural change in the virus, inhibited viral entry, suppressed synthesis of negative-strand RNA.	Ting et al. 2018
AgNMs including spherical AgNPs, two types of Ag nanowires, Ag colloids	NPs: < 20 nm, Nanowires: 60 and 400 nm, Colloids: 10 nm	Transmissible gastroenteritis virus (TGEV) a significant coronavirus	Decreased cell apoptosis caused by TGEV infection by activation of p38/mitochondria-caspase-3 signaling in stem cells.	Lv et al. 2014
Cationically modified chitosan nano/microspheres (HTCC-NS/MS)	60 nm ~2.5 μm	Human coronavirus NL63 (HCoV-NL63) and OC43 (HCoV-OC43)	Direct attractive interactions of HTCC with the virus.	Ciejka et al. 2017
Seven different carbon quantum dots (CQDs): First generation CQDs: CQDs-1, CQDs-2, CQDs-3, CQDs-4 Second generation CQDs: CQDs-5, CQDs-6, CQDs-7	CQDs-1 (4.5 ± 0.2 nm), CQDs-2 (5.5 ± 0.3 nm), CQDs-3 (6.3 ± 0.4 nm), CQDs-4 (6.5 ± 0.4 nm), CQDs-5 (9.2 ± 0.3 nm), CQDs-6 (7.6 ± 0.2 nm), CQDs-7 (8.0 ± 0.2 nm)	Human coronavirus 229E (HCoV-229E)	Due to the interaction of the functional group of CQDs with HCoV-229E entry receptor, viral entry is inhibited. Viral replication inhibition was also seen.	Łoczechin et al. 2019

FDA approved iron oxide nanoparticles (Fe_2O_3 and Fe_3O_4) (IONPs)	-	SARS-CoV-2	Interaction of IONPs with SARS-CoV-2 S1-RBD that leads to ROS release and conformational changes of viral structural proteins inhibits viral entry into host cells, limits viral replication and infection.	Abo-Zeid et al. 2020
Layered double hydroxide (LDH), an inorganic nanoparticle intercalated with the shRNA-plasmid	-	SARS-CoV-2	Conjugated nanoparticles with shRNA with specific sequence target viral RNA to degrade them (proposed hypothesis).	Acharya et al. 2020
saRNA lipid nanoparticles	-	SARS-CoV-2	High cellular responses, as characterized by IFN-γ production, upon re-stimulation with SARS-CoV-2 peptides.	Mckay et al. 2020
Graphene nanoplatelet and graphene oxide	-	SARS-CoV-2	Cotton and polyurethane materials functionalized with bidimensional graphene nanoplatelets trap SARS-CoV-2 and have the potential to reduce the spread of COVID-19.	De Maio. et al. 2020
Quantum dots (semiconductor nanomaterials)	1-10 nm	SARS-CoV-2	Therapeutic molecules functionalized or coated onto the surface of QDs increase the drug release profile, subsequently targeting COVID-19 that inhibit the activity of the SARS-CoV-2 virus.	Manivannan and Ponnuchamy 2020
Zinc nanoparticles	-	SARS CoV-2	Zn^{2+} showed antiviral activity through inhibiting SARS-CoV RNA polymerase.	Skalny et al. 2020

(Contd.)

Table 1: *(Contd.)*

Nanoparticle	Size (diameter)	CoV Type	Mechanism (in-vivo/in-vitro data)	Reference
BNT162b1 lipid-nanoparticles (vaccine candidate)	-	SARS-CoV-2	Encodes the trimerized receptor-binding domain (RBD) of the spike glycoprotein of SARS-CoV-2.	Mulligan et al. 2020
PVP-coated silver nanoparticles	-	Respiratory syncytial virus (RSV)	Interference of viral attachment to host cell.	Sun et al. 2008
Nanoparticle composite TPNT1	-	SARS-CoV-2	Block viral entry by inhibiting the binding of SARS-CoV-2 spike proteins to ACE2 receptor and interfering with the syncytium formation.	Chang et al. 2020
Silver nanoparticles	10 nm	SARS-CoV-2	Inhibits viral entry step *via* disrupting viral integrity.	Jeremiah et al. 2020

Houston, prepared a respiratory mask with a hydrophobic coating that can effectively prevent SARS-CoV-2. MVX nano-surgical mask, a product of MVX Prime Ltd, is acclaimed to kill 99.9% of all viruses that come into contact. Wakamomo, a Vietnamese company, has developed a mask using nanobiotechnology to stop COVID-19 as the used GECIDE nano-fabric has been tested for inactivation of human Coronavirus up to 99% (Nanotechnology Products Database 2020).

3.2.2 Nanotechnology-Based Sanitizers

EnvisionSQ, a Canadian manufacturing company, has developed a 'NanoCleanSQ' solution, a sanitizer with an excellent adhesive property that can stick to any surface and only goes after washing with water. Laboratory testing assured its' capacity to eliminate 99.9999% common bacteria and enveloped viruses, which include viruses inducing the renowned SARS-CoV-2. Similarly, Malaysian company SHEPROS has developed 'Nano Silver Hand sanitizer', a silver nanoparticle-based solution that effectively works against coronavirus with 99% elimination efficiency (Nanotechnology Products Database 2020). A list of some other nanoparticles-based products is indicated in Table 2.

3.2.3 Nanotechnology in COVID-19 Infection Detection

To win the battle against SARS-CoV-2, one of the major strategies adopted is to keep uninfected persons away from those who are infected. There is also a critical need to urgently design a highly standardized template for quick screening and detection of the virus whilst avoiding false negative/positive results. Currently, three diagnostic techniques include RT-PCR (reverse-transcription polymerase chain reaction), gene sequencing, POC (point-of-care diagnostic approach) that detects antibodies against SARS-CoV-2 in the patient sample using a lateral flow immunoassay, and CT (chest computed tomography) (Ai et al. 2020, Kim 2020, Wang et al. 2020). Moitra et al. (2020) claimed a selective naked-eye detection of SARS-CoV-2 within 10 minutes. In their report, thiol-modified ASO capped AuNPs agglomerate selectively in the presence of its target RNA sequence of SARS-CoV-2 and show a change in surface plasmon resonance with a redshift of ~40 nm in the absorbance spectra. Again, colloidal AuNPs-based lateral-flow assay is also used for rapid detection of IgM antibodies, specific antibodies present in blood against the SARS-CoV-2 virus (Huang et al. 2020). Furthermore, a lateral flow immunoassay using lanthanide-doped NPs has been developed for rapid and sensitive detection of anti-SARS-CoV-2 immunoglobulin G (IgG) in human serum with positive-identification permission in suspicious cases (Chen et al. 2020). A field-effect transistor (FET) based biosensor produced by coated graphene sheets with a specific antibody against SARS-CoV-2 is being used for the rapid detection of the virus in human nasopharyngeal swab specimens (Seo et al. 2020). Chiral zirconium quantum dots, a new class of nanocrystals, are used for the optical detection of coronavirus. The fluorescence properties of immune-conjugated QD-MP NPs, separated by an external magnetic field allow biosensing of coronavirus with a limit of detection of 79.15 EID/50 μL (Ahmed et al. 2018). A list of some nanoparticles-based detectors is indicated in Table 3.

Table 2: Some nanotechnology-based products with potential for prevention of the spread of COVID-19 virus

Product Name	Nanoparticles Used	Manufacturing Company	Purpose Served
Virucidal graphene-based composite ink	Silver-graphene oxide-based nanoparticles	ZEN Graphene Solutions Ltd, Canada.	SARS-CoV-2 virucidal composite ink that can be applied to the fabrics, masks, and personal protective equipment (PPE) for increased protection.
G+ Fibrics	Graphene nanoparticles	Directa Plus PLC, UK.	The non-toxic and bactericidal fabric used in the production of gowns, gloves, masks, PPE.
The New Clean	Mineral nano-crystals	Nano Touch Materials, LLC, USA.	Disinfectant that is used to clean elevator buttons, door handles, back of the phone, mats, and even skin.
PRELYNX Portal	Nano-polymer vapor	123Design, USA.	Medical scanner and sanitizer designed to stop the spread of SARS-CoV-2.
Liquid Guard	Not mentioned	Nano-Care Deutschland AG, Germany.	Disinfectant with broad-spectrum antimicrobial activity against various pathogens including SARS-CoV-2.
Mack Antonoff HVAC	Different nanoparticle-based filters	Mack Antonoff HVAC, USA.	Virus removal, clean air filtration system to combat SARS-CoV-2.
Silver Nano Colloid	Nano-sized silver colloids with an average size of 50 nm	Danesh Gostaran Azar Sakhtar, Iran.	Disinfectants are used in different fields as an industry, agriculture, cosmetics and packaging to prevent infections including SARS-CoV-2.

Source: Nanotechnology Products Database.

Table 3: Some nanotechnology-based products used to detect SARS-CoV-2

Product Name	Nanoparticle Used	Manufacturing Company	Purpose Served
ROS detector in sputum sample (RDSS)	Multi-wall carbon nanotube	Nano Hesgar Sazan Salamat Arya, Iran.	ROS/H_2O_2 electrochemical system consists of multi-wall carbon nanotube, used for CoVs detection.
SARS-CoV-2 rapid POC CE-IVD test	Gold (Au) nanoparticle/ nanopowder	NanoComposix, USA.	A lateral flow assay using AuNPs to detect SARS-CoV-2.
SARS-CoV-2 rapid test cassette	Gold (Au) nanoparticle/ nanopowder	SureScreen Diagnostics Ltd., UK.	Rapid detection of coronavirus through IgG, IgM antibody detection.
SAFER-sample kit	Not mentioned	Lucence Diagnostic Pte Ltd., Singapore.	Rapid diagnosis of RNA viral infection such as SARS-CoV-2.
SARS-CoV-2 point-of-need diagnostic test	Gold (Au) nanoparticle/ nanopowder	Mologic Ltd., UK.	Rapid test for exposure to the virus providing results within 10 minutes.
MinION sequencer	Not mentioned	Oxford Nanopore Technologies Ltd., UK.	A portable, real-time device for DNA and RNA sequencing and rapid detection of SARS-CoV-2.

Source: Nanotechnology Products Database.

4. Nanotechnology-based Therapeutic Approach against SARS-CoV-2

4.1 Nanotechnology in SARS-CoV-2 Vaccine Development

Vaccines are one of the most constructive and cost-efficient ways to combat infectious diseases (Remy et al. 2014). Till now different companies all across the world are trying their best to develop a vaccine that is suitable for the SARS-CoV-2 pandemic. Currently, there are over 169 vaccine candidates under development, with 26 of them in the clinical trial phase (World Health Organization, Coronavirus disease data 2020). Tai et al. (2020), in their recent studies, reported the activity of a novel receptor-binding domain (RBD)-based mRNA vaccine against SARS-CoV-2 using lipid nanoparticles (LNPs). The lipid nanoparticles detect whether the designed S1 and RBD mRNAs stably express antigen in specific cells. A particular advantage of these sorts of vaccines is the stimulation of antibody and CD^{4+} T cell response, and they additionally evoke CD^{8+} cytotoxic T cell response which completely exterminates virus (Pardi et al. 2018, Smith et al. 2020). The collaboration between BioNTech and Pfizer has developed a potential mRNA-based COVID-19 vaccine BNT162b2, which is expected to enter its phase-3 clinical trial soon.

DNA vaccines are another type of nucleic acid-based vaccine that offers higher stability than the mRNA vaccine. DNA vaccine incorporates a DNA plasmid molecule that encodes one or more than one antigen (Liu 2019). Mediphage Bioceuticals, a company based in Canada, has developed a DNA vaccine that activates the body's natural immune response to fight against SARS-CoV-2. Entos Pharmaceuticals, another Canadian company, has developed a vaccine that is on track for its phase-1 clinical trial.

Subunit vaccine, another type of vaccine, is reported to have activity against SARS-CoV-2. An adjuvant is required to erect a potent immune response from this type of vaccine (Moyle and Toth 2013). Till now different research has been conducted on 'the molecular clamp technology', a synthetic polymerase to develop a subunit vaccine (Takashima et al. 2011). The University of Queensland's potential SARS-CoV-2 vaccine was developed in collaboration with Viroclinics Xplore, a Netherlands-based company, that used molecular clamp technology which locks the spike protein into a shape that allows the immune system to recognize and then neutralizes the virus. NVX-CoV2373 vaccine established by Novavax Inc. involved a recombinant protein nanoparticle technology platform to generate antigens derived from the coronavirus spike protein (Nanotechnology Products Database 2020). A list of some nanoparticles-based vaccine candidates is indicated in Table 4.

5. Carbon-Based Nanomaterials against SARS-CoV-2

Carbon-based nanomaterials can be a promising solution to this emerging crisis. Starting from primary protective materials, detection processes to the discovery of vaccines, carbon nanoparticles are now being used in the hope that they might combat SARS-CoV-2. Different companies have developed graphene-based filter masks using the antiviral property of graphene nanoparticles. As mentioned before, an American company, LIGC applications limited manufactured a Guardian G-volt mask using graphene that is 99% effective against particles over 0.3 micrometers. Again graphene-enhanced protective face mask manufactured by Versarien plc provides a minimum filtration of 95%. Graphene-based composite ink is another product applied to fabrics to enhance virucidal activity (Nanotechnology Products Database 2020).

As is used in the process of infection prevention, carbon-based nanoparticles have shown great progress in the section of infection detection. The 2D hexagonally arranged, thickly layered single-based carbon atom, graphene is being used in bio-sensing platforms (Li et al. 2020a). Properties like high ionic mobility and great surface area allow them to be great bio-sensors (Peña-Bahamonde et al. 2018). A field-effective transistor (FET), using a bio-sensing platform that has been developed recently, has successfully detected SARS-CoV-2 from human nasopharyngeal swabs (Seo et al. 2020).

Graphene, carbon nanotubes, and nanodiamonds hold a great capacity to activate the immune system (Pescatori et al. 2013, Orecchioni et al. 2017, Fusco et al. 2020).

Table 4: Nanotechnology-based possible vaccine candidates against SARS-CoV-2

Vaccine Candidate	Manufacturing Company	Mechanism
DNA-based vaccine	Mediphage Bioceuticals, Inc Canada.	Activate body's immune response to protect against SARS-CoV-2
TNX-1800	Tonix Pharmaceuticals Holding Corp., USA.	A potential horsepox vaccine that expresses protein from the virus that causes COVID-19.
VLP (virus-like particle) vaccine	Medicago Inc., Canada.	VLP is the first step in the development of the COVID-19 vaccine which is further analyzed for antibody development.
GV-MVA-VLPTM	GeoVax, Inc., USA.	Stimulates both humoral and cellular arms of the immune system to recognize, prevent, and control the target infection.
Fusogenix DNA vaccine	Entos Pharmaceuticals, Canada.	Plasmid DNA in vaccines allows optimized payload encoding multiple protein epitopes to stimulate the body's natural antibody production which prevents SARS-CoV-2 infection.
Arcturus' STARR™ Tecnology based vaccine	Arcturus Therapeutics Ltd., USA.	Self-replicating RNA-based vaccine triggers rapid and prolonged antigen expression within the host cell which builds protective immunity against SARS-CoV-2.
1c-SApNP vaccine	Ufovax, LLC, USA.	As a VLP, the nanoparticle vaccine induces the immune system to rapidly generate antibodies that deactivate the SARS-CoV-2.
The University of Queensland's potential COVID-19 vaccine.	Viroclinics Xplore, Netherlands.	The molecular clamp technology-based vaccine locks the spike protein into a shape that allows the immune system to recognize and then neutralize the virus.
Ad5-nCoV vaccine	Cansino Biologics Inc., China.	The genetically engineered vaccine carries the replication-defective adenovirus type 5 as a vector to express SARS-CoV-2 spike protein.
mRNA- 1273 vaccine	Moderna, Inc., USA.	The mechanism is to instruct the patients' cells to produce a protein that could prevent, cure, and treat the disease.
NVX-CoV2373 vaccine	Novavax, Inc., USA.	The proprietary recombinant protein nanoparticle technology platform generates antigens derived from the coronavirus spike protein.

Source: Nanotechnology Products Database.

Graphene oxide with functionalized amino groups activates STAT1/IRF1 interferon signaling in T-cells and monocytes that produces more T-cell chemoattractants, macrophage 1 (M1) 1/T-helper 1 (Th1) polarizing immune system with less toxicity (Orecchioni et al. 2017). Developing an effective vaccine against SARS-CoV-2 is the best solution to look out for. In this context, nanomaterial-based vaccine adjuvants can be of great help (Zhu et al. 2014). Ma et al. (2015) stated that GO nanomaterial excites an inflammasome biosensor (NLRP3)-dependent expression of IL-1β in macrophages. This characteristic can be used while developing vaccines.

6. Conclusion

With the current outbreak of the SARS-CoV-2 pandemic, it has become imperative to proffer urgent and veritable solutions. Nanoparticles that have been in use for diverse human welfare purposes for a very long time can be a suitable alternative if the potential is well harnessed. Nanoparticles have a diameter that is similar to the size of SARS-CoV-2, the causative virus for COVID-19 disease. From different studies, it has been seen that nanoparticles can easily mimic the characteristics of the SARS-CoV-2, which makes it a suitable candidate for antiviral therapy or therapeutic carrier. Starting from PPE to vaccines, nanoparticles are being used in different sectors as a promising weapon to fight against SARS-CoV-2. In this review, we have highlighted the antiviral activity of nanoparticles alongside their potential prospects in the nanotechnology-based treatment process.

References

Abo-Zeid, Y., Ismail, N.S., McLean, G.R. and Hamdy, N.M. 2020. A molecular docking study repurposes FDA approved iron oxide nanoparticles to treat and control COVID-19 infection. Eur. J. Pharm. Sci. 153: 105465.

Acharya, R. 2020. Prospective vaccination of COVID-19 using shRNA-plasmid-LDH nanoconjugate. Med. Hypoth. 143: 110084.

Ahmed, S.R., Kang, S.W., Oh, S., Lee, J. and Neethirajan, S. 2018. Chiral zirconium quantum dots: a new class of nanocrystals for optical detection of coronavirus. Heliyon, 4(8): e00766.

Ai, T., Yang, Z., Hou, H., Zhan, C., Chen, C., Lv, W., Tao, Q., Sun, Z. and Xia, L. 2020. Correlation of chest CT and RT-PCR testing in coronavirus disease 2019 (COVID-19) in China: a report of 1014 cases. Radiology 200642.

Cavalli, R., Donalisio, M., Civra, A., Ferruti, P., Ranucci, E., Trotta, F. and Lembo D. 2009. Enhanced antiviral activity of Acyclovir loaded into β-cyclodextrin-poly (4-acryloylmorpholine) conjugate nanoparticles. J. Control. Release 137(2): 116-122.

Chan, W.C.W. (2020). Nano research for COVID-19. ACS Nano, 14: 3719-3720.

Chang, S.-Y., Huang, K.-Y., Chao, T.-L., Kao, H.-C., Pang, Y.-H., Lu, L., Chiu, C.-L., Huang, H.-C., Cheng T.-J.R., Fang, J.-M. and Yang, P.-C. 2021 Nanoparticle composite TPNT1 is effective against SARS-CoV-2 and influenza viruses. Sci. Rep. 11(1): 8692.

Chen, Z., Zhang, Z., Zhai, X., Li, Y., Lin, L., Zhao, H., Bian, L., Li, P., Yu, L., Wu, Y. and Lin, G. 2020. Rapid and sensitive detection of anti-SARS-CoV-2 IgG, using lanthanide-doped nanoparticles-based lateral flow immunoassay. Anal. Chem. 92(10): 7226-7231.

Ciejka, J., Wolski, K., Nowakowska, M., Pyrc, K. and Szczubiałka, K. 2017. Biopolymeric nano/microspheres for selective and reversible adsorption of coronaviruses. Materials Sci Eng: C 76: 735-742.

Cojocaru, F.D., Botezat, D., Gardikiotis, I., Uritu, C.M., Dodi, G., Trandafir, L., Rezus, C., Rezus, E., Tamba, B.-I. and Mihai, C.-T. 2020. Nanomaterials designed for antiviral drug delivery transport across biological barriers. Pharmaceutics, 12(2): 171.

Cui, J., Li, F. and Shi, Z.L. 2019. Origin and evolution of pathogenic coronaviruses. Nat. Rev. Microbiol. 17(3): 181-192.

De Maio, F., Palmieri, V., Babini, G., Augello, A., Palucci, I., Perini, G., Salustri, A., Spilman, P., Spirito, M.D., Sanguinetti, M., Delogu, G., Rizzi, L.G., Cesareo, G., Soon-Shiong, P., Sali and M. and Papi, M. 2021. Graphene nanoplatelet and graphene oxide functionalization of face mask materials inhibits infectivity of trapped SARS-CoV-2. medRxiv. iScience 24(7): 102788.

Du Toit, A. 2020. Outbreak of a novel coronavirus. Nat. Rev. Microbiol 18(3): 123-123.

Du, T., Liang, J., Dong, N., Lu, J., Fu, Y., Fang, L., Xiao, S. and Han, H. 2018. Glutathione-capped Ag_2S nanoclusters inhibit coronavirus proliferation through blockage of viral RNA synthesis and budding. ACS Appl. Materials Interfaces 10(5): 4369-4378.

Fusco, L., Avitabile, E., Armuzza, V., Orecchioni, M., Istif, A., Bedognetti, D., Ros, T.D. and Delogu, L.G. 2020. Impact of the surface functionalization on nanodiamond biocompatibility: a comprehensive view on human blood immune cells. Carbon, 160, 390-404.

Gaurav, C., Marc, J.M., Sourav, K., Vianni, C., Deepa, G. and Sergio, O. 2020. Nanotechnology for COVID-19: therapeutics and vaccine research. ACS Nano. 14(7): 7760-7782.

Haggag, E.G., Elshamy, A.M., Rabeh, M.A., Gabr, N.M., Salem, M., Youssif, K.A. and Abdelmohsen, U.R. 2019. Antiviral potential of green synthesized silver nanoparticles of Lampranthus coccineus and Malephora lutea. Int. J. Nanomed 14: 6217.

Hang, X., Peng, H., Song, H., Qi, Z., Miao, X. and Xu, W. 2015. Antiviral activity of cuprous oxide nanoparticles against hepatitis C virus in vitro. J. Virol. Methods 222: 150-157.

https://www.pfizer.com/news/press-release/press-release-detail/pfizer-and-biontech-propose-expansion-pivotal-covid-19.

Huang, C., Wen, T., Shi, F.J., Zeng, X.Y. and Jiao, Y.J. 2020. Rapid detection of IgM antibodies against the SARS-CoV-2 virus via colloidal gold nanoparticle-based lateral-flow assay. ACS Omega 5(21): 12550-12556.

Jeremiah, S.S., Miyakawa, K., Morita, T., Yamaoka, Y. and Ryo, A. 2020. Potent antiviral effect of silver nanoparticles on SARS-CoV-2. Biochem. Biophys. Res. Commun. 533(1): 195-200.

Kim, H. 2020. Outbreak of novel coronavirus (COVID-19): What is the role of radiologists? Eur. Radiol. 30(6): 3266-3267.

Kumar, R., Nayak, M., Sahoo, G.C., Pandey, K., Sarkar, M.C., Ansari, Y., Das, V.N.R., Topno, R.K., Bhawna, Madhukar, M. and Das, P. 2019. Iron oxide nanoparticles based antiviral activity of H1N1 influenza A virus. J. Infect. Chemother. 25(5): 325-329.

Li, Q., Guan, X., Wu, P., Wang, X., Zhou, L., Tong, Y. et al. 2020. Early transmission dynamics in Wuhan, China of novel coronavirus-infected pneumonia. N. Engl. J. Med. 382: 1199-1207.

Li, S., Ma, L., Zhou, M., Li, Y., Xia, Y., Fan, X., Cheng, C. and Luo, H. 2020. New opportunities for emerging 2D materials in bioelectronics and biosensors. Curr. Opin. Biomedl. Engineering 13: 32-41.

Liu, C., Yang, Y., Gao, Y., Shen, C., Ju, B., Liu, C., Tang, X., Wei, J., Ma, X., Zhu, Y., Liu, W., Xu, S., Liu, Y., Yuan, J., Wu, J., Liu, Z., Zhang, Z., Liu, L., Wang, P. and Zhang, P. 2020. Viral architecture of SARS-CoV-2 with post-fusion spike revealed by Cryo-EM. Biorxiv. Structure, 28(11): 1218-1224.e4.

Liu, D.X., Fung, T.S., Chong, K.K.L., Shukla, A. and Hilgenfeld, R. 2014. Accessory proteins of SARS-CoV and other coronaviruses. Antiviral Res. 109: 97-109.

Liu, M.A. 2019. A comparison of plasmid DNA and MRNA as vaccine technologies. Vaccines 7(2): 37.

Łoczechin, A., Séron, K., Barras, A., Giovanelli, E., Belouzard, S., Chen, Y.-T., Metzler-Nolte, N., Boukherroub, R., Dubuisson, J. and Szunerits, S. 2019. Functional carbon quantum dots as medical countermeasures to human Coronavirus. ACS Appl. Mater. Interfaces 11(46): 42964-42974.

Lv, X., Wang, P., Bai, R., Cong, Y., Suo, S., Ren, X. and Chen, C. 2014. Inhibitory effect of silver nanomaterials on transmissible virus-induced host cell infections. Biomaterials, 35(13): 4195-4203.

Ma, J., Liu, R., Wang, X., Liu, Q., Chen, Y., Valle, R.P., Zuo, Y.Y., Xia, T. and Liu, S. 2015. Crucial role of lateral size for graphene oxide in activating macrophages and stimulating pro-inflammatory responses in cells and animals. ACS Nano, 9(10): 10498-10515.

Manivannan, S. and Ponnuchamy, K. 2020. Quantum dots as a promising agent to combat COVID-19. Appl. Organometallic Chem. 34(10): e5887.

Masters, P.S. 2006. The molecular biology of coronaviruses. Adv. Virus Res. 66: 193-292.

McKay, P.F., Hu, K., Blakney, A.K., Samnuan, K., Brown, J.C., Penn, R., Zhou, J., Bouton, C.R., Rogers, P., Polra, K., Lin, P.J.C., Barbosa, C., Tam, Y.K., Barclay, W.S. and Shattock, R.J. 2020. Self-amplifying RNA SARS-CoV-2 lipid nanoparticle vaccine candidate induces high neutralizing antibody titers in mice. Nature Comm. 11(1): 1-7.

Medhi, R., Srinoi, P., Ngo, N., Tran, H.V. and Lee, T.R. 2020. Nanoparticle-based strategies to combat COVID-19. ACS Appl. Nano. Mater. acsanm.0c01978.

Moitra, P., Alafeef, M., Dighe, K., Frieman, M. and Pan, D. 2020. Selective naked-eye detection of SARS-CoV-2 mediated by N gene targeted antisense oligonucleotide capped plasmonic nanoparticles. ACS Nano 14(6): 7617–7627.

Mori, Y., Ono, T., Miyahira, Y., Nguyen, V.Q., Matsui, T. and Ishihara, M. 2013. Antiviral activity of silver nanoparticle/chitosan composites against H1N1 influenza A virus. Nanoscale Res. Lett. 8(1): 93.

Mortola, E. and Roy, P. 2004. Efficient assembly and release of SARS coronavirus-like particles by a heterologous expression system. FEBS Lett. 576(1-2): 174-178.

Moyle, P.M. and Toth, I. 2013. Modern subunit vaccines: development, components, and research opportunities. ChemMedChem, 8(3): 360-376.

Mulligan, M.J., Lyke, K.E., Kitchin, N., Absalon, J., Gurtman, A., Lockhart, S. et al. 2020. Phase 1/2 study of COVID-19 RNA vaccine BNT162b1 in adults. Nature 586(7830): 589-593.

Nanotechnology Products Database. COVID-19. 2020. Available in: https://product.statnano.com/.

Neuman, B.W., Kiss, G., Kunding, A.H., Bhella, D., Baksh, M.F., Connelly, S., Droese, B., Klaus, J.P., Makino, S., Sawicki, S.G., Siddell, S.G., Stamou, D.G., Wilson, I.A., Kuhn, P. and Buchmeier, M.J. 2011. A structural analysis of M protein in coronavirus assembly and morphology. J. Str. Biol. 174(1): 11-22.

Novel Coronavirus (2019-nCoV) situation reports 2 World Health Organization (WHO).

Orecchioni, M., Bedognetti, D., Newman, L., Fuoco, C., Spada, F., Hendrickx, W. et al. 2017. Single-cell mass cytometry and transcriptome profiling reveal the impact of graphene on human immune cells. Nat. Commun. 8: 1109.

Pardi, N., Hogan, M.J., Porter, F.W. and Weissman, D. 2018. mRNA vaccines—a new era in vaccinology. Nat. Rev. Drug. Discov. 17(4): 261.

Peña-Bahamonde, J., Nguyen, H.N., Fanourakis, S.K. and Rodrigues, D.F. 2018. Recent advances in graphene-based biosensor technology with applications in life sciences. J. Nanobiotechnol 16(1): 1-17.

Pescatori, M., Bedognetti, D., Venturelli, E., Menard-Moyon, C., Bernardini, C., Muresu, E., Piana, A., Maida, G., Manetti, R., Sgarrella,F., Bianco, A. and Delogu, L.G. 2013. Functionalized carbon nanotubes as immunomodulator systems. Biomaterials, 34: 4395-4403.

Rabiee, N., Bagherzadeh, M., Ghasemi, A., Zare, H., Ahmadi, S., Fatahi, Y., Dinarvand, R., Rabiee, M., Ramakrishna, S., Shokouhimehr, M. and Varma, R.S. 2020. Point-of-use rapid detection of SARS-CoV-2: nanotechnology-enabled solutions for the covid-19 pandemic. Int. J. Mol. Sci. 21(14): 5126.

Remy, V., Largeron, N., Quilici, S. and Carroll, S. 2014. The economic value of vaccination: why prevention is wealth. Value in Health 17(7): A450.

Seo, G., Lee, G., Kim, M.J., Baek, S.H., Choi, M., Ku, K.B., Lee, C.S., Jun, S., Park, D., Kim, H.G., Kim, S.J., Lee, J.O., Kim, B.T., Park, E.C. and Kim S.I. 2020. Rapid detection of COVID-19 causative virus (SARS-CoV-2) in human nasopharyngeal swab specimens using field-effect transistor-based biosensor. ACS Nano 14(4): 5135-5142.

Skalny, A.V., Rink, L., Ajsuvako, O.P., Aschner, M., Gritsenko, V.A., Alekseenko, S.I., Svistunov, A.A., Petrakis, D., Spandidos, D.A., Aaseth, J., Tsatsakis, A. and Tinkov, A.A. 2020. Zinc and respiratory tract infections: perspectives for COVID-19 (Review). Int. J. Mol. Med. 46(1): 17-26.

Smith, T.R., Patel, A., Ramos, S., Elwood, D., Zhu, X., Yan, J. et al. 2020. Immunogenicity of a DNA vaccine candidate for COVID-19. Nat. Commun 11(1): 2601.

Staroverov, S.A., Vidyasheva, I.V., Gabalov, K.P., Vasilenko, O.A., Laskavyi, V.N. and Dykman, L.A. 2011. Immunostimulatory effect of gold nanoparticles conjugated with transmissible gastroenteritis virus. Bull. Exp. Biol. Med. 151(4): 436.

Sun, L., Singh, A.K., Vig, K., Pillai, S.R. and Singh, S.R. 2008. Silver nanoparticles inhibit replication of respiratory sincitial virus. J. Biomed. Biotechnol. 4: 149-158.

Tai, W., Zhang, X., Drelich, A., Shi, J., Hsu, J.C., Luchsinger, L., Hillyer, C.D., Tseng, C.T.K., Jiang, S. and Du, L. 2020. A novel receptor-binding domain (RBD)-based mRNA vaccine against SARS-CoV-2. Cell Res. 30(10): 932-935.

Takashima, Y., Osaki, M., Ishimaru, Y., Yamaguchi, H. and Harada, A. 2011. Artificial molecular clamp: a novel device for synthetic polymerases. Angewandte Chemie International Edition 50(33): 7524-7528.

Ting, D., Dong, N., Fang, L., Lu, J., Bi, J., Xiao, S. and Han, H. 2018. Multisite inhibitors for enteric coronavirus: antiviral cationic carbon dots based on curcumin. ACS Appl. Nano. Materials 1(10): 5451-5459.

Tooze, J., Tooze, S. and Warren, G. 1984. Replication of coronavirus MHV-A59 in sac-cells: determination of the first site of budding of progeny virions. Eur. J. Cell. Biol. 33(2): 281-293.

Udugama, B., Kadhiresan, P., Kozlowski, H.N., Malekjahani, A., Osborne, M., Li, V.Y.C., Li, V.Y.C., Chen, H., Mubareka, S., Gubbay, J.B. and Chan, W.C.W. 2020. Diagnosing COVID-19: the disease and tools for detection. ACS Nano 14(4): 3822-3835.

Vert, M., Doi, Y., Hellwich, K.H., Hess, M., Hodge, P., Kubisa, P., Rinaudo, M. and Schué, F. 2012. Terminology for biorelated polymers and applications (IUPAC Recommendations 2012). Pure Appl. Chem. 84(2): 377-410.

Vijayakumar, S. and Ganesan, S. 2012. Gold nanoparticles as an HIV entry inhibitor. Curr. HIV Res. 10(8): 643-646.

Wang, W., Xu, Y., Gao, R., Lu, R., Han, K., Wu, G. and Tan, W. 2020. Detection of SARS-CoV-2 in different types of clinical specimens. Jama 323(18): 1843-1844.

WHO. 2020. World Health Organization. The push for COVID-19 vaccine.

WHO. 2020b. World Health Organization. Water, sanitation, hygiene, and waste management for the COVID-19 virus: interim guidance.

World Health Organization. Coronavirus disease 2019 (COVID-19): Situation report, 59. 2020. Available online: https://www.who.int/emergencies/diseases/novel-coronavirus-2019/situation-reports (accessed on 20 September 2020).

Wu, D., Wu, T., Liu, Q. and Yang, Z. 2020. The SARS-CoV-2 outbreak: what we know. Int. J. Infect. Dis. 94: 44-48.

Zhou, P., Yang, X.L., Wang, X.G., Hu, B., Zhang, L., Zhang, W. et al. 2020. A pneumonia outbreak associated with a new coronavirus of probable bat origin. Nature, 579(7798): 270-273.

Zhu, M., Wang, R. and Nie, G. 2014. Applications of nanomaterials as vaccine adjuvants. Human Vaccines Immunother. 10(9): 2761-2774.

Zu, Z.Y., Jiang, M.D., Xu, P.P., Chen, W., Ni, Q.Q., Lu, G.M. and Zhang, L.J. 2020. Coronavirus disease 2019 (COVID-19): a perspective from China. Radiology 200490.

Recent Developments on the Removal of Antibiotics Using Carbon-Based Nanocomposites: A Mystic Approach in Nanoscience

Anita Kongor[1]*, Mohd Athar[2], Manoj Vora[1], Keyur Bhatt[3] and Vinod Jain[1]

[1] Department of Chemistry, School of Sciences, Gujarat University, Navrangpura, Ahmedabad - 380009
[2] Physics Department, University of Cagliari, 09042, Monserrato (CA), Italy
[3] Mehsana Urban Institute of Science, Department of Chemistry, Ganpat University, Kherva, Gujarat - 384012, India

1. Introduction

The term 'antibiosis' was coined by Vuillemin in 1889; much later the noun 'antibiotic' was first used by Waksman in 1942 who quoted 'Out of the earth shall come thy salvation' (Finch et al. 2010). Although medicinal use of plants/herbs or natural molecules was significant throughout the history of mankind, when did the use of antibiotics gain special interest? (Bérdy 2012) This discovery was indeed needed that witnessed significant remarks in the later 1940s for the urgent problems of chemotherapy. Though antibiotics are life-saving medicines, their self-medication and triggering of malpractice in clinical decisions or indiscriminate use of antibiotics are some of the medical negligences that have started playing with the body's immune system. With the synthetic progress and bio-medical discoveries, the inappropriate use of antibiotics has posed a burning issue, especially its antagonistic effects to the security of aquatic species and on human health. Particularly, the hypersensitive effect is a major concern in case of elderly seniors and persons with less-capable immune response. An infectious challenge and risk of causing therapy complication alert the health-related quality-of-life outcomes.

Since the mid-1990s, there have been reported cases in which antibiotics acted as a source of organic contaminants in surface water (Watkinson et al. 2009, Chen et al. 2015). Antibiotics have frequently been detected in surface water and wastewater

Corresponding author: anitakongorchem@outlook.com

at concentrations generally ranging between 0.01 and 1.0 µg/L (Ben et al. 2019). Due to this, these contaminants enter the major sources of drinking water. Figure 1 gives a schematic representation of different sources of antibiotics. As per our survey, there are five key reasons why the removal of antibiotics from wastewater is obligatory: (1) they are found in low concentration limits; (2) they are highly polar; (3) as per the author's knowledge, until now, no agency or organization has approved any concentration permissible limits; (4) the traditional treatment technology does not meet the pace with the increasing antibiotic load in the environment; and (5) lack of general awareness of antibiotic contamination.

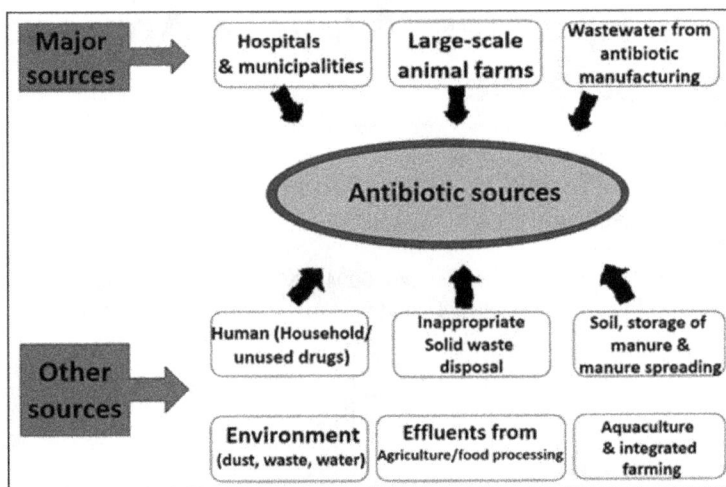

Figure 1: Sources of antibiotics.

The growing population and inappropriate use of antibiotics poses risk to aquatic organisms, which leads to the development of antibiotic-resistance bacteria. Mutiyar et al. evaluated a total load of pharmaceutical compounds released from a sewage treatment plant, representing the first measurement of antibiotic residues in aqueous environments in Delhi, India (Mutiyar and Mittal 2014). Recoveries were varied from 25.5-108.8%, and the maximum concentration of ampicillin was found in wastewater influents (104.2 ± 98.11 µgL^{-1}) and effluents (12.68 ± 8.38 µgL^{-1}). Thus, it is likely that antibiotics will be released into the environment via various pathways.

The widespread occurrence of water and unexpected diversity in climate changes is exposing the human health and sustainability of the ecosystems. Major concerns about antibiotic pollution have led research communities to lay novel experimental protocols and allocate serious water-quality problems. The root of nano-science in the disciplines of chemistry and physics is considered more traditional. Several researchers have encountered many outstanding solutions with precise and functional environmental tools. However, there have been different opportunities and risks of nanotechnologies. There are several breakthrough applications for nano-science-related products, which have been translated from the research laboratories to consumer use. Some of the significant milestones include targeted drug therapeutics, nanostructured bone replacements, nano-scale body imaging, and on-site healthcare diagnostics (Roco et al. 2011). Likewise, as far as the pollutant removal performance of nanomaterials is concerned, many elaborative studies have been reported to evaluate the sustainability of the synthesized nanomaterials (Xuandong Wang et al. 2019, Engel and Chefetz 2016, Chauhan and Agnihotri 2019, Li et al. 2019, Chen et al. 2020). When using nanomaterials, a significant consideration is that nano-sized materials can only have active experimental services and should not harm the bio-system when released to the environment (Du et al. 2019, Xin Wang et al. 2019, Peng et al. 2020, T. Wang et al. 2019, Rai et al. 2018, Andreani et al. 2020, Matysiak et al. 2016). For a decade, nanotechnology is one such technique that has attracted researchers to achieve better performance for water treatment in comparison to existing conventional methods.

Currently, the global scenario has changed upside down leaving questions such as what is of more importance, i.e. the prevention of the survival of microbes or to refocus on the metabolic power of microbes against antibiotics? Or is it to find preventive measures to tackle antibiotic contamination? Each question depends on the subject specialization one proceeds with; for example, a chemist will find an experimental way to prevent antibiotic pollution and a biologist may find a microbiological phenomenon to build medical applications. Therefore, before identifying measures to mitigate the issues of antibiotic removal, let us refine our knowledge about different groups of antibiotics, the beneficiary role of nanomaterials as an emerging solution, and lastly gain knowledge about the recent technologies adopted to gain success against its adverse effects.

Antibiotics are categorized under different groups based on the World Health Organization (WHO) and Anatomical Therapeutic Chemical (ATC) classification system, wherein antibiotics have been classified as antibacterial for systemic use (ATC code: J01). Others such as antimycotic for systemic use (ATC code: J02), antiviral (ATC code: J05), and drugs for tuberculosis (ATC code: J04) treatment were excluded, as were antibiotics for topical use (Hsia et al. 2019, WHO-Oslo 2008). Furthermore, antibiotics have also been classified into the four current AWaRe categories: 'Access', 'Watch', 'Reserve', and 'Not Recommended'. Antibiotics not noted in the AWaRe classification were further categorized into one of the four groups based on the guidance of members and advisors of the twenty-first-century expert committee on the selection and use of essential medicines. The AWaRe antibiotic classification thus provides a useful framework for exploring antibiotic consumption patterns (Klein et al. 2021).

Till 1959-1960, the synthesis and study of microbial antibiotics was the subject of research for many chemists, but later a decisive change was witnessed with the discovery of semisynthetic penicillin (Béahdy 1974). After this era, further classifications were reported according to their mechanism of action, physicochemical properties, management regulation, chemical structure, origin, and spectrum of activity. The key to main antibiotic families has been classified as shown in Table 1. It is interesting to note that the earlier classification was taken into consideration because of higher complexity and complete lack of uniform denomination. However, inclusive information about the most primary classification of antibiotics with few representative examples has been included in a tabular format.

Table 1: Classification of antibiotics

Classification Type	Sub-Classification Type	Representative Examples	Reference
(A) Primary classification based on chemical structure	**With the following principal constituent:**		(Balsalobre et al. 2019) (Q. Li et al. 2019)
	(a) Sugar	Labilomycin, bambermycin, macarbomycins	
	(b) Macrocyclic lactone ring	12-Membered ring (Methymycin); 14-Membered ring without sugar (Albocycline, cineromycin B); Other large macrolactones (Botrycidin, cryptomycins)	
	(c) Quinone skeleton	Naphthoquinone derivatives (Biflorin, chymaphillin); Benzoquinone derivatives (Auranthioglioclodin, fumigatin)	
	(d) Amino acid, peptide	Simple amino acids (Azaserine, Anticapsin); Homopeptides (Distamycin, Ilamycin) Heteromer peptides (Amphomycin, Siomycin)	
	(e) Nitrogen-containing heterocyclic systems	Pyrrolnitrin, Anisomycin	
	(f) Oxygen-containing heterocyclic systems	Simple furans (Asarinin); Small lactones (Acetomycin)	
	(g) Alicyclic skeleton	Cycloalkane derivatives (Caldariomycin); Oligoterpene antibiotics (Cephalospori)	
	(h) Aromatic skeleton	Benzene compounds (Miconidin) Nonbenzoid aromatic compounds (Thujaplicin)	

(B) Commonly used classes of antibiotics		
(i) Aliphatic chain	Simple alkanes (Lipoxamycin) Aliphatic compounds with sulfur contents (Fluopsin); with phosphorus content (Phosphonomycin)	
With the following basic functional group/atom:		
(a) Beta-lactams	Penicillins, Monobactam, Carbapenems and Cephalosporins	(Hashmi 2019)
(b) Macrolides	Erythromycin, Azithromycin, Clarithromycin and Clindamycin	(Du et al. 2019)
(c) Tetracyclines	*Short-acting for 6-8 hrs* (Tetracycline, Oxytetracycline), *Intermediate acting for 12 hrs* (Democycline, Methacycline) and *Long acting for 15-18 hrs* (Doxycycline, Minocycline)	(Sadick 2001)
(d) Quinolones	*1st generation* (Nalidixic acid, Cinoxacin), *Fluoroquinolones:* *2nd generation* (Norfloxacin, ciprofloxacin, ofloxacin), *3rd generation* (Sparfloxacin, Grepafloxacin) and *4th generation* (Moxifloxacin, Gatifloxacin, Difloxacin)	(Andersson and MacGowan 2003)
(e) Aminoglycosides	*Systemic:* (Mycin sub-category includes: Streptomycin, Kanamycin, Tobramycin; Micin sub-category includes: Amikacin, Gentamicin,) and *Topical* (Neomycin and Framycetin)	(Gonzalez and Spencer 1998)
(f) Sulfonamides	Antimicrobial (Sulfamethoxazole, Sulfamethizole, Sulfamoxole, Sulfamethazine) and Nonantimicrobial (Methzolamide, Furosemide, Sulfasalazine, Celecoxib, Sumatriptan)	(Hassanein 2019)

Let us have an overview of the use of nanomaterials, which have been briskly explored for the removal of antibiotics, and other organic contaminants present in wastewaters. Nowadays, researchers have drawn their attention toward the use of biogenic nanoparticles or nanoparticles synthesized using waste materials. This approach is simply pre-eminent as it involves balancing a stable ecosystem with individual life forms. For example, Gopal et al. reported the synthesis of bentonite-supported bimetallic nanocomposites using the pomegranate rind extract, which was further employed for the removal of tetracycline in contaminated natural water systems (Gopal et al. 2020). Low cost and biogenic palladium nanoparticles were prepared using waste *E. coli* cells by He et al. which were found to degrade approximately 87.7% of ciprofloxacin. Moreover, for continuous reductive degradation of ciprofloxacin, Pd(0) hydrogen hydride must be constantly generated and the H_2 needs to be bubbled in the solution (He et al. 2020).

Particularly, there have been potential reports of carbon-based nanomaterials for pollution remediation issues. Gao et al. reported chicken feather waste for synthesizing nitrogen-doped carbon nanotubes and explored its high catalytic activity and good reusability in the reduction of 4-nitrophenol (Gao et al. 2014). It has been speculated that nanocomposites prepared using carbonaceous material and metal nanoparticles have synergistic effects to improve the removal of organic contaminants in wastewaters. For instance, biogenic silver nanoparticles supported on carbonaceous material could exhibit 91% of methylene blue removal from water on nine hours of contact time (Vilchis-Nestor et al. 2014). There are consistent potential benefits of using carbon-based nanomaterials for the removal of antibiotics in comparison to other support materials, such as:

(a) Serve key molecular-level chemical interactions, electrostatic and covalent interaction, hydrogen bonding, π-π interaction, and hydrophilic effects as important contributors to bond specific polar antibiotic molecules.
(b) No effect of media, such as acidic or alkaline solution, which protects the encapsulated materials/entities from degradation.
(c) Apart from chemical interactions, carbon-based nanomaterials, such as graphene derivatives, possess, excellent physical adsorption capabilities.
(d) Cost-effective, stable, and ease of separation.
(e) The use of carbon-based nanotubes and graphene oxides display antimicrobial properties.
(f) Surface modification leads to the ease of separation and removal of carbon nanocomposites can be enhanced, which makes it a potential aspirant for decontamination applications.
(g) Short filtration times and improved filtration efficiency (above 90%) of hybrid carbon nanomaterial-based membranes

Therefore, a closer overview of the removal of antibiotics using carbon-based nanocomposites and the recent methods adopted so far have been emphasized. Special emphasis is laid on the membrane-based technologies adopted for the removal of antibiotics. Very briefly, membrane-based separation is one of the environmentally friendly processes, which has gained significant interest in recent times. The membrane processes used in water treatment can be classified based on the pore size, driving forces, type of membrane, and average permeability. To name a few, microfiltration, ultrafiltration, nanofiltration, reverse osmosis, electro-dialysis, electro-deionization, and membrane degasification are membrane-based separation techniques (Singh et al. 2016). Additionally, recent developments in green chemistry and membrane-based separation/removal as a reigning technology will be recorded to gain the newest and state-of-the-art insights based on carbon-based nanomaterials. This will accomplish the purpose of this chapter and publicize the epidemiological knowledge of post-antibiotic consequences, highlighting the importance of its ongoing surveillance.

2. Carbon-Based Nanocomposites: A General Introduction

Nanoparticles are described as dispersion of inorganic substances which fall in the size range of 1-100 nanometers (nm) in diameter. Nanocomposites are the combination of solid materials with more than one phase differing in properties on a nanometer scale (less than 100 nm). Carbon-based nanocomposites are one of the leading materials, which are chemically inert at the nano-scale level and are highly hydrophobic. These materials have been widely used for the remediation of environmental pollutants, such as heavy metals and other organic contaminants (Figure 2). They are employed in industrial applications, such as supercapacitors, sensors, electromagnetic absorbers, photovoltaic cells, photodiodes, and optical limiting devices (Ates et al. 2017). Carbon nanocomposites are particularly structural carbon-based materials, which include fullerenes, activated carbon, carbon-based nanofibers, carbon black nanoparticles, carbon nanotubes, carbide-derived carbon, graphene, and its derivatives such as graphene oxide, nano-diamonds, and carbon-based quantum dots. The extraordinary features of these materials are as follows: lightweight, high surface area, high mechanical strength, electron mobility, thermal conductivity, chemical versatility, and relatively higher biosafety. Fundamentally,

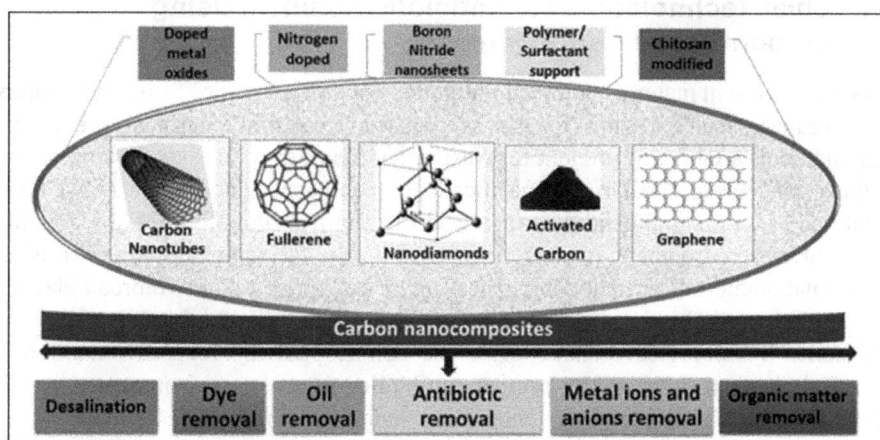

Figure 2: Potential of carbon-based nanocomposites for environmental remediation.

carbon-based nanomaterials have been classified on the basis of their dimensions, such as zero-dimension carbon-based nanomaterial include graphene quantum dots, one-dimension include carbon nanotubes and graphene nano-ribbons, two-dimension include graphene and few-layer graphene (<10 layers), and three-dimension include graphite (Crevillen et al. 2019). These nanocomposites have improved physical properties and unprecedented flexibility which can be applied for the manufacturing of electronic and fabricated devices.

The synthesis of carbon nanocomposites mainly include the following methods: (a) Melt Blending involves the fabrication of carbon nanomaterial via blending into a molten polymer matrix under intense shearing; (b) *In Situ* Polymerization involves initial swelling of carbon nanomaterial by a monomer, and upon addition of the initiator, the polymerization begins by heat or light irradiation (Díez-Pascual 2020); and (c) Solution Mixing/Solvent Processing in which a solution of carbon nanomaterial is simply mixed with a polymer-solvent at room temperature followed by ultrasonication under identical conditions (Díez-Pascual and Díez-Vicente 2016). Although solution mixing is meant to produce high-quality carbon nanocomposites, the melt blending method is quite simple and can be opted for large-scale production (Choudhary and Gupta 2001).

According to literature reports, graphene has been prepared by mechanical exfoliation, epitaxial growth, chemical vapor deposition, and chemical exfoliation (Huang and Ruoff 2020, Z. Gao et al. 2020, Maitra et al. 2012). However, amongst these methods, mechanical exfoliation renders a convenient method for providing high-quality graphene samples. Carbon nanotubes are synthesized through the vaporization of graphite samples using two main methods, namely electric arc-discharge and laser ablation (Moses et al. 2018, Eatemadi et al. 2014, Lee et al. 2006). Using the chemical vapor deposition method, high-purity carbon nanotubes for commercial purposes where the carbon-containing vapor is passed over supported metal catalyst nanoparticles, such as nickel, cobalt, and iron in a furnace (Yu et al. 2016).

2.1 Chief Technologies for Antibiotic Removal Using Carbon-Based Nanocomposite

There are different materials adopted for the removal of antibiotics from wastewaters at laboratory scale. Owing to the specific physical and chemical properties, nanomaterials are more proficient for wastewater treatment processes in comparison to other conventional materials. Specifically, carbon-based nanomaterials have been identified as a potential candidate to ensure the removal of antibiotics as far as the safety of the environment is concerned. Technically, the removal processes using carbon nanomaterials or nanocomposites can be categorized into two broad classes, that is biotic and abiotic methods. The biotic pathway trails the biodegradation of antibiotics by biological means, while the abiotic pathway focuses on different chemical/physical processes to remove antibiotics from the environment. Explicit information on both biotic and non-biotic/abiotic methods for the removal of antibiotics is given in the following section.

2.2 Biotic and Non-Biotic/Abiotic Methods

Although antibiotics have a shorter life span, yet their hydrophobic and lipophilic properties make them persistent in the water bodies. Therefore, an in-depth study is crucially needed to understand the removal of antibiotics, especially using different abiotic or non-biotic methods. This takes into account, hydrolysis, oxidation, reduction, sorption, and photolysis reactions to aid the removal of antibiotics. Abiotic methods have been elucidated as one of the finest eco-friendly methods for the elimination of antibiotics. Generally, the optimum performance of smaller size nanomaterials or nanocomposites is considerably enhanced due to surface area. Applications of hybrid carbon-based membranes are particularly suited to remove different environmental pollutants issues, such as dyes, organic matter, antibiotics, toxic ions, and biomolecules. Interestingly, the integration of carbon nanomaterials into a variety of membrane materials, such as ceramic and polymeric types can considerably increase the surface membrane characteristics. However, research on using polymeric-type membranes is limited due to many downside facts, such as poor tolerances to pH-specific reagents. Some of the benefits of modified membrane holding nano-features include more a number of contacts with the porous membrane surface, improvement of antifouling properties of the membranes (Wang et al. 2018), improved photocatalytic properties to degrade organic contaminants, and increase chemical and mechanical stability (Homaeigohar and Elbahri 2017). Blending or fabrication of carbon-based nanocomposites onto different membrane structures has been well demonstrated to exhibit all of the aforementioned properties (L. Ma et al. 2017, Song et al. 2018). Moreover, the remarkable conductive, mechanical, and antibacterial properties of carbon-based nanomaterials have been widely used to fabricate novel membranes for water treatment and desalination with advanced characteristics (Manawi et al. 2016, H.Y. Yang et al. 2013). Furthermore, it is noted that the use of multifunctional carbon nanocomposites in the active layers of membranes has addressed different challenges of wastewater treatment processes. The tabular representation of carbon-based nanomaterial membrane, type of removal method of antibiotics, a model antibiotic used, and carbon nanomaterial size based on the last five years (2015-2020) has been listed in Table 2.

Amongst several interesting examples, a greener approach such as microwave-induced catalyst is considered amongst the most efficient techniques to remove antibiotic residues from polluted water. Liu et al. reported a new microwave-induced oxidation catalyst for degrading chlortetracycline using the nanocomposites of Fe_3O_4 and carbon materials (S. Liu et al. 2018). The resulting composite has a porous structure with a high specific surface area and the established strong interactions between Fe_3O_4 and carbon nanotube promoted the regeneration of the catalyst and extended its lifecycle. Magnetite-based carbon nanocomposites demonstrate an environmentally friendly method for the removal of antibiotics. In the same context, another illustration of the synthesis of magnetic fullerenes via one-step catalytic thermal dissociation of PET bottle wastes and its application as adsorbent for the removal of ciprofloxacin antibiotic (Elessawy et al. 2020). The presence of the iron phase in the nanocomposites strongly enhanced adsorption capacity and adsorbent recycling. Alternatively, another low-cost carbon nanocomposites have been

Table 2: Removal of various antibiotics using carbon-nanocomposites based membranes

Carbon–Nanocomposites-Based Membrane	Removal Method	Model Antibiotic	Removal Efficiency	Reference
Coal-based carbon-membrane coated with nano Sb-SnO$_2$	Electrocatalytic oxidation	Tetracycline	96.5% after 6 hours operation	(Z. Liu et al. 2017)
Graphene modified e-Fenton catalytic membrane	Combination of electro-Fenton and membrane filtration process	Florfenicol	90%	(Jiang et al. 2018)
Carbon nanotube@nitrogen doped carbon/Al$_2$O$_3$ membrane	Membrane filtration with peroxy-monosulfate (PMS) activation	Sulfamethoxazole	65% after 100 minutes duration operation	(H. Ma et al. 2020)
Graphene oxide (GO)/poly(vinylidenefluoride) (PVDF) electrospun nanofibrous membranes(ENM)	Adsorptive-filtration	Tetracycline	17.92 mg/g with 1.5 wt% of GO (GO$_{1.5}$/PVDF) ENM	(Park et al. 2018)
All-carbon nanofiltration membrane that consists of MWCNTs interposed between GO nanosheets.	Adsorption process – vacuum filtration	Tetracycline	99.23%	(Guohai Yang et al. 2018)
Two-dimensional boron nitride nanosheets/GO nanocomposite membrane	Adsorption process – vacuum filtration	Tetracycline	96.1%	(Guohai Yang et al. 2019)
Carbon-polymeric flat sheet ultrafiltration membranes	Adsorption and filtration	Metronidazole	–	(Nadour, Boukraa and Benaboura 2019)
Carbon membrane coated with nano-TiO$_2$	Electrocatalytic oxidation and membrane separation	Tetracycline	100%	(Z. Liu et al. 2016)

mentioned in a very recent report, where ferroferric oxide nanoparticles assisted powdered activated carbon was prepared for tetracycline removal (Zhou et al. 2020). The regeneration ability and magnetic separation property of the iron-based carbon nanocomposites cause feasible adsorbent property in comparison to neat carbon nanomaterial. A significant study demonstrated that carbonaceous nanocomposites could be used as a cost-effective and potential sorbent for environmental remediation. The study has been demonstrated to remove sulfamethazine using carbonaceous nanocomposites synthesized pyrolysis of rice straw biomass as a cheap precursor material obtained from agricultural residues (Zhang et al. 2016). Another advantage of this method was that there were not any effects of harsh aging on antibiotic sorption in the presence of soil. Sulfamethazine possesses ionizable acidic groups and aromatic rings, which make them highly polar, and therefore for its feasible removal, carbon materials have been doped with nitrogen to provide polar surface and basic sites. One such example was reported by G. Xu et al. where a novel adsorbent based on nitrogen-doped flower-like porous carbon nanostructures was proposed for the removal of sulfamethazine (Xu et al. 2019). The N-doped carbon nanostructures could exhibit a maximum monolayer adsorption capacity of 610 mg g^{-1} within five minutes of adsorption equilibrium.

The applicability of multi-walled carbon nanotube (MWCNT) loaded iron metal-organic framework (MIL-53(Fe)) composite-based adsorbent was studied for the removal of tetracycline hydrochloride, oxytetracycline hydrochloride, and chlortetracycline hydrochloride (Xiong et al. 2018). This study exemplifies that the combination of metal-organic framework materials with MWCNT could enhance the adsorption capacity for antibiotic adsorption. Another effective technology was reported for the removal of ciprofloxacin through the modification of Fe_3O_4–MoO_3 by activated carbon (AC) (Mahmoud et al. 2021). The adsorption of ciprofloxacin onto the assembled nanocomposites, Fe_3O_4-MoO_3 and Fe_3O_4-MoO_3-AC was found to be 80-71.9% and 93-90.5%, respectively. One interesting example of using novel adsorbent, Fe_3O_4/activated carbon/chitosan, and a magnetic-activated carbon/chitosan nanocomposite were used for the removal of ciprofloxacin, erythromycin, and amoxicillin (Danalıoğlu et al. 2017). The use of magnetic counterparts in the making of carbon nanocomposites is very crucial as it eliminates the filtration step and imparts easy separation.

There had been no literature reports about the synthesis of carbon-based nanocomposites using a green route, liquid-phase reduction method until Wang et al. developed novel carbon spheres supported by nanoscale zero-valent iron. Carbon spheres as a support material can effectively prevent nanoscale zero-valent iron particles from agglomeration to maintain the high surface area and reactivity. Thus, this synergistic effect improved the metronidazole removal efficiency (Xiangyu Wang, Du and Ma 2016).

Li et al. demonstrated an efficient degradation of tetracycline hydrochloride using a coupled membrane bioreactor/microbial fuel cell system with granular activated carbon as an expanded dynamic membrane cathode. The degradation was based on synergetic bio-electrochemical catalytic reaction and adsorption. The method adopted was energy-saving and environmentally friendly enabling 90% of tetracycline removal (Y. Li et al. 2017). Hayati et al. reported an interesting and novel integration of photocatalytic and ultrasonic methods for sulfadiazine degradation and removal. Herein, magnesium oxide was coated on carbon nanotubes in an ultrasound-assisted hydrothermal method and used as a heterogeneous catalyst. Biodegradability studies were also carried out which served the purpose to decrease the toxicity of sulfadiazine and after the sono-photocatalytic treatment it was found that its biodegradability was increased (Hayati et al. 2020).

3. Issues and Challenges

The removal of antibiotics from water systems using carbon-based nanocomposites can be applied to practical aspects, such as point-of-use filters. Taking note of carbon nanotubes (CNTs), it was first considered that they are very expensive and could not be used for large-scale applications (H.Y. Yang et al. 2013, Ncibi and Sillanpää 2015). However, literature experimental studies have demonstrated that its continuous production could be possible to mass-produce high-quality CNTs at a lower cost (Upadhyayula et al. 2009). Another significant step of analysis is the stability of adsorbent-based carbon nanocomposites in various pH environments. In other examples, a strong influence of pH was mentioned, whereby it was reported that the speciation of metronidazole and sulfamethoxazole was hindered during the adsorption process. The reason being at pH 6, metronidazole was found in its neutral form, with the lowest value for solubility whereas sulfamethoxazole because of its speciation at pH 6 is more soluble and much less hydrophobic. However, the observed trend may also be related to the size of antibiotics (Teixeira et al. 2019). Recently, multi-walled carbon nanotubes based electrochemical membrane filtration was used or degrading antibiotics, including sulfamethoxazole, ciprofloxacin, and amoxicillin in both single and mixed systems (Tan et al. 2020). In the case of the degradation of ciprofloxacin, the degradation efficiency was affected by solution pH. The structure and interactions between the carbon-based nanomaterials and surfactants should provide proper binding sites for antibiotics adsorption techniques. In the experiments of the electrochemical membrane filtration process, Tan et al. found that among the targeted pollutants, including sulfamethoxazole, ciprofloxacin, and amoxicillin, and sodium dodecylbenzene sulfonate has the most significant effect on the degradation of ciprofloxacin while it exhibited the weakest effect on amoxicillin. Understanding the effects of these parameters on water treatment processes is critical to being able to effectively remove antibiotics from wastewater (Zeidman et al. 2020, Hong et al. 2020). Intriguingly, it has been shown that the sorption behavior largely depends on multiple factors, such as surface area, pore volume, pore size distribution, and surface functionalities of the carbon nanomaterials (Zhang et al. 2020). Although many newer approaches have been witnessed for the removal of antibiotics, there is an urgent need to improve the degradation efficiency of antibiotics. Membrane-based filtration techniques have been corroborated with adsorption and catalytic-based removal methods. However, to practice or translate such lab-scale experiments on a commercial scale is still an exclusive area of research.

Furthermore, a lot needs to be studied to pre-identify the target antibiotic molecule and understand its removal mechanism using carbon nanocomposites. This directs to the selectivity of the surface of the functionalized carbon nanocomposites toward particular antibiotics. After fine-tuning the nanocomposites selectivity using covalently modified end groups, specific kinds of antibiotics can be easily targeted before heading for the removal procedure (Rashid and Ralph 2017). It is very important to study the influence of salt ionic strength when carbonaceous nanocomposites are used for the removal of antibiotics. For example, Zhang et al. reported that the adsorption of sulfamethazine increased with an increase in NaCl concentration because of the electrostatic screening of the surface charge by the counter ion species added (C. Zhang et al. 2016).

4. Future Outlook and Perspectives

Over the last two decades, antibiotics have been frequently found in various water streams. In this chapter, the knowledge and material-based progress has been discussed with a nearly improved understanding of synthetic as well as application-based dimensions of carbon-based nanocomposites. Depending on the current scenario in dealing with various diseases, it is for sure that the 'era of chemists' will be followed by the 'era of microbiologists'. The right approach will be to merge the teamwork of experts on nanomaterials with the researchers of biology to enter a new phase of a geotechnical system. It is anticipated that the innovations in carbonaceous materials in conjugation with nanotechnology will be quite frequent in the coming years. The most recent approach has made use of the antibiotic-free concept using nanotechnology, which is the most advanced approach to mitigate the issue of antibiotic contamination (Wang et al. 2020). The suitable and fascinating properties offer new vistas to young researchers to work in a newer direction in the study of antibiotic removal. Therefore, carbon nanocomposites can be iterated to meet the following significant requirements needed for presently used removal methods. They are:

(a) Stable method due to its development with a combination of different materials, such as inorganic nanomaterials (titanium oxide, silver, iron, and iron oxides), chitosan, and polymeric materials.
(b) Due to extraordinary conductivity and electron mobility-like characteristics, it can function as an electrocatalyst.
(c) Important from a surface chemistry point of view, as more accessible adsorption sites provide higher adsorption rates for biocompatible applications.
(d) Efficacy is increased due to the presence of high electron donor capacity. This is needed for most antibiotic compounds possessing acidic and ionizable groups.
(e) Positive impact on the environment as the practical protocol steps require less chemical mass and energy.

6. Concluding Remarks

Nanomaterials have crossed their early phase of development and many reliable tools with new ideas have driven open discoveries since the last few years. Out of which, the ongoing research and development progress with respect to carbon nanocomposites for the removal of antibiotics have flagged to be accountable for the welfare of present and future ecological safety (Tufa 2015). Though a plethora of literature has been published, practical application of carbon nanocomposites in an industrial context is rare. To sum up, the authors wish to propose an economical model whereby environment-friendly nanomaterials are configured to remove pollutants from wastewater.

Acknowledgment

Anita Kongor gratefully acknowledges the financial assistance provided by the Council of Scientific & Industrial Research (CSIR), New Delhi, for Research Associateship Fellowship (File No. 09/70 (0073) 2020 EMR-I). Manoj Vora gratefully acknowledges the financial support received from UGC in the form of Rajiv Gandhi National Fellowship, RGNF-JRF (RGNF-2017-18-SC-GUJ-47048). The authors also acknowledge UGC Infonet and Information and Library Network (INFLIBNET) (Ahmedabad) for e-journals.

Conflicts of Interest

The authors declare no conflict of interest.

References

Andersson, Monique I. and MacGowan, Alasdair P. 2003. Development of the quinolones. J. Antimicrob. Chemother. 51(Suppl. 1): 1–11.
Andreani, Tatiana, Paula M.V. Fernandes, Verónica Nogueira, Vera V. Pinto, Maria José Ferreira, Maria Graça Rasteiro, Ruth Pereira and Pereira, Carlos M. 2020. The critical role of the dispersant agents in the preparation and ecotoxicity of nanomaterial suspensions. Environmen. Sci. Pollut. Res. 27: 19845–19857.
Ates, Murat, Eker, Aysegul Akdogan and Eker, Bulent. 2017. Carbon nanotube-based nanocomposites and their applications. J. Adhes. Sci. Technol. 31: 1977–1997.
Balsalobre, Luz, Ana Blanco and Teresa Alarcón. 2019. Beta-lactams. Antibiot. Drug Resist. 57–72.
Béahdy, János. 1974. Recent developments of antibiotic research and classification of antibiotics according to chemical structure. Adv. Appl. Microbiol. 18: 309–406.
Ben, Yujie, Fu, Caixia, Hu, Min, Liu, Lei, Wong, Ming Hung and Zheng, Chunmiao. 2019. Human health risk assessment of antibiotic resistance associated with antibiotic residues in the environment: a review. Environ. Res. 169: 483–493.
Bérdy, János. 2012. Thoughts and facts about antibiotics: where we are now and where we are heading. J Antibiot. 65: 385–395.
Chauhan, Anjali, Sillu, Devendra and Agnihotri, Shekhar. 2019. Removal of pharmaceutical contaminants in wastewater using nanomaterials: a comprehensive review. Curr. Drug Metabol. 20: 483-505.
Chen, Jianqiu, Fengzhu Zheng and Ruixin Guo. 2015. Algal feedback and removal efficiency in a sequencing batch reactor algae process (SBAR) to treat the antibiotic cefradine. PLoS ONE 10(7): 1–11.
Chen, Lirong, Wenrui Feng, Jian Fan, Kai Zhang and Zhenchao Gu. 2020. Removal of silver nanoparticles in aqueous solution by activated sludge: mechanism and characteristics. Sci. Total. Environ. 711: 135155.

Choudhary, Veena and Gupta, Anju. 2001. In Tech Polymer_carbon_nanotube_nanocomposites. Pdf.

Crevillen, Agustín G., Escarpa, Alberto and García, Carlos D. 2019. Carbon-based Nanomaterials in Analytical Chemistry. pp. 1–36. The Royal Society of Chemistry.

Danalıoğlu, Selen Tuğba, Şahika Sena Bayazit, Özge Kerkez Kuyumcu and Mohamed Abdel Salam. 2017. Efficient removal of antibiotics by a novel magnetic adsorbent: magnetic activated carbon/chitosan (MACC) nanocomposite. J. Mol. Liq. 240: 589–596.

Díez-Pascual, Ana M. and Angel L. Díez-Vicente. 2016. Poly(propylene fumarate)/ polyethylene glycol-modified graphene oxide nanocomposites for tissue engineering. ACS Appl. Mater. Interface 8: 17902–17914.

Díez-Pascual, Ana M. 2020. Carbon-based polymer nanocomposites for high-performance applications. Polymers. 12: 872.

Du, Jingjing, Yuyan Zhang, Ruilin Guo, Fanxiao Meng, Yucong Gao, Chuang Ma and Hongzhong Zhang. 2019. Harmful effect of nanoparticles on the functions of freshwater ecosystems: insight into nanoZnO-polluted stream. Chemosphere 214: 830–838.

Eatemadi, Ali, Hadis Daraee, Hamzeh Karimkhanloo, Mohammad Kouhi, Nosratollah Zarghami, Abolfazl Akbarzadeh, Mozhgan Abasi, Younes Hanifehpour and Sang Woo Joo. 2014. Carbon nanotubes: properties, synthesis, purification, and medical applications. Nanoscale Res. Lett. 9: 1–13.

Elessawy, Noha A., Mohamed Elnouby, Gouda, M.H., Hesham A. Hamad, Nahla A. Taha, Gouda, M. and Mohamed S. Mohy Eldin. 2020. Ciprofloxacin removal using magnetic fullerene nanocomposite obtained from sustainable PET bottle wastes: adsorption process optimization, kinetics, isotherm, regeneration and recycling studies. Chemosphere 239: 124728.

Engel, Maya and Benny Chefetz. 2016. Removal of triazine-based pollutants from water by carbon nanotubes: impact of dissolved organic matter (DOM) and solution chemistry. Water Res. Vol. 106. Elsevier Ltd.

Finch, R.G., Greenwood, David, Norrby, S.R. and Whitley, R.J. 2010. Antibiotic and Chemotherapy e-book. ISBN: 978-0-7020-4064-1. SAUNDERS Elsevier Limited.

Gao, Lei, Ran Li, Xuelin Sui, Ren Li, Changle Chen and Qianwang Chen. 2014. Conversion of chicken feather waste to N-doped carbon nanotubes for the catalytic reduction of 4-nitrophenol. Environ. Sci. Technol. 48(17): 10191–10197.

Gao, Zhaoli, Sheng Wang, Joel Berry, Qicheng Zhang, Julian Gebhardt, William M. Parkin, Jose Avila, Hemian Yi, Chaoyu Chen, Sebastian Hurtado-Parra, Marija Drndić, Andrew M. Rappe, David J. Srolovitz, James M. Kikkawa, Zhengtang Luo, Maria C. Asensio, Feng Wang and A.T. Charlie Johnson. 2020. Large-area epitaxial growth of curvature-stabilized ABC trilayer graphene. Nature Commun. 11(1): 1–10.

Gopal, G., Sankar, H., Natarajan, C. and Mukherjee, A. 2020. Tetracycline removal using green synthesized bimetallic nZVI-Cu and bentonite supported green nZVI-Cu nanocomposite: a comparative study. J. Environ. Manage. 254: 109812.

Gonzalez, L.S. and Spencer, J.P. 1998. Aminoglycosides: a practical review. American Fam. Physic. 58: 1811–1820.

Hashmi, M.Z. ed. 2019. Antibiotics and Antimicrobial Resistance Genes in the Environment. Volume 1. *In*: Advances in Environmental Pollution Research Series. Elsevier.

Hassanein, Mahmoud Moussa. 2019. Sulfonamides: far from obsolete. Int. J. Contemp. Pediatr. 6: 2740.

Hayati, Farzan, Ali Akbar Isari, Bagher Anvaripour, Moslem Fattahi and Babak Kakavandi. 2020. Ultrasound-assisted photocatalytic degradation of sulfadiazine using MgO@CNT heterojunction composite: effective factors, pathway and biodegradability studies. Chem. Eng. J. 381: 122636.

He, Peipei, Tianyu Mao, Anming Wang, Youcheng Yin, Jinying Shen, Haoming Chen and Pengfei Zhang. 2020. Enhanced reductive removal of ciprofloxacin in pharmaceutical wastewater using biogenic palladium nanoparticles by bubbling H2. RSC Adv. 10: 26067–26077.

Homaeigohar, Shahin and Mady Elbahri. 2017. Graphene membranes for water desalination. NPG Asia Mater. 9: e427–e427.

Hong, Nian, Qin Cheng, Ashantha Goonetilleke, Erick R. Bandala and An Liu. 2020. Assessing the effect of surface hydrophobicity/hydrophilicity on pollutant leaching potential of biochar in water treatment. J. Indus. Eng. Chem. 89: 222–232.

Hsia, Yingfen, Brian R. Lee, Ann Versporten, Yonghong Yang, Julia Bielicki, Charlotte Jackson, Jason Newland, Herman Goossens, Nicola Magrini and Mike Sharland. 2019. Use of the WHO access, watch and reserve classification to define patterns of hospital antibiotic use (AWaRe): an analysis of paediatric survey data from 56 countries. The Lancet Glob. Health 7: e861– e871.

Huang, Ming and Ruoff, Rodney S. 2020. Growth of single-layer and multilayer graphene on Cu/Ni alloy substrates. Acc. Chem. Res. 53: 800–811.

Jiang, Wen Li, Xue Xia, Jing Long Han, Yang Cheng Ding, Muhammad Rizwan Haider and Ai Jie Wang. 2018. Graphene modified electro-fenton catalytic membrane for in situ degradation of antibiotic florfenicol. Environ. Sci. Technol. 52: 9972–9982.

Klein, Eili Y., Maja Milkowska-Shibata, Katie K. Tseng, Mike Sharland, Sumanth Gandra, Céline Pulcini and Ramanan Laxminarayan. 2021. Assessment of WHO antibiotic consumption and access targets in 76 countries, 2000–15: an analysis of pharmaceutical sales data. Lancet Infect. Dis. 21: 107–115.

Lee, Yang Doo, Hyeon Jae Lee, Jong Hun Han, Jae Eun Yoo, Yun Hi Lee, Jai Kyeong Kim, Sahn Nahm and Byeongr Kwon Ju. 2006. Synthesis of double-walled carbon nanotubes by catalytic chemical vapor deposition and their field emission properties. J. Phys. Chem. B 110: 5310–5314.

Li, Qi, Xijuan Chen, Xin Chen, Yan Jin and Jie Zhuang. 2019. Cadmium removal from soil by fulvic acid-aided hydroxyapatite nanofluid. Chemosphere 215: 227–233.

Li, Yihua, Lifen Liu and Fenglin Yang. 2017. Destruction of tetracycline hydrochloride antibiotics by FeOOH/TiO2 granular activated carbon as expanded cathode in low-cost MBR/MFC coupled system. J. Membrane Sci. 525 (November 2016): 202–209.

Liu, Shiyuan, Lefu Mei, Xiaoliang Liang, Libing Liao, Guocheng Lv, Shuaifei Ma, Shiyao Lu, Amr Abdelkader and Kai Xi. 2018. Anchoring Fe3O4 nanoparticles on carbon nanotubes for microwave-induced catalytic degradation of antibiotics. ACS Appl. Mater. Inter. 10: 29467–29475.

Liu, Zhimeng, Mengfu Zhu, Zheng Wang, Hong Wang, Cheng Deng and Kui Li. 2016. Effective degradation of aqueous tetracycline using a nano-TiO2/carbon electrocatalytic membrane. Materials 9: 1–14.

Liu, Zhimeng, Mengfu Zhu, Lei Zhao, Cheng Deng, Jun Ma, Zheng Wang, Hongbin Liu and Hong Wang. 2017. Aqueous tetracycline degradation by coal-based carbon electrocatalytic filtration membrane: effect of nano antimony-doped tin dioxide coating. Chem. Engineer. J. 314: 59–68.

Ma, Huanran, Guanlong Wang, Zhiming Miao, Xiaoli Dong and Xiufang Zhang. 2020. Integration of membrane filtration and peroxymonosulfate activation on CNT@nitrogen doped carbon/Al2O3 membrane for enhanced water treatment: insight into the synergistic mechanism. Sep. Pur. Technol. 252: 117479.

Ma, Lining, Xinfa Dong, Mingliang Chen, Li Zhu, Chaoxian Wang, Fenglin Yang and Yingchao Dong. 2017. Fabrication and water treatment application of carbon nanotubes (CNTs)-based composite membranes: a review. Membranes 7: 16.

Mahmoud, Mohamed E., Shaimaa R. Saad, Abdel Moneim El-Ghanam and Rabah Hanem A. Mohamed. 2021. Developed magnetic Fe_3O_4–MoO_3-AC nanocomposite for effective removal of ciprofloxacin from water. Mater. Chem. Phys. 257: 123454.

Maitra, Urmimala, Ramakrishna Matte, H.S.S., Prashant Kumar and Rao, C.N.R. 2012. Strategies for the synthesis of graphene, graphene nanoribbons, nanoscrolls and related materials. Chimia 66: 941–948.

Manawi, Yehia, Victor Kochkodan, Muataz Ali Hussein, Moe A. Khaleel, Marwan Khraisheh and Nidal Hilal. 2016. Can carbon-based nanomaterials revolutionize membrane fabrication for water treatment and desalination? Desalination 391: 69–88.

Matysiak, Magdalena, Lucyna Kapka-Skrzypczak, Kamil Brzóska, Arno C. Gutleb and Marcin Kruszewski. 2016. Proteomic approach to nanotoxicity. J. Proteom. 137: 35–44.

Moses, Joseph Christakiran, Ankit Gangrade and Biman B. Mandal. 2018. Carbon nanotubes and their polymer nanocomposites. pp. 145-175. *In*: Nanomaterials and Polymer Nanocomposites: Raw Materials to Applications. Elsevier Inc. ISBN: 978-0-12-814615-6. India. Niranjan Karak (Ed.). Advanced Polymer and Nanomaterial Laboratory. Center for Polymer Science and Technology, Department of Chemical Sciences, Tezpur University, Tezpur, India.

Mutiyar, Pravin K. and Mittal, Atul K. 2014. Occurrences and fate of selected human antibiotics in influents and effluents of sewage treatment plant and effluent-receiving river Yamuna in Delhi (India). Environ. Monitor. Assess. 186: 541–557.

Nadour, Meriem, Fatima Boukraa and Ahmed Benaboura. 2019. Removal of diclofenac, paracetamol and metronidazole using a carbon-polymeric membrane. J. Environm. Chem. Engin. 7: 103080.

Ncibi, Mohamed Chaker and Mika Sillanpää. 2015. Optimized removal of antibiotic drugs from aqueous solutions using single, double and multi-walled carbon nanotubes. J. Hazard. Mater. 298: 102–110.

Park, Jeong Ann, Aram Nam, Jae Hyun Kim, Seong Taek Yun, Jae Woo Choi and Sang Hyup Lee. 2018. Blend-electrospun graphene oxide/poly(vinylidene fluoride) nanofibrous membranes with high flux, tetracycline removal and anti-fouling properties. Chemosphere 207: 347–356.

Peng, Zan, Xiaojuan Liu, Wei Zhang, Zhuotong Zeng, Zhifeng Liu, Chang Zhang, Yang Liu, Binbin Shao, Qinghua Liang, Wangwang Tang and Xingzhong Yuan. 2020. Advances in the application, toxicity and degradation of carbon nanomaterials in environment: a review. Environ. Int. 134: 105298.

Rai, Prabhat Kumar, Vanish Kumar, Sang Soo Lee, Nadeem Raza, Ki Hyun Kim, Yong Sik Ok and Daniel C.W. Tsang. 2018. Nanoparticle-plant interaction: implications in energy, environment and agriculture. Environ. Int. 119: 1–19.

Rashid, Md Harun Or and Stephen F. Ralph. 2017. Carbon nanotube membranes: synthesis, properties, and future filtration applications. Nanomaterials 7(5): 99.

Roco, Mihail C., Mirkin, Chad A. and Hersam, Mark C. 2011. Nanotechnology research directions for societal needs in 2020: summary of international study. J. Nanopart. Res. 13: 897–919.

Sadick, Neil S. 2001. Systemic Antibiotic Agents. Dermatol. Clin. 19: 1–21.

Singh, R. and Hankins, N. eds. 2016. Emerging Membrane Technology for Sustainable Water Treatment. ISBN: 978-0-444-63312-5. Imprint: Elsevier Science.

Song, Na, Xueli Gao, Zhun Ma, Xiaojuan Wang, Yi Wei and Congjie Gao. 2018. A review of graphene-based separation membrane: materials, characteristics, preparation and applications. Desalination 437: 59–72.

Tan, Ting Yuan, Zhuo Tong Zeng, Guang Ming Zeng, Ji Lai Gong, Rong Xiao, Peng Zhang, Biao Song, Wang Wang Tang and Xiao Ya Ren. 2020. Electrochemically enhanced simultaneous degradation of sulfamethoxazole, ciprofloxacin and amoxicillin from aqueous solution by multi-walled carbon nanotube filter. Sep. Pur. Technol. 235 (September 2019).

Teixeira, S., Delerue-Matos, C. and Santos, L. 2019. Application of experimental design methodology to optimize antibiotics removal by walnut shell based activated carbon. Sci. Total Environ. 646: 168–176.

Tufa, Ramato Ashu. 2015. Perspectives on environmental ethics in sustainability of membrane based technologies for water and energy production. Environ. Technol. Innov. 4: 182–193.

Upadhyayula, Venkata K.K., Shuguang Deng, Mitchell, Martha C. and Smith, Geoffrey B. 2009. Application of carbon nanotube technology for removal of contaminants in drinking water: a review. Sci. Total Environ. 408: 1–13.

Vilchis-Nestor, A.R., Trujillo-Reyes, J., Colín-Molina, J.A., Sánchez-Mendieta, V. and Avalos-Borja, M. 2014. Biogenic silver nanoparticles on carbonaceous material from sewage sludge for degradation of methylene blue in aqueous solution. Int. J. Environ. Sci. Technol. 11: 977–986.

Wang, Ting, Jingjing Wu, Shaoxin Xu, Chenguang Deng, Lijun Wu, Yuejin Wu and Po Bian. 2019. A potential involvement of plant systemic response in initiating genotoxicity of Ag-nanoparticles in Arabidopsis Thaliana. Ecotoxicol. Environ. Safety 170: 324–330.

Wang, Xiangyu, Yi Du and Jun Ma. 2016. Novel synthesis of carbon spheres supported nanoscale zero-valent iron for removal of metronidazole. Appl. Surf. Sci. 390: 50–59.

Wang, Xin, Jingke Song, Jianfu Zhao, Zhongchang Wang and Xuejiang Wang. 2019. In-situ active formation of carbides coated with NP–TiO_2 nanoparticles for efficient adsorption-photocatalytic inactivation of harmful algae in eutrophic water. Chemosphere 228: 351–359.

Wang, Xuandong, Renli Yin, Lixi Zeng and Mingshan Zhu. 2019. A review of graphene-based nanomaterials for removal of antibiotics from aqueous environments. Environ. Pollut. 253: 100–110.

Wang, Ying, Ming Yong, Song Wei, Yuqing Zhang, Wei Liu and Zhiping Xu. 2018. Performance improvement of hybrid polymer membranes for wastewater treatment by introduction of micro reaction locations. Prog. Nat. Sci. 28(2): 148–159.

Wang, Yue, Yannan Yang, Yiru Shi, Hao Song and Chengzhong Yu. 2020. Antibiotic-free antibacterial strategies enabled by nanomaterials: progress and perspectives. Adv. Mater. 32(18): 1904106.

Watkinson, A.J., Murby, E.J., Kolpin, D.W. and Costanzo, S.D. 2009. The occurrence of antibiotics in an urban watershed: from wastewater to drinking water. Sci. Total Environ. 407: 2711–2723.

World Health Organization. 2008. WHO Collaborating Centre for Drug Statistics Methodology. Anatomical therapeutic chemical (ATC) classification index-including defined daily doses (DDDs) for plain substances. Oslo: WHO-Oslo [cited 2008 Oct].

Xiong, Weiping, Guangming Zeng, Zhaohui Yang, Yaoyu Zhou, Chen Zhang, Min Cheng, Yang Liu, Liang Hua, Jia Wan, Chengyun Zhou, Rui Xu and Xin Li, 2018. Adsorption of tetracycline antibiotics from aqueous solutions on nanocomposite multi-walled carbon nanotube functionalized MIL-53(Fe) as new adsorbent. Sci. Total Environ. 627: 235–244.

Xu, Guiju, Beibei Zhang, Xiaoli Wang, Na Li, Lu Liu, Jin Ming Lin and Ru Song Zhao. 2019. Nitrogen-doped flower-like porous carbon nanostructures for fast removal of sulfamethazine from water. Environm. Pollut. 255.

Yang, Guohai, Dan Dan Bao, Daqing Zhang, Cheng Wang, Lu Lu Qu and Hai Tao Li. 2018. Removal of antibiotics from water with an all-carbon 3D nanofiltration membrane. Nanoscale Res Lett. 13.

Yang, Guohai, Daqing Zhang, Cheng Wang, Hong Liu, Lulu Qu and Haitao Li. 2019. A novel nanocomposite membrane combining Bn nanosheets and Go for effective removal of antibiotic in water. Nanomaterials 9(3): 386

Yang, Hui Ying, Zhao Jun Han, Siu Fung Yu, Kin Leong Pey, Kostya Ostrikov and Rohit Karnik. 2013. Carbon nanotube membranes with ultrahigh specific adsorption capacity for water desalination and purification. Nature Commun. 4.

Yu, Fei, Yong Li, Shen Han and Jie Ma. 2016. Adsorptive removal of antibiotics from aqueous solution using carbon materials. Chemosphere 153: 365–385.

Zeidman, Ahdee B., Oscar M. Rodriguez-Narvaez, Jaeyun Moon and Erick R. Bandala. 2020. Removal of antibiotics in aqueous phase using silica-based immobilized nanomaterials: a review. Environ. Technol. Innov. 20: 101030.

Zhang, Chen, Cui Lai, Guangming Zeng, Danlian Huang, Chunping Yang, Yang Wang, Yaoyu Zhou and Min Cheng. 2016. Efficacy of carbonaceous nanocomposites for sorbing ionizable antibiotic sulfamethazine from aqueous solution. Water Res. 95: 103–112.

Zhang, Meng, Shu Tao and Xilong Wang. 2020. Interactions between organic pollutants and carbon nanomaterials and the associated impact on microbial availability and degradation in soil: a review. Environ. Sci. Nano 7: 2486–2508.

Zhou, Jiahui, Fang Ma and Haijuan Guo. 2020. Adsorption behavior of tetracycline from aqueous solution on ferroferric oxide nanoparticles assisted powdered activated carbon. Chem. Eng. J. 384: 123290.

CHAPTER

14

Scope and Challenges for Green Synthesis of Functional Nanoparticles

Raktima Chatterjee[1a], Shinjini Sarkar[1a], Amit Kumar Dutta[2], Yuksel Akinay[3]*, Saumya Dasgupta[1a]* and Madhumita Mukhopadhyay[1a,b]*

[1a]Department of Chemistry, Amity Institute of Applied Sciences (AIAS), Amity University, Kolkata - 700156, W.B. India
[1b]Department of Material Science & Technology, Maulana Abul Kalam Azad University of Technology (MAKAUT), West Bengal, Simhat, Haringhata - 741249, Nadia, West Bengal, India
[2] Department of Chemistry, Bangabasi Morning College, 19, Raj Kumar Chakraborty Sarani, Kolkata - 700009, W.B. India
[3] Department of Mining, Engineering Faculty, Van Yuzuncu Yil University, Turkey

1. Nanotechnology and Biosynthesis

Nanotechnology is primarily the application of science and the innovations optimized related to the arena of any materials on the atomic scale. This specific arena reopens the imbibed properties of an already known system, e.g. metal, composite, metal hybrid, etc., in nano-metric order. During such transformation, the electron charge density waves termed plasmons are generated, thereby perturbing the intensity

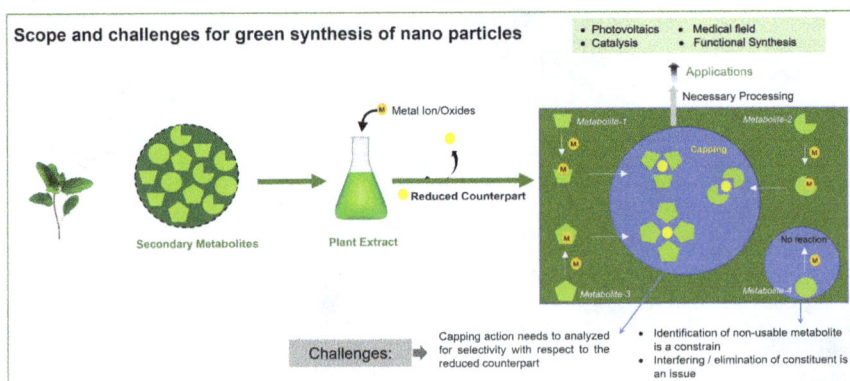

Corresponding authors: mmukhopadhyay@kol.amity.edu; madhubanerji@gmail.com; sdasgupta@kol.amity.edu; yukselakinay@yyu.edu.tr

of incident and reflected light in proportionate to the mass. The newer/improved properties enable the nanoparticles to significantly contribute in the area of chemicals, electronics, agriculture, drug-gene delivery, space industry, renewable energy sector, optical devices, single-electron transistors, light emitters, photo-electrochemical applications, catalysis, etc. (Parashar et al. 2009, Parashar et al. 2009, Bankar et al. 2010). These nanoparticles possess defined optical, chemical and mechanical properties, which enable them as smart materials for specific applications.

In this context, the important aspect in this research arena is dependent on the methods for synthesis of nanoparticles as a function of its variable physical and chemical properties, like size, morphology, composition, etc. (Bankar et al. 2010, Kasthuri et al. 2009, Nagajyothi and Lee 2011). The most popular chemical synthesis routes of such nanoparticles include chemical reduction, physicochemical reduction, radiolysis, electrochemical routes, UV-initiated photoreduction, microemulsion technique, photoinduced reduction, electrochemical synthetic route, irradiation route, etc. (Iravani et al. 2014). Polymeric matrix-like Poly (methyl vinyl etherco-maleic anhydride) is engaged by the researchers to be used as both reducing and stabilizing agents (7). The synthesized nanoparticles (NPs) are found to be stable with their properties intact at room temperature for up to 30 days. Morphological, FTIR studies, further establish the formation of nano-sized particles consisting of a layer of polymer (5-8 nm) [poly (methyl vinyl etherco-maleic anhydride)] coating encircling them (Maity et al. 2011). Medina et al. further reported that the NPs (10.2-13.7 nm) are crystalline with FCC lattice with the least surface energy as a spherical entity (Maity et al. 2011). Though the chemical means of nanoparticle synthesis is effective and quantitative, it bears certain major disadvantages of being costly, toxic and potentially hazardous. The chemical technique consists of a difficult separation procedure, high pressure and energy requirement (Iravani et al. 2014). The techniques are bio-incompatible due to the use of dangerous compounds (hydrazine or potassium bitartrate). These chemicals cause carcinogenicity, genotoxicity and cytotoxicity (Iravani et al. 2014).

In addition, the chemical synthesis route requires external capping agents for the stabilization of nano-size particles (Iravani et al. 2014). Literature reports suggest that some researchers prefer to use physical methods for nanoparticle synthesis as solvent contamination can be avoided. For studies like toxicity measurement (for the long term) for the synthesized nano-particle, physical approaches like the laser ablation process are one of the widely used techniques (Jung et al. 2006). The primary advantage of this process is the absence of a solution phase owing to which pure and uncontaminated metal colloids can be prepared in comparison to the evaporation condensation process (Tsuji et al. 2002). However, such physical methods have limited applications due to the absence of solvent phase and optimum tailoring of particle size is an issue. In such circumstances, the implementation of an environment-friendly process for the synthesis of functional nano-particles is in demand since the application arena is widely spread. This enforces the engagement of bio-sources as the active ingredient for reducing agents for NPs synthesis. According to the reports, in general, the bio-cells (in either producers or consumers, like simple bacterial cells to fungi and plants) are consisting of bio-reductant as well as particle stabilizing agents. The primary research lies in the identification of these functional reagents for such bio-nanosynthesis (Mohanpuria et al. 2008). The primary concern to using bio-synthetic route is to optimize the critical protocols, like operating temperature, nature and variety of organism, identification of eco-friendly catalyst from either plant/animal sources, optimum growth conditions of the enzyme and cell type, the existence of any inheritable properties, etc. (Gahlawat and Choudhury 2019). A general schematic based on interconnection among bionanoscience and nanotechnology is shown in Figure 1.

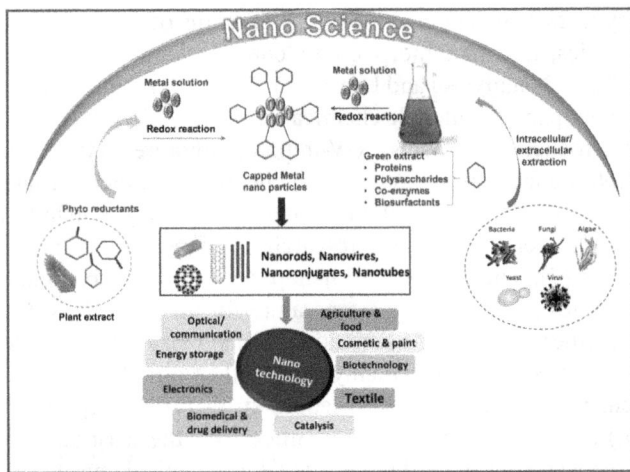

Figure 1. General scheme for interconnection among nanoscience and bionanotechnology.

The subsequent sections will provide an outline regarding the scope of synthesis of NPs from bio-origins (Section 2 and Section 3). Newer biosystems have been identified and undertaken for selective and functional NP synthesis and are elaborated in Section 4 along with the tools of characterizations for such systems. Finally, the correlation of biosynthesized functional nanomaterials is connected with technology, i.e. application. The application section has been divided into biotechnology, like gene therapy and material science aspect viz. sensors, catalyst, electrochemical devices, etc., in Section 5. Section 5 also highlights the concept of green toxicology with a primary focus on the influence of pollutants, antagonistic influence, etc. caused by different means onto bio-life. Any chemical frequently used in industry, medicine or naturally occurring may exhibit certain hazards to humans and the environment under certain circumstances, which are measured by physicochemical properties. The primary concept according to medieval physician Paracelsus is that the "dose makes the poison", i.e. every chemical is toxic at respective doses (Borzelleca 2000). Mechanistic toxicology is defined as the subject which deals with the study of toxic (i.e. dose dependence of chemicals) and elucidates the associated mechanism through which the influence is exerted. Since the effect of any chemical is absolutely concentration/dose-dependent, a detailed study on toxicology in terms of either descriptive or clinical effects is needed. For chemicals used in therapeutic cases, the potential toxicity is to analyze in clinical toxicology. On the other hand, the effect of toxicity of similar drugs, on organisms other than humans, is taken care by ecological toxicologists. On such note, green toxicologists primarily design rules to redox the bio-toxicity of the chemicals for their usage either as a drug or any other means as per the application. This is intrinsically correlated with the central nervous system and circulatory system of the body which is controlled by the blood-brain barrier. The primary aim of the drug design is to protect such intrinsic BBB from the influence of xenobiotics. The effectivity of any chemical within the biosystem depends upon the concentration/dose and can be correlated using dose-response relation. The threshold response of any drug/chemical can be well studied from the dose-response curve as shown in Figure 2. The response curve (interaction of xenobiotics with receptors) acts as a calibration plot that helps in the estimation of the usable and maximal efficacy of the chemical under study.

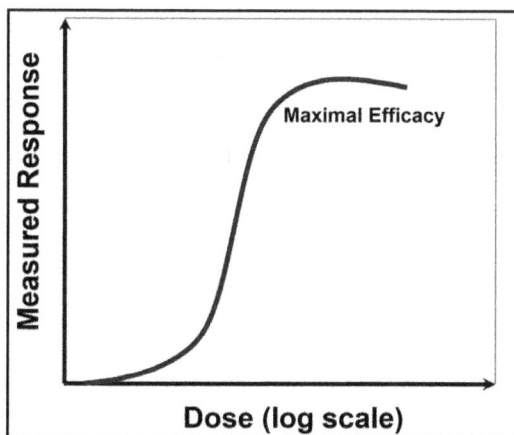

Figure 2. Relationship among measured response against applied dose of any chemical/drug.

In any biosystem, the influence of chemicals can only be accessed either as assistive or as a harmful drug based on the site of transportation through the ADME process viz absorption, distribution, metabolism and excretion (details discussed in Section 5.1). The ADME process is correlated with the dose-response curve to study the influence of any drug on biosystem (Dose exposure → ADME → site concentration). Determination for flux (usage) of any chemical is studied through: (a) concentration of drug as a function of time; (b) extent of bioabsorption of the drug within the cell; (c) chemical interaction of the induced drug with cell nutrients assisted through bio catalyst; and (d) rate of excretion from the organism. The role of biosynthesis of nanoparticles is undertaken nowadays for designing chemicals for biodegradability, which can reduce the toxicity/risk (Risk = Intrinsic hazard + Exposure) by reducing exposure time or reforming the toxic to moderately non-toxic version (Borzelleca 2000). According to the reports, this can be done by incorporating molecular characteristics which facilitates biodegradation and acts judicious strategic sector for reduction of risk factors. The incorporation of biosynthetic agents enables the advancement in mechanistic toxicology. Designing chemicals for exploiting soft spots required detailed information about the factors responsible for the kinetics and dynamics of toxicology. A scheme correlating the green toxicogenomics with bio-nanoparticles is shown in Figure 3.

Figure 3a describes the role of bio-nanoparticles (BNP) in gene therapy, wherein the mutual interaction between DNA and BNPs allows selective gene delivery within the nucleus for necessary medication. On the other hand, the BNP acting as a drug tend to function either by biofilm formation and disrupting the cell membrane of active germs through oxidative stress generation or by disconnecting the electron transport chain. The proposed mechanism for such action is shown in Figure 3b. Reports suggest that the bio-nanoparticles (metallic, organic, metallic oxide, bimetallic, etc.) essentially avoid the auto-resistive response of bio-cell toward the targeted chemical and is also associated with medicinal effect against bacteria, microbes, etc. (Baptista et al. 2018). A tabulation regarding the influence of bio-nanoparticles on the demolition of multiple drug resistance pathogens is given in Table 1 (Baptista et al. 2018). The database of the influence of nanoparticles on targeted bacteria along with its mechanism of action is as described in Table 1, which has been elaborated by Baptista et al. (2018). Finally, the conclusive remarks in Section 6 enlist the major challenges that reopen a new arena of the present research and future aspect for the same.

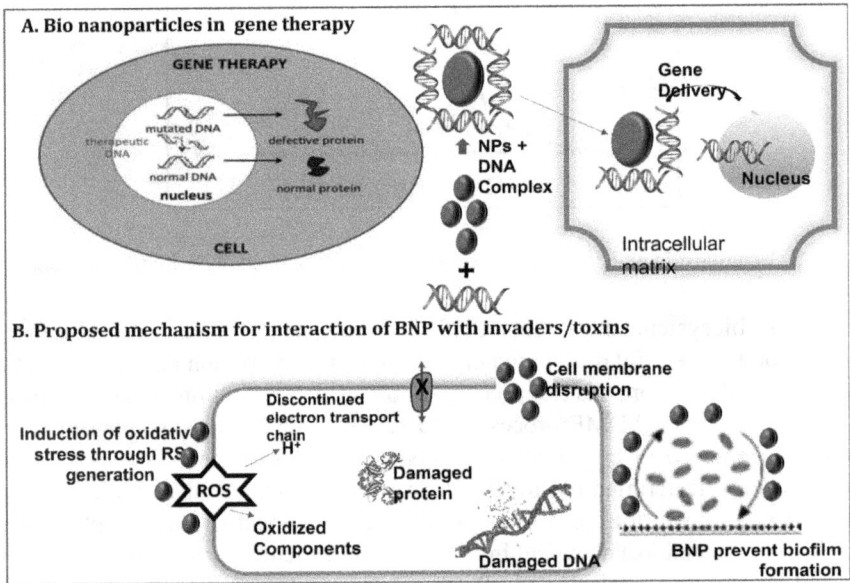

Figure 3. (A) Schematic for role of bionanoparticles toward gene therapy and (B) Proposed mechanism for function of BNP in toxigenomics (Baptista et al. 2018).

2. Phyto Nanoscience: Challenges and Prospects

For the synthesis of functional nanoparticles, 'Phyto' or plant sources act as an effective and easy reactant source. The dependence on metal salts for the growth of such producers has been already established from literature reports. The reports

Table 1: Influence of bionanoparticles on pathogens and mechanism of action (Baptista et al. 2018)

Sl. No.	Nanoparticle Type	Targeted Bacteria	Mechanism of Action
1	Au	*S. aureus* (Vancomycin-resistant)	Interaction with Valinomycin.
		S. aureus (Methicillin-resistant)	Photothermal therapy with ROS generation.
		E. coli, K. pneumoniae (Cefotaxime-resistant)	Damage of genetic.
		S. aureus, E. coli, P. aeruginosa (Ampicillin-resistant)	Combination with ampicillin. Lead to entry into the bacterial cell.
		Streptococcus bovis, S. Epidermidis (Kanamycin-resistant)	Destruction of invader cell wall.
		Proteus mirabilis, A. baumannii (Carbapenems-resistant)	Disturb the osmotic balance and disrupt the integrity of the bacterial cell wall.
		P. aeruginosa (Biofilm formation)	Interaction with cell surface.
		E. coli, P. aeruginosa, S. aureus (MDR)	Penetration through biofilm layers and interaction with cellular components.
		E. coli, K. pneumoniae, E. cloacae (MDR)	Photodynamic therapy/photothermal therapy.
2	Ag	*E. coli, P. aeruginosa* (Ampicillin-resistant)	Combination with ampicillin leads to entry into the bacterial cell. Inhibition of cell wall synthesis, protein synthesis and nucleic acid synthesis.
		S. aureus, E. coli, P. aeruginosa, K. pneumoniae, E. Faecalis (Erythromycin-resistant)	Cell surface damage and loss of the chain integrity.
		S. pneumoniae (Teicoplanin-resistant)	ROS generation, cellular uptake of silver ions, cascade of intracellular reaction.

(Contd.)

Table 1: (*Contd.*)

Sl. No.	Nanoparticle Type	Targeted Bacteria	Mechanism of Action
		E. coli (MDR)	ROS generation.
		A. baumannii (MDR)	Attach to the cell wall leading to structural changes in the permeability of the cell membrane.
		P. aeruginosa, E. coli (MDR)	Combination with antibiotics.
		S. epidermidis Conjugation with AMP. Mycobacterium smegmatis Vibrio fluvialis, P. aeruginosa (MDR/Biofilm formation)	Conjugation with AMP.
3	ZnO	K. pneumoniae (Ampicillin-carbenicillin-resistant)	ROS generation and disruption of bacterial cell wall.
4	CuO	E. coli, S. aureus (MDR)	ROS generation.
		S. aureus, P. aeruginosa (MDR)	Modulation of nitrogen metabolism.
5	Cu	S. aureus (Methicillin-resistant)	Copper ions release and subsequently bind with DNA leading to disorder of helical structure.
6	Fe$_3$O$_4$	E. coli, S. aureus, P. aeruginosa (MDR)	Radiofrequency (RF) coupled with magnetic core shell nanoparticles lead to RF-mediated physical perturbation of cell membranes and bacterial membrane dysfunction. Penetrate the membrane and interference in the electron transfer.
7	Al$_2$O$_3$	S. aureus (Methicillin-resistant)	Disruption of bacterial cell wall and ROS generation.

#	Type	Organism	Mechanism
8	Au/Ag bimetallic	*Enterococcus* (Vancomycin-resistant)	Theranostic system for SERS and a PDT.
		E. coli, S. aureus, E. faecalis, P. aeruginosa (Biofilm formation)	Disruption of bacterial cell wall and inactivate the proteins and enzymes for ATP production.
		B. subtilis E. coli, K. pneumoniae, S. aureus (MDR)	Combination with antibiotics.
		S. aureus, Micrococcus Luteus (MDR)	
9	Au/Pt bimetallic	*E. coli* (MDR)	Damage of the inner membrane, increase intracellular ATP level.
10	Cu/Ni bimetallic	*S. aureus, E. coli, S. Mutans* (MDR)	Adsorption of ions to the bacteria cells.

state that the intracellular nanoparticle synthesis includes the growth of the plant in organic media saturated with minerals, important required metal ions and hydroponic solution, etc. (Jha and Prasad 2010). In such processes, nature directs the *in-situ* synthesis of metal/metal complex nanoparticles for the growth and development of plants (Jha and Prasad 2010). The bioreduction of metal salts like magnesium, potassium etc., during assimilation within plants is an example of green biosynthesis, wherein both the bio reductants and capping agents are biosources. In a similar manner, extracellular techniques for nanoparticle synthesis initiates with the usage of extract prepared after boiling/processing of leaves, roots, stems, etc. According to the literature studies, based on the availability of bioactive reagents fruits, stems, seeds, latex, bark and roots are employed for the biosynthesis of nanoparticles and will be described in consequent sections. The research in this arena lies in the identification of suitable reducing agents for metals as well as capping agents for optimization of the generated particles. Selectivity of plant and its part with respect to the metal is the primary objective of such research content. Table 2 describes a brief description of different plants and a list of nanoparticles associated with such synthesis.

Nanoparticle synthesis from plants sources is easy to identify during the reaction course because of the associated color transformation in the visible wave length regime (from light greenish to brownish). Modulation of solution temperature and concentration of hydrogen ion enables variation in the features of synthesized nanoparticles (like shape and size). Song and Kim studied the synthesis of Ag nano particles using an extract from five variable leaves (Song and Kim 2008). The efficient and effective leaf extract is reported to be of Magnolia which generates Ag nanoparticles of 15 to 500 nm (with a UV absorbance peak at 430 nm). Kasthuri et al. (2008) showed the stabilization/capping effect of Ag nanoparticles using Phyllanthin extract through cyclic voltammetry studies (which shows a negative shift of cathodic current) and the existence of —OCH_3 group of the extract studied through FTIR. On such note, Bankar et al. (2010) employed banana peel, Safaepour et al. (2009) used Geranium graveolens and Bar et al. (2009), Parashar et al. (2009), Khandel et al. (2018), Nagajyothi et al. (2011) and Khatoon et al. (2018) used latex extract, Mentha piperita leaf extract and rhizome extract of Dioscorea Batatas for the synthesis of Ag nanoparticles. The silver nano particles (spherical and flower shapes) capped with Dioscorea Batatas are examined for medicinal activity, i.e. as a potent antimicrobial drug in contradiction to *Saccharomyces cerevisiae, Bacillus subtilis, E. coli, Candida albicans* and *S. aureus*. In such a case, the synthesized Ag nanoparticles are reported to exhibit maximum antimicrobial activity against *Saccharomyces cerevisiae* and *Candida albicans* showed the least activity for *Escherichia coli*.

Spherical ZnO nanoparticles (7-29 nm) synthesized from the gum/polymeric matrix of *C. procera* are reported by Singh et al. (2011). Such gum extract is further reported to be an effective bio-reductant and stabilizing agent for ZnO nanoparticles. Similarly, the proteins and some amino acids of Glycine max (soybean) leaf extract are reported to be one of the bio-reductants for the production of Pd nanoparticles (Petla et al. 2012). The FTIR study supported the capping action of such amino acid surfactants of Glycine max which retain the particle size of the resultant Pd nanoparticle. Gopalkrishnan et al. (2012) confirmed the reducing action of water-soluble carbohydrates present within *Tridax procumbens*, which promotes the

Table 2: List of variable plants and the associated metal nanoparticles (Song and Kim 2008, Kasthuri et al. 2008, Singh et al. 2011, Singh et al. 2011, Petla et al. 2012, Shankar et al. 2004, Philip 2009, Zhan et al. 2011, Gardea-Torresedey et al. 2003, Shankar et al. 2004, Armendariz et al. 2004, Ankamwar et al. 2005, Huang et al. 2007, Li et al. 2007, Bar et al. 2009, Yadav and Rai 2011, Chandran et al. 2006, Loo et al. 2012, Herrera-Becerra et al. 2008, Lee et al. 2011, Sharma et al. 2009)

Sl. No.	Name of the Plants	Nanoparticles Synthesized
1	• *Medicago sativa* (Alfalfa, leaf) • *Phyllanthus amarus* (Schumacher) • *Geranium indicum* (Alfalfa, leaf) • Tamarind (leaf) • *C. camphora* (camphor, dried leaf)) • *Magnolia kobus* • *Pinus densiflora* (Japanese red pine) • *Diopyros kaki* (Oriental persimmon) • *C. annum* (capsicum, extract from fruit) • *H. antidysenterica* (kurchi, leaf extract) • *U. fasciata* (ethyl acetate extract) • *C. zeylanicum* (cinnamon, dermas part) • *Camellia sinensis* (tea shrub) • *J. curcas* (gum and seed)	Silver
2	*Aloe vera* (bioextract from leaf) • *Avena sativa* (oat, biomass)	Gold (Au)
3	*Alfa alfa* (powder milled)	Iron (Fe)
4	*M. kobus* (leaf extract) • *T. procumbens* (daisy, leaf extract)	Copper (Cu)
5	*Physalis alkekengi* (plant extract) • *C. procera* (latex)	Zinc (Zn)
6	*G. max* (leaf extract)	Palladium (Pd)
7	*Salvadora persica* (tooth brush tree)	Ag-Ni bimetallic nanoparticles
8	*Azadirachta indica* (neem, leaf) • *Agaricus bisporus* (mushroom, pulp)	Au-Ag nanoparticles
9	*Cacumen platycladi* (Chinese medicinal herb)	Au-Pd nanoparticles
10	*Coriander sativam* (coriander)	ZnO nanoparticles

synthesis of CuO nanoparticles. Furthermore, polymerization with polyaniline is found to prevent the self-oxidation of such transition metal oxide which showed significant antibacterial activity of 100% at 30 μg.cm^{-3} (Gopalkrishnan et al. 2012). Medicinal *alfalfa* milled powder has been used as an effective reducing agent for FeO^{2+} to FeO (~ 10 nm) (Gopalkrishnan et al. 2012). Optimizing the pH at a basic range of 10, further resulted in a smaller FeO dimension of ~4 nm. The main difficulty in using this plant extract is the co-existence of dense biomass which restricts the microscopic observation of experimental nanoproducts in the range of ~10 nm. Phyto nanoscience technique can avoid multistep processes during the culture of cells in the case of bio-cells acting as the bio-reducing agents (as will be discussed in section Section 3). In addition, the phyto-based synthesis process offers faster reaction time and better caping of the formed nanoparticles.

Bimetallic Nanoparticles

Synthesis of bimetallic nanoparticles involves competitive reduction using a plant extract in a solution of two/more metal ion solutions. Based on the Nernst equation (Equation 1), the metal ion having higher reduction potential tend to reduce faster, which is further tailored by the formal potential ratio developed during *in-situ* synthesis.

$$DE = DE^\circ = \frac{Rt}{nF} \ln Q_r \qquad (1)$$

where DE° is the standard electrode potential, the formal potential is represented by DE (potential at the temperature of interest), R, T, n, F and Q_r are universal gas constant, operating temperature, <u>electrons</u> transferred in <u>half-reaction</u>, Faraday's constant and the quotient of the cell reaction, respectively. In case of considerable difference in the magnitude of reduction potential, core-shell morphology is expected for such bimetallic nanoparticles. In the solution of gold and silver metal ions, it is expected that Ag ions bear the tendency to reduce later and thereby form a shell (adsorbed) onto the already formed Au nanoparticles. Shankar et al. (2004) and Song and Kim et al. (2008) proposed hydrogen bonding and electrostatic interaction as the operative forces among the capping agents and the generated bimetallic nanoparticles. Though successive reduction process of two metal ions should generate core-shell type functional particle, in practice the formal potential difference allows the formation of alloyed bimetallic nanoparticles. This is further favored by an increase in reaction rate during the competitive reduction process. Therefore, synthesis of core-shell Au@Ag nanoparticle required optimization of solution properties (discussed in Section 2.1) so as to favor successive reduction and inhibit the simultaneous reaction. An important factor controlling the morphology of bimetallic particles of nano-dimension is the ionic ratio among the interacting metal ions. An insight is exemplified in Table 3 wherein Sheny et al. (2011) reported that at 9:1 ionic ratio,

Table 3: List of variable plants and the associated metal nanoparticles (Riaz et al. 2020, Tamuly et al. 2013, Pandey et al. 2015, Elemike et al. 2019, Mondal et al. 2011, Sheny et al. 2011, Zhan et al. 2011, Zhang et al. 2010)

Nanoparticle	Morphology of Bimetallic Nanoparticle	Ionic Ratio
Au-Ag	Alloy (Tamuly et al.)	1:1
Ag-Au	Ag as the shell surrounding around Au (Shankar et al. and Song et al.)	1:1
Au-Ag	Alloy (Sheney et al. and Mondal et al.)	1:1
		2:1
		3:1
		4:1
		9:1
Au-Ag	Alloy (Mondal et al.)	1:3
Au-Pd	Alloy (Zhang et al.)	1:1
Ag-Ni	Heterogenous matrix (Riaz et al.)	1:1

Au-Ag alloyed particle is observed from TEM and SPR band, whereas 1:1 and 1:2 of Au and Ag generates core-shell Ag@Ag nanoparticles. Irrespective of involved metal ions, alloy bimetallic nanoparticles are larger than core-shell particles.

However, Philip et al (2009) fail to obtain the core-shell morphology of the Au-Ag system even at the stoichiometry of 4:1. Riaz et al. (2020) employed *Salvadora persica* plant extract to synthesize Ag-Ni bimetallic nanoparticles which exhibit significant antioxidant potential. These heterogeneous nanoparticles are studied for the removal of Cr (VI) and Congo red dye from the aquatic source. The optimization of pH and other reaction parameters are discussed in Section 2.1, respectively. The bimetallic nanoparticles are found to be more selective and efficient for various applications with the bottleneck of morphology optimization and control of the concurrent reduction reactions. Hence, the present section establishes the significant role of a plant or Phyto sources toward the green synthesis of metallic/bimetallic and metal oxide nanoparticles for variable applications.

2.1 Factors Influencing the Phytosynthesis

This subsection primarily focuses on the factors influencing the biosynthesis of nanoparticles. The factors can be identified as (a) solution pH, (b) temperature, (c) reaction time and (d) dose of used biomass. The pH of the solution affects the nucleation center by fulfilling a proportionate relation. The rate of reduction is found to get accelerated with an increase in the population of such nucleation sites because it triggers the activity of the functional groups within the biomass extract (Armendariz et al. 2004). Consequently, the morphology and the dimension of the synthesized nanoparticle get perturbed with change in pH (1-11) as observed by Satish Kumar et al. (2009) in biosynthesizing nanoparticles of Ag with *C. zeylanicum*. In acidic pH, ellipsoidal Ag nanoparticles are obtained, whereas they changed to large spherical particles at higher hydrogen ion concentrations. It is also reported that variation in the concentration of hydrogen ions can optimize variable shapes of the resultant nanoparticles viz. hexagonal, tetrahedral, icosahedral multi-twinned, irregular and rod-shaped, etc.

Low pH of ~2 promotes aggregation (25-85 nm) of Au nanoparticles synthesized by *Avenasativa*, whereas a pH of 3-4 allows the accessibility of acidic functional groups, like —COOH and —OH for binding with gold and prevent nucleation to produce much smaller particles (Armendariz et al. 2004). This observation is further supported by Zhan et al. (2011) that high pH accelerates the reduction kinetics chloroaurate ions and thereby boost the homogenous nucleation and reduces the anisotropic growth forming many nano-sized Au particles upon using *Cacumen platyclade* extract. However, lower pH tends to slow the reduction and form heterogeneous nucleation and secondary nucleation to form large aggregates of Au particles. For the bimetallic synthesis of Au-Ag alloy, lower pH of 3.3 reduces the reduction rate of Ag (require 48 hours) compared to Au. An increase in the pH of the medium to basic range (10.8) enables simultaneous reduction as evident from monomodal plasmon resonance peak and get capped by hydroxyl groups of the plant extract (Jacob et al. 2012).

Alike pH, the temperature is another crucial factor in synthesis of the nanoparticle. In biosynthesis mode using plants, three temperatures are significant, namely (a) drying temperature, (b) extraction temperature and duration and (c) reaction temperature (Mohamad et al. 2013). The antioxidant content of the plant part is found to depend on the drying temperature (temperature employed for drying the plant source after cleaning treatment) and reduces slowly with fast-drying as reported by Naithani et al. (2006). Compared to slow drying using solar light or shade (Mittal et al. 2012, Ali et al. 2011), Sasidharan et al. (2010) used the fast oven drying technique to remove moisture from the plant source. The former technique is too slow, whereas the latter reduces the antioxidant content significantly. Finally, the freeze-drying process is established to be an efficient drying technique that helps in the sustenance of bio-reductant within the source for bio-nanosynthesis (Zaino et al. 2009, Silva et al. 2007). Chen et al. (2011) suggested that properly dried citrus peel extracts generate even more phenolic content compared to fresh peels. Researchers have established that the polyphenol content of the plant extract is dependent on the extraction temperature (at which the reactive functional species from the extract is eliminated) and duration of the same (Sardsaengjun et al. 2010, Machado et al. 2013). An increase in the working temperature has been found to favor the extraction of polyphenols for reaction and helps in accelerating the solubility of solute and the extent of diffusion of the solvent. However, the phenol compound tends to get decomposed at a higher temperature regime of ~90 °C (Sousa et al. 2008). Finally, an increase in reaction temperature (during the bio-reductant process using plant extract) is reported to form agglomerated particles by accelerating the nucleation rate. Sneha et al. (2010) found mainly nano-triangles of Au by using Piper beetle leaf at 200°C (operating temperature), which changes to nanoplatelets of 5-500 nm at 30-40°C. Another important observation is noted with the bimetallic synthesis of Au-Ag using *Anacardium occidentale* by Sheny et al. (2011). Greater leaf extract is required at room temperature (~2.5 ml), which reduces to ~0.6 ml at 100°C. This is owing to the fact that higher temperature tends to increase the reaction kinetics of the involved active species and stabilize the nanoparticles with a larger size. The primary bottleneck in this regard is to synthesize stable lower dimension nanoparticles with active capping and moderate reaction kinetics.

Compared to the microbe-mediated synthesis of bio-nanoparticles, plant catalyzed synthesis is more rapid and demarcated. The shape, size and yield of nanoparticles produced are also affected by the reaction time of biosynthesis (i.e. incubation of suspension medium). A time period of ~2 days is required for the synthesis of Au nanoparticles using *Coriandrum sativum* compared to 4 days required by micro-organisms (Nazeruddin et al. 2014). Li et al. (2007) studied the synthesis of silver nanoparticles of ~10 nm (for 5 hours reaction) using *Capsicum annuum* leaf, which tends to form polycrystalline as well as spherical particles. Similar reactant pair when allowed to react for ~13 hours synthesize 25-40 nm silver nanoparticles which proves the dependence on the reaction time. The bio-reducing process is influenced by the biomass content of the plant source and is termed as loading. Plant sources consist of many bioactive agents within which both bio-reductant and anti-sintering agents are required for nanoparticle synthesis. In such an aspect, optimizing the necessary bio-content is crucial. The medicinal plant named aloe vera can be engaged for the synthesis of Au nanoparticles using the extract of leaf and emerged as nanogold triangles at low extract loading (Chandran et al. 2006). Upon increasing the extract dose, the morphological study showed the transformation of the formed nano-triangles to spherical shapes. Therefore, during R&D, the slow dosage needs to be enforced with a gradual increase so that the change in morphology of the resultant nanoparticle can be tracked and tailored accordingly. A general scheme highlighting the mentioned parameters required for bio-reduction of metals along with capped nanoparticle synthesis is shown in Figure 4 for better correlation.

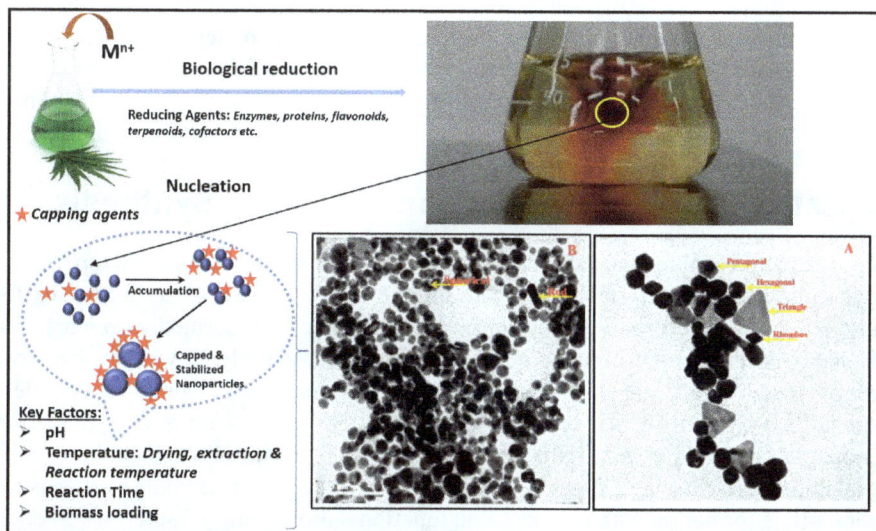

Figure 4. Schematic for factors involved in bio reduction of metal ions; A and B: Transmission electron microscopy images of Au nanoparticles synthesized using Portulaca grandiflora. Variable shapes are observed as marked in the figure. Reprinted with permission from Vijayaraghavan and Ashokkumar 2017. Plant-mediated biosynthesis of metallic nanoparticles: A review of literature, factors affecting synthesis, characterization techniques and applications. Journal of Environmental Chemical Engineering 5: 4866-4883 from Elsevier.

2.2 Role of Biological Agents Toward Nanoparticle Synthesis

The interesting area of the Phytosynthesis of nanoparticles is in the identification of the exact mechanism of green reduction and capping. Huang et al. reported the bio-reducing nature of *C. camphora* extract toward the synthesis of silver nanoparticles (Huang et al. 2007). Similarly, aldoses, as well as ketones sugar, are reported to act as the reducing agents toward chloroaurate ions by Shankar et al. (2007). Jatropha latex is employed for the synthesis of Ag nanoparticles, wherein a cyclic octapeptide (curcacycline A) and cyclic nonapeptide (curcacycline B) act as the bio-reductant and capping agent along with an anti-coagulating agent enzyme (curacin) (Bar et al. 2009). In such bioreactive agents discussed specific bonds (i.e. functional groups) of saturated and unsaturated carbon-carbon and carbon-oxygen bond accomplished the function of green reductant and stabilizing agent. Vibrational spectroscopy reports the completion of either Ag/Au nanoparticle formation using vibrational spectra. After the formation of Ag/Au nanoparticles from Phyllanthin extract, methoxy functional group (—OCH$_3$) is reported to form (1,088 cm^{-1}) from the FTIR study (Bar et al. 2009). As already reported in other cases, multiple shapes of Ag/Au nanoparticles can be synthesized by variation in the dosage of the plant extract. Such nanoparticles are based on the loading of Phyllanthin extract, variable shapes of the obtained nanomaterials are obtained and are employed for a specific application. In this connection, a hypothetical mechanism is predicted by Jha and Prasad (2010) in which either phytochemicals of Cycas leaf-like polyphenols or metallothioneins/ phytochelatin, etc., are proposed to act as the bio-reducing agent. The redox reaction with the respective metal ions is proposed to comprise a two-step process involving quercetin via an intermediate complex present in *Opuntia ficus-indica* plant extract (Gade et al. 2010). Therefore, spectroscopic tools act as the key characterization in the identification of bioactive agents responsible for bio-reduction and capping.

3. Extremophiles: Applications in Green Synthesis

The organisms capable of sustaining extreme environmental conditions of temperature, pressure are termed *extremophiles*. The unique enzymes used by such species for functioning in such forbidden environments are collectively termed *extremozyme*s. Bio-nanoparticle synthesis using such extremophiles, viz. algae, viruses, bacteria, actinomycetes and yeast is an eco-friendly technique that is comparatively less expensive, toxic-free which enables optimization of various traits of synthesized nanoparticles (Singh et al. 2016). Apart from being a green route and enabling easy unsalability, these biological agents are capable of undergoing template synthesis for nanoparticle formation with reproducible functional morphology. Table 1 (in Section 1) has exemplified the drug effect of synthesized bio-nanoparticles on specific strains of microorganisms. Table 4 lists the variety of nanoparticles that could be synthesized from microorganisms. Broadly, the following classification provides a broader view on the variability of nanoparticles synthesized from microorganisms as reported in the literature:

(a) metallic and bimetallic nanoparticles;
(b) nanoparticles of either magnetic or non-magnetic metal oxides;
(c) sulfide particles;
(d) carbonate nanoparticles;
(e) phosphate nanoparticles, etc.

Biosynthesis of nanoparticles as tabulated in Table 4 can be classified in terms of microorganisms which act as the nano-factories are (a) bacteria and actinomycetes, (b) fungi and yeast, (c) algae and (d) virus.

Bacteria and Actinomycetes

Bacteria are the class of extremophiles that possess several defense mechanisms, like intracellular sequestration, efflux pumps, aliovalent metal oxidation states and precipitation at the extracellular level to cope up with the harsh and toxic environment

Table 4: Synthesis of variable nanoparticles from extremophiles (Fayaz et al. 2010, Vigneshwaran et al. 2005, Sneha et al. 2010, Fayaz et al. 2009, Li et al. 2011)

Nanoparticle	Name of Microorganisms	Dimension of Nanoparticles (nm)	Shape of Nanoparticles	Culturing Temperature (°C)
Ag	*T. viride* (outside the cell)	2-50	Spherical	27
	Phaenerochaete chrysosporium (extracellular)	10^1 to 10^2	Tomb-like structure	37
	Corynebacterium glutamicum (extracellular)	5-40	Not regular in shape	30
	Trichoderma viride (extracellular)	~3	N.A.	10-40
	Aspergillus fumigatus	4-20	Spherical	20
Au	*Enterobacter* sp. (intracellular)	2-5	Spherical	30
	Rhodococcus sp.(intracellular)	10	Spherical	37
	Shewanella oneidensis (extracellular)	~10	Round	30
	Plectonema boryanum (intracellular)	12-25	Box-like	25-100
	P. boryanum	10-6	Octahedral	25
	Yarrowia lipolytica (extracellular)	14	Three angular	30
	Brevibacterium casei (intracellular)	10-50	Spherical	37
Pt	*Shewanella algae* (intracellular)	5	N.A.	25
Hg	*Verticillium* sp.	25	Spherical	25
Se	*Shewanella sp.* (extracellular)	180	Globular	30

(Contd.)

Table 4: (*Contd.*)

Nanoparticle	Name of Microorganisms	Dimension of Nanoparticles (nm)	Shape of Nanoparticles	Culturing Temperature (°C)
Au-Ag alloy	*Fusarium oxysporum* (extracellular)	8-14	Spherical	25
Pd	*Desulfovibrio desulfuricans* (extracellular)	50	Spherical	35
Uranium (VI), Technetium (VII), Chromium (VI), Cobalt (III), Manganese (IV)	*Pyrobaculum Islandicum* (extracellular)	N.A.	Spherical	100
Fe₃O₄	*Shewanella Oneidensis*	~ 45	Rectangular, four or six membered	28
	Yeast cells	N.A.	Nanopowders	36
	HSMV-1 (intracellular)	113	Pellet-shaped	63
	QH-2 (intracellular)	81 ± 23 × 58 ± 20	Rectangular	22-26
Sb₂O₃	*Saccharomyces Cerevisiae* (intracellular)	2-10	Spherical	25-60
TiO₂	*Lactobacillus* sp. (extracellular)	2-10	Spherical	25
BaTiO₃	*Lactobacillus* sp. (extracellular)	20-80	Tetragonal	25
	Fusarium oxysporum (extracellular)	4-5	Spherical	25
ZrO₂	*Fusarium oxysporum* (extracellular)	3-11	Spherical	25

CdS	*Schizosaccharomyces pombe* and *Candida glabrata* (intracellular)	1-.1.5	Hexagonal lattice	N.A.
	Rhodobacter sphaeroides (intracellular)	8	Hexagonal lattice	N.A.
	Fusarium oxysporum (extracellular)	5-20	Spherical	
FeS	Sulfate-reducing bacteria	2	Spherical	N.A.
$SrCO_3$	*Fusarium oxysporum* (extracellular)	10-50	Needle like	27
$Zn_3(PO_4)_2$	Yeast (extracellular)	10-80 × 80-200	Rectangular	25
CdSe	*Fusarium oxysporum* (extracellular)	9-15	Spherical	10

(Iravani 2014). This defense mechanism could be reutilized for acting as the biosource for nanoparticle synthesis through an extracellular or intracellular mechanism. Silver resistant strain of *Pseudomonas stutzeri* (AG259) is capable of producing silver nanoparticles (size 200 nm) using nicotinamide adenine dinucleotide enzyme that oxides to NAD^+ and supplies electron (Joerger et al. 2001). Intercellular reduction of multiple targets, like Ag, Ni, Co, Pd, Li, Rh, Fe and Pt, is feasible using the bacterium *Pseudomonas aeruginosa* (Srivastava and Constanti 2012), wherein external agents for stabilization is not required (hence, pH is not a susceptible parameter). Biosynthesis of Pd, Pt and Te was successfully carried out by Ahmed et al. (2018) and Zonaro et al. (2017) using *Shewanella loihica* PV-4 and *Ochrobactrum* sp. The former generates 2-7 nm Pd and Pt nanoparticles using biofilms of *S. loihica* which showed significant catalytic activity for degradation of methyl orange dye. *Ochrobactrum* sp. serves as an excellent source as a bio-reductant for toxic tellurite oxyanions and converts them to functional Te nanoparticles. Saravaran et al. (2018) have synthesized Ag nanoparticles (41-62 nm) using *Bacillus brevis* which shows remarkable activity against multidrug-resistant strains of *S. aureus* and *S. typhi*. *Bacillus subtilis* bacteria separated from the Hatti gold mine of India was utilized for the synthesis of Au nanoparticles as they are found to be resistant to aurum ions (Srinath et al. 2018). This biocatalyst (Au ions) is reported to decompose other significant dyes, which are not environment friendly (Srinath et al. 2018). Apart from terrestrial bacteria, marine microbial cultures are also explored as the nano-factories as established by Malhotra et al. (2013) using *Stenotrophomonas* for the biosynthesis of Au and Ag nanoparticles. Certain low molecular weight proteins of such *Stenotrophomonas* are found to promote bio-synthesis (Malhotra et al. 2013). Literature reports have also established the function of the use of extracellular polymeric substances (EPS) of certain bacteria that can function as the bio-reductant and stabilizing agents of Au, Ag, Cu, etc (Section 3.1). *Actinomycetes* are unicellular, gram-positive bacteria that consist of extracellular enzymes, secondary metabolites and high protein capable of acting as bio-reductant and capping agents in both intracellular and extracellular mode. The enzymes present in the cell wall of Actinobacteria *Rhodococcus* NCIM 2891 were used as the primary active reagents for the biosynthesis of Ag nanoparticles (10 nm) by Otari et al. (2012). Similarly, acidophilic actinobacteria, *Streptacidiphilus durhamensis*, is reported to accelerate the production of ~50 nm Ag nanoparticles, which further proved to be lively in contradiction of *P. mirabilis, P. aeruginosa* and *Staphylococcus aureus* (Buszewski et al. 2016).

Fungi and Yeast

Biosynthesis of metal/metal oxide nanoparticles is also reported to be synthesized from fungi owing to its higher productivity and capability as a bio-reductant. The binding ability of fungi with the respective target metal ion is reported to be higher which is another advantage of such biosynthesis (Singh et al. 2016). Bioaccumulation activity toward the metal ions (intended to reduce) are higher for fungi. Since the problems of external treatments using detergents, other frequencies are not required during extracellular treatment, it is undertaken more frequently. In another report, *S. radiatum* is found to be an effective bio-reductant for the production of white-rot fungus along with the extracellular formation of silver nanoparticles (10-40 nm).

This is extracted from the Eturnagaram forest of Warangal. Both the pathogenic strains of gram-negative and positive are reported to get successfully attacked and evaded with this synthesized nano-Ag (Metuku et al. 2014). *Colletotrichum* sp. is reported by Suryavanshi et al. for the synthesis of Al_2O_3 nanoparticles which need further functionalization using special oils (extracted from the *C. medica* and *E. globulus*) (Suryavanshi et al. 2017). Such nano-functionalized oil is effective against foodborne pathogens.

In addition to fungi, yeasts can cause the detoxification of metal ions via a certain inherent mechanism. The primary mechanism(s) responsible for such detoxification are complex chelation, intracellular sequestration, precipitation by bio-means, etc. Such facts can be explained based on their inherent property of toxic absorption nature. Biomimicking synthesis of Ag nanoparticles is reported using a variety of ascomycetous yeast *Yarrowia lipolytica* (Apte et al. 2013). The reports establish that the yeast cells successfully produce melanin pigment and is accountable for the bio-mineralization of active metal ions which undergoes successive bio-reduction. On a similar note, 20-80 nm circular Ag nanoparticles are synthesized from *Candida utilis* NCIM 3469. The synthesized Ag particles have been tested against pathogens and showed significant activity (Waghmare et al. 2015). Significant antiproliferative activity against *K. pneumoniae* and *S. aureus* is observed from 2-10 nm Ag nanoparticles synthesized from *Candida lusitania* from the gut of a termite (Eugenio et al. 2016). The yeast cells inherit numerous mechanisms to be meal tolerant and hence the produced nanoparticles possess selectivity and can be synthesized with optimized characteristic features.

Algae

Algae is also considered to be a significant nano-factory for the production of nanoparticles. Unicellular microalga, *Chlorella vulgaris* were first dried and engaged to reduce silver metal ions (~ 5-15 nm) by Ferreira et al. (2017) which act as the antimicrobial agents. Similar algae and marine alga, *Sargassum bovinum* (from the Persian Gulf) were engaged for the synthesis of Palladium nano (5-10 nm circular and mono dispersed particles) within 10 minutes reaction time (Arsiya et al. 2017). Aqueous extracts of *S. plagiophyllum* (a marine alga) are reported to produce silver oxide and silver nanoparticles which can be also synthesized with a different morphology using *Caulerpa racemose* (Dhas et al. 2014, Edison et al. 2016).

ZnO is one of the widely used materials in the field of optoelectronics, drug chemistry, the food industry, etc. However, functionalization of ZnO including off stoichiometric compositions requires a costly, time-consuming synthesis procedure. In this context, Rajesh Kumar et al. (2018) utilized marine seaweeds (brown colored) alga like *Turbinaria conoides*, *Padina tetrastromatica* and *Sargassum muticum* extract engaged by Sanaeimehr et al. (2018). This ZnO is examined for its effectiveness against human liver cancer. The marine algae are of immense interest owing to the presence of the biologically active compound and secondary metabolites for which they participate in the bio-redox reaction and as a stabilizing agent. The reported bio-medicinal activity of such algae is reported to be in the area of antiviral agents, anticancer agents, antidiabetic, antioxidants and cardioprotective.

Viruses

The significant property of providing a protein capsid outer wall for interaction with metal ions is the primary advantage of using a virus for bio-nanosynthesis of metal ions (Kobayashi et al. 2012). Intra modification of viruses can be done so as to enable them as a template for patterning and deposition or treated so as to be the 3D carries for drug delivery. For the synthesis of bio nanoparticles for application in nanocomposites which are important materials for bioengineering in drug delivery, like cancer, myalo type ailments, etc. Semiconductor ZnS and CdS nanoparticles are attempted to be synthesized using a biogenic route (explained in Section 4). Compared to all viruses, plant viruses are reported to be technologically safer tools owing to their stability (structural and biochemical), easy processing, non-toxicity and non-pathogenicity. An interesting study is undertaken for size optimization of the nanoparticles using selective plant extracts, like tobacco *mosaic virus* (TMV), *Nicotiana benthamiana, Musa pradisiaca, Avena sativa* and *bovine papilloma virus* (BPV) (Love et al. 2017). The presence of only TMV is found to significantly reduce the size of nanoparticles with capping action. *Potato virus X* is studied for cancer ailment by employing a doxorubicin drug by Le et al. (2017). The said virus can promote the synthesis of elongated filamentous nanoparticle which bears more penetration power for drug action compared to spherical ones (Le et al. 2017). The major drawback of such viruses is that they require the involvement of host organisms for protein expression. The relevant BNP synthesis from such viruses is, therefore, under-developed and limited for large-scale application. Apart from significant advantages, the main bottlenecks for virus-enabled biosynthesis are scaling-up processes, the extent of reproducibility, complete study on synthesis mechanism and control on size and monodispersity.

3.1 Mechanism of Nanoparticle Development by Extremophiles

Nanoparticle synthesis through microorganism source is quite different from that in plants/plant parts whose probable mechanism. To date there are many schools of thought regarding the same; however, three primary mechanisms can be highlighted in terms of the available reports like (a) reductase enzyme and proteins, (b) role of redox mediators and (c) electron shuttle quinones.

Role of Proteins

The primary objective in engaging the virus as nano-factories is to utilize their defense mechanism against toxic metal and reduce them to protect own selves. The involvement of proteins as one of the proposed defense tools is found to involve (a) metal reduction to lower valence state, (b) formation of specific and selective complex and (c) intracellular or extracellular dissimilatory oxidation (Tanzil et al. 2016). Reports have clarified that in response to such metal ion toxicity, the natural defense process of microorganisms tends to enforce NADH-based nitrate reductase enzyme to get self-oxidation to nitrite and even to nitrogenous gases. The examination of the culture proves the formation of such an oxidized product from which the formation of reduced metal nanoparticles can be correlated. Researchers

(Srivastava and Constanti 2012) have even studied the stability of the synthesized non-aggregated Ag nanoparticles for months without the addition of any capping agent. Furthermore, antimicrobial activity is observed for such nanoparticles against infections, like urinary tract infection producing traces of isolates such as *K. pneumoniae, Bacillus* sp., *S. aureus, E. coli*, etc.

Another very interesting report by Hamedi et al. establishes that among the four phases of bacterial growth viz. lag, logarithmic, stationery and death, the highest synthesis rate of nanoparticles is reported in the third stage wherein the extracellular enzyme nitrate activity is extreme (Hamedi et al. 2017). Moreover, the enhanced C:N ratio induces the generation of the small and narrow size particle distribution of nanoparticles. In the extracellular mechanism of reduction, the electron donators are identified to be the conductive bacterial external appendages pili and surface proteins (Cologgi et al. 2011). This is proved in the bio-mineralization of uranium nanoparticles reported by *Geobacter sulfurreducens* using pilin-inducing and noninducing conditions. Extracellular reduction of uranium ions is only preferred by the strain having pilling supplementation, else the biosynthesis resulted inside the periplasmic space. The hypothesis of Vasylevskyi and his group established that the initial reduction process of metal ions ($M^{m+} - M^0$) is an endogenic process, whereas further aggregation to M_n is an exorgenic technique which is reported using the photoinduced electron transfer through tetrapeptide/Ag^+ solution. The subsequent sections will report such bio-nanosynthesis using certain redox mediators and other components which are present within these extremophiles and perform the dual function of reductant and capping agent.

Redox Mediators

Redox mediators, i.e. electron shuttle quinones enable electron transportation using NADH, cytochrome *c*, ubiquinol, etc. The ubiquinol is also termed as oxide/ superoxide catalyst and such redox proteins are associated with the respective redox proteins (Bewley et al. 2013). The mode of operation involves an extracellular synthesis of metal nanoparticles as reported, wherein Shewan ellaoneidensis MR-1 is used along with oxide of iron compounds as an effective acceptor of the electron in respiration in absence of oxygen involving *c*-type cytochromes as the redox mediator. Electrons are proposed to get transported across the cell membrane to the surface through a metal-reducing pathway (Mtr). The proteins components of such Mtr pathway can be mentioned as *CymA, MtrA, MtrB, MtrC* and *OmcA* which help in the transportation of electrons. The route for electron transportation is reported to be from quinol at the inner membrane (IM) across the outer membrane to the ferric oxide surface. Such path incorporates periplasmic space as the medium (Gahlawat and Choudhury 2019).

Role of Protein

CymA first tends to oxidize the quinol within the inner-membrane and help in the transportation of electrons to MtrA involving other proteins through a periplasmic membrane. According to the reports, the decaheme cytochrome C protein (*MtrA*) is found to exist within porin-like protein (*MtrB*) and in the outer membrane portion.

MtrC and *OmcA* actually function as bio-reductants (terminally placed) and reduce the (Fe(III) oxides) and transfer electrons through their exposed heme part (Gahlawat and Choudhury 2019). Followed by this, the electrons are transferred from *MtrC* (outer-membrane) to *OmcA*, wherein *MtrB* and *MtrA* act as the catalyst. Apart from these allotted functions, the cytochrome C-proteins like *OmcA* and *MtrC* are translocated from the outer membrane to the cell surface. This translocation is carried out by the type II bacterial secretion pathway. Hence, it may be stated that the mentioned multiheme complexes promote the extracellular reduction of metal ions. They intend to allow electron transportation from the inner to the outer cell membrane and catalyzes such bio reduction through periplasm.

Role of Exopolysaccharides

The exopolysaccharides (EPS) are bioactive agents of bacteria which serve the purpose of environmental protection, adherence to surface and cell to cell interactions. They generally function in the extracellular mode in their function of bioreducing and capping. Functional groups aldehyde (rhamnose as well as pyranose sugars) and hemiacetal groups present within the exopolysaccharides of *Escherichia coli* biofilm acts as the primary reducing for Ag ions and get self-oxidized to carboxyl groups (Kang et al. 2014) and are established from FTIR and C_{13} NMR study.

Detoxification of aqueous sources from Cd using Pseudomonas aeruginosa JP-11 is also an example of nano synthesis by EPS of biosource. These EPS are made active through the chemical association of sulfur groups (owing to high stability constants) so as to accelerate the adsorption of cadmium ions. They tend to reduce Cd by initiating self-attachment in an aqueous solution and generate nanoparticles of 20-40 nm dimension of CdS nanoparticles (Raj et al. 2016). Pullulan polymer is used as the effective reducing agent for Au ions by Choudhury et al. which showed first-order kinetics and resulted in reaction acceleration at 100°C. The basic structure of pullulan is unaltered as proved through vibrational spectroscopy with only oxidation of side-chain aliphatic alcoholic groups during such bio-reduction. A proposed schematic for the mechanism following the redox mediators and exopolysaccharides is given in Figure 5 (Choudhury et al. 2014).

3.2 Size and Morphology Optimization for Targeted Application

Details on the variants of extremophiles involved in the synthesis of biomolecules are described in Sections 3 and 3.1. Each reactant species can further retreat in terms of the reaction conditions so as to optimize the traits of the synthesized nanoparticles. The primary factor reported is the adhesive chemicals present onto the cell wall of such microorganisms which induces biomineralization. Secondary factors mentioned are pH, temperature, the involved reactant and the medium stoichiometry (Klaus-Joerger et al. 2001). The detailed study on the growth kinetics during the formation of Ag nanoparticles upon employing *Morganella psychrotolerans* is undertaken by Ramanathan et al. (2011). At an initial growth temperature of 20°C, only spherical nanoparticles of 2-5 nm are observed which eventually resulted in a mixture of phases like triangular and hexagonal in the form of nanoplates along with spherical ones at 25°C. Reducing the temperature within 15-20°C generates

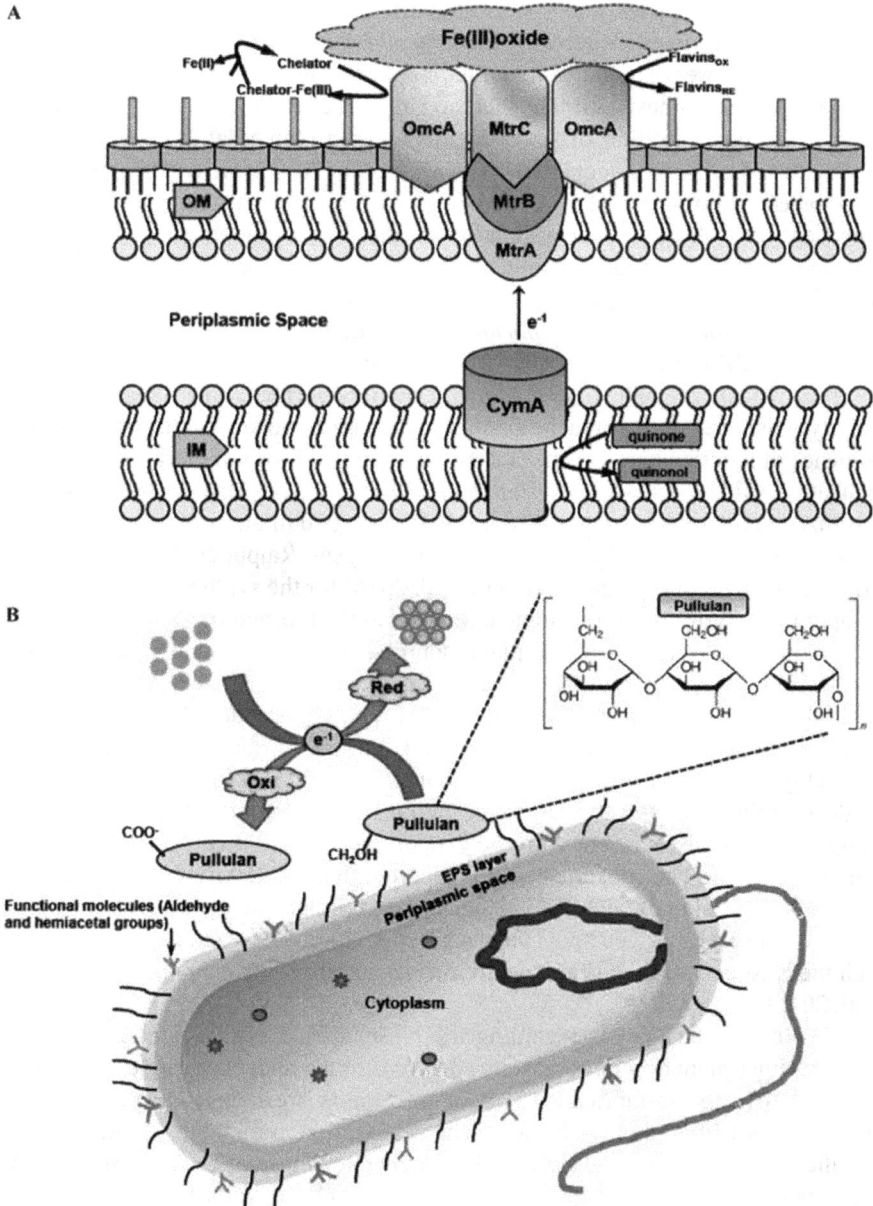

Figure 5. (A) The pathway for metal-reducing electron transfer of *S. oneidensis* through extracellular mechanism. (B) Scheme for the biomineralization of gold ions by pullulan exopolysaccharide. Reprinted with permission from Shi, L., K.M. Rosso, T.A. Clarke, D.J. Richardson, J.M. Zachara and J.K. Fredrickson 2012. Molecular underpinnings of Fe(III) oxide reduction by *Shewanella oneidensis* MR-1. Front. microbiol 3: 50 with permission from RSC and Choudhury, A.R., A. Malhotra, P. Bhattacharjee and G. Prasad. 2014. Facile and rapid thermo-regulated biomineralization of gold by pullulan and study of its thermodynamic parameters. Carbohydr Polym 106: 154-159 from Elsevier.

a mixture of nanoplates and spherical particles. Surprisingly, the population of spherical nanoparticles reduced significantly at 4°C but caused much bigger (70-100 nm) spherical cores. The influence of temperature, metal ion concentration and pH on nanoparticle synthesis is established by Yumei et al. (2017) by using *Arthrobacter* sp. A lower concentration of silver nitrate solution (1 mM at pH ~7-8) generates face-centered cubic Ag nanoparticles (9 to 72 nm) at 70°C which aggregates (72 nm) at a concentration of 3 mM. The reaction is found to get terminated at a pH below 5 and above 8. A detailed investigation on the process parameters like hydrogen ion concentration, solution parameters, the time required for reaction on the particle size of synthesized nanoparticles using fungal route is studied by Bhargava et al. (2016) using *Cladosporium oxysporum* for Au nanoparticles. At 1:5 biomass to water ratio and 1 mM concentration (pH = 7), the maximum yield of Au nanoparticles is observed which is capable of degrading the activity of rhodamine B through catalyst activity. Mishra et al. reported the synthesis of aurum nanoparticles by optimizing the influence of reaction parameters, like temperature, the time required for reaction and incubation using *Trichoderma viride* and *H. lixii* species (Mishra et al. 2014). The optimized temperature is reported to be 30 °C using 10 minutes of reaction time that serves as a strong biocatalyst and antimicrobial agent. Rajput et al. studied multiple strains of *F. oxysporum* (fungi) (Rajput et al. 2016) for the synthesis of nanoparticles by optimization of involved parameters affecting the morphology and size.

Similar to the usage of plant sources, functional nanoparticles can be synthesized using extremophiles and are discussed in Section 3.1. Higher variability of carbohydrates constitutes EPSs, like sugars consisting of single units of D-galactose and glucose, D-mannose, L-fucose, L-rhamnose, D-galacturonic acid, D-glucuronic acid, D form of mannuronic acid, L-guluronic acid, N-acetyl-D-glucosamine and N-acetyl-D-galactosamine. A large number of noncarbohydrate components are also present with the exopolysaccharides, like carboxyl, pyruvate, sulfate and phosphate substituents. It is reported that the metal ion upon interacting with such EPSs containing reducing sugar and encircled/chelated. Chelation being a thermodynamically driven spontaneous process, the redox reaction progress at ease with the least activation barrier and are further stabilized by capping agents (Freitas et al. 2011).

Finally, the challenges regarding the bio-mediated synthesis of nanoparticles through either plant or microorganism sources is in the understanding of mechanistic approach toward bio-fabrication. Literature reports are being daily updated with proof concepts based on advanced characterization tools, but identification of specific agents, their isolation, probability of reaction in isolation, etc. are required to understand the biochemical pathway for the development of tailor-made nanoparticles. Bulk scale production of such nanoparticles is another restriction of such processes maintaining the controlled size and morphology. Bulk bioproduction is reported to be affected by numerous factors, like cost, energy requirement, low nanoparticles yield and multimodal distribution of size. Lastly, the major shortcomings or stringent properties that need attention are clearance of nanostructures *in-vivo*, dispersal profile, release kinetics and evaluation of the biocompatibility.

4. Biogenic Synthesis of Nanoparticles

Green Synthesis from Vitamins

Vitamins have been identified by researchers to act as bio-reductant agents. Morphologically optimized nanowires and nanospheres have been synthesized using Vitamin B2 as the reducing and capping agents (Nadagouda and Varma 2006). This role is also used to treat certain tumor cells using such vitamins. Reports have also established the active role of Vitamin C along with chitosan polymer for the bio-reduction of many metal ions. The ascorbic acid is proposed to act as the capping and reducing agent along with chitosan as the stabilizing species. The synthesis is dependent on the concentration of stabilizing agent which is proportional to metal ion concentration (Malassis et al. 2016). Another report exhibits the synthesis of Ag nanoparticles using Vitamin C in *Desmodium triflorum*. In this case, the protons generated from glycolysis coherently work along with NAD to act as the reducing agent and promote the redox reaction (Ahmad et al. 2011).

Green Synthesis from Enzymes

The involvement of enzymes as the bio-reductant for nanoparticle synthesis is one of the interesting findings owing to their interesting well-defined structure, purity, presence of growth accelerating chemicals, etc. Bimetallic Fe/Pd nanoparticles are synthesized using specific enzymes within polymer multilayer-assembled membranes held through electrostatic interactions (Gour et al. 2019). Manivasagan et al. employed extracellular amylase for reducing $AuCl_4$ wherein the enzyme is generated using fermentation of *Bacillus licheniformis* at pH 8 (Gour et al. 2019). Another attempt has utilized *E. coli* species for the synthesis of Au nanoparticles by extracting reductase enzyme (sulfite) which in turn bears significant activity against human pathogenic (known as antifungal) (Gour et al. 2019). Reports have been also published for the extraction of natural enzymes and bio-molecules from herb/plant-based extract viz. corn cob, Cocos nucifera coir, wheat and rice bran, fruit seeds and peels, palm oil etc., for the synthesis of functional nanoparticles (Gour et al. 2019). Compared to the reduction of Ag ions using a chemical mode of synthesis using sodium borohydride enzymes from beet juice showed enhanced catalytic activity. The former method/technique tends to the transformation of 4-nitrophenol to 4-aminophenol. Size optimization of Ag^0 is found to be effectively tailored by changing the concentration of beet juice (Gour et al. 2019). The coupling of Au nanoparticles with a redox enzyme is reported to effectively act as the efficient electron conductor/transmitter in a sensor application (Gour et al. 2019). Such electron transmitter functions between the biocatalyst and the electrode present within the assembly.

Green Synthesis from Wastes

Waste management is an important and serious issue in today's world. Waste is being classified based upon its reusability, decomposition, safety, etc. so that the involved chemical/bio-hazard can be eliminated. The bio-reductants necessary for nanoparticle synthesis are now reported to be existent within the waste also. Based on the literature reports, the selection of bio-reductants is extracted from the following: (a) biodegradable, (b) agricultural, (c) domestic, (d) kitchen, (e) agricultural, (f) industrial waste, (g) coffee wastewater and (h) green waste.

The monodispersed Au particles (6-20 nm) synthesized using mango peel showed an interesting feature of non-toxicity event at a considerably higher concentration of 160 μg ml^{-1}. The possible motive proposed is owing to the presence of bio capping agents and their interactions, the toxic effect of Au nanoparticles is minimized (Yan et al. 2014). The wine industry acts as the largest biosource of bio-reductants in the form of grape wastes. The spherical and polygonal-shaped Ag nanoparticles are successfully synthesized from grape seed extract and analyzed through FTIR to study the presence of —OH functional group, cyclobenzene, etc. Such nanoparticles are found to possess significant activity against gram-negative and gram-positive bacteria (Xu et al. 2015). In the line of such study, 7 nm Ag particles can be synthesized

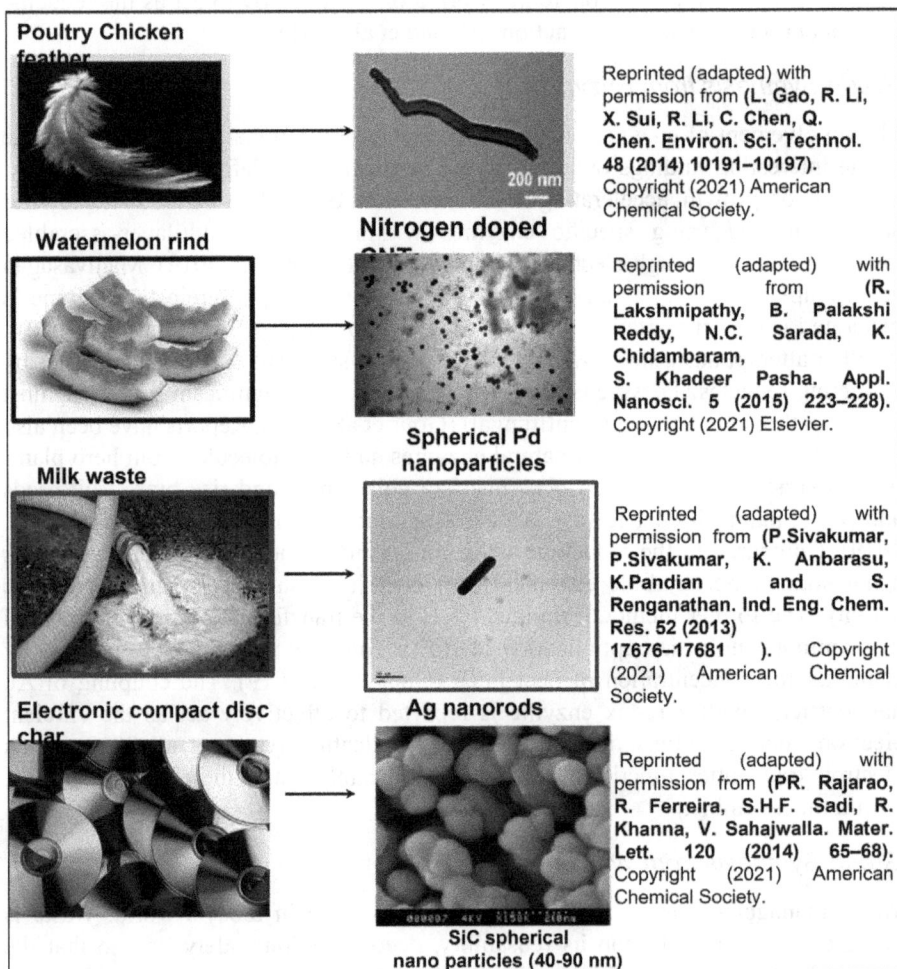

Figure 6. Examples of green synthesis of some nanoparticles from waste products.

from orange peel extract using microwave-assisted methods (Kahrilas et al. 2014). The TEM images clearly showed the formation of an organic layer surrounding the formed nanoparticles. Chicken eggshell membrane being a protein of double-layered membrane comprises glycoproteins. It is reported that among these glycoproteins' collagen of type I, V and X are present along with glycine, alanine and uronic acid.

ESM is also found to be useful for acting as the template for macroporous materials, extracting heavy metals and biosensing of enzyme immobilized matrix of Au-ESM. An account of some nanoparticles derived from waste products is shown in Figure 6.

The redox reaction involving Au^{3+} to Au and oxidized counterpart of ESM are studied using absorption and fluorescence spectroscopy (Sharma et al. 2019). Annona squamosa or custard apple, banana peel, etc. are engaged as effective bio-reductant and capping agents for the synthesis of Ag and Mn_3O_4 (super capacitive properties) particles. It is reported by researchers that the primary bio-reductant (for Ag ions) present in the custard apple peel are water-soluble ketone and hydroxyl groups. Similarly, cellulose, lignin, proteins, pectin, etc. could be efficiently used for green synthesis which is in conjugation termed as biopolymers. Hydroxyapatite nanoparticles have been prepared using pectin as the bio-reductant present in banana peel (Boethling et al. 2007). Among these, the synthesized Mn_3O_4 are studied to have high electrochemical reversibility. Furthermore, the superior stability is further confirmed by the presence of 93% specific capacitance tested for more than 2,000 cycles of charge-discharge cycle.

5. Applications of Biosynthesized Nanoparticles

This section highlights the recent progress of biosynthesized nanoparticles in biomedical research, including gene and cancer therapy, bioimaging, drug delivery, antimicrobial and biosensing applications and as alternative therapeutic agents over conventional treatment strategies. Application is also highlighted in the arena of environmental remediation, biosensors, catalysts, ferrofluids, alternate energy sector and magnetic storage media.

5.1 Gene Therapy and Toxicogenomics

Due to the present COVID-19 pandemic situation, virus infection is posed to be a serious problem to bio-life. When a human body cell is infected by a virus, virus gene or virus DNA do rapidly replicate to produce a large number of replicas. The virus DNA enters into the host cell protein synthesis process and disturbs the normal cell function. The act of inserting a gene to treat a diseased cell is termed 'gene therapy' in which no external drug is induced or surgery is performed (Rapti et al. 2011, Pouton and Seymour 2001). The identified mutated genes are replaced with healthy genes/genetic material, such as correct DNA sequence and the normal function is tried to get induced. The major challenge in this arena of gene therapy research is the selection of competent delivery vectors (Dizaj et al. 2014) having lower cytotoxicity levels and enhanced transfection efficiencies. This selection is important in finalizing a target gene for specific tissues or cells.

In this research arena, newer vectors of non-viral nature that can overcome physiological blockades are increasing thereby enriching the nanotechnology sector (Morachis et al. 2012). In this arena, the research has shifted from the usage of chemical non-viral vectors to natural bio-materials for gene delivery. This can be accounted for by the specific nature of such bio-based non-viral vectors. In order to protect the entrapped molecule from deterioration and for escape from the reticulo endothelial system, optimization of size, shape, porosity and other properties of the synthesized nanoparticles are undertaken. In this category, certain compounds are studied like gold, carbon nanotubes, silica, supra-molecular system, fullerenes, quantum dots, etc. (Katragadda et al. 2010). Calcium phosphate (Olton et al. 2011) particles are studied to be biocompatible and easily degraded and effectively engaged as an important role in endocytosis and get readily absorbed owing to their high binding affinity. Silica (Kneuer et al. 2000) particle is first functionalized as aminosilanes (owing to less toxicity) and is utilized as a gene delivery vehicle. Bio nanoparticles from gold sources exhibit a high tissue penetration effect. These particles have strong absorption of light near the infra-red region (Tiwari and Lee 2013). The photothermal effect is used to modify the surface of Au nanoparticles for transfecting to the cell and helps to control the release profile of the gene. This results in lowering the toxicity of the lipoplexes during transfection with Au during *in vitro* studies with a problem of interference with the cell. Carbon nanotubes (cylindrical fullerenes) (Ramos-Perez et al. 2013, Shi Kam et al. 2004) and fullerenes (soluble carbon molecules) along with other supramolecular systems has been applied *in-vitro* for some animal models. Binding with DNA or bio-protein can further be facilitated by coating such nanoparticles. The strategy is to enhance the surface area by engaging such nanoparticles and in addition, they can easily surpass the cellular barriers. This can increase the transfection efficiency of the bio-module. However, further research and development is required to study long-term safety, size and shape, functionalization of system/chemical under concern on the transfection efficiency to enhance their quantifiable application.

Bio-nanoparticles find immense application as drug carriers in neuro/brain-related diseases wherein the conventional drugs fail to bypass the blood-brain barrier. The concept is that such functional nanoparticles are capable of crossing the BBB and delivering the drug to the target site for the remission of diseases like brain tumors, Alzheimer's and Parkinson's. The main property of such nanoparticles is the capability of getting accumulated and interacting with the affected cell, thereby increasing the permeability and retention (EPR) and could be utilized in drug delivery (Laquintana et al. 2009). Another important property of such bio-nanoparticles required to act as effective drug carriers is mono-dispersity, devoid of toxicity, easy complex tendency with ligands, etc. Among all metallic nanoparticles, gold, dendrimers, quantum dots, polymer gels, ZnO or iron oxide nanoparticles are reported to be safer with fewer toxic effects for drug action (Khan et al. 2014) in cancer therapy compared to the conventional drugs. In addition, for molecular imaging purposes, inorganic nanoparticles are highly applied (Das et al. 2019) as the contrast agent. Some typical imaging techniques are magnetic resonance imaging (MRI), positron emission tomography (PET), computed tomography (CT), ultrasound and optical imaging (Narayanan et al. 2012). According to the reports, nanorods of gold can exhibit tunable absorption in a low wavelength region of both visible and near-infrared region which makes them potential applications in the field of biosensing, photothermal therapy and gene delivery.

The significance of toxicology with bio-nanomaterials is already established in Section 1. The correlation between biological activity and chemical structure has been always a source of curiosity for researchers. QSAR models are employed for predicting the potential activity of a chemical to act as a drug by studying and tailoring the extent of toxicity (Selassie et al. 2002). A description of molecular descriptors required for such screening is tabulated in Table 5.

Table 5: Molecular descriptors required for screening of chemicals for toxicology

Parameters	Lipophilicity (as Log P)
Topological parameters	Aqueous solubility
	Henry's law constant
	Reaction rate
	Topological parameters and molecular connectivity
Electrical parameters	Organizational fragments
	Hammett constants
	pKa
	Dipole moment
Steric parameters	Structural fragments
	Molecular orbitals
	Steric parameters Taft constants
	Molecular weight, volume and surface area

Highly persistent along with bio-accumulative chemicals are generally considered as environmental hazards. Therefore, separate protocols are documented by national and international regulatory agencies and are practiced to introduce specific guidelines and standards for characterizing these attributes of the chemical (United Nations Globally Harmonized System of Classification and Labelling of Chemicals 2011). Persistent chemicals tend to remain within the environment for decades and hence evaluated through half-lives in variable environmental media. Those chemicals which can resist degradation owing to their slow transformation are termed recalcitrant or refractory pollutants. Transportation of chemicals can result through enzyme metabolism (biotic route) or abiotically (e.g., hydrolysis, photolysis and oxidation) or through a combination of both routes. Design for biodegradability is an important aspect for consideration, wherein systematic paths are designed for increasing the rates of biodegradability of persistent matter to harmless reagents which would minimize the deadliness by restricting the exposure time and thereby termed as risk reduction strategy. In addition, the specific molecular features affect the biodegradation rates as given in Figure 7 (Boethling et al. 2007).

In this connection accumulation of bio-products results when the extent of absorption is higher compared to the rate of excretion. The synthesized nanoparticles

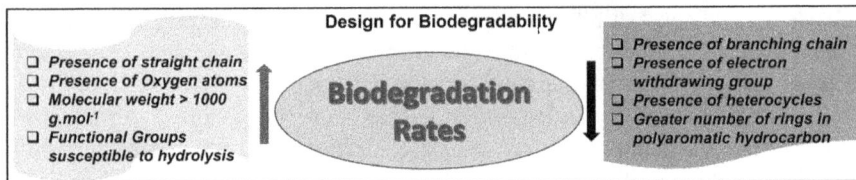

Figure 7. Representation of design for biodegradability of chemicals for toxicogenomics.

that bear the tendency to get bioaccumulated are lipophilic and are stored within the fatty tissues (Legierse et al. 1999). Hence, the bioaccumulation factor (BAF) is proposed which can be defined as per Equation 2:

$$BAF = \frac{\text{Concentration of a chemical within biosystem at steady state}}{\text{Concentration of that chemical in the environment}} \quad (2)$$

However, the BAF factor to date has the limitation to be applied in presence of an aqueous medium only. Toxicogenomics being a discipline of pharmacology that deals with the interaction of gene and protein activity with cell/tissue in response to exposure to toxic substances, optimization of BAF are very significant.

Finally, the process of risk assessment is required which is further delineated into noncancer and cancer methods. Noncancer assessment of risk is used to identify points of departure (POD), including no observed (adverse) effect level (NO(A)EL), lowest observed (adverse) effect level (LO(A)EL) or the benchmark dose (BMD) for critical toxicological endpoints for the most sensitive and appropriate species. The point of departure is calculated from the variable test data obtained from animal species. However, for cancer risk assessment it is considered that a certain chemical can theoretically initiate carcinogenesis. This is followed by the generation of models to result in a point that estimates the associated confidence interval and known as cancer slope factors (CSFs) which represent the extent of carcinogen. Hence, the efficacy of such bio-nanoparticles is studied using BMD parameters and also suing exposure point concentrations (EPCs).

5.2 Bio-Nanoparticles in Catalysis and Renewable Energy

The bio-synthesized functional nanoparticles synthesized also bear an application toward environmental remediation, such as in wastewater treatment, degradation of organic dyes and pollutants, biosensors, catalysts, ferrofluids and in magnetic storage media (Huang et al. 2014, Kuang et al. 2013, Smuleac et al. 2011, Kavitha et al. 2013). A photovoltaic solar cell is one of the next-gen energy carriers owing to its renewable nature and high energy capacity. Ternary oxide, spinel oxides (ferrites $M^{II}Fe_2^{III}O_4$) or other $M^{II}M_2^{III}O_4$ semiconductors, ABO_3 are promising materials because of their narrow bandgap energy (~1.6-1.7 eV but not over 2.2 eV) that can expand the range of available energy bandgaps from near-ultraviolet to visible regime. The mentioned compositions are attempted to synthesize through bioprocess to produce functional nanoparticles which show a large absorption coefficient and low toxicity. Researchers (Nagatani et al. 2015) have synthesized wurtzite-derived β-CuGaO$_2$, β-AgGaO$_2$, maintaining the β-NaFeO$_2$ structure which has narrow bandgap energy. Figure 8

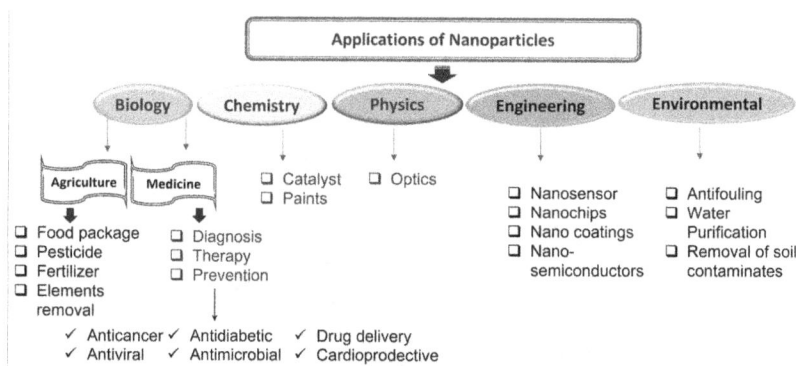

Figure 8. Applications of bionanoparticles.

represents the applications of nanoparticles in various fields. The application in the arena of catalysis is interesting and significant with respect to chemical industries and pharmacology. Decomposition of peroxide to oxygen is catalyzed effectively using Au, Ag and Pd. Au nanoparticles are effectively produced using Gnidia glauca flower by Ghosh et al. which exhibit significant catalytic properties in the reduction of 4-nitrophenol to 4-aminophenol by $NaBH_4$ in the aqueous phase (Ghosh et al. 2012). The nanosized metal nanoparticles, such as Au, Ag and Pt, are applied for various commercial personal care products, such as shampoo, soap, detergent, anti-aging creams and perfumes (Kumar and Yadav 2009). Applications of nanoparticles are emerging in crop protection and agriculture.

5.3 Green Synthesis of Carbon-Based Nanomaterials

Another important arena of green synthesis is in the utilization of carbon-based systems viz. graphene, which comprises a multivariant application. Carbon-based materials, such as graphene (2D), meso-carbons, carbon nanotubes (CNTs), carbon dots, helical carbon nanostructures, have attracted great attention due to their unique properties, including high mechanical and thermal properties, good thermal stability, biocompatibility, non-toxicity (Ismail 2019, Bandeira et al. 2020, Coros et al. 2020). Recently, graphene has received intense interest for different applications, like high mechanical, electrical conductivity, thermal properties, etc. Graphene, possessing high mechanical, electrical conductivity and thermal properties, is a hexagonally arranged two-dimensional (2D) form of sp^2-hybridized carbon having a single atom thick. Besides graphene, the carbon quantum dots (CDs) characterized as a subclass of nanoparticles and having a spherical size of < 10 nm is the new type of fluorescent carbon material (Novoselov et al. 2012, Akinay and Kizilcay 2020, Yao et al. 2019). Various synthesis methods have been performed on the preparation of these carbon-based nanomaterials. But in the last decades, the green synthesis of these materials has been a new focus. The environment-friendly and low-cost methods are the new research field for the green synthesis of nanoparticles in which plant extracts are used as the dopant and reduction agents as already discussed above. In this section, the green synthesis of graphene and carbon quantum dot materials is reviewed and reported. The primary principle of green synthesis is also depicted in Figure 9.

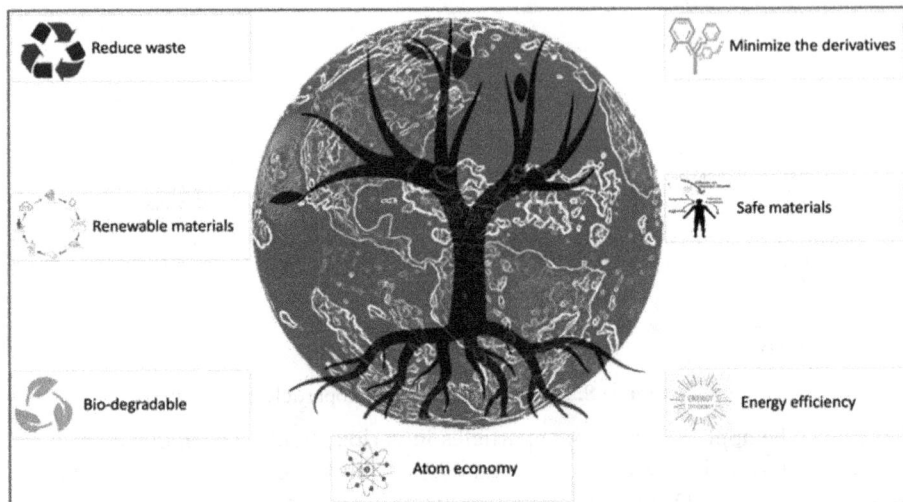

Figure 9. The principle of green synthesis of nanoparticles.

The principle of green synthesis of nanoparticles can be classified as follows:

- Prevent wastes: To reduce and minimize hazardous wastes during the synthesis process.
- Use of renewable materials: Prefer renewable chemicals.
- Minimize the derivatives: It is important to use fewer derivatives as a stabilizer.
- Bio-degradable products: Using and synthesizing biodegradable and non-toxic products.
- Safe materials: Using safe materials that are not hazardous for humans and the environment.
- Energy efficiency: Avoid using high energy (optimize the temperature and ambient).
- Atom economy: Improving reaction efficiency.

The synthesis of graphene can be classified into two different methods, namely top-down method and the bottom-up method (Lim et al. 2018, Choi et al. 2010).

Top-Down Methods

It is well known that graphite is the combination of many sheets bonded with each other by van der Waals forces. The top-down method focuses on building atomic graphene layers from the graphite stack by overcoming these van der Waals forces. In this method, different approaches are used to create graphene layers by applying an attack to graphite powders. The attack that forms graphene sheets is done by mechanical exfoliation, arc discharge, liquid-phase exfoliation (LPE), chemical exfoliation, oxidative exfoliation and chemical synthesis methods. The mechanical exfoliation method is divided into two main categories in which normal force synthesis methods and three-roll machines. When the external mechanical force reaches a sufficient amount (> 300 nN/μm²), the weak van der Waals bond energy can be exceeded and single atomic layers are formed. On the other hand, the chemical exfoliation method uses alkali ions solution to disperse the graphite layer. In this process, the interlayer spacing distance (up to 3.34 Å) is increased by reducing van der Waal bond energy (2 eV.nm²) (Lee et al. 2019). However, the biggest challenge of the top-down approach is its low yield, which is not suitable for high amount production.

Bottom-Up Methods

The bottom-up method uses carbon molecule sources as a precursor to synthesize graphene layers. The chemical vapor deposition (CVD) method, thermal pyrolysis, chemical synthesis and laser-assisted synthesis are the most used methods for the bottom-up approach. This method provided a large surface area, high purity, and defect-free graphene sheets. However, this method requires high production costs and complex synthesis steps. The schematic representation of Top-Down and Bottom-Up methods are given in Figure 10.

Green Synthesis of Graphene

As briefly explained above, bottom-up and top-down methods result in a large number of toxic substances (such as hydrazine, sulfur compounds and sodium borohydride), energy consumption and high costs. Therefore, recently the green synthesis methods of graphene have been extensively studied using different biological reducing agents, such as plant extracts, microorganisms, covering agents, polysaccharides, enzymes, amino acids, microbes, alcoholic derivatives and so on. This green synthesis method provides cost-efficiency, environment-friendly products, non-toxic derivatives, high product yields and easy processability. The overview of the green synthesis process for graphene and graphene oxide compounds is summarized in Table 6.

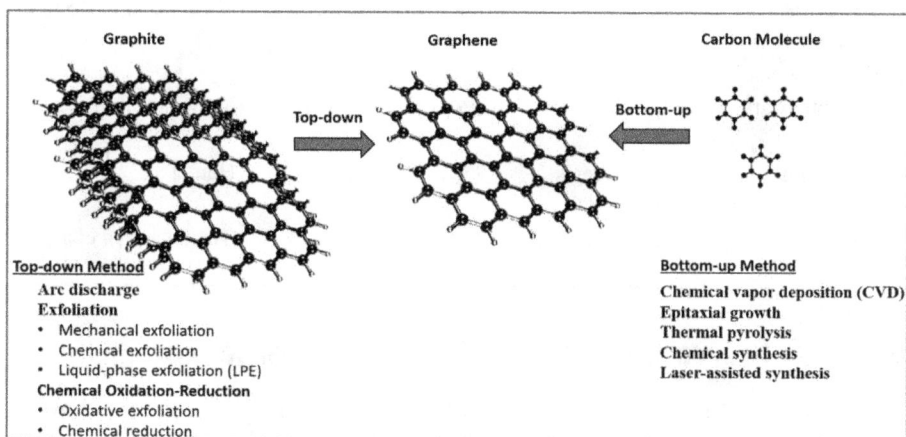

Figure 10. The schematic representation of top-down and bottom-up methods.

Table 6: Green synthesize process for graphene and graphene oxide compounds

Reducing Agent	Method	Reaction Conditions	Reference
Sugars: glucose, fructose and sucrose	Stirring	Stirred for 60 minutes at 95°C	(Zhu et al. 2010)
Pomegranate juice	Hummer's method	0–5°C for 15 minutes	(Tavakoli et al. 2015)
Bacillus marisflavi biomass	Hummer's and Offeman's method	35 minutes at 35°C	(Gurunathan et al. 2013)
Rose water	Hummer's method	35 minutes at 35°C	(Haghighi and Tabrizi 2013)
Grapes (*Vitis vinifera*)	Hummer's method	1 hour at 35°C	(Khodabakhshi et al. 2014)
Cellulose	Solvent system	at 120°C	(Wei et al. 2017)
Urea	Hummer's method	1 hour at 35°C	(Chamoli et al. 2018)
Bacillus marisflavi biomass	Hummer's and Offeman's method	30 minutes at 35°C	(Gurunathan et al. 2014)
Lignin	Exfoliation in water	1 hour at 35°C	(Wang et al. 2019)
Sugar beet leaves	Exfoliation in PVA	130°C for 3 hours	(Attia et al. 2018)
Mangifera indica, Ficus religiosa or Polyalthia longifolia leaf	Hummer's method	30 minutes at 35°C	(Chamoli et al. 2016)
Camphor leaves (*Cinnamomum camphora*)	One-step pyrolysis	1,200°C for 4 minutes	(Shams et al. 2015)
Eucalyptus bark extract	Hummer's method	80-85°C for 24 hours	(Manchala et al. 2019)
Artemisinin	Hummer's method	95°C for 24 hours	(Hou et al. 2018)
Aloe vera (AV)	Hummer's method	95°C for 24 hours	(Bhattacharya et al. 2017)

6. Conclusive and Future Direction Remarks

In recent years, biosynthesis of nanoparticles has gained significant attention due to the associated advantages with the technique, advanced property of the particles and application ease. Researchers have undertaken producers (plant sources) and consumer sources, like bacteria, fungi, yeast, microalgae, seaweeds, viruses, etc., as the potential synthesizers which bear bio-reducing agents as well as the capping agents. These biocapping retains the nanomorphology and prevent particle agglomeration and toxicity. On the basis of broad scientific content, produced nanoparticles can be classified as (a) metal, (b) metal oxide, (c) bimetallic, (d) metal complex, (e) doped perovskites nanoparticles, etc. (of the central metal ion Ag, Pd, Au, Cu, Pt and oxides, like zinc oxide, cuprous oxide, titanium dioxide, etc.) and (f) carbon-based particles. Compared to extremophile origin, biosynthesis using phyto/plants is more accepted owing to the inexpensive, environmentally friendly process, simple one-step reaction, safer technique and devoid of any post-processing complexities. The factors affecting such biosynthesis (discussed in detail in Section 2 and Section 3) are found to be dependent on the concentration of hydrogen ion, dosage of the drug (chemical), reaction and drying temperature, the time required for the reaction and microbe strain, growth media, etc. The nanoparticles synthesized via such green synthesis are characterized by X-ray diffraction, morphological studies, like SEM, TEM, AFM, UV-spectrophotometry and thermal studies like DSC, TG, XPS and mass spectroscopy. Magnetic, semiconducting and conducting are studied using SQUID, impedance spectra, etc. The primary objective(s) undertaken by researchers in this field are (a) understanding and optimizing the factors affecting the nanoparticle synthesis using plants/animal sources and (b) application of these nanoparticles in a specific arena, like catalysis, agricultural, renewable energy sector, medicinal, etc.

In this connection, the primary lacuna remains in the understanding of the exact mechanism towards the bioreduction undertaken by the biosources, stability of the formed particles and identification of capping agent. The application of such nanoproducts is still limited in the field of catalysis and drug. Hence, advanced research needs to be undertaken in the arena of the drug industry using such nanoparticles synthesized from bio-source. In addition, the yield of the nanomaterials also needs to be enhanced many folds which possess another shortcoming in terms of technological application so that biosynthetic routes can compete well with either chemical or physical techniques. Furthermore, more concentrated research is to be undertaken regarding the involvement and optimization of parameters affecting the bio-nanoparticles synthesis and functionalization.

Acknowledgment

The authors (MM, SD) acknowledge Amity University, Kolkata, India for infrastructural and financial support. YA acknowledges Van Yuzuncu Yil University, Mining Department, Turkey.

References

Ahmad, N., Sharma, S., Singh, V.N., Shamsi, S.F., Fatma, A. and Mehta, B.R. 2011. Biosynthesis of silver nanoparticles from Desmodium triflorum: a novel approach towards weed utilization. Biotechnol. Res. Int. 2011: 454090.

Ahmed, E., Kalathil, S., Shi, L., Alharbi, O. and Wang, P. 2018. Synthesis of ultra-small platinum, palladium and gold nanoparticles by Shewanella loihica PV-4 electrochemically active biofilms and their enhanced catalytic activities. J. Saudi Chem. Soc. 22: 919-929.

Akinay, Y. and Kizilcay, A.O. 2020. Computation and modeling of microwave absorbing CuO/graphene nanocomposites. Polymer Composites 41: 227-232.

Ali, D.M., Thajuddin, N., Jeganathan, K. and Gunasekaran, M. 2011. Plant extract mediated synthesis of silver and gold nanoparticles and its antibacterial activity against clinically isolated pathogens. Colloids and Surf B 85: 360-365.

Ankamwar, B., Chaudhary, M. and Sastry, M. 2005. Gold nanotriangles biologically synthesized using tamarind leaf extract and potential application in vapor sensing. Synth. React. Inorg. Metal-Org. Nano-Met Chem. 35: 19-26.

Apte, M., Sambre, D., Gaikawad, S., Joshi, S., Bankar, A., Kumar, A.R. and Zinjarde, S. 2013. Psychrotrophic yeast Yarrowia lipolytica NCYC 789 mediates the synthesis of antimicrobial silver nanoparticles via cell-associated melanin. AMB Expres. 3: 32.

Armendariz, V., Gardea-Torresdey, J.L., Herrera, I., Jose-Yacaman, M., Peralta-Videa, J.R. and Santigo, P. 2004. Size controlled gold nanoparticles formation by Avena sativa biomass: use of plants in nanobiotechnology. J. Nanoparticle Res. 6: 377-382

Arsiya, F., Sayadi, M.H. and Sobhani, S. 2017. Green synthesis of palladium nanoparticles using Chlorella vulgaris. Mater. Lett. 186: 113-115.

Attia, N.F., Park, J. and Oh, H. 2018. Facile tool for green synthesis of graphene sheets and their smart free-standing UV protective film. Appl. Surf Sci. 458: 425-430.

Bandeira, M., Giovanela, M., Roesch-Ely, M., Devine, D.M. and da Silva, J.C. 2020. Green synthesis of zinc oxide nanoparticles: a review of the synthesis methodology and mechanism of formation. Sustain. Chem. Pharm. 15: 100223.

Bankar, A., Joshi, B., Kumar, A.R. and Zinjarde, S. 2010. Banana peel extract mediated novel route for the synthesis of silver nanoparticles. Col. Surfaces Physicochem. Eng. Asp. 368: 58-63.

Baptista, P.V., McCusker, M.P., Carvalho, A., Ferreira, D.A., Mohan, N.M., Martins, M. and Fernandes, A.R. 2018. Nano-strategies to fight multidrug resistant bacteria – a battle of the titans. Front Microbiol. 9: 1441 (1-26).

Bar, H., Bhui, D.K., Sahoo, G.P., Sarkar, P., De, S.P. and Misra, A. 2009. Green synthesis of silver nanoparticles using latex of Jatropha curcas. Colloids Surf. A Physicochem. Eng. Asp. 339: 134-139.

Bewley, K.D., Ellis, K.E., Firer-Sherwood, M.A. and Elliott, S.J. 2013. Multi-heme proteins: nature's electronic multi-purpose tool. Biochim. Biophys. Acta, Bioenerg. 1827: 938-948.

Bhargava, A., Jain, N., Khan, M.A., Pareek, V., Dilip, R.V. and Panwar, J. 2016. Utilizing metal tolerance potential of soil fungus for efficient synthesis of gold nanoparticles with superior catalytic activity for degradation of rhodamine B. J. Environ. Manage. 183: 22-32.

Bhattacharya, G., Sas, S., Wadhwa, S., Mathur, A., McLaughlin, J. and Roy, S.S. 2017. Aloe vera assisted facile green synthesis of reduced graphene oxide for electrochemical and dye removal applications. RSC Adv. 7: 26680-26688.

Boethling, R.S., Sommer, E. and Fiore, D.D. 2007. Designing small molecules for biodegradability. Chem. Rev. 107: 2207–2227.

Borzelleca, J.F. 2000. Paracelsus: herald of modern toxicology. Toxicol. Sci. 53: 2-4.

Buszewski, B., Railean-Plugaru, V., Pomastowsk, P., Rafinska, K., Szultka-Mlynska, M., Golinska, P., Wypij, M., Laskowski, D. and Dahm, H. 2016. Antimicrobial activity of biosilver nanoparticles produced by a novel Streptacidiphilus durhamensis strain. J. Microbiol. Immunol. Infect. 51: 45-54.

Chamoli, P., Das, M.K. and Kar, K.K. 2018. Urea-assisted low temperature green synthesis of graphene nanosheets for transparent conducting film. J. Phy. Chem. Solids, 113: 17-25.

Chamoli, P., Sharma, R., Das, M.K. and Kar, K.K. 2016. Mangifera indica, Ficus religiosa and Polyalthia longifolia leaf extract-assisted green synthesis of graphene for transparent highly conductive film. RSC Adv. 6: 96355-96366.

Chandran, S.P., Ahmad, A., Chaudhary, M., Pasricha, R. and Sastry, M. 2006. Synthesis of gold nanotriangles and silver nanoparticles using aloe vera plant extract. Biotechnol. Prog. 22: 577-583.

Chen, M.L., Yang, D.J. and Liu, S.C. 2011. Effects of drying temperature on the flavonoid, phenolic acid and antioxidative capacities of the methanol extract of citrus fruit (Citrus sinensis (L.) Osbeck) peels. Int. J. Food Sci. Technol. 46: 1179-1185.

Choi, W., Lahiri, I., Seelaboyina, R. and Kang, Y.S. 2010. Synthesis of graphene and its applications: a review. Crit. Rev. Solid State Mater. Sci. 35: 52-71.

Choudhury, A.R., Malhotra, A., Bhattacharjee, P. and Prasad, G. 2014. Facile and rapid thermo-regulated biomineralization of gold by pullulan and study of its thermodynamic parameters. Carbohydr. Polym. 106: 154-159.

Cologgi, D.L., Lampa-Pastirk, S., Speers, A.M., Kelly, S.D. and Reguera, G. 2011. Extracellular reduction of uranium via Geobacter conductive pili as a protective cellular mechanism. Proc. Natl. Acad. Sci. USA 108: 15248-15252.

Coros, M., Pogacean, F., Turza, A., Dan, M., Berghian-Grosan, C., Pana, I.O. and Pruneanu, S. 2020. Green synthesis, characterization and potential application of reduced graphene oxide. Physica. E 119: 113971.

Das S., Kotcherlakota, R. and Patra, C.R. 2019. Noninvasive imaging techniques of metal nanoparticles and their future diagnostic applications. *In*: Shukla, A. (eds). Medical Imaging Methods. Springer, Singapore.

Dhas, T.S., Kumar, V.G., Karthick, V., Angel, K.J. and Govindaraju, K. 2014. Facile synthesis of silver chloride nanoparticles using marine alga and its antibacterial efficacy. Spectrochim. Acta Part A 120: 416-420.

Dizaj, S.M., Jafari, S. and Khosroushahi, A.Y. 2014. A sight on the current nanoparticle-based gene delivery vectors. Nanoscale Res. Lett. 9: 252-1-9.

Edison, T.N.J.I., Atchudan, R., Kamal, C. and Lee, Y.R. 2016. Caulerpa racemosa: a marine green alga for eco-friendly synthesis of silver nanoparticles and its catalytic degradation of methylene blue. Bioprocess Biosyst. Eng. 39: 1401-1408.

Elemike, E.E., Onwudiwe, D.C., Fayemi, O.E. and Botha, T.L. 2019. Green synthesis and electrochemistry of Ag, Au, and Ag–Au bimetallic nanoparticles using golden rod (Solidago canadensis) leaf extract. Appl. Phy. A 25: 42.

Eugenio, M., Muller, N., Frases, S., Almeida-Paes, R., Lima, L.M.T., Lemgruber, L., Farina, M., de Souza, W. and SantAnna, C. 2016. Yeast-derived biosynthesis of silver/silver chloride nanoparticles and their antiproliferative activity against bacteria. RSC Adv. 6: 9893-9904.

Fayaz, A.M., Balaji, K., Kalaichelvan, P.T. and Venkatesan, R. 2009. Fungal based synthesis of silver nanoparticles – an effect of temperature on the size of particles. Col. Surf. B 74: 123-126.

Fayaz, A.M., Balaji, K., Girilal, M., Yadav, R., Kalaichelvan, P.T. and Venketesan, R. 2010. Biogenic synthesis of silver nanoparticles and their synergistic effect with antibiotics: a study against gram-positive and gram-negative bacteria. Nanomedicine: Nanotech., Bio. and Medicine 6: 103-109.

Ferreira da Silva, V., ConzFerreira, M.E., Lima, L.M.T., Fraśes, S., de Souza, W. and Sant'Anna, C. 2017. Green production of microalgae-based silver chloride nanoparticles with antimicrobial activity against pathogenic bacteria. Enzyme Microb. Technol. 97: 114-121.

Freitas, F., Alves, V.D. and Reis, M.A.M. 2011. Advances in bacterial exopolysaccharides: from production to biotechnological applications. Trends Biotechnol. 29: 388-398.

Gade, A., Gaikwad, S., Tiwari, V., Yadav, A., Ingle, A. and Rai, M. 2010. Biofabrication of silver nanoparticles by Opuntia ficus-indica: in vitro antibacterial activity and study of the mechanism involved in the synthesis. Curr. Nanosci. 6: 370-375.

Gahlawat, G. and Choudhury, A.R. 2019. A review on the biosynthesis of metal and metal salt nanoparticles by microbes. RSC Adv. 9: 12944-12967.

Gardea-Torresedey, J.L., Gomez, E., Jose-Yacaman, M., Parsons, J.G., Peralta-Videa, J.R. and Tioani, H. 2003. Alfalfa sprouts: a natural source for the synthesis of silver nanoparticles. Langmuir 19: 1357-1361.

Ghosh, S., Patil, S., Ahire, M., Kitture, R., Gurav, D.D., Jabgunde, A.M., Kale, S., Pardesi, K., Shinde, V., Bellare, J., Dhavale, D.D. and Chopade, B.A. 2012. Gnidia glauca flower extract mediated synthesis of gold nanoparticles and evaluation of its chemocatalytic potential. J. Nanobiotechnol. 10: 17-1-9.

Gopalkrishnan, K., Ramesh, C., Raghunathan, V. and Thamilselvan, M. 2012. Antibacterial activity of Cu$_2$O nanoparticles on E. cole synthesized from Tridas procumbens leaf extract and surface coating with polyaniline. Dig. J. Nanomater. Biostruct. 7: 833-839.

Gour, A. and Jain, N. Kumar. 2019. Advances in green synthesis of nanoparticles. Inter. J. Artif. Cells Nanomed. Biotechnol. 47: 844-851.

Gurunathan, S., Han, J.W., Eppakayala, V. and Kim, J.H. 2013. Green synthesis of graphene and its cytotoxic effects in human breast cancer cells. Int. J. Nanomedicine 8: 1015-1027.

Gurunathan, S., Haan, J.W., Kim, E., Kwon, D.N., Park, J.K. and Kim, J.H. 2014. Enhanced green fluorescent protein-mediated synthesis of biocompatible graphene. J. Nanobiotech. 12: 41-1-16.

Haghighi, B. and Tabrizi, M.A. 2013. Green-synthesis of reduced graphene oxide nanosheets using rose water and a survey on their characteristics and applications, RSC Adv. 3: 13365-13371.

Hamedi, S., Ghaseminezhad, M., Shokrollahzadeh, S. and Shojaosadati, S.A. 2017. Controlled biosynthesis of silver nanoparticles using nitrate reductase enzyme induction of filamentous fungus and their antibacterial evaluation. Artif. Cells Nanomed. Biotechnol. 45: 1588-1596.

Herrera-Becerra, R., Zorrilla, C., Rius, J.L. and Ascencio, A. 2008. Electron microscopy characterization of biosynthesized iron oxide nanoparticles. Appl. Phys. A 91: 241-246.

Hou, D., Liu, Q., Wang, X., Quan, Y., Qiao, Z., Yu, L. and Ding, S. 2018. Facile synthesis of graphene via reduction of graphene oxide by artemisinin in ethanol. J. Materiomics 4: 256-265.

Huang, J., Li, Q., Sun, D., Liu, Y., Su, Y., Yang, X., Wang, H., Wang, Y., Shao, W. and He, J.N. 2007. Biosynthesis of silver and gold nanoparticles by novel sundried Cinnamomum camphora leaf. Nanotechnology 18: 105-106.

Huang, L., Weng, X., Chen, Z., Megharaj, M. and Naidu, R. 2014. Synthesis of iron-based nanoparticles using oolong tea extract for the degradation of malachite green. Spectrochim Acta A Mol. Biomol. Spectrosc. 117: 801-804.

Iravani, S. 2014. Bacteria in nanoparticle synthesis: current status and future prospects. Int. Sch. Res. Notices. 2014: 1-18.

Iravani, S., Korbekandi, H., Mirmohammadi, S.V. and Zolfaghari, B. 2014. Synthesis of silver nanoparticles: chemical, physical and biological methods. Res. Pharm. Sci. 9: 385-406.

Ismail, Z. 2019. Green reduction of graphene oxide by plant extracts: a short review. Ceramics International 45: 23857-23868.

Jacob, J., Mukherjee, T. and Kapoor, S. 2012. A simple approach for facile synthesis of Ag, anisotropic Au and bimetallic (Ag/Au) nanoparticles using cruciferous vegetable extracts. Mater. Sci. Eng: C 32: 1827-1834.

Jha, A.K. and Prasad, K. 2010. Green synthesis of silver nanoparticles using Cycas leaf. Int. J. Green Nanotechnol. 1: 110-117.

Joerger, T.K., Joerger, R., Olsson, E. and Granqvist, C.G. 2001. Bacteria as workers in the living factory: metal-accumulating bacteria and their potential for materials science. Trends Biotechnol. 19: 15-20.

Jung, J., Oh, H., Noh, H., Ji, J. and Kim, S. 2006. Metal nanoparticle generation using a small ceramic heater with a local heating area. J. Aerosol. Sci. 37: 1662-1670.

Kahrilas, G.A., Wally, L.M., Fredrick, S.J., Hiskey, M., Prieto, A.L. and Owens, J.E. 2014. Microwave-assisted green synthesis of silver nanoparticles using orange peel extract. ACS Sustain. Chem. Eng. 2: 367-376.

Kang, F., Alvarez, P.J. and Zhu, D. 2014. Microbial extracellular polymeric substances reduce Ag+ to silver nanoparticles and antagonize bactericidal activity. Environ. Sci. Technol. 48: 316-322.

Kasthuri, J., Kanthiravan, K. and Rajendiran, N. 2009. Phyllanthin-assisted biosynthesis of silver and gold nanoparticles: a novel biological approach. J. Nanoparticle Res. 11: 1075-1085.

Katragadda, C.S., Choudhury, P.K. and Murthy, P. 2010. Nanoparticles as non-viral gene delivery vectors. Indian J. Pharm. Edu. Res. 44: 109-111.

Kavitha, A.L., Prabu, H.G., Babu, S.A. and Suja, S.K. 2013. Magnetite nanoparticles-chitosan composite containing carbon paste electrode for glucose biosensor application. J. Nanosci. Nanotechnol. 13: 98-104.

Khan, A.K., Rashid, R., Murtaza, G. and Zahra, A. 2014. Gold nanoparticles: synthesis and applications in drug delivery. Trop. J. Pharm. Res. 13: 1169-1177.

Khandel, P., Yadaw, R.K., Soni, D.K., Kanwar, L. and Sathi, S.K. 2018. Biogenesis of metal nanoparticles and their pharmacological applications: present status and application prospects. J. Nanostruct. Chem. 8: 217-254.

Khatoon, A., Khan, F., Ahmad, N., Shaikh, S., Mohd. S., Rizvi, D., Shakil, S., Al-Qahtani, M.H., Abuzenadah, A.M., Tabrez, S., Ahmed, A.B.F., Alafnan, A., Islam, H., Iqbal, D. and Dutta, R. 2018. Silver nanoparticles from leaf extract of Mentha piperita: eco-friendly synthesis and effect on acetylcholinesterase activity. Life Sciences 209: 430-434.

Khodabakhshi, S., Karami, B., Eskandari, K., Hoseini, S.J. and Rashidi, A. 2014. Graphene oxide nanosheets promoted regioselective and green synthesis of new dicoumarols. RSC Adv. 4: 17891-17895.

Klaus-Joerger, T., Joerger, R., Olsson, E. and Granqvist, C.G. 2001. Bacteria as workers in the living factory: metal-accumulating bacteria and their potential for materials science. Trends Biotechnol. 19: 15-20.

Kneuer, C., Sameti, M., Bakowsky, U., Schiestel, T., Schirra, H., Schmidt, H. and Lehr, C.M. 2000. A nonviral DNA delivery system based on surface modified silica-nanoparticles can efficiently transfect cells in vitro. Bioconjug. Chem. 11: 926-932.

Kobayashi, M., Tomita, S., Sawada, K., Shiba, K., Yanagi, H., Yamashita, I. and Uraoka, Y. 2012. Chiral meta-molecules consisting of gold nanoparticles and genetically engineered tobacco mosaic virus. Opt. Express 20: 24856-24863.

Kuang, Y., Wang, Q., Chen, Z., Megharaj, M. and Naidu, R. 2013. Heterogeneous Fenton-like oxidation of monochlorobenzene using green synthesis of iron nanoparticles. J. Colloid Interface Sci. 410: 67-73.

Kumar, V. and Yadav, S.K. 2009. Plant-mediated synthesis of silver and gold nanoparticles and their applications. J. Chem. Technol. Biotechnol. 84: 151-157.

Laquintana, V., Trapani, A., Denora, N., Wang, F., Gallo, J.M. and Trapani, G. 2009. New strategies to deliver anticancer drugs to brain tumors. Expert Opin. Drug Deliv. 6: 1017-1032.

Le, D.H., Lee, K.L., Shukla, S., Commandeur, U. and Steinmetz, N.F. 2017. Potato virus X, a filamentous plant viral nanoparticle for doxorubicin delivery in cancer therapy. Nanoscale 9: 2348-2357.

Lee, H.J., Lee, G., Jang, N.R., Yun, J.H., Song, J.Y. and Kim, B.S. 2011. Biological synthesis of copper nanoparticles using plant extract. Nanotechnology 1: 371-374.

Lee, X.J., Hiew, B.Y.Z., Lai, K.C., Lee, L.Y., Gan, S., Thangalazhy-Gopakumar, S. and Rigby, S. 2019. Review on graphene and its derivatives: synthesis methods and potential industrial implementation. Journal of the Taiwan Institute of Chemical Engineers 98: 163-180.

Legierse, K.C.H.M., Verhaar, H.J.M., Vaes, W.H.J., De Bruijn, J.H.M. and Hermens, J.L.M. 1999. An analysis of the time dependent acute aquatic toxicity of organophosphorus pesticides: the Critical Target Occupation (CTO) model. Environ. Sci. Technol. 33: 917-925.

Li, S., Shen, Y., Xie, A., Yu, X., Oiu, L., Zhang, L. and Zhang, O. 2007. Green synthesis of silver nanoparticles using Capsicum annuum L. extract. Green Chem. 9: 852-858.

Li, X., Xu, H., Zhe-Sheng Chen and Chen, G. 2011. Biosynthesis of nanoparticles by microorganisms and their applications. Journal of Nanomaterials 2011, 270974.

Lim, J.Y., Mubarak, N.M., Abdullah, E.C., Nizamuddin, S. and Khalid, M. 2018. Recent trends in the synthesis of graphene and graphene oxide based nanomaterials for removal of heavy metals—a review. Journal of Industrial and Engineering Chemistry 66: 29-44.

Loo, Y.Y., Chieng, W.B., Nishibuchi, M. and Radu, S. 2012. Synthesis of silver nanoparticles by using tea leaf extract from Camellia sinensis. Inter. J. Nanomed. 7: 4263-4267.

Love, A.J., Talianski, M.E., Chapman, S.N. and Shaw, J. 2017. US Pat. No. 9,688,964, U.S. Patent and Trademark Office, Washington, DC.

Machado, S., Pinto, S., Grosso, T., Nouws, H., Albergaria, T. and Delerue-Matos, C. 2013. Characterization of green zero-valent iron nanoparticles produced with tree leaf extracts. Sci. Total Environ. 445: 1-8.

Maity, D., Kanti Bain, M., Bhowmick, B., Sarkar, J., Saha, S., Acharya, K., Chakraborty, M. and Chattopadhyay, D. 2011. In situ synthesis, characterization, and antimicrobial activity of silver nanoparticles using water soluble polymer. J. Appl. Polym. Sci. 122: 2189-2196.

Malassis, L., Dreyfus, R., Murphy, R.J., Hough, L.A., Donnio, B. and Murray, C.B. 2016. One-step green synthesis of gold and silver nanoparticles with ascorbic acid and their versatile surface post functionalization. RSC Adv. 6: 33092-33100.

Malhotra, A., Dolma, K., Kaur, N., Rathore, Y.S., Mayilraj, S. and Choudhury, A.R. 2013. Biosynthesis of gold and silver nanoparticles using a novel marine strain of Stenotrophomonas. Bioresour. Technol. 142: 727-731.

Manchala, S., Tandava, V.S.R.K., Jampaiah, D., Bhargava, S.K. and Shanker, V. 2019. Fabrication of a novel $ZnIn_2S_4/g-C_3N_4$/graphene ternary nanocomposite with enhanced charge separation for efficient photocatalytic H_2 evolution under solar light illumination. Photochem. Photobiol. Sci. 18: 2952-2964.

Metuku, R.P., Pabba, S., Burra, S., Gudikandula, K. and Charya, M.S. 2014. Biosynthesis of silver nanoparticles from Schizophyllum radiatum HE 863742.1: their characterization and antimicrobial activity. Biotech. 4: 227-234.

Mishra, A., Kumari, M., Pandey, S., Chaudhry, V., Gupta, K. and Nautiyal, C. 2014. Biocatalytic and antimicrobial activities of gold nanoparticles synthesized by Trichoderma sp. Bioresour. Technol. 166: 235-242.

Mittal, A.K., Kaler, A. and Banerjee, U.C. 2012. Free radical scavenging and antioxidant activity of silver nanoparticles synthesized from flower extract of Rhododendron dauricum. Nano Biomed. Eng. 4: 118-124.

Mohamad, N.A.N., Arham, N.A., Jai, J. and Hadi, A. 2013. IEEE International Conference on Control System, Computing and Engineering, 29 Nov.-1 Dec. 2013, Penang, Malaysia.

Mohanpuria, P., Rana, N.K. and Yadav, S.K. 2008. Bio-synthesis of nanoparticles: technological concepts and future applications. J. Nanoparticle Res. 10: 507-517.

Mondal, S., Roy, N., Laskar, R.A., Sk, I., Basu, S., Mandal, D. and Begum, N.A. 2011. Biogenic synthesis of Ag, Au and bimetallic Au/Ag alloy nanoparticles using aqueous extract of mahogany (Swietenia mahogani JACQ.) leaves. Colloids. and Surf. B 82: 497-504.

Morachis, J.M., Mahmoud, E.A., Sankaranarayanan, J. and Almutairi, A. 2012. Triggered rapid degradation of nanoparticles for gene delivery. J. Drug Deliv. 2012: 1-9.

Nadagouda, M.N. and Varma, R.S. 2006. Green and controlled synthesis of gold and platinum nanomaterials using vitamin B2: density-assisted self-assembly of nanospheres, wires and rods. Green Chem. 8: 516-518.

Nagajyothi, P.C. and Lee, K.D. 2011. Synthesis of plant mediated silver nanoparticles using Dioscorea batatas rhizome extract and evaluation of their antimicrobial activities. J. Nanomater. 22: 3303-3305.

Nagatani, H., Suzuki, I., Kita, M., Tanaka, M., Katsuya, Y., Sakata, O., Miyoshi, S., Yamaguchi, S. and Omata, T. 2015. Structural and thermal properties of ternary narrow-gap oxide semiconductor; wurtzite-derived β-CuGaO$_2$. Inorg. Chem. 54: 1698-1704.

Naithani, V., Nair, S. and Kakkar, P. 2006. Decline in antioxidant capacity of Indian herbal teas during storage and its relation to phenolic content. Food Res. Int. 39: 176-181.

Narayanan, S., Sathy, B.N. and Mony, U. 2012. Biocompatible magnetite/gold nanohybrid contrast agents via green chemistry for MRI and CT bioimaging. ACS Appl. Mater. Interfaces 4: 251-260.

Nazeruddin, G.M., Prasad, N.R., Prasad, S.R., Shaikh, Y.I., Waghmare, S.R. and Adhyapak, P. 2014. Coriandrum sativum seed extract assisted in situ green synthesis of silver nanoparticle and its anti-microbial activity. Ind. Crops Prod. 60: 212-216.

Novoselov, K.S., Fal, V.I., Colombo, L., Gellert, P.R., Schwab, M.G. and Kim, K. 2012. A roadmap for graphene. Nature 490: 192-200.

Olton, D.Y.E., Close, J.M., Sfeir, C.S. and Kumta, P.N. 2011. Intracellular trafficking pathways involved in the gene transfer of nano-structured calcium phosphate-DNA particles. Biomaterials 32: 7662-7670.

Otari, S., Patil, R., Nadaf, N., Ghosh, S. and Pawar, S. 2012. Green biosynthesis of silver nanoparticles from an actinobacteria Rhodococcus sp. Mater. Lett. 72: 92-94.

Pandey, P.C., Singh, R. and Pandey, Y. 2015. Controlled synthesis of functional Ag, Ag–Au/Au–Ag nanoparticles and their Prussian blue nanocomposites for bioanalytical applications. RSC Adv. 5: 49671-49679.

Parashar, U.K., Saxena, P.S. and Shrivastava, A. 2009. Bioinspired synthesis of silver nanoparticles. Dig. J. Nanomater. Biostruct. 4: 159-166.

Parashar, V., Parashar, R., Sharma, B. and Pandey, A.C. 2009. Parthenium leaf extract mediated synthesis of silver nanoparticles: a novel approach towards weed utilization. Dig. J. Nanomater. Bios. 4: 45-50.

Petla, R.K., Vivekanandhan, S., Misra, M., Mohanty, A.K. and Satyanarayana, N. 2012. Soybean (Glycine max) leaf extract based green synthesis of palladium nanoparticles. J Biomater Nanobiotechnol. 3: 14-19.

Philip, D. 2009. Biosynthesis of Au, Ag and Au-Ag nanoparticles using edible mushroom extract. Spectrochim Acta A Mol. Biomol. Spectrosc. 73: 374-381.

Pouton, C.W. and Seymour, L.W. 2001. Key issues in non-viral gene delivery. Adv. Drug Deliv. Rev. 46: 187-203.

Raj, R., Dalei, K., Chakraborty, J. and Das, S. 2016. Extracellular polymeric substances of a marine bacterium mediated synthesis of CdS nanoparticles for removal of cadmium from aqueous solution. J. Colloid. Interface Sci. 462: 166-175.

Rajeshkumar, S. 2018. Synthesis of zinc oxide nanoparticles using algal formulation (Padina tetrastromatica and Turbinaria conoides) and their antibacterial activity against fish pathogens. Res. J. Biotechnol. 13: 15-19.

Rajput, S., Werezuk, R., Lange, R.M. and McDermott, M.T. 2016. Fungal isolate optimized for biogenesis of silver nanoparticles with enhanced colloidal stability. Langmuir 32: 8688-8697.

Ramanathan, R., O'Mullane, A.P., Parikh, R.Y., Smooker, P.M., Bhargava, S.K. and Bansal, V. 2011. Bacterial kinetics-controlled shape-directed biosynthesis of silver nanoplates using Morganella psychrotolerans. Langmuir 27: 714-719.

Ramos-Perez, V., Cifuentes, A., Coronas, N., Pablo, A. and Borrós, S. 2013. Modification of carbon nanotubes for gene delivery vectors. Methods Mol. Biol. 1025: 261-268

Rapti, K., Chaanine, A.H. and Hajjar, R.J. 2011. Targeted gene therapy for the treatment of heart failure. Can. J. Cardiol. 27: 265-283.

Riaz, T., Mughal, P., Shahzadi, T., Shahid, S. and Athar Abbasi, M. 2020. Green synthesis of silver nickel bimetallic nanoparticles using plant extract of Salvadora persica and evaluation of their various biological activities. Mat. Res. Express 6: 1250.

Safaepour, M., Shahverdi, A.R., Shahverdi, H.R., Khorramizadeh, M. and Gohari, A.R. 2009. Green synthesis of small silver nanoparticles using geraniol and its cytotoxicity against fibrosarcoma-Wehi 164. Avicenna J. Med. Biotechnol. 1: 111-115.

Sanaeimehr, Z., Javadi, I. and Namvar, F. 2018. Antiangiogenic and antiapoptotic effects of green-synthesized zinc oxide nanoparticles using Sargassum muticum algae extraction. Cancer Nanotechnol. 9: 3.

Saravanan, M., Barik, S.K., Mubarak Ali, D., Prakash, P. and Pugazhendhi, A. 2018. Synthesis of silver nanoparticles from Bacillus brevis (NCIM 2533) and their antibacterial activity against pathogenic bacteria. Microb. Pathog. 116: 221-226.

Sardsaengjun, C. and Jutiviboonsuk, A. 2010. Mangiferin in leaves of three Thai Mango (Mangifera indica L.) varieties. Thai Pharmaceut. Health. Sci. J. 5: 14-17.

Sasidharan, S., Nilawatyi, R., Xavier, R., Latha, L.Y. and Amala, R. 2010. Wound healing potential of Elaeis guineensis Jacq leaves in an infected albino rat model. Molecules 15: 3186-3199.

Satishkumar, M., Sneha, K., Won, S.W., Cho, C.W., Kim, S. and Yun, Y.S. 2009. Cinnamon zeylanicum bark extract and powder mediated green synthesis of nano-crystalline silver particles and its bactericidal activity. Colloids Surf. B Biointer. 73: 332–338.

Selassie, C.D., Mekapati, S.B. and Verma, R.P. 2002. QSAR: then and now. Curr. Med. Chem. 2: 1357-1379.

Shams, S.S., Zhang, L.S., Hu, R., Zhang, R. and Zhu, J. 2015. Synthesis of graphene from biomass: a green chemistry approach. Mater. Lett. 161: 476-479.

Shankar, S.S., Ahmad, A., Pasricha, R. and Sastry, M. 2003. Bio reduction of chloroaurate ions by Geranium leaves and its endophytic fungus yields gold nanoparticles of different shapes. J. Mater. Chem. 13: 1822-1826.

Shankar, S.S., Ahmad, A., Rai, A. and Sastry, M. 2004. Rapid synthesis of Au, Ag and bimetallic Au core-Ag shell nanoparticles using neem (Azadirachta Indica) leaf broth. J. Colloid. Interface Sci. 275: 496-502.

Sharma, D., Kanchi, S. and Bisetty, K. 2019. Biogenic synthesis of nanoparticles: a review. Arab. J. Chem. 12: 3576-3600.

Sharma, V.K., Yngard, R.A. and Lin, Y. 2009. Silver nanoparticles: green synthesis and their antimicrobial activities. Adv. Colloid. Interface Sci. 30: 83-96.

Sheny, D.S., Mathew, T. and Philip, D. 2011. Phytosynthesis of Au, Ag and Au-Ag bimetallic nanoparticles using aqueous extract and dried leaf of Anacardium occidentale. Spectrochim Acta Part A 79: 254-262.

Shi Kam, N.W., Jessop, T.C., Wender, P.A. and Dai, H. 2004. Nanotube molecular transporters: internalization of carbon nanotube-protein conjugates into Mammalian cells. J. Am. Chem. Soc. 126: 6850-1

Shi, L., Rosso, K.M., Clarke, T.A., Richardson, D.J., Zachara, J.M. and Fredrickson, J.K. 2012. Molecular underpinnings of Fe(III) oxide reduction by Shewanella oneidensis MR-1. Front. Microbiol. 3: 50.

Silva, E.M., Rogez, H. and Larondelle, Y. 2007. Optimization of extraction of phenolics from Inga edulis leaves using response surface methodology. Sep. Purif. Technol. 55: 381-387.

Singh, M., Manikandan, S. and Kumarguru, A.K. 2011. Nanoparticles: a new technology with wide applications. Res. J. Nanosci. Nanotechnol. 1: 1-11.

Singh, P., Kim, Y.J., Zhang, D. and Yang, D.C. 2016. Biological synthesis of nanoparticles from plants and microorganisms. Trends Biotechnol. 34: 588-599.

Singh, R.P., Shukla, V.K., Yadav, R.S., Sharma, P.K., Singh, P.K. and Pandey, A.C. 2011. Biological approach of zinc oxide nanoparticles formation and its characterization. Adv. Mater. Lett. 2: 313-317.

Smuleac, V., Varma, R., Sikdar, S. and Bhattacharyya, D. 2011. Green synthesis of Fe and Fe/Pd bimetallic nanoparticles in membranes for reductive degradation of chlorinated organics. J. Memb. Sci. 379: 131-137.

Sneha, K., Sathishkumar, M., Mao, J., Kwak, I.S. and Yun, Y.S. 2010. Corynebacterium glutamicum-mediated crystallization of silver ions through sorption and reduction processes. Chem. Eng. J. 162: 989-996.

Sneha, K., Sathishkumar, M., Kim, S. and Yun, Y.S. 2010. Counter ions and temperature incorporated tailoring of biogenic gold nanoparticles. Proc. Biochem. 45: 1450-1458.

Song, J.Y. and Kim, B.S. 2008. Biological synthesis of bimetallic Au/Ag nanoparticles using Persimmon (Diopyros kaki) leaf extract. Korean J. Chem. Eng. 25: 808-811.

Song, J.Y. and Kim, B.S. 2008. Rapid biological synthesis of silver nanoparticles using plant leaf extracts. Bioprocess Biosyst. Eng. 44: 1133-1138.

Sousa, A., Ferreira, T.C., Barros, L., Bento, A. and Pereira, T.A. 2008. Antioxidant potential of traditional stoned table olives "Alcaparras": influence of the solvent and temperature extraction conditions. LWT – Food Sci. Technol. 41: 739-745.

Srinath, B., Namratha, K. and Byrappa, K. 2018. Eco-friendly synthesis of gold nanoparticles by Bacillus subtilis and their environmental applications. Adv. Sci. Lett. 24: 5942-5946.

Srivastava, S.K. and Constanti, M. 2012. Room temperature biogenic synthesis of multiple nanoparticles (Ag, Pd, Fe, Rh, Ni, Ru, Pt, Co and Li) by Pseudomonas aeruginosa SM1. J. Nanopart. Res. 14: 831-840.

Suryavanshi, P., Pandit, R., Gade, A., Derita, M., Zachino, S. and Rai, M. 2017. Colletotrichum sp.- mediated synthesis of sulphur and aluminium oxide nanoparticles and its in vitro activity against selected food-borne pathogens. LWT – Food Sci. Technol. 81: 188-194.

Tamuly, C., Hazarika, M., Borah, S.C., Das, M.R. and Boruah, M.P. 2013. In situ biosynthesis of Ag, Au and bimetallic nanoparticles using Piper pedicellatum C.DC: green chemistry approach. Colloids and Surf. B 102: 627-634.

Tanzil, A.H., Sultana, S.T., Saunders, S.R., Shi, L., Marsili, E. and Beyenal, H. 2016. Biological synthesis of nanoparticles in biofilms. Enzyme Microb. Technol. 95: 4-12.

Tavakoli, F., Salavati-Niasari, M. and Mohandes, F. 2015. Green synthesis and characterization of graphene nanosheets. Mater. Res. Bull. 63: 51-57.

Tiwari, P.K. and Lee, Y.S. 2013. Gene delivery in conjunction with gold nanoparticle and tumor treating electric field. J. Appl. Phys. 114: 054902-1-5.

Tsuji, T., Iryo, K., Watanabe, N. and Tsuji, M. 2002. Preparation of silver nanoparticles by laser ablation in solution: influence of laser wavelength on particle size. Appl. Surf. Sci. 202: 80-85.

United Nations Globally Harmonized System of Classification and Labelling of Chemicals (GHS), 2011. 2nd Edn, United Nations, New York and Geneva.

Vigneshwaran, N., Kathe, A.A., Varadarajan, P.V., Nachane, R.P. and Balasubramanya, R.H. 2006. Biomimetics of silver nanoparticles by white rot fungus, Phaenerochaete chrysophobia. Colloid. and Surf. B. 53: 55-59.

Vijayaraghavan, K. and Ashokkumar, T. 2017. Plant-mediated biosynthesis of metallic nanoparticles: a review of literature, factors affecting synthesis, characterization techniques and applications. J. Environ. Chem. Eng. 5: 4866-4883.

Waghmare, S.R., Mulla, M.N., Marathe, S.R. and Sonawane, K.D. 2015. Eco-friendly production of silver nanoparticles using Candida utilis and its mechanistic action against pathogenic microorganisms. Sonawane. Biotech. 5: 33-38.

Wang, S., Hu, Z., Shi, J., Chen, G., Zhang, Q., Weng, Z. and Lu, M. 2019. Green synthesis of graphene with the assistance of modified lignin and its application in anticorrosive waterborne epoxy coatings. Appl. Surf. Sci. 484: 759-770.

Wei, X., Huang, T., Yang, J.H., Zhang, N., Wang, Y. and Zhou, Z.W. 2017. Green synthesis of hybrid graphene oxide/microcrystalline cellulose aerogels and their use as superabsorbent. J. Hazard Mater. 335: 28-38.

Xu, H., Wang, L., Su, H., Gu, L., Han, T., Meng, F. and Liu, C. 2015. Making good use of food wastes: green synthesis of highly stabilized silver nanoparticles from grape seed extract and their antimicrobial activity. Food Biophys. 10: 12-18.

Yadav, A. and Rai, M. 2011. Bioreduction and mechanistic aspects involved in the synthesis of silver nanoparticles using Holarrhena Antidysenterica. J. Bionanosci. 5: 70-73.

Yan, D., Zhang, H., Chen, L., Zhu, G., Wang, Z., Xu, H. and Yu, A. 2014. Super capacitive properties of Mn_3O_4 nanoparticles bio-synthesized from banana peel extract. RSC Adv. 4: 23649-23652.

Yao, B., Huang, H., Liu, Y. and Kang, Z. 2019. Carbon dots: a small conundrum. Trends in Chemistry 1: 235-246.

Yumei, L., Yamei, L., Qiang, L. and Jie, B. 2017. Rapid biosynthesis of silver nanoparticles based on flocculation and reduction of an exopolysaccharide from Arthrobacter sp. B4: its antimicrobial activity and phytotoxicity. J. Nanomater. 2017: 1-8.

Zaino, M.K., Abdul-Hamid, M.A., Abu Bakar, F. and Pak Dek, S. 2009. Effect of different drying methods on the degradation of selected flavonoids in Centella asiatica. Int. Food Res. J. 16: 531-537.

Zhan, G., Huang, J., Lin, L., Lin, W., Emmanuel, K. and Li, Q. 2011. Synthesis of gold nanoparticles by Cacumen Platycladi leaf extract and its simulated solution: toward the plant-mediated biosynthetic mechanism. J. Nanoparticle Res. 13: 4957-4968.

Zhan, G., Huang, J., Du, M., Abdul-Rauf, I., Ma, Y. and Li, Q. 2011. Green synthesis of Au–Pd bimetallic nanoparticles: single-step bio reduction method with plant extract. Mater. Lett. 65: 2989-2999.

Zhang, G., Kuang, Y., Liu, J., Cui, Y., Chen, T. and Zhou, H. 2010. Fabrication of Ag/Au bimetallic nanoparticles by UPD-redox replacement: application in the electrochemical reduction of benzyl chloride. Electrochem. Commun. 12: 1233-1236.

Zhu, C., Guo, S., Fang, Y. and Dong, S. 2010. Reducing sugar: new functional molecules for the green synthesis of graphene nanosheets. ACS Nano 4: 2429-2437.

Zonaro, E., Piacenza, E., Presentato, A., Monti, F., Dell'Anna, R., Lampis, S. and Vallini, G. 2017. Ochrobactrum sp. MPV1 from a dump of roasted pyrites can be exploited as bacterial catalyst for the biogenesis of selenium and tellurium nanoparticles. Microb. Cell Fact. 16: 215.

Part IV: Wastewater Treatment and Environmental Applications

A Brief Review on Carbon-Based Nanosystems for Detection of Pollutants in Industrial Wastewater and Living Cells and Their Subsequent Removal Strategies

Shubham Roy[1], Souravi Bardhan[1], Nur Amin Hoque[2] and Sukhen Das[1]*

[1] Department of Physics, Jadavpur University, Kolkata - 700032, India
[2] Department of Marine Information Technology, Zhejiang University,
Zhoushan - 316021, PR China

1. Introduction

In ancient days, people were aware of water pollution. Shreds of evidence suggest that Indus Valley (2600-1700 BCE) was far advanced in wastewater management (Lofrano and Brown 2010). They used separate channels for drainage water, which was passed through tapered terra-cotta pipelines into small sumps (Feo et al. 2014). The sumps collected the solid wastes using the sedimentation process and the freshwater was further used from above. Similarly, Egyptian, Greek, and Roman civilizations were also equipped with their wastewater management systems (Angelakis and Spyridakis 2010).

Things changed when the Roman era collapsed. It is known as the sanitary Dark Age (476-1800) (Lofrano and Brown 2010). During this period, the culture of wastewater treatment and water resource management was abandoned (Yannopoulos et al. 2017). Lack of concern regarding hygiene and health along with rumors spreading and convictions probably caused this. Urbanization and the evolution of technology during the mid-nineteenth century bought industrial wastes into the game (Geels 2006). Nowadays, water pollution is occurring in numerous pathways, and among them industrial water pollution is causing severe harm to the environment. Various industries, like textile, leather, steel, chemical, fertilizer, etc., are discharging hazardous contaminants directly into the water through pipelines and small canals

Corresponding author: sdasphysics@gmail.com

(Yadav et al. 2019). Industrial contaminants, such as chromium, arsenic, lead and mercury, are enormously harmful to human health even at lower doses (Duruibe et al. 2007) and some of them, iron, nitrate and sulfate can also damage water bodies above a critical limit (Du et al. 2020). Long-time exposure to these contaminants results in various diseases, such as chronic diarrhea, various bowel syndromes and improper pigmentation of the skin (Fernandez-Luqueno et al. 2013), besides several carcinogenic diseases (Fasinu and Orisakwe 2013) like cirrhosis of the liver, lung cancer, skin cancer, etc., which could be minimized if the proper wastewater management system is employed. Recent advancements in materials science, especially nanotechnology, enable us with various nanostructures capable of determining contaminants rapidly and removing them from water in a facile manner (Liu et al. 2019). Thus, comprehensive and exhaustive knowledge is much needed to implement proper wastewater management systems on-site.

The present study is contemplated to discuss different carbon-based nanomaterials for wastewater treatment. Initially, this work includes a brief overview of various carbonaceous nanostructures and their applications in numerous fields. Detection of pollutants in wastewater is the preliminary step toward wastewater management. Hence, special attention has been given to fluorometric sensing using carbon nanotechnology. Numerous applications of bioimaging and biosensing using carbonaceous nanomaterials have been incorporated in this work to study the toxicological aspects of industrial contaminants. Adsorption-based remediation and modern piezo-catalysis techniques employing various carbon-based nanostructures and polymer-nanocomposites have been discussed here in a detailed manner, which portrays the importance of carbonaceous nanomaterials in this field.

2. Carbon-Based Nano Systems: A New Era in Nanotechnology

Carbon nanomaterials are gaining importance for applications in various fields, such as superconductors (Tang et al. 2001), electronic materials (Dai et al. 2012), reinforcement material in composites (Esawi and Farag 2007), drug delivery systems (Hosnedlova et al. 2019), biosensors and biomedical applications (Mohajeri et al. 2019). Carbon nanomaterials exhibit a unique combination of physicochemical properties, such as remarkably high electrical conductivity (Chen and Dai 2013), mechanical strength (Hayashi et al. 2007), optical absorption and emission properties (Dekaliuk et al. 2014) and thermal stability (Bannov et al. 2020). These unique properties of carbon nanomaterials significantly vary with their dimension and structure. The synthesis of such carbon nanostructures is quite facile and industrially feasible (Huczko et al. 2014).

Earlier studies show that fluorescent carbon dot (CD) is a dimensionless promising imaging and fluorometric sensing agent for the detection of various contaminants (Molaei 2020) in aqueous media. Surprisingly, the size-dependent tuneability of optical emission enhances the applicability of this novel fluorometric probe (Kailasa et al. 2019). This zero-dimensional carbon nanomaterial is one of the newest and most promising candidates in the carbon nanotechnology era.

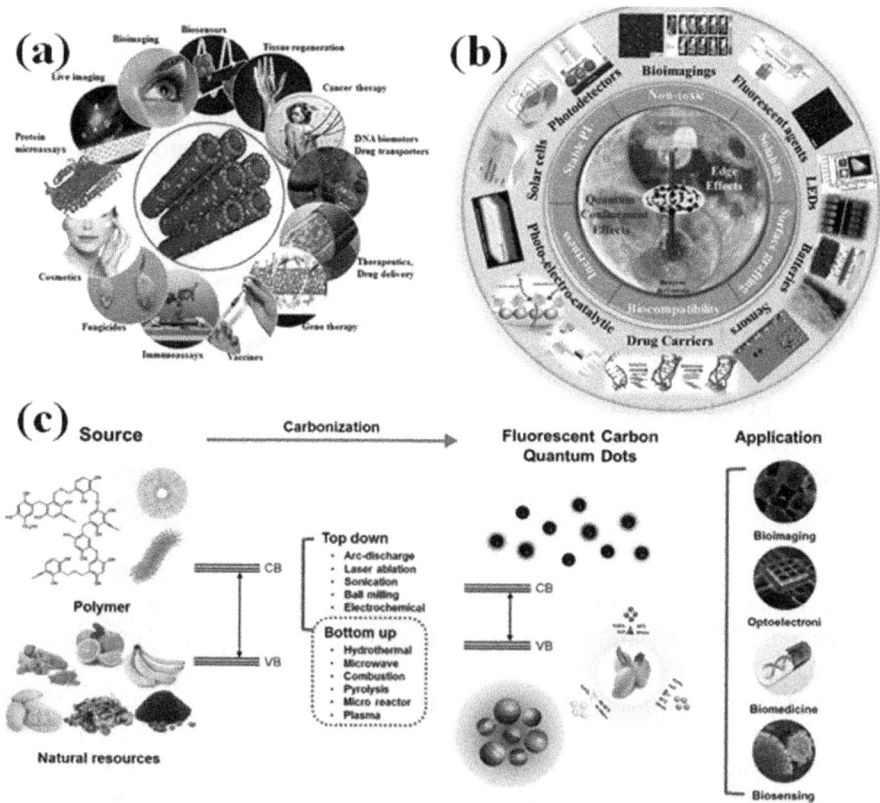

Figure 1. Applications of various carbon-based nanostructures in modern days showing: (a) CNT as a next generation advanced nanomaterial. Reproduced from Merum et al. 2017. Copyright 2017 with permission from Elsevier. (b) Some fluorometric and colorimetric applications of GQD. Reproduced from Tian et al. 2018. Copyright 2018 with permission from Elsevier. (c) Synthesis and numerous physicochemical applications of carbon quantum dots. Reproduced from Mazrad et al. 2018. Copyright 2018 with permission from Royal Society of Chemistry

On the other hand, 1D carbon nanotube (CNT) is an excellent electronic material having a great contribution to piezoelectric nanogenerators and electrodes (Vilatela and Marcilla 2015). High electrical conductivity and electrical stability make this material desirable for the electrode (Liu et al. 2012) and capacitor applications. A few studies have been initiated with CNT and CNT polymer nanocomposites, which show an enhanced thermoelectric figure (An et al. 2017) of merit upon CNT addition. The 2D graphene is quite popular in adsorption-based wastewater treatment due to its high adsorption capacities (Yap et al. 2018). Graphene is a lightweight and hard nanomaterial having a high surface area (Jiang and Fan 2014). It has uniform porosity, a large aspect ratio and hollow pores, making it a promising adsorbent for wastewater treatment applications.

Recently, researchers are working on the waste-based synthesis of carbon nanomaterials. Thus, various biomaterials and waste products have been used extensively to synthesize functionalized carbon nanostructures (Deng et al. 2016). In reality, the use of such natural materials and waste products not only minimizes the production cost but also sustainably decreases environmental pollution.

3. Water Quality Assessment with Carbon-Based Nanosystems

3.1 Limitations of Conventional Systems and Rise of Fluorometric Probes

Water quality assessment is generally based on chemical, physical and bacteriological parameters. Escalating technological demand liberates enormous amounts of metals, ions and complexes into the water causing physical and chemical damage to the water quality. Hence, water quality assessment, i.e. detection of such foreign elements is necessary to combat industrial water pollution.

Several techniques have been employed, such as atomic absorption spectroscopy, cyclic voltammetric method, potentiometry, light scattering method and liquid chromatography for water quality assessment (Conti et al. 2002, Ammann 2002). But these conventional methods have their limitations. In reality, these methods are site and sample, specifically having limited range and long measurement times (Bardhan et al. 2020). Henceforth, recently developed fluorometric sensors are gaining importance for their fast and accurate response (Bandi et al. 2020).

A fluorophore is a fluorescent chemical compound or a part of any chemical compound that can re-emit light upon light excitation (Bachmann et al. 2006). Generally, the interaction between fluorophore and contaminant alters the overall fluorescence property of the fluorophore (Xu et al. 1996). This simple photophysical phenomenon is known as fluorometric sensing. These fluorometric sensors are way more sensitive (Wang et al. 2018) than other conventional sensors, including colorimetric sensors. The fluorescence intensity is measured directly without any reference beam, which makes this technique selective and sensitive.

3.2. Carbonaceous Nanomaterials as Fluorometric Probes: Theory and Pathways

Carbon nanostructures are a class of novel nanomaterials that are widely used in various fluorometric applications, such as fluorometric sensors, biomarkers and bioimaging agents (Pirsaheb et al. 2019). Predominantly, carbon dot and graphene quantum dot show significant fluorescence emission than other carbon-based nanosystems. The emission peak position of these quantum dots is related to their excitation wavelength (Zhu et al. 2015), which may result from the wide distribution of differently sized particles, different emissive traps and surface chemistry that make these carbonaceous nanostructures fluorescent. An extensive study of functionalized CD and GQD shows that the incorporation of nitrogen and sulfur creates C=N and C=S moieties, respectively (Ouyang et al. 2019), which results in surface functionalization of these nanodots.

Thus, it incorporates new energy levels and emits better fluorescence than the undoped quantum dots. Surprisingly, CNT show promising near-infrared (NIR) fluorescence when encapsulated into a polymer matrix (Wolfbeis 2015) and becomes a striking fluorescent probe for metal ions, proteins and small biological species, including glucose, H_2O_2, ATP and NO (Kruss et al. 2013).

Carbonaceous nanomaterials undergo numerous pathways that are involved in fluorescence-based contaminant detection; among them Inner Filter Effect (IFE), Photoinduced Electron Transfer (PET) and Förster Resonance Energy Transfer (FRET) routes are quite popular.

3.2.a Inner Filter Effect (IFE)

It is observed that when fluorophores interact with foreign elements (quencher in this case), such as ions, molecules, complexes with a relatively higher concentration and some quenchers 'block' the emission of the fluorophore (Chen et al. 2018). In this case, some fluorophores are less accessible than others and result in fluorometric quenching (turn off). In other words, fluorescence quenching suggests the presence of the targeted quencher in the sample (Zheng et al. 2013).

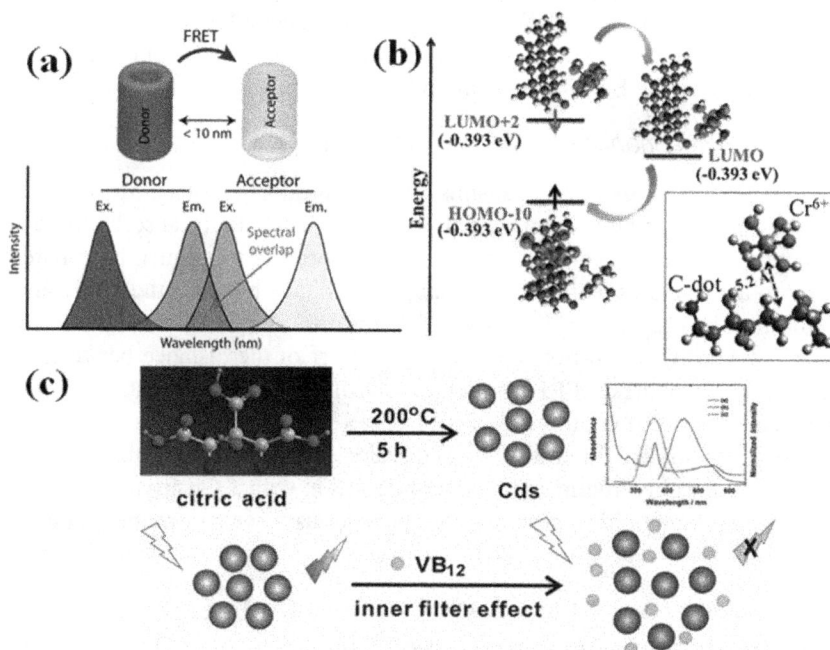

Figure 2. Different fluorometric sensing pathways depicting: (a) FRET based sensing technique. Reproduced from Broussard and Green 2017. Copyright 2017 with permission from Elsevier. (b) DFT based determination of PET based sensing with CDs. Reproduced from Bardhan et al. 2020. Copyright 2020 with permission from Royal Society of Chemistry. (c) Inner Filter Effect using carbon dots showing fluorescence quenching. Reproduced from Ding et al. 2018. Copyright 2018 with permission from Elsevier.

IFE is a steady-state phenomenon that usually occurs in the ground state. Studies suggest that carbon dot undergoes IFE in various cases to detect silver nanoparticles in cosmetics, biomarkers like alkaline phosphatase, hexavalent chromium and ascorbic acid in wastewater (Zheng et al. 2013). IFE can happen in other carbon nanostructures, like CNT and graphene quantum dot (GQD). Zhao et al. (2010) reported the behavior of Bovine Serum Albumin adsorbed onto CNT through IFE. Similarly, GQD depicts its efficacy in detecting artemisinin, cyanide and blood glucose using the inner filter effect (Zhu et al. 2019).

3.2.b Photoinduced Electron Transfer (PET)

In PET, donors are nearby (<10 nm) with acceptors get excited when photons are irradiated (Roy et al. 2019). This donor-acceptor pair in an excited state collides with each other and electrons get transported to the acceptor moiety, causing a rapid alteration of the fluorescence intensity. This phenomenon is known as photoinduced electron transfer. Fluorescence lifetimes get changed in this phenomenon as the PET occurs in the excited state. Lan et al. in their 2015 paper proposed a carbon dot-based turn-on sensor (sensitivity 84 nM) for hydrogen peroxide following the PET mechanism. In this case, the carbon dot acts as an electron acceptor. A similar phenomenon has been reported by Pan et al. showing detection of mercury ions using PET-based technique. Carbon nanotube undergoes PET as well (Baskaran et al., 2005). Phthalocyanines-CNT complex shows significant photoinduced electron transfer as reported by Ballesteros et al. in their 2007 paper.

3.2.c Förster Resonance Energy Transfer (FRET)

FRET is the radiationless energy transmission from the donor molecule to the acceptor molecule. The donor molecule is the fluorophore that initially absorbs the energy and the acceptor is the molecule to which the energy is subsequently transferred. In order to achieve a decent FRET system, the distance between the donor and the acceptor must be less than 5-10 nm (Mondal et al. 2013). The efficacy of this energy transfer is inversely proportional to the sixth power of the distance between donor and acceptor, which makes FRET extremely sensitive to a little alteration in distance (Dennis and Bao 2008). Currently, FRET-based sensing probes are quite popular due to their sensitivity. Bu et al. (2014) developed a carbon dot-gold nanoparticle FRET system for polybrominated biphenyl detection with a sensitivity of 0.039 μg/ml. FRET-based ammonia sensor with carbon dot as a sensing probe (Ganiga and Cyriac 2016) has also been reported recently showing a 3 ppm detection limit.

3.3 Carbonaceous Fluorometric Probes for Detection of Contaminants in Wastewater

Industrial wastewater largely contains heavy metal ions and metal complexes. Thus, detection of industrial heavy metals is necessary to assess the water quality. Fluorescent carbon-based nanosystems have been widely used in the past few years to detect heavy metal ions in wastewater.

3.3.a Carbon Quantum Dot (CD)

In their recent paper, Bardhan et al. (2020) showed nitrogenous CD doped in natural microcline nanoparticles can detect Cr(VI) and Fe(III) simultaneously. The detection limit, in this case, is very low. Besides, Ma et al. (2019) reported a novel CD nanocomposite for Cu(II) ion detection. A dual-mode colorimetric and fluorometric sensor based on CD has also been proposed by Kalaiyarasan et al. (2019) for copper ion detection with a detection limit of 3.8 nM. Low dimensionality causes a higher probability of these quantum dots to agglomerate and lose their fluorescence. This could be eliminated by employing the *in-situ* growth technique of CD in other nanostructures. Roy et al. (2020) proposed this unique technique and detected Cr(VI) in real-life wastewater using CD-doped boehmite nanoparticles. The detection limit was found to be 58 nM in this case. Other heavy metal ions, such as mercury, arsenic, selenium, zinc and lead, have also been detected using CD in quite a similar manner (Devi et al. 2019), making it a promising fluorometric agent for industrial usage.

3.3.b Graphene Quantum Dot (GQD)

Promising biocompatibility, fluorescence and chemical stability makes GQD a potential candidate for fluorometric sensing. GQD is capable to detect hazardous heavy metal ions, like As(V) and Hg(II) using the FRET mechanism (Chini et al. 2019) when attached to carbon dots with very low detection limits. Nitrogen-doped GQD was used to sense Fe(III) ions with a detection limit of 90 nM (Ju and Chen 2014). Similarly, sulfur-doped GQD was found to be a promising fluorometric probe for Ag(I) ions in water (Bian et al. 2017). Shtepliuk et al. (2017) developed a theoretical strategy to detect Cd, Hg and Pb in water using GQD-based nanoprobe. They used density functional methods (DFT) to understand the sensing capability of this novel probe.

3.3.c Carbon-Based Fluorescent Nanofibers and Polymers

Recently, emphasis has been given to optically active carbonaceous polymers and nanofibers for fluorometric detection of industrial contaminants in wastewater. Primarily, Hg(II), Ni(II), Cu(II), Al(III) and Fe(III) ions have been detected (Ramdzan et al. 2020) using these fluorescent nanosystems. Polymers like, Polycaprolactone (PCL), polyaniline (PANI), poly(vinyl alcohol) (PVA) and ethyl cellulose (EC), have been widely used to fabricate innovative optically active fluorometric probes (Terra et al. 2017) for this purpose. In several reports, it has been shown that multiple heavy metal detection is possible using these fluorometric probes (Kailasa et al. 2019, Bandi 2020). Wang et al. (2014) detected Cr(III) and Cu(II) simultaneously using electrospun 1,4-DHAQ-doped cellulose nanofiber films. Similarly, Wei et al. in their 2014 paper showed PVA nanofibrous membrane modified with spirolactam-rhodamine derivatives for simultaneous detection of Fe(III), Cr(III) and Hg(II). These polymeric sensing probes are way more stable and reusable than aqueous nano-sensors. In reality, these polymers do not inherit fluorescence properties. Thus, loaded with fluorophores (Tao et al. 2017) such as dyes, conjugated polymers, carbon-based nanomaterials, and nanoparticles produce fluorescent detection systems.

4. Toxicity Assessment of Industrial Wastewater Using Carbon-Based Nanostructures

Carbon nanoparticles having the benefit of biocompatibility with additional fluorescence quality created a large domain in the field of toxicological and nano theranostics study (Molaei 2019). Both *in-vitro* and *in-vivo* applications of such nanomaterials are widely explored (Molaei 2019) to seek new avenues for advanced bio-sensing and bio-monitoring of industrial wastewater. Although most of the toxicologic studies involve short-term indicators of toxicity or alteration of cellular functions or interfere with metabolic pathways, carbon nanostructures have reported overcoming such drawbacks and can be modified accordingly for even sub-chronic exposure assays or various pathologic reactions for *in-vivo* experiments without causing a secondary modification of host cells (Hurt et al. 2006).

Bio-sensing and bio-imaging techniques involving the detection of specific molecules or pollutants in wastewater have colossal significance in environmental safety and health care, hence there is a huge demand for detecting pollutants at the cellular or genetic level (Gao et al. 2014). The fluorescence technique emerged as one of the feasible, time-saving and reliable methods to fulfill such prerequisites. The intrinsic fluorescence properties of carbon-based nanostructures, especially nitrogen or sulfur-doped CDs and GQDs, make them potential candidates for bio-sensors and bio-imaging agents.

Moreover, these quantum dots can be conjugated with various small molecules, antibodies and nanoparticles to detect cancer cells, conduct targeted sensing and can be used as biomarkers (Simpson et al. 2018). They can even be useful in detecting the presence of various molecules or heavy metals (Deshmukh et al. 2018), such as $Hg(II)$, $Pb(II)$, $Cu(II)$, $Fe(III)$, $Cr(VI)$ within live cells, thus possessing a great role in biomagnification study. Even the electroactive properties of graphene facilitate the synthesis of electro-chemiluminescent and fluorescent biosensors (Pumera 2011).

Another benefit of carbon nanostructure, especially CDs is that they have remarkable photostability and high resistance toward metabolic degradation, making them preferable for *in-vivo/in-vitro* bioimaging applications with relatively lower chances of cytotoxicity as compared to other quantum dots (CdSe, CdTe) (Li et al. 2017). Other than fluorescence turn-on and turn-off biosensors, various carbon nanostructure-based colorimetric biosensors have exhibited promising potential toward targeted detection at very low concentrations compared to conventional colorimetric biosensors developed from silver or gold nanoparticle or metallic QDs (Sabela et al. 2017).

Bhunia et al. 2013 chemically designed functionalized carbon nanoparticle-based nanoprobes with tunable emission paving the path for the advancement of carbon nanostructure-based bioimaging probes. Recent researches to explore new fluorescent markers in cell imaging proved carbon nanoparticles as a suitable candidate due to their high fluorescence and low cytotoxicity (Song et al. 2018) since they can instantly detect the presence of a very low concentration of specific pollutants or heavy metals in human cell or microorganism by the fluorescent turn-on-off method (Pirsaheb et al. 2019, Sharma et al. 2016).

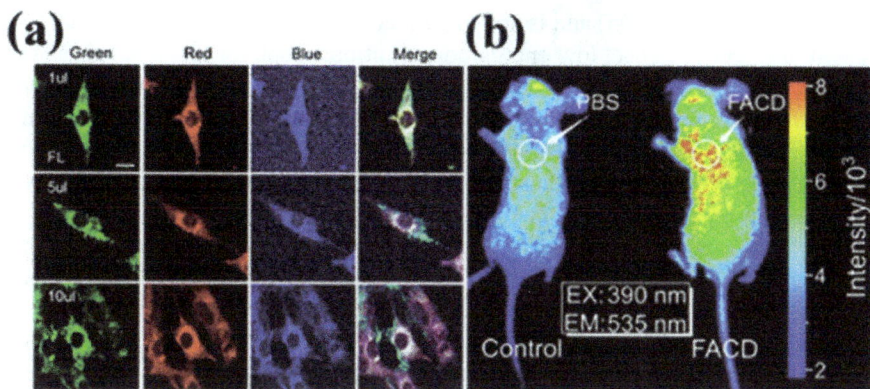

Figure 3. Bioimaging studies using carbon based nano systems showing: (a) N-isopropylacrylamide passivated carbon dots for multicolor *in-vitro* bioimaging applications. Reproduced from Kim et al. 2018. Copyright 2018 with permission from Elsevier. (b) Aspirin based CDs for *in-vivo* bioimaging in mice. Adapted from Xu et al. 2016. Copyright 2016 with permission from American Chemical Society

In the quest for fabricating next-generation probes for *in-vivo* fluorescent imaging, Yang et al. (2009) were the first to inject mice with carbon dots to explore feasibility. After the success, further similar investigations were performed by coupling carbon nanomaterials with other active molecules for various *in-vivo* applications, such as cancer diagnostics (Luo et al. 2013) which can validate the *in-vitro* assessments on cancer cells.

Hence, there are future endeavors of development of more advanced carbon nanostructure-based biosensors and modifying the bio-imaging techniques to detect even minute quantity of specific molecule or heavy metal even in live cells and animal models to combat environmental pollution and maintain health safety along with the determination of pathway for biomagnification, thus preventing health hazards from industrial water pollution.

5. Remediation of Industrial Wastewater Using Carbon-Based Nanostructures

Removal of toxic pollutants and heavy metals, both from the wastewater and living cells has attracted huge attention because of its high significance in healthcare and the environment. Various techniques for pollutant removal, such as adsorption, ion exchange, reverse osmosis, coagulation, electrochemical and solvent extraction, were reported (Dąbrowski et al. 2004) but most of them suffered drawbacks. A few of these techniques are costly. Particularly, ion exchange suffers recyclability and electrical techniques require high energy consumption (Fiyadh et al. 2019).

5.1 Adsorption-Based Removal of Heavy Metals Using Carbonaceous Nanomaterials

The adsorption-based removal of contaminants has widely proved to be effective

(Wingenfelder et al. 2005) and is now practiced extensively because of its simplicity, ease of use, convenience, low energy consumption, scope for regeneration and high level of effectiveness even when pollutants are present at low concentration. Carbon nanomaterials undergo both adsorption processes, which makes them promising adsorbents for heavy metal removal from wastewater. Although various organic and inorganic adsorbents and bio-adsorbents are still used, carbon-based nanomaterials are now gaining popularity due to their surface properties, high stability and inertness and in many cases due to their porous structure (Khezami and Capart 2005). Moreover, the incorporation of various functional groups on their surface or tailoring their physical and chemical properties enhanced their sorption capacities (Yoon et al. 2007). The adsorptive capacity of these carbon-based nanomaterials was evaluated using isotherm models like pseudo-second-order rate equation, Langmuir, Freundlich, Redlich-Peterson and Dubinin-Radushkevich models (Xu et al. 2018).

5.1.a Carbon Nanotubes

CNTs have emerged as a potential next-generation adsorbent because of their remarkable structural, physical and chemical properties and very high adsorption capacity which attributes to the presence defects, functional groups, outside surfaces, grooves and interstitial regions between nanotube bundles (Thostenson et al. 2001). Although they are highly stable, they are not a good adsorbent themselves, hence requires surface functionalization, such as the acid modification or grafting of functional groups like $-NH_2$ or $-SH$, increasing the number of pre-existing groups like $-COOH$ or $-OH$ or by coating with specific nanomaterials to achieve high selectivity and excellent adsorbing capacity of heavy metals. Various reports proved that functionalized multi-walled CNTs can effectively tackle toxic heavy metals, like As^{3+}, Hg^{2+}, Cd^{2+}, Cr^{6+}, Pb^{2+}, Ni^{2+} in an aqueous medium (Fiyadh et al. 2019, Abbas et al. 2016). The metal ions interact with the CNT surface and are adsorbed via various mechanisms, mainly by physical adsorption, sorption-precipitation, electrostatic attraction lowering the metal ion concentration from an aqueous medium. The mechanism followed and affinity of functionalized CNTs toward heavy metal ions depend on various factors, like pH, temperature, surface area and charge, concentration and ionic strength and hence various affinity orders are reported in different studies. Rao et al. (2007) reported the following affinity order $Cd^{2+} < Cu^{2+} < Zn^{2+} < Ni^{2+} < Pb^{2+}$, while Stafiej and Pyrzynska (2007) reported $Mn^{2+} < Zn^{2+} < Co^{2+} < Pb^{2+} < Cu^{2+}$. In most of these cases, the adsorption model follows the Langmuir equation, which applies to adsorption on homogeneous surfaces, while in some cases it follows Freundlich equations, which mainly represent adsorption on the heterogeneous surface (Abbas et al. 2016). CNTs are also reported to successfully remove organic pollutants such as dyes from wastewater due to their tendency to interact by non-covalent forces like hydrogen bonding, π–π stacking and van der Waals forces (Gupta and Saleh 2013).

5.1.b Graphene and Its Derivatives

Graphene and graphene-based nanostructure are used for the past few years as an economical adsorbent for wastewater treatment because of the desirable surface properties and functional sites (Tabish et al. 2018), which can be further enhanced by

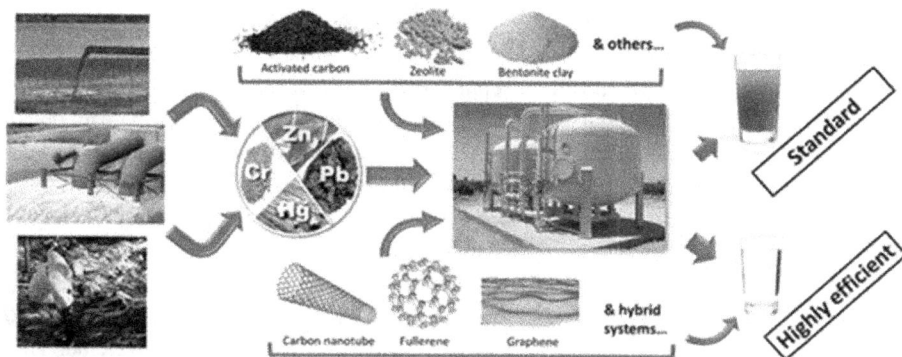

Figure 4. A brief comparison between conventional adsorbents and futuristic nano-adsorbents showing their efficacy in real life condition. Reproduced from Burakov et al. 2018. Copyright 2018 with permission from Elsevier.

the oxidation method. The negative charge on graphene or graphene oxide surface facilitates the electrostatic attraction of heavy metal ions. Moreover, the presence of functional groups like –COOH or –OH attributes to ion exchange or surface complex formation. Although graphene has poor adsorption selectivity, the specificity can be improved by functionalization, adsorbent thickness maintenance and desired ambient conditions, which also increases the adsorption capacity, thus facilitating heavy metal removal. Affinity order for heavy metals was found to be $Ca^{2+} \approx Sr^{2+} < Zn^{2+} \approx Ni^{2+} \approx Cd^{2+} < Cu^{2+} < Pb^{2+}$, which can be correlated with their electronegativity and standard reduction potential (Peng et al. 2017). In the case of anionic pollutants of wastewater, like fluoride, chromate, arsenate and arsenite, as reported by Reyes Bahena et al. (2002), under low pH ligand exchange reactions can take place as the anions can displace –OH groups from graphene surface.

5.1.c Carbon-Based Polymer Nanocomposites

Scientists are now giving more emphasis on carbon nanostructure-doped polymer adsorbents (Zhao et al. 2018) for their multiple benefits, like practicability, feasible regeneration, mechanical stability and durability, adjustable surface functional groups, benign nature and the polymeric matrix provides greater surface area for adsorption or specific interaction. An obvious feature of carbon-polymer nanocomposites for heavy metal removal is their high sorption ability (Wu et al. 2011) and reduction of the potential risk of release of pollutants back to the environment after they interact with the polymer matrix. Modifications of such polymeric membranes with improved regeneration capacity and multiple-use facilities to curb wastage have become a new challenge for scientists. Presently, the removal of heavy metals from living cells using carbon-based nanostructures is gaining popularity, mainly due to its biocompatibility.

5.2 Piezo-Catalytic Removal of Contaminants from Aqueous Media

Removal of organic pollutants from wastewater is of utmost importance. Various strategies like photocatalysis and membrane separation have been widely introduced to combat this. But lack of selectivity, high cost in membrane separation and rapid electron-hole pair recombination are the common difficulties that limit semiconductor-based photocatalysis and membrane separation.

Piezo-catalysis is a relatively new concept of eliminating these drawbacks demonstrating unique catalytic properties as a result of the creation of the built-in electric field by the dipole polarization. This dipolar polarization delivers a driving force to move the photoinduced charge carriers enabling their separation (Liang et al. 2019).

The era of piezocatalysis has begun with piezoelectric crystals, like $BaTiO_3$, $PbTiO_3$ and $NaNbO_3$ (Wang et al. 2019). Later, Lin et al. (2014) incorporated zirconium into the lead titanate crystals and achieved better piezocatalytic activity. Similarly, Sharma et al. (2020) co-doped cerium and calcium into standard barium titanate crystals to clean dye/pharmaceutical wastewater. ZnO/Al_2O_3 nanosheet-based ultrasonic catalyst was prepared by Nie et al. (2020) very recently, which is capable of removing methyl orange dye from an aqueous medium.

Carbon and its nanostructures are unexplored in this field until now. A few works report the enhancement of piezocatalytic activity when carbon nanomaterials were added to piezo-electric materials. Chen et al. in their recent paper reported novel barium titanate/carbon hybrid nanocomposites for piezocatalytic degradation of dye in wastewater (Chen et al. 2020). In this work, the authors achieved the best results for 2% carbon-doped barium titanate crystal, which rapidly eliminates around 75.5% of rhodamine-B dye under ultrasonic vibration in wastewater. This enhancement can be ascribed to the action of carbon's charge transfer, which promotes the effective separation of the piezoelectrically induced charges in this case. Thus, the application of carbonaceous nanostructures in piezocatalytic removal of pollutants is still a futuristic domain, which could be a game-changer in the next few years.

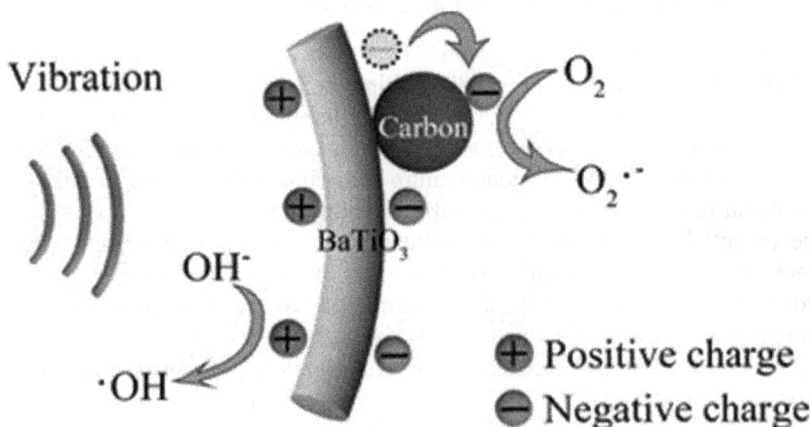

Figure 5. The working mechanism of $BaTiO_3$/carbon piezocatalytic nanocomposite showing charge separation process under vibration. Reproduced from Chen et al. 2020. Copyright 2020 with permission from Elsevier.

6. Prospective and Summary - Conclusions and Outlook

In this review, the importance of wastewater treatment has been discussed in great depth with various solution strategies. Initially, the role of materials science, especially, carbon-nanotechnology in wastewater treatment is discussed and two distinct (sensing of industrial contaminants and their removal) branches of wastewater remediation have been elaborated. The development of carbon nanotechnology enables us with numerous quantum dots (CD and GQD) and carbon nanostructures (CNT and graphene); those are having exceptionally remarkable physicochemical properties to detect and remove industrial contaminants from wastewater. In reality, fluorometric detection of carbonaceous nanostructures has been discussed and their potential applicability of toxicological analysis has been depicted, which could be the next-generation sensor technology for wastewater treatment plants.

Apart from that, plausible remediation strategies involving classical adsorption-based contaminant removal and novel piezocatalytic removal have also been discussed in a detailed manner. Carbon nanomaterials with high aspect ratios have already emerged as promising adsorbent materials, but carbon nanomaterials mediated piezocatalysis is a relatively new approach, which could be thoroughly studied in near future.

In summary, carbon nanostructures have emerged as next-generation wastewater monitoring and remediation materials for combating environmental pollution and wastewater-related health hazards.

Acknowledgments

The authors would like to thank the Department of Physics, Jadavpur University, for extending their facilities.

Funding Details

S.D. would like to acknowledge DST-SERB (Grant No. EEQ/2018/000747) for funding.

Conflict of Interests

The authors declare no conflict of interest.

References

Abbas, A., Al-Amer, A.M., Laoui, T., Al-Marri, M.J., Nasser, M.S., Khraisheh, M. and Atieh, M.A. 2016. Heavy metal removal from aqueous solution by advanced carbon nanotubes: critical review of adsorption applications. Sep. Purif. Technol. 157: 141-161.

Ammann, A.A. 2002. Speciation of heavy metals in environmental water by ion chromatography coupled to ICP–MS. Anal. Bioanal. Chem. 372(3): 448-452.

An, C.J., Kang, Y.H., Song, H., Jeong, Y. and Cho, S.Y. 2017. High-performance flexible thermoelectric generator by control of electronic structure of directly spun carbon nanotube webs with various molecular dopants. J. Mater. Chem. A 5(30): 15631-15639.

Angelakis, A.N. and Spyridakis, D.S. 2010. A brief history of water supply and wastewater management in ancient Greece. Water Science and Technology: Water Supply 10(4): 618-628.

Bachmann, L., Zezell, D.M., Ribeiro, A.D.C., Gomes, L. and Ito, A.S. 2006. Fluorescence spectroscopy of biological tissues—a review. Appl. Spectrosc. Rev. 41(6): 575-590.

Ballesteros, B., de la Torre, G., Ehli, C., Aminur Rahman, G.M., Agulló-Rueda, F., Guldi, D.M. and Torres, T. 2007. Single-wall carbon nanotubes bearing covalently linked phthalocyanines–photoinduced electron transfer. J. Am. Chem. Soc. 129(16): 5061-5068.

Bandi, R., Dadigala, R., Gangapuram, B.R., Sabir, F.K., Alle, M., Lee, S.H. and Guttena, V. 2020. N-doped carbon dots with pH-sensitive emission, and their application to simultaneous fluorometric determination of iron (III) and copper (II). Microchim. Acta 187(1): 30.

Bannov, A.G., Popov, M.V. and Kurmashov, P.B. 2020. Thermal analysis of carbon nanomaterials: advantages and problems of interpretation. J. Therm. Anal. Calorim. 142(1): 349-370.

Bardhan, S., Roy, S., Chanda, D.K., Ghosh, S., Mondal, D., Das, S. and Das, S. 2020. Nitrogenous carbon dot decorated natural microcline: an ameliorative dual fluorometric probe for Fe^{3+} and Cr^{6+} detection. Dalton Trans. 49(30): 10554-10566.

Baskaran, D., Mays, J.W., Zhang, X.P. and Bratcher, M.S. 2005. Carbon nanotubes with covalently linked porphyrin antennae: photoinduced electron transfer. J. Am. Chem. Soc. 127(19): 6916-6917.

Bhunia, S.K., Saha, A., Maity, A.R., Ray, S.C. and Jana, N.R. 2013. Carbon nanoparticle-based fluorescent bioimaging probes. Sci. Rep. 3(1): 1-7.

Bian, S., Shen, C., Qian, Y., Liu, J., Xi, F. and Dong, X. 2017. Facile synthesis of sulfur-doped graphene quantum dots as fluorescent sensing probes for Ag^+ ions detection. Sens. Actuators, B 242: 231-237.

Broussard, J.A. and Green, K.J. 2017. Research techniques made simple: methodology and applications of Förster Resonance Energy Transfer (FRET) microscopy. J. Invest. Dermatol. 137(11): 185-191.

Bu, D., Zhuang, H., Yang, G. and Ping, X. 2014. An immunosensor designed for polybrominated biphenyl detection based on fluorescence resonance energy transfer (FRET) between carbon dots and gold nanoparticles. Sens. Actuators, B 195: 540-548.

Burakov, A.E., Galunin, E.V., Burakova, I.V., Kucherova, A.E., Agarwal, S., Tkachev, A.G. and Gupta, V.K. 2018. Adsorption of heavy metals on conventional and nanostructured materials for wastewater treatment purposes: a review. Ecotoxicol. Environ. Saf. 148: 702-712.

Chen, L., Jia, Y., Zhao, J., Ma, J., Wu, Z., Yuan, G. and Cui, X. 2021. Strong piezocatalysis in barium titanate/carbon hybrid nanocomposites for dye wastewater decomposition. J. Colloid Interface Sci. 586: 758-765.

Chen, S., Yu, Y.L. and Wang, J.H. 2018. Inner filter effect-based fluorescent sensing systems: a review. Anal. Chim. Acta 999: 13-26.

Chen, T. and Dai, L. 2013. Carbon nanomaterials for high-performance supercapacitors. Mater. Today 16(7-8): 272-280.

Chini, M.K., Kumar, V., Javed, A. and Satapathi, S. 2019. Graphene quantum dots and carbon nano dots for the FRET based detection of heavy metal ions. Nano-Struct. Nano-Objects 19: 100347.

Conti, M.E., Tudino, M.B., Muse, J.O. and Cecchetti, G. 2002. Biomonitoring of heavy metals and their species in the marine environment: the contribution of atomic absorption spectroscopy and inductively coupled plasma spectroscopy. Res. Trends Appl. Spectros. 4: 295-324.

Dąbrowski, A., Hubicki, Z., Podkościelny, P. and Robens, E. 2004. Selective removal of the heavy metal ions from waters and industrial wastewaters by ion-exchange method. Chemosphere 56(2): 91-106.

Dai, L., Chang, D.W., Baek, J.B. and Lu, W. 2012. Carbon nanomaterials for advanced energy conversion and storage. Small 8(8): 1130-1166.

De Feo, G., Antoniou, G., Fardin, H.F., El-Gohary, F., Zheng, X.Y., Reklaityte, I., Butler, D., Yannopoulos, S. and Angelakis, A.N. 2014. The historical development of sewers worldwide. Sustainability 6(6): 3936-3974.

Dekaliuk, M.O., Viagin, O., Malyukin, Y.V. and Demchenko, A.P. 2014. Fluorescent carbon nanomaterials: "quantum dots" or nanoclusters? Phys. Chem. Chem. Phys. 16(30): 16075-16084.

Deng, J., You, Y., Sahajwalla, V. and Joshi, R.K. 2016. Transforming waste into carbon-based nanomaterials. Carbon 96: 105-115.

Dennis, A.M. and Bao, G. 2008. Quantum dot–fluorescent protein pairs as novel fluorescence resonance energy transfer probes. Nano Lett. 8(5): 1439-1445.

Deshmukh, M.A., Shirsat, M.D., Ramanaviciene, A. and Ramanavicius, A. 2018. Composites based on conducting polymers and carbon nanomaterials for heavy metal ion sensing. Crit. Rev. Anal. Chem. 48(4): 293-304.

Devi, P., Rajput, P., Thakur, A., Kim, K.H. and Kumar, P. 2019. Recent advances in carbon quantum dot-based sensing of heavy metals in water. TrAC, Trends Anal. Chem. 114: 171-195.

Ding, L., Yang, H., Ge, S. and Yu, J. 2018. Fluorescent carbon dots nanosensor for label-free determination of vitamin B12 based on inner filter effect. Spectrochim. Acta, Part A 193: 305-309.

Du, L., Wu, A., Liu, G., Li, H., Yu, B. and Wang, X. 2020. Green autofluorescence eleocytes from earthworm as a tool for detecting environmental iron pollution. Ecol. Indic. 108: 105695.

Duruibe, J.O., Ogwuegbu, M.O.C. and Egwurugwu, J.N. 2007. Heavy metal pollution and human biotoxic effects. Int. J. Phys. Sci. 2(5): 112-118.

Esawi, A.M. and Farag, M.M. 2007. Carbon nanotube reinforced composites: potential and current challenges. Mater. Design 28(9): 2394-2401.

Fasinu, P.S. and Orisakwe, O.E. 2013. Heavy metal pollution in sub-Saharan Africa and possible implications in cancer epidemiology. Asian Pac. J. Cancer Prev. 14(6): 3393-3402.

Fernandez-Luqueno, F., López-Valdez, F., Gamero-Melo, P., Luna-Suárez, S., Aguilera-González, E.N., Martínez, A.I., García-Guillermo, M.D.S., Hernandez-Martinez, G., Herrera-Mendoza, R., Álvarez-Garza, M.A. and Pérez-Velázquez, I.R. 2013. Heavy metal pollution in drinking water – a global risk for human health: a review. Afr. J. Environ. Sci. Technol. 7(7): 567-584.

Fiyadh, S.S., AlSaadi, M.A., Jaafar, W.Z., AlOmar, M.K., Fayaed, S.S., Mohd, N.S., Hin, L.S. and El-Shafie, A. 2019. Review on heavy metal adsorption processes by carbon nanotubes. J. Clean. Prod. 230: 783-793.

Ganiga, M. and Cyriac, J. 2016. FRET based ammonia sensor using carbon dots. Sens. Actuators, B 225: 522-528.

Gao, X., Ding, C., Zhu, A. and Tian, Y. 2014. Carbon-dot-based ratiometric fluorescent probe for imaging and biosensing of superoxide anion in live cells. Anal. Chem. 86(14): 7071-7078.

Geels, F.W. 2006. The hygienic transition from cesspools to sewer systems (1840–1930): the dynamics of regime transformation. Research Policy 35(7): 1069-1082.

Gupta, V.K. and Saleh, T.A. 2013. Sorption of pollutants by porous carbon, carbon nanotubes and fullerene – an overview. Environ. Sci. Pollut. Res. 20(5): 2828-2843.

Hayashi, T., Kim, Y.A., Natsuki, T. and Endo, M. 2007. Mechanical properties of carbon nanomaterials. Chemphyschem 8(7): 999-1004.

Hosnedlova, B., Kepinska, M., Fernandez, C., Peng, Q., Ruttkay-Nedecky, B., Milnerowicz, H. and Kizek, R. 2019. Carbon nanomaterials for targeted cancer therapy drugs: a critical review. Chem. Rec. 19(2-3): 502-522.

Huczko, A., Dąbrowska, A., Łabędź, O., Soszyński, M., Bystrzejewski, M., Baranowski, P. Bhatta, R., Pokhrel, B., Kafle, B.P., Stelmakh, S. and Gierlotka, S. 2014. Facile and fast combustion synthesis and characterization of novel carbon nanostructures. Phys. Status Solid B 251(12): 2563-2568.

Hurt, R.H., Monthioux, M. and Kane, A. 2006. Toxicology of carbon nanomaterials: status, trends, and perspectives on the special issue. Carbon 44(6): 1028-1033.

Jiang, L. and Fan, Z. 2014. Design of advanced porous graphene materials: from graphene nanomesh to 3D architectures. Nanoscale 6(4): 1922-1945.

Ju, J. and Chen, W. 2014. Synthesis of highly fluorescent nitrogen-doped graphene quantum dots for sensitive, label-free detection of Fe (III) in aqueous media. Biosens. Bioelectron. 58: 219-225.

Kailasa, S.K., Ha, S., Baek, S.H., Kim, S., Kwak, K. and Park, T.J. 2019. Tuning of carbon dots emission color for sensing of Fe^{3+} ion and bioimaging applications. Mater. Sci. Eng. C 98: 834-842.

Kalaiyarasan, G. and Joseph, J. 2019. Efficient dual-mode colorimetric/fluorometric sensor for the detection of copper ions and vitamin C based on pH-sensitive amino-terminated nitrogen-doped carbon quantum dots: effect of reactive oxygen species and antioxidants. Anal. Bioanal. Chem. 411(12): 2619-2633.

Khezami, L. and Capart, R. 2005. Removal of chromium (VI) from aqueous solution by activated carbons: kinetic and equilibrium studies. J. Hazard. Mater. 123(1-3): 223-231.

Kim, M.C., Yu, K.S., Han, S.Y., Kim, J.J., Lee, J.W., Lee, N.S., Jeong, Y.G. and Kim, D.K. 2018. Highly photoluminescent N-isopropylacrylamide (NIPAAM) passivated carbon dots for multicolor bioimaging applications. Eur. Polym. J. 98: 191-198.

Kruss, S., Hilmer, A.J., Zhang, J., Reuel, N.F., Mu, B. and Strano, M.S. 2013. Carbon nanotubes as optical biomedical sensors. Adv. Drug Deliv. Rev. 65(15): 1933-1950.

Lan, M., Di, Y., Zhu, X., Ng, T.W., Xia, J., Liu, W., Meng, X., Wang, P., Lee, C.S. and Zhang, W. 2015. A carbon dot-based fluorescence turn-on sensor for hydrogen peroxide with a photo-induced electron transfer mechanism. ChemComm 51(85): 15574-15577.

Li, M., Yu, C., Hu, C., Yang, W., Zhao, C., Wang, S., Zhang, M., Zhao, J., Wang, X. and Qiu, J. 2017. Solvothermal conversion of coal into nitrogen-doped carbon dots with singlet oxygen generation and high quantum yield. Chem. Eng. J. 320: 570-575.

Liang, Z., Yan, C.F., Rtimi, S. and Bandara, J. 2019. Piezoelectric materials for catalytic/ photocatalytic removal of pollutants: recent advances and outlook. Appl. Catal. B 241: 256-269.

Lin, H., Wu, Z., Jia, Y., Li, W., Zheng, R.K. and Luo, H. 2014. Piezoelectrically induced mechano-catalytic effect for degradation of dye wastewater through vibrating Pb $(Zr_{0.52}Ti_{0.48})O_3$ fibers. Appl. Phys. Lett. 104(16): 162907.

Liu, L., Luo, X.B., Ding, L. and Luo, S.L. 2019. Application of nanotechnology in the removal of heavy metal from water. In: Nanomaterials for the Removal of Pollutants and Resource Reutilization. pp. 83-147. Elsevier.

Liu, X.M., Dong Huang, Z., Woon Oh, S., Zhang, B., Ma, P.C., Yuen, M.M. and Kim, J.K. 2012. Carbon nanotube (CNT)-based composites as electrode material for rechargeable Li-ion batteries: a review. Compos. Sci. Technol. 72(2): 121-144.

Lofrano, G. and Brown, J. 2010. Wastewater management through the ages: a history of mankind. Sci. Total Environ. 408(22): 5254-5264.

Luo, P.G., Sahu, S., Yang, S.T., Sonkar, S.K., Wang, J., Wang, H., LeCroy, G.E., Cao, L. and Sun, Y.P. 2013. Carbon "quantum" dots for optical bioimaging. J. Mater. Chem. B 1(16): 2116-2127.

Ma, X., Lin, S., Dang, Y., Dai, Y., Zhang, X. and Xia, F. 2019. Carbon dots as an "on-off-on" fluorescent probe for detection of Cu (II) ion, ascorbic acid, and acid phosphatase. Anal. Bioanal. Chem. 411(25): 6645-6653.

Mazrad, Z.A.I., Lee, K., Chae, A., In, I., Lee, H. and Park, S.Y. 2018. Progress in internal/ external stimuli responsive fluorescent carbon nanoparticles for theranostic and sensing applications. J. Mater. Chem. B 6(8): 1149-1178.

Merum, S., Veluru, J.B. and Seeram, R. 2017. Functionalized carbon nanotubes in bio-world: applications, limitations and future directions. Mater. Sci. Eng. B 223: 43-63.

Mohajeri, M., Behnam, B. and Sahebkar, A. 2019. Biomedical applications of carbon nanomaterials: drug and gene delivery potentials. J. Cell. Physiol. 234(1): 298-319.

Molaei, M.J. 2019. A review on nanostructured carbon quantum dots and their applications in biotechnology, sensors, and chemiluminescence. Talanta 196: 456-478.

Molaei, M.J. 2020. Principles, mechanisms, and application of carbon quantum dots in sensors: a review. Anal. Methods 12(10): 1266-1287.

Mondal, S., Das, T., Ghosh, P., Maity, A., Mallick, A. and Purkayastha, P. 2013. FRET-based characterisation of surfactant bilayer protected core-shell carbon nanoparticles: advancement toward carbon nanotechnology. ChemComm 49(69): 7638-7640.

Nie, Q., Xie, Y., Ma, J., Wang, J. and Zhang, G. 2020. High piezo-catalytic activity of ZnO/ Al_2O_3 nanosheets utilizing ultrasonic energy for wastewater treatment. J. Clean. Prod. 242: 118532.

Ouyang, Z., Lei, Y., Chen, Y., Zhang, Z., Jiang, Z., Hu, J. and Lin, Y. 2019. Preparation and specific capacitance properties of sulfur, nitrogen co-doped graphene quantum dots. Nanoscale Res. Lett. 14(1): 219.

Pan, X., Zhang, Y., Sun, X., Pan, W. and Wang, J. 2018. A green emissive carbon-dot-based sensor with diverse responsive manners for multi-mode sensing. Analyst 143(23): 5812-5821.

Peng, W., Li, H., Liu, Y. and Song, S. 2017. A review on heavy metal ions adsorption from water by graphene oxide and its composites. J. Mol. Liq. 230: 496-504.

Pirsaheb, M., Mohammadi, S. and Salimi, A. 2019. Current advances of carbon dots based biosensors for tumor marker detection, cancer cells analysis and bioimaging. TrAC Trends Analyt. Chem. 115: 83-99.

Pirsaheb, M., Mohammadi, S., Salimi, A. and Payandeh, M. 2019. Functionalized fluorescent carbon nanostructures for targeted imaging of cancer cells: a review. Microchim. Acta 186(4): 1-20.

Pumera, M. 2011. Graphene in biosensing. Mater. Today 14(7-8): 308-315.

Ramdzan, N.S.M., Fen, Y.W., Anas, N.A.A., Omar, N.A.S. and Saleviter, S. 2020. Development of biopolymer and conducting polymer-based optical sensors for heavy metal ion detection. Molecules 25(11): 2548.

Rao, G.P., Lu, C. and Su, F. 2007. Sorption of divalent metal ions from aqueous solution by carbon nanotubes: a review. Sep. Purif. Technol. 58(1): 224-231.

Reyes Bahena, J.L., Robledo Cabrera, A., López Valdivieso, A. and Herrera Urbina, R. 2002. Fluoride adsorption onto α-Al_2O_3 and its effect on the zeta potential at the alumina-aqueous electrolyte interface. Sep. Sci. Technol. 37(8): 1973-1987.

Roy, S., Bardhan, S., Chanda, D.K., Roy, J., Mondal, D. and Das, S. 2020. In situ-grown Cdot-wrapped boehmite nanoparticles for Cr (VI) sensing in wastewater and a theoretical probe for chromium-induced carcinogen detection. ACS Appl. Mater. Interfaces 12(39): 43833-43843.

Roy, S., Pal, K., Bardhan, S., Maity, S., Chanda, D.K., Ghosh, S., Karmakar, P. and Das, S. 2019. Gd (III)-doped boehmite nanoparticle: an emergent material for the fluorescent sensing of Cr (VI) in wastewater and live cells. Inorg. Chem. 58(13): 8369-8378.

Sabela, M., Balme, S., Bechelany, M., Janot, J.M. and Bisetty, K. 2017. A review of gold and silver nanoparticle-based colorimetric sensing Assays. Adv. Eng. Mater. 19(12): 1700270.

Sharma, M., Halder, A. and Vaish, R. 2020. Effect of Ce on piezo/photocatalytic effects of $Ba_{0.9}Ca_{0.1}Ce_xTi_{1-x}O_3$ ceramics for dye/pharmaceutical waste water treatment. Mater. Res. Bull. 122: 110647.

Sharma, V., Saini, A.K. and Mobin, S.M. 2016. Multicolour fluorescent carbon nanoparticle probes for live cell imaging and dual palladium and mercury sensors. J. Mater. Chem. B 4(36): 6154-6154.

Shtepliuk, I., Caffrey, N.M., Iakimov, T., Khranovskyy, V., Abrikosov, I.A. and Yakimova, R. 2017. On the interaction of toxic heavy metals (Cd, Hg, Pb) with graphene quantum dots and infinite graphene. Sci. Rep. 7(1): 1-17.

Simpson, A., Pandey, R.R., Chusuei, C.C., Ghosh, K., Patel, R. and Wanekaya, A.K. 2018. Fabrication characterization and potential applications of carbon nanoparticles in the detection of heavy metal ions in aqueous media. Carbon 127: 122-130.

Song, J., Ma, Q., Zhang, S., Liu, H., Guo, Y. and Feng, F. 2018. S, N-Co-doped carbon nanoparticles with high quantum yield for metal ion detection, IMP logic gates and bioimaging applications. New J. Chem. 42(24), 20180-20189.

Stafiej, A. and Pyrzynska, K. 2007. Adsorption of heavy metal ions with carbon nanotubes. Sep. Purif. Technol. 58(1): 49-52.

Tabish, T.A., Memon, F.A., Gomez, D.E., Horsell, D.W. and Zhang, S. 2018. A facile synthesis of porous graphene for efficient water and wastewater treatment. Sci. Rep. 8(1): 1-14.

Tang, Z.K., Zhang, L., Wang, N., Zhang, X.X., Wen, G.H., Li, G.D., Wang, J.N., Chan, C.T. and Sheng, P. 2001. Superconductivity in 4 angstrom single-walled carbon nanotubes. Science 292(5526): 2462-2465.

Tao, S., Zhu, S., Feng, T., Xia, C., Song, Y. and Yang, B. 2017. The polymeric characteristics and photoluminescence mechanism in polymer carbon dots: a review. Mater. Today Chem. 6: 13-25.

Terra, I.A., Mercante, L.A., Andre, R.S. and Correa, D.S. 2017. Fluorescent and colorimetric electrospun nanofibers for heavy-metal sensing. Biosensors 7(4): 61.

Thostenson, E.T., Ren, Z. and Chou, T.W. 2001. Advances in the science and technology of carbon nanotubes and their composites: a review. Compos. Sci. Technol. 61(13): 1899-1912.

Tian, P., Tang, L., Teng, K.S. and Lau, S.P. 2018. Graphene quantum dots from chemistry to applications. Mater. Today Chem. 10: 221-258.

Vilatela, J.J. and Marcilla, R. 2015. Tough electrodes: carbon nanotube fibers as the ultimate current collectors/active material for energy management devices. Chem. Mater. 27(20): 6901-6917.

Wang, C., Li, Z., Pan, Z. and Li, D. 2018. Development and characterization of a highly sensitive fluorometric transducer for ultra low aqueous ammonia nitrogen measurements in aquaculture. Comput. Electron. Agric. 150: 364-373.

Wang, S., Wu, Z., Chen, J., Ma, J., Ying, J., Cui, S., Yu, S., Hu, Y., Zhao, J. and Jia, Y. 2019. Lead-free sodium niobate nanowires with strong piezo-catalysis for dye wastewater degradation. Ceram. Int. 45(9): 11703-11708.

Wang, Z.X. and Ding, S.N. 2014. One-pot green synthesis of high quantum yield oxygen-doped, nitrogen-rich, photoluminescent polymer carbon nanoribbons as an effective fluorescent sensing platform for sensitive and selective detection of silver (I) and mercury (II) ions. Anal. Chem. 86(15): 7436-7445.

Wei, Z., Zhao, H., Zhang, J., Deng, L., Wu, S., He, J. and Dong, A. 2014. Poly(vinyl alcohol) electrospun nanofibrous membrane modified with spirolactam–rhodamine derivatives for visible detection and removal of metal ions. RSC Adv. 4(93): 51381-51388.

Wingenfelder, U., Hansen, C., Furrer, G. and Schulin, R. 2005. Removal of heavy metals from mine waters by natural zeolites. Environ. Sci. Technol. 39(12): 4606-4613.

Wolfbeis, O.S. 2015. An overview of nanoparticles commonly used in fluorescent bioimaging. Chem. Soc. Rev. 44(14): 4743-4768.

Wu, A., Jia, J. and Luan, S. 2011. Amphiphilic PMMA/PEI core–shell nanoparticles as polymeric adsorbents to remove heavy metal pollutants. Colloids Surf. A 384(1-3): 180-185.

Xu, C., Williams, R.M., Zipfel, W. and Webb, W.W. 1996. Multiphoton excitation cross-sections of molecular fluorophores. Bioimaging 4(3): 198-207.

Xu, J., Cao, Z., Zhang, Y., Yuan, Z., Lou, Z., Xu, X. and Wang, X. 2018. A review of functionalized carbon nanotubes and graphene for heavy metal adsorption from water: preparation, application, and mechanism. Chemosphere 195: 351-364.

Xu, X., Zhang, K., Zhao, L., Li, C., Bu, W., Shen, Y., Gu, Z., Chang, B., Zheng, C., Lin, C. and Sun, H. 2016. Aspirin-based carbon dots, a good biocompatibility of material applied for bioimaging and anti-inflammation. ACS Appl. Mater. Interfaces 8(48): 32706-32716.

Yadav, V.B., Gadi, R. and Kalra, S. 2019. Clay based nanocomposites for removal of heavy metals from water: a review. J. Environ. Manage. 232: 803-817.

Yang, S.T., Cao, L., Luo, P.G., Lu, F., Wang, X., Wang, H., Meziani, M.J., Liu, Y., Qi, G. and Sun, Y.P. 2009. Carbon dots for optical imaging in vivo. J. Am. Chem. Soc. 131(32): 11308-11309.

Yannopoulos, S., Yapijakis, C., Kaiafa-Saropoulou, A., Antoniou, G. and Angelakis, A.N. 2017. History of sanitation and hygiene technologies in the Hellenic world. J. Water Sanit. Hyg. Dev. 7(2): 163-180.

Yap, P.L., Kabiri, S., Tran, D.N. and Losic, D. 2018. Multifunctional binding chemistry on modified graphene composite for selective and highly efficient adsorption of mercury. ACS Appl. Mater. Interfaces 11(6): 6350-6362.

Yoon, H., Ko, S. and Jang, J. 2007. Nitrogen-doped magnetic carbon nanoparticles as catalyst supports for efficient recovery and recycling. Chem. Commun. (14): 1468-1470.

Zhao, G., Huang, X., Tang, Z., Huang, Q., Niu, F. and Wang, X. 2018. Polymer-based nanocomposites for heavy metal ions removal from aqueous solution: a review. Polym. Chem. 9(26): 3562-3582.

Zhao, X., Liu, R., Chi, Z., Teng, Y. and Qin, P. 2010. New insights into the behavior of bovine serum albumin adsorbed onto carbon nanotubes: comprehensive spectroscopic studies. J. Phys. Chem. B 114(16): 5625-5631.

Zheng, M., Xie, Z., Qu, D., Li, D., Du, P., Jing, X. and Sun, Z. 2013. On-off-on fluorescent carbon dot nanosensor for recognition of chromium (VI) and ascorbic acid based on the inner filter effect. ACS Appl. Mater. Interfaces 5(24): 13242-13247.

Zhu, S., Song, Y., Zhao, X., Shao, J., Zhang, J. and Yang, B. 2015. The photoluminescence mechanism in carbon dots (graphene quantum dots, carbon nanodots, and polymer dots): current state and future perspective. Nano Res. 8(2): 355-381.

Zhu, S., Yan, X., Sun, J., Zhao, X.E. and Wang, X. 2019. A novel and sensitive fluorescent assay for artemisinin with graphene quantum dots based on inner filter effect. Talanta 200: 163-168.

A Recent Review on Synthesis, Potential Environmental Applications and Socio-Economic Impact of Waste-Derived Carbon Nanotubes

Sakshi Kabra Malpani[1], Ajay Kumar[2], Rena Hada[3] and Deepti Goyal[4]*

[1] Department of Chemistry, Jyoti Nivas College Autonomous, Bengaluru, Karnataka, India
[2] Institute of Biomedical Sciences and Department of Mechanical and Electromechanical Engineering, National Sun Yat-sen University, Taiwan
[3] Department of Chemistry, Ganpat University, Kherva, Mahesana, Gujarat, India
[4] Department of Applied Chemistry, School of Vocational Studies & Applied Sciences, Gautam Buddha University, Greater Noida, UP, India.

1. Introduction

Nanomaterials owing to their remarkable physical, chemical, mechanical and thermal properties have been playing a substantial role in revolutionizing the industrial growth of this decade. Carbon, the third most abundant element on Earth's crust, exists in different allotropic forms and is extensively utilized in the synthesis of carbon-based nanomaterials (CBNs). Among many CBNs as shown in Figure 1, carbon nanotubes (CNTs) are admirable and promising aspirants in diverse applications, like fuel cells, batteries, supercapacitor electrodes, conducting films, photovoltaic cells, biosensors, drug discovery, cancer therapy, etc. (Banerjee et al. 2015, Huang et al. 2012, Hwang et al. 2017, Z. Li et al. 2010, Yang et al. 2011, Zhu et al. 2013) and more precisely in environment based applications viz. adsorption of heavy metal ions, removal of toxic dyes, wastewater remediation, ozonation of organic pollutants, etc. (Duan et al. 2020, Konicki and Pełech 2019, Restivo et al. 2021, Yin et al. 2020).

CNTs consist of cylindrical graphene sheets with both ends capped by a fullerene-like structure. Depending on the number of carbon layers, they can be divided into two main groups – single-walled CNTs (SWCNTs) and multi-walled CNTs (MWCNTs). Sometimes, double-walled carbon nanotubes (DWCNTs) have also been considered as a separate class of CNTs. CNTs are gaining much research

Corresponding author: deeptiskjain@yahoo.com

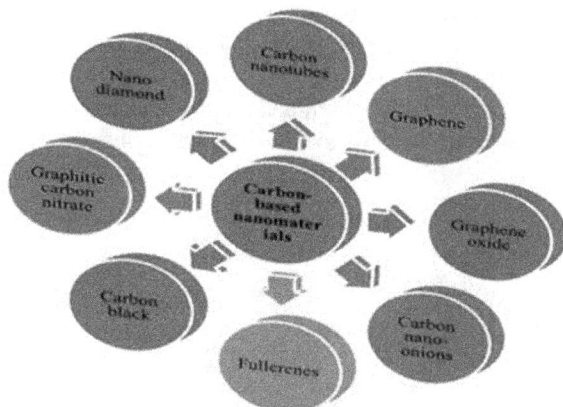

Figure 1: Various types of carbon-based nanomaterials.

attention because of their unique 1D-layered structure and fascinating properties, like high porosity, hollow structure, hydrophobicity, large specific surface area, light-weight, great flexibility and elasticity, good sorbent capacity, etc. Various conventional methods, like arc discharge, chemical vapor deposition (CVD) and laser ablation, have been constantly used in the synthesis of CNTs. However, such methods demand expensive feedstock resources, complicated instruments, an excessive amount of harsh chemicals and acidic waste generation, which confined massive use of CNTs on a commercial scale. Therefore, there is a pressing need to explore cost-efficient, environment benign green methods as sustainable alternatives in CNTs production. This research will set an example in facilitating circular economy by up-cycling of various wastes (industrial, agricultural, plastic, etc.) into value-added products and thus enhancing our socio-economic standard. It would also provide a complementary solution to some underlying problems, like waste utilization, clean manufacturing of CNTs, energy and resource conservation, reduced pollution, etc.

This chapter gives a deep insight into CNTs, their properties, conventional and green methods of synthesis. Potential applications of waste-derived CNTs in various environment facets are also discussed. It also emphasizes the role of CNTs in promoting social, economic and technological progress and proposes future perspectives in this regard. The goal of this chapter is to evaluate and document interdisciplinary research in waste up-cycling, facile, green synthesis of CNTs, their environment-based applications and socio-economic impact reported during this decade (2010-2020).

2. Conventional Methods for CNTs' Synthesis

Synthesis methods play an important role in the production of carbon nanotubes with desired properties. There are many techniques used to produce SWCNTs and MWCNTs, including arc discharge, laser vaporization and chemical vapor deposition techniques that mainly occur in the gaseous phase. Initially, arc discharge and laser ablation techniques were used to fabricate CNTs which require high temperature but these methods were then replaced by low-temperature CVD technique. Due to low temperature, the properties of CNTs such as diameter, length, density, purity and orientation can be controlled (Abbasi et al. 2014, Awasthi et al. 2011, Engels et al. 2011, Hata 2016, Ren et al. 2015, Sharma et al. 2015, Tang et al. 2013). The various conventional methods used for CNT synthesis are described in Figure 2.

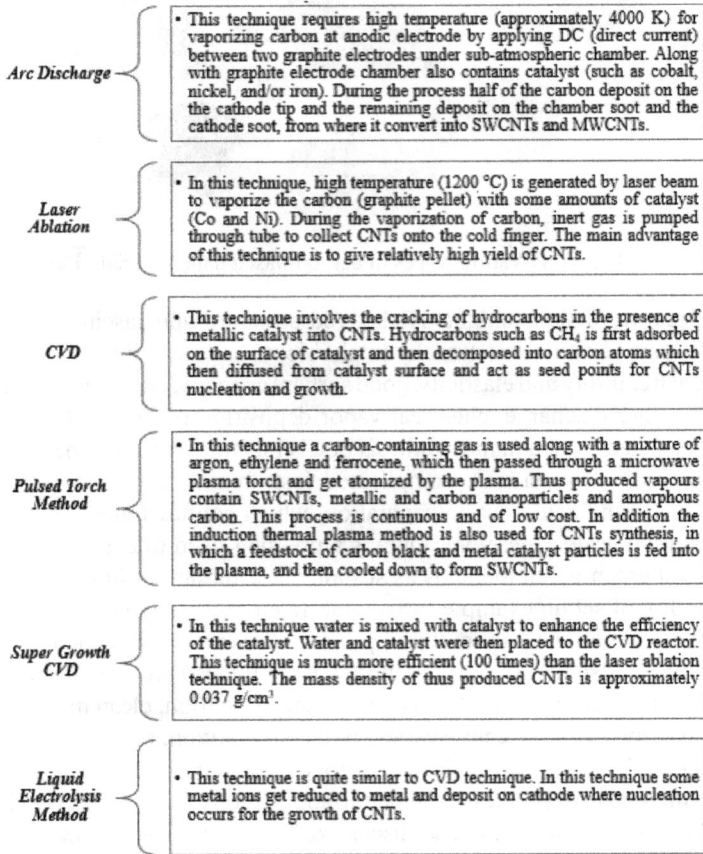

Arc Discharge — • This technique requires high temperature (approximately 4000 K) for vaporizing carbon at anodic electrode by applying DC (direct current) between two graphite electrodes under sub-atmospheric chamber. Along with graphite electrode chamber also contains catalyst (such as cobalt, nickel, and/or iron). During the process half of the carbon deposit on the the cathode tip and the remaining deposit on the chamber soot and the cathode soot, from where it convert into SWCNTs and MWCNTs.

Laser Ablation — • In this technique, high temperature (1200 °C) is generated by laser beam to vaporize the carbon (graphite pellet) with some amounts of catalyst (Co and Ni). During the vaporization of carbon, inert gas is pumped through tube to collect CNTs onto the cold finger. The main advantage of this technique is to give relatively high yield of CNTs.

CVD — • This technique involves the cracking of hydrocarbons in the presence of metallic catalyst into CNTs. Hydrocarbons such as CH_4 is first adsorbed on the surface of catalyst and then decomposed into carbon atoms which then diffused from catalyst surface and act as seed points for CNTs nucleation and growth.

Pulsed Torch Method — • In this technique a carbon-containing gas is used along with a mixture of argon, ethylene and ferrocene, which then passed through a microwave plasma torch and get atomized by the plasma. Thus produced vapours contain SWCNTs, metallic and carbon nanoparticles and amorphous carbon. This process is continuous and of low cost. In addition the induction thermal plasma method is also used for CNTs synthesis, in which a feedstock of carbon black and metal catalyst particles is fed into the plasma, and then cooled down to form SWCNTs.

Super Growth CVD — • In this technique water is mixed with catalyst to enhance the efficiency of the catalyst. Water and catalyst were then placed to the CVD reactor. This technique is much more efficient (100 times) than the laser ablation technique. The mass density of thus produced CNTs is approximately 0.037 g/cm³.

Liquid Electrolysis Method — • This technique is quite similar to CVD technique. In this technique some metal ions get reduced to metal and deposit on cathode where nucleation occurs for the growth of CNTs.

Figure 2: Different conventional methods for CNTs synthesis.

3. Green Synthesis of CNTs Using Different Waste Materials

To make the process environmentally friendly and reduce the cost of CNTs production, several green synthesis methods can be used. Low-valued waste materials, such as agricultural wastes, plastic wastes, renewable feedstock and industrial wastes, can act as carbon precursors for CNTs production (Kumar et al. 2016). In the following sections, synthesis methods are discussed thoroughly in terms of the types of wastes. Figure 3 presents an overview of different waste materials, which are being used to synthesize CNTs.

Wastes

- Agricultural wastes
  - Sugar cane bagasse
  - Gum wood
  - Rice byproducts
  - Saw dust
  - Fruit waste
- Plastic wastes
  - Polyethylene (PE)
  - Polypropylene (PP)
  - Poly vinyl alcohol (PVA)
  - Polyterthalete (PET)
- Renewable feedstocks
  - Biomass
  - Vegetable oil
  - Plant derivatives
  - Poultry products
- Industrial wastes
  - Waste tyres
  - Fly ash (FA)
  - Soot particles
  - Paper sludge

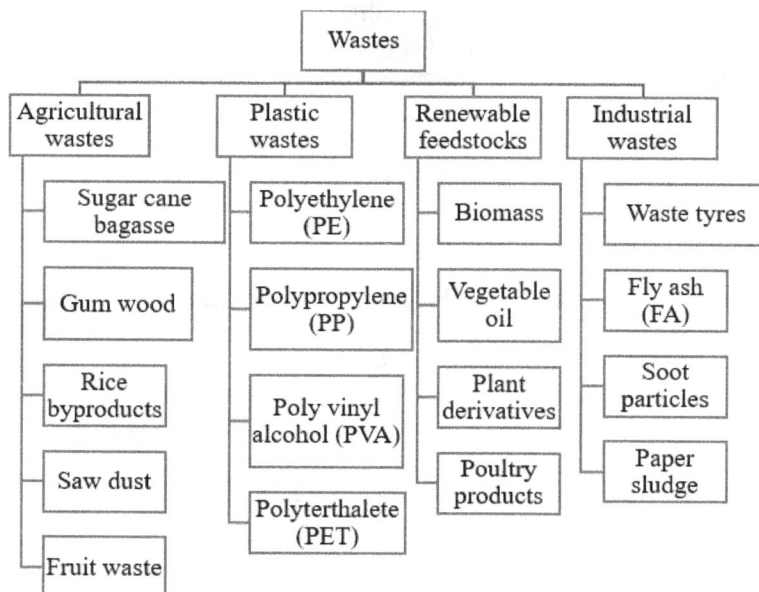

Figure 3: Waste materials for CNTs' synthesis.

3.1 Synthesis of CNTs from Agricultural Wastes

In recent years, agro wastes have been attracted much attention in the synthesis of carbonaceous nanomaterials, such as CNTs, graphene and carbon nanofibers. Agricultural wastes, such as rice by-products, sugar cane bagasse, sawdust, fruits and waste leaves, consisting of cellulose, hemicellulose and lignin (a rich source of carbon) have been widely studied for the synthesis of CNTs. A group of researchers has synthesized CNTs with 50 nm diameter using wood sawdust. The process was done at 750 °C under optimized reaction conditions (Bernd et al. 2017). The reaction was performed in a designed reactor in an environmentally friendly manner.

Besides using wood fiber, Wang et al. (2015) conducted the synthesis of graphenated CNTs starting from rice husk waste via microwave plasma irradiation method. They found that the obtained graphenated CNTs mainly composed of 2-6 layers of graphene standing on the sidewalls of CNTs with 10 μm length and 50-200 nm of diameter. A similar study was reported on microwave synthesis of CNTs using rice husk ash. The results emphasize the use of Te plasma-induced from microwaves, which is responsible for increasing the rate of decomposition of rice husk in the presence of the catalyst. Furthermore, the prepared CNTs were characterized by TEM, FESEM, TGA and Raman spectroscopy to ensure their size and shape (Wang et al. 2015).

In another study, MWCNTs were synthesized via catalytic conversion of sugarcane bagasse at high temperatures. The MWCNTs thus produced were 50 μm in length and 20-50 nm in diameter (Alves et al. 2011). In a similar study, bamboo charcoal was used to synthesize CNTs via the CVD method at very high temperatures (Zhu et al. 2012). The study revealed that the calcium silicate present in the bamboo charcoal may be responsible for the growth of MWCNTs. In recent research, CNTs have been synthesized by metal-loaded rice straw via a hydrothermal process. The study showed that the CNT was produced with a high yield having a high specific surface area and 22-66 nm diameter (Fathy 2017).

Shi et al. (2014) studied the microwave-induced synthesis of MWCNTs using Gumwood as a carbon source. The reaction was performed over SiC catalyst at microwave power 300 W and temperature 500°C for 30 minutes. The CNTs thus produced were of 50 nm diameter (Shi et al. 2014).

Another study on the synthesis of CNTs/CNFs from sawdust has been reported by Bernd et al. (2017) In this study the pyrolysis of sawdust has been done in a tubular reactor using zinc as a reducing agent and the ferrocene or Fe/Mo/MgO as a catalyst. The reaction was performed at 750 °C for 3 hours. Thus, the CNTs that were produced were of different shapes and sizes (Bernd et al. 2017).

In a recent investigation, a mixture of SWCNTs and MWCNTs has been synthesized via catalytic hydrothermal pyrolysis of rice straw over Fe-Ni/Al$_2$O$_3$ catalyst at 830 °C for 60 minutes. The results revealed that the CNTs were produced with a 3.8-47 nm diameter and relatively higher surface areas ranging from 10.1-188 m^2/g (Lotfy et al. 2018).

Recently bamboo-shaped MWCNTs were obtained from cotton stalks via catalytic hydrothermal pyrolysis. The authors performed the whole process over alumina-supported tri-metal oxide catalyst Fe-Ni-Cu/Al$_2$O$_3$ at 830 °C for 60 minutes. They got MWCNTs of different diameters ranging from 50-100 nm. The surface area was 110 m^2/g (Azzam et al. 2019).

In a most recent study, environmentally friendly synthesis of MWCNTs with 7 nm diameter via pyrolysis of chickpea peel at 400 °C in an aqueous medium was reported. Thus, prepared MWCNTs were showed fluorescence in the cytoplasm of 22RV1 human prostate carcinoma cell line without exerting any sign of cytotoxicity. The MWCNTs also exhibited remarkable cytocompatibility in human immortalized prostate epithelial RWPE1 cells (Singh et al. 2020).

3.2 Synthesis of CNT from Plastic Wastes

Plastic waste is considered the most substantial and challenging non-biodegradable waste material in the world. Globally, it is now included as the major component in solid waste with an annual production of about 150 million tons. To utilize this abundant waste, many countries have initiated safe disposal at the household level and its recycling. Some most common plastic materials used in everyday life are polypropylene (PP), polystyrene (PS), polyethylene (PE), polyvinyl alcohol (PVA), polyethylene terephthalate (PET) and nylon. Efforts are being made not only to constrain the generation of plastic wastes and associated spoiling but also to find ways of utilizing them sustainably. These polymeric wastes can be recycled and reused in different environmental applications, such as for the synthesis of CBNs (graphenes and CNTs) composites, polymers, construction materials, paper, etc. Although the synthesis of these materials is time and energy-consuming due to high carbon content, economic, eco-friendly and self-sustaining production of CNTs can be still achieved.

Numerous studies have been done for the synthesis of CNTs via low-cost, energy-saving and green processes. The green synthesis of CNTs from waste plastics is reported on different reactors, such as quartz tube, autoclave, crucible and muffle furnace, moving bed, fluidized bed, etc. (Bazargan and McKay 2012). In all the processes, the conversion of large molecules into small carbon precursors via pyrolysis is done (Zhuo and Levendis 2014). Amongst several plastic wastes, PP and PE have been widely used for the green synthesis of CNTs (Gong et al. 2012). Along with this other waste polyolefins such as PVA and PET were also successfully used for CNTs synthesis. A group of researchers has studied different conversion processes for waste plastics to CNTs and categorized them into two classes viz:

(i) One-pot conversion in which solid plastic wastes, such as PP, PE, PVA, PS, PVC, PET, etc., in presence of the catalyst gets converted into CNTs under *in-situ* conditions.

(ii) Stepwise conversion in which plastic wastes first decomposed in the initial step. Thus, obtained decomposed waste in its gaseous phase is then mixed with catalysts to produce CNTs.

The conversion of different plastic wastes at different reaction conditions (catalyst, reactor, % yield, etc.) as well as corresponding references are summarized in Table 1.

Table 1: Plastic wastes for the synthesis of CNTs

Type of Plastic Wastes	Catalyst	Reactor	CNTs Yield%	Reference
PE	Stainless steel meshes	Quartz tube	>10	(Zhuo et al. 2010)
PP	Activated carbon and Ni_2O_3	Quartz tube	50	(Gong et al. 2012)
LDPE and PP	Ni-Mo/Al_2O_3 catalyst	-	58	(Aboul-Enein et al. 2018)
PP and PE	Ni/Al-SBA-15 catalysts	-	74.1	(Yang et al. 2016)
PP	Ni/Mo/MgO catalyst	Silica crucible	>45.8	(Bajad et al. 2015)
PP	Stainless steel 316 tube	Stainless steel 316 tube	42	(Tripathi et al. 2017)
PP	Ni catalyst	-	20	(Mishra et al. 2012)
PP	Ni/Ca-Al catalyst	-	>10	(Wu et al. 2012)
PVA	FA	Quartz tube	NR	(Nath and Sahajwalla 2012)
PET	Catalyst-free	Arc discharge	39	(Joseph Berkmans et al. 2014)

3.3   Synthesis of CNTs from Renewable Feedstocks

In the recent scenario for minimization of environmental problems, there is an urgent requirement of sustainable resources for the synthesis of carbon-based nanomaterials (CBNs) (Titirici et al. 2015). Such renewable materials have been widely explored for the development of many advanced CBNs, such as nanospheres, nanofibers, carbon dots, graphenes and CNTs. Likewise, the use of several renewable precursors, such as raw biomasses, juices, fruit peels, bio-derived hydrocarbons and other bio-based materials, have been well reported for the fabrication of CBNs (Huang et al. 2014, Mewada et al. 2013). The synthesis of both SW and MWCNTs using renewable precursors as the carbon source has received the most attention. The renewable resource-derived CNTs exhibit similar properties to conventional CNTs and thus in this review, we have included the growth of CNTs from renewable feedstocks.

Different materials, such as plant oils, biodiesel and solid biomasses, like grass and leaves, plant-sourced materials and animal-sourced materials, such as chicken fat and feathers, have been used for CNTs production. The selection and purity of these renewable resource-based carbon precursors, selection of catalyst and optimization of experimental conditions are the main criteria for better fabrication of CNTs (Wu et al. 2013). Various renewable feedstocks used for the fabrication of CNTs are summarized in Table 2.

3.4   Synthesis of CNTs Using Industrial Waste

Industrial waste generation has increased enormously around the world in recent decades, and there are no signs of it slowing down. By 2050, worldwide municipal solid waste generation is expected to have increased by roughly 70% to 3.4 billion metric tonnes. To reduce it, a solid waste management system should be developed following 3R Technologies (Reduce, Reuse and Recycle) by a variety of techniques for resource recovery, waste minimization and proper disposal of solid waste. Many research groups all around the world are engaging in the reuse and recycling of industrial wastes (Hada et al. 2020, Malpani and Rani 2019). Carbon-containing waste can be a good feedstock for the generation of advanced CBNs like CNTs.

It is well known that automobiles are the fastest growing mode of transportation throughout the world; as scrap tires are not biodegradable and thus they are a serious source of environmental problems. It contains approximately 88% of carbon (Murillo et al. 2006). So, several researchers focused on how to make use of scrap rubber in the generation of power, diesel fuel, carbonaceous materials, etc. (Xu et al. 2020). Pyrolysis of waste tires is one of the important methods for waste tires disposal (Lewandowski et al. 2019). Yang et al. used scrap tires as a carbon source for the synthesis of short hollow thick-walled CNTs using the CVD method in the presence of cobalt catalyst at 400 °C in the N_2 atmosphere (Yang et al. 2012). Zhang and Williams utilized waste truck tires as a carbon source for the production of CNTs and H_2 gas with a Ni/Al_2O_3 catalyst. They used the pyrolysis-catalytic steam reforming process at a temperature range of 700-900 °C (Zhang and Williams 2016). Recently Mwafy also got success to convert pyrolyzed waste tires into MWCNT of 40 nm using CVD at 850 °C with ferrocene as a catalyst (Mwafy 2020).

Table 2: Fabrication of CNTs using different renewable feedstock materials

Renewable Feedstocks	Carbon Precursor	Reaction Conditions	CNTs Diameter	Reference
Biomass	Waste leaves	Dried, crumbled and preheated at 250 °C/1 hour in the air and washed with alcohol, calcined at 500 °C in air.	40-77 nm	(Qu et al. 2013)
	B. juncea plant	Dried, crumbled and mixed with 90% ethanol at 70 °C/120 minutes and filtered, mixed with 250 mL of HNO_3, heated at 400 °C/5 minutes.	80 nm	(Qu et al. 2012)
Vegetable oils	Camphor oil	Oil + ferrocene catalyst, heated above 800 °C.	0.63-1.14 nm	(TermehYousefi et al. 2014)
	Sesame oil	Oil + ferrocene catalyst, heated at 900 °C	30-60 nm	(Kumar, Singh and Tiwari 2016)
	Sunflower oil	Oil + ferrocene catalyst heated at 800-850 °C.	20-40 nm	(Yadav et al. 2011)
	Castor oil	Oil + ferrocene catalysts heated at 800 °C under an inert atmosphere.	15-40 nm	(Awasthi et al. 2011)
	Neem oil	Oil + ferrocene + Fe catalyst	15-20 nm	(Kumar, Tiwari, and Srivastava 2011)
	Palm oil	Oil + Fe/Co impregnated catalyst, heated at 750 °C	30 nm (with Fe), 90 nm (with Co)	(Suriani et al. 2013)
	Palm oil	Oil + ferrocene catalyst, heated at 750-950 °C	20-70 nm	(Robaiah et al. 2018)
Plant derivatives	Biodiesel oil of Jatropha curcas	Oil + Fe/Co/Mo catalysts well dispersed on a silica/alumina substrate	16-24 nm	(Karthikeyan and Mahalingam 2010)

(Contd.)

Table 2. (*Contd.*)

Renewable Feedstocks	Carbon Precursor	Reaction Conditions	CNTs Diameter	Reference
	Jatropha-derived biodiesel	Biodiesel + ferrocene and ferrocene-acetonitrile catalysts	20-50 nm ᵃCAN nanotubes 30-60 nm	(Kumar et al. 2013)
	Methyl ester of Helianthus annuus oil	Silicon + oil + ferrocene catalyst	10 and 30 nm	(Angulakshmi et al. 2013)
	Polyisocyanate crosslinked cellulose acetate	Polyisocyanate crosslinked cellulose acetate + NiCl₂ catalyst and fumed silica as a template heated at 750 °C	24-38 nm	(Dubrovina et al. 2014)
Poultry products	Chicken fat oil	Chicken fat oil + ferrocene catalyst heated at 750 °C	18 and 78 nm	(Suriani et al. 2013)
	Chicken fat	Chicken fat heated at 470 °C-900 °C	19.8-31.7 nm	(Gao et al. 2014)

ᵃCAN - Carbon-acrylonitrile

The combustion of coal in power stations generates a huge amount of FA as a by-product. The management of this huge amount of FA consequently becomes a global concern. FA is generally disposed of as landfill in the fulfilment of dams and lagoons. Research on recycling and reuse of FA as filler and catalyst/support has established an eco-friendly technology for value-added products and composites (Hada et al. 2020, Malpani and Rani 2019). FA was used as catalytic support for CNTs growth by CVD due to having ideal compositions (SiO_2, Al_2O_3 and Fe_2O_3). Salah et al. (2016) reported an economical method for the large-scale synthesis of CNTs from carbon-rich-FA using low-pressure CVD. C_2H_2 gas was taken as a precursor and carbon-rich-FA was used as a co-precursor of carbon along with a catalyst and this method produced CNTs of 20-40 nm with uniform length and diameter (Salah et al. 2016).

Paper sludge is also an economic and environmental problem for paper and board industries. Several researchers utilized paper sludge as adsorbent, fertilizers, building materials, etc. Paper mill sludge (PMS) was first utilized to prepare Fe, N and S co-doped CNTs by Zhang et. al. (2019) The CNT/nanoporous carbon nanocomposites were synthesized by pyrolysis of the mixture of PMS and melamine at 900 °C/4 hours in N_2 atmosphere (Zhang et al. 2019). Zhou et. al. recently published their study wherein the pyrolysis technique was adopted by Zhang et al. used to upgrade industrial paper sludge into biochar with N doped CNTs in the presence of melamine at 900 °C/2 hours in the N_2 atmosphere (Zhou et al. 2020).

Printed circuit board (PCB) waste was first utilized to convert PCB into CNTs by pyrolysis process at 600 °C/30 minutes in an inert atmosphere of N_2 by Quan et. al. (2010). In this process, pyrolysis oil obtained from PCB by pyrolysis was used as a carbon source and ferrocene as a catalyst for CNTs production.

Kowthaman et al. (2020) studied the synthesis of CNTs from engine soot particles by laser ablation vaporization method. A mixture of pretreated soot particles, ethanol and water was irradiated using Nd:YAG laser operating at 10 Hz of repetition rate and pulse duration of 10 ns for 90 minutes under constant stirring of 500 rpm (Kowthaman and Arul Mozhi Selvan 2020).

4. Environment Applications

Research related to the potential environmental applications of CNTs is presented in Table 3.

5. Socio-Economic Facets

This chapter also outlines socio-economic facets of waste-derived CNTs and their environment-based applications along with their technological advancements. It attempts to critically evaluate synergy between society, environment and economy, which leads to total sustainable development that is quite essential for a developing nation like India. Reduced demands on limited natural resources, up-cycling of wastes, minimal use of environmental sources for waste dumping, cost-cutting, reduced ecological pollution, clean and green manufacturing of CNTs are the important parameters that should be focused on while developing effective strategies for the well-being of society as well as economy. Designing green methods to synthesize

Table 3: Environmental applications of CNTs

Process	Materials	Application	Reference
Adsorption/remediation	CNTs	Wastewater treatment and water remediation	(Yin et al. 2020)
	CNTs	Heavy metal removal/environmental remediation	(Duan et al. 2020)
	CNT/calcium alginate composites	Copper removal	(Y. Li et al. 2010)
	Sulfur-coated magnetic CNT	Mercury(II) removal	(Fayazi 2020)
	MWCNT/TiO$_2$	Phenol and cyanide adsorption	(Kariim et al. 2020)
	SWCNT	Adsorption of hazardous synthetic organic chemicals	(Ghosh et al. 2019)
	CNT, graphene sheet, metal oxide	Adsorption of organic pollutants	(Awad et al. 2020)
	Magnetite SWCNT, MWCNT	Adsorption of mercury ions	(Alijani and Shariatinia 2018)
	MWCNT, SWCNT and [a]HCNT	Removal of ethylbenzene	(Bina et al. 2012)
	Functionalized CNT	Remediation of phenol	(Moradi et al. 2018)
	N-doped MWCNT	Nickel ions removal	(Balog et al. 2020)
	[b]Mod-MWCNT	Cationic dye removal	(Konicki and Pełech 2019)
	MWCNT/ZnO	Cr(VI) ions adsorption	(Murali et al. 2020)
	Magnetic MWCNT	Anionic dyes adsorption	(Zhao et al. 2020)
	[c]ZIF-8/hydroxylated MWCNT nanocomposite	Adsorption of phosphate ions from water using ZIF-8/hydroxylated MWCNT nanocomposite	(Wang et al. 2020)
	MWCNT	Sorbents for nuclear waste management	(Sengupta and Gupta 2017)
Catalysis	CNT catalyst	Ozonation of 4-nitrobenzaldehyde.	(Santos et al. 2020)
	Bi/CNT catalyst	Non-mercury catalytic acetylene hydrochlorination	(Lian et al. 2020)
	Zn-CNTs-Cu catalyst	*In-situ* generation of H$_2$O$_2$	(Fu et al. 2021)

	Cu-anchored CNT catalyst	Reduction of 4-nitrophenol	(Yang et al. 2020)
	Ball-milled MWCNT catalyst	Ozonation of organic pollutants	(Restivo et al. 2020)
	Activated MWCNT catalyst support	Liquid phase hydrogenation of unsaturated aldehydes	(Machado et al. 2010)
	CNT heterogeneous catalyst	Hydrogenation, dehydrogenation, CO/H_2 conversion, oxidation, ammonia synthesis and decomposition, CNT synthesis, dehalogenation	(Yan et al. 2015)
	CNT supported electrocatalysts	Fuel cells	(Luo et al. 2015)
Membrane filtration	MWCNT/Al_2O_3 membrane, SWCNT film, CNT membrane	Filtration of air samples: dHEPA and high performing air filters, ultra-low penetration air filters,	(Rashid and Ralph 2017)
	SWCNT/MWCNT hybrid filter, MWCNT filter	Filtration of bacteria and virus particles like *E. coli*, MS2 bacteriophage virus particles, and *Salmonella typhimurium*	(Rashid and Ralph 2017)
	Functionalized CNT buckypaper, CNT paper	Filtration of organic compounds using pervaporation techniques such as humic acid, ethylene glycol, organophosphate bioremediation	(Rashid and Ralph 2017)
	ZnO/MWCNTs nanocomposites, eGO-CNT hybrid, CNT/Polyacrylonitrile	Membrane filtration in industrial wastewater treatment and drinking water purification, micropollutant removal	(Rashid and Ralph 2017)
Environmental sensing	CNT, MWCNT	Chemical and physical property sensor (gas, molecule and chemical element, humidity, strain, pressure, etc.) biosensor for drugs	(Bezzon et al. 2019)

[a]HCNT-hybrid CNT, [b]Mod-Modified, [c]ZIF-zeolitic imidazolate framework, [d]HEPA-High efficiency particulate air, [e]GO-reduced graphene oxide

CNTs from different types of wastes could be an important part of the solid waste management (SWM) plan of low and middle-income countries where a major part of such wastes are dumped openly. Effects of socio-economic factors, public health, environment laws and cost-effectiveness are the key points in designing a long term integrated approach of SWM plan, which not only works in the area of the betterment of society but also improves its financial status by job creation, reducing overall cost, etc. (Kamran et al. 2015). This improved SWM plan and its effective implementation will increase social awareness, reduce expenditure on waste disposal and clean and economical production of important CNTs, improve quality living, financial benefits, waste utilization and can also increase job opportunities for common people. Government should impose strict laws and regulations in the direction of building circular economy models that could meet our social, economic, sustainable, environmental, technical and resource-oriented goals.

6. Conclusions

This chapter highlighted recent progress in the up-cycling of various types of wastes in the green synthesis of CNTs and their environment-based applications. It also casts light on socio-economic facets of CNTs production through sustainable recycling of wastes. CNTs with many remarkable properties are high in demand in the current scenario of the industrial revolution, and their synthesis from wastes not only complies with all the principles of green chemistry but also alleviates the problem of disposal of these wastes to some extent. It will also help in the improvement of social and economic standards, including the creation of new job avenues. It also aims at promoting, encouraging future possibilities, research and study in this topic.

Conflict of Interest

The authors declare no conflict of interests.

Reference

Abbasi, E., Aval, S., Akbarzadeh, A., Milani, M., Nasrabadi, H., Joo, S., , S.W., Hanifehpour, Y., Nejati-Koshki, K., and Pashaei-Asl, R. 2014. Dendrimers: synthesis, applications, and properties. Nanoscale Res. Lett. 9: 1–10.

Aboul-Enein, A.A., Awadallah, A.E., Abdel-Rahman, A.A.-H. and Haggar, A.M. 2018. Synthesis of multi-walled carbon nanotubes via pyrolysis of plastic waste using a two-stage process. Fullerenes, Nanotub. Carbon Nanostruc. 26: 443–450.

Alijani, H. and Shariatinia, Z. 2018. Synthesis of high growth rate SWCNTs and their magnetite cobalt sulfide nanohybrid as super-adsorbent for mercury removal. Chem. Eng. Res. Des. 129: 132–149.

Alves, J.O., Zhuo, C., Levendis, Y. A. and Tenório, J.A.S. 2011. Catalytic conversion of wastes from the bioethanol production into carbon nanomaterials. Appl. Catal. B Environ. 106: 433–444.

Angulakshmi, V.S., Rajasekar, K., Sathiskumar C. and Karthikeyan, S. 2013. Growth of vertically aligned carbon nanotubes on a silicon substrate by a spray pyrolysis method. New Carbon Mater. 28: 558.

Awad, A.M., Jalab, R., Benamor, A., Nasser, M.S., Ba-Abbad, M.M., El-Naas, M. and Mohammad, A.W. 2020. Adsorption of organic pollutants by nanomaterial-based adsorbents: an overview. J. Mol. Liq. 301: 112335.

Awasthi, K., Kumar, R., Raghubanshi, H., Awasthi, S., Pandey, R., Singh, D., Yadav, T.P. and Srivastava, O.N. 2011. Synthesis of nano-carbon (nanotubes, nanofibres, graphene) materials. Bull. Mater. Sci. 34: 607–614.

Azzam, E.M.S., Fathy, N.A., El-Khouly, S.M. and Sami, R.M. 2019. Enhancement of the photocatalytic degradation of methylene blue dye using fabricated CNTs/TiO$_2$/AgNPs/ Surfactant nanocomposites. J. Water Process Eng. 28: 311–321.

Bajad, G.S., Tiwari, S.K. and Vijayakumar, R.P. 2015. Synthesis and characterization of CNTs using polypropylene waste as precursor. Mater. Sci. Eng. B 194: 58–77.

Balog, R., Manilo, M., Vanyorek, L., Csoma, Z. and Barany, S. 2020. Comparative study of Ni(II) adsorption by pristine and oxidized multi-walled N-doped carbon nanotubes. RSC Adv. 10: 3184–3191.

Banerjee, S.S., Todkar, K.J., Khutale, G.V., Chate, G.P., Biradar, A.V., Gawande, M.B., Zboril, R. and Khandare, J.J. 2015. Calcium phosphate nanocapsule crowned multiwalled carbon nanotubes for pH triggered intracellular anticancer drug release. J. Mater. Chem. B 3: 3931–3939.

Bazargan, A. and McKay, G. 2012. A review – synthesis of carbon nanotubes from plastic wastes. Chem. Eng. J. 195–196.

Bernd, M.G.S., Bragança, S.R., Heck, N. and Filho, L.C.P.D.S. 2017. Synthesis of carbon nanostructures by the pyrolysis of wood sawdust in a tubular reactor. J. Mater. Res. Technol. 6: 171–177.

Bezzon, V.D.N., Montanheiro, T.L.A., De Menezes, B.R.C., Ribas, R.G., Righetti, V.A.N., Rodrigues, K.F., and Thim, G.P. 2019. Carbon nanostructure-based sensors: a brief review on recent advances. Adv. Mater. Sci. Eng. 2019: 1–21.

Bina, B., Pourzamani, H., Rashidi, A. and Amin, M.M. 2012. Ethylbenzene removal by carbon nanotubes from aqueous solution. J. Environ. Public Health 2012: 1–8.

Duan, C., Ma, T., Wang, J. and Zhou, Y. 2020. Removal of heavy metals from aqueous solution using carbon-based adsorbents: a review. J. Water Process. Eng. 37: 101339.

Dubrovina, L., Naboka, O., Ogenko, V., Gatenholm, P. and Enoksson, P. 2014. One-pot synthesis of carbon nanotubes from renewable resource: cellulose acetate. J. Mater. Sci. 49: 1144–1149.

Engels, V., Geng, J., Jones, G.M., Elliott, J.A., Wheatley, A.E.H. and Boss, S.R. 2011. Cobalt catalyzed carbon nanotube growth on graphitic paper supports. Curr. Nanosci. 7: 315–322.

Fathy, N.A. 2017. Carbon nanotubes synthesis using carbonization of pretreated rice straw through chemical vapor deposition of camphor. RSC Adv. 7: 28535–28541.

Fayazi, M. 2020. Removal of mercury(II) from wastewater using a new and effective composite: sulfur-coated magnetic carbon nanotubes. Environ. Sci. Pollut. Res. 27: 12270–12279.

Fu, T., Gong, X., Guo, J., Yang, Z. and Liu, Y. 2021. Zn-CNTs-Cu catalytic in-situ generation of H$_2$O$_2$ for efficient catalytic wet peroxide oxidation of high-concentration 4-chlorophenol. J. Hazard. Mater. 401: 123392.

Gao, L., Hu, H., Sui, X., Chen, C. and Chen, Q. 2014. One for two: conversion of waste chicken feathers to carbon microspheres and (NH$_4$)HCO$_3$. Environ. Sci. Technol. 48: 6500–6507.

Ghosh, S., Ojha, P.K. and Roy, K. 2019. Exploring QSPR modeling for adsorption of hazardous synthetic organic chemicals (SOCs) by SWCNTs. Chemosphere 228: 545–555.

Gong, J., Liu, J., Wan, D., Chen, X., Wen, X., Mijowska, E., Jiang, Z., Wang, Y. and Tang, T. 2012. Catalytic carbonization of polypropylene by the combined catalysis of activated carbon with Ni_2O_3 into carbon nanotubes and its mechanism. Appl. Catal. A Gen. 449: 112–120.

Hada, R., Goyal, D., Yadav, V. Singh, Siddiqui, N. and Rani, A. 2020. Synthesis of NiO nanoparticles loaded fly ash catalyst via microwave assisted solution combustion method and application in hydrogen peroxide decomposition. Materials Today: Proceed. 28: 119–123.

Hata, K. 2016. A super-growth method for single-walled carbon nanotube synthesis. Synth. English Ed. 9: 167–179.

Huang, H., Xu, Y., Tang, C.J., Chen, J.R., Wang, A.J. and Feng, J.J. 2014. Facile and green synthesis of photoluminescent carbon nanoparticles for cellular imaging. New J. Chem. 38: 784–789.

Huang, Z.D., Zhang, B., Oh, S.W., Bin Zheng, Q., Lin, X.Y., Yousefi, N. and Kim, J.K. 2012. Self-assembled reduced graphene oxide/carbon nanotube thin films as electrodes for supercapacitors. J. Mater. Chem. 22: 3591–3599.

Hwang, Y., Park, S.-H. and Lee, J. 2017. Applications of functionalized carbon nanotubes for the therapy and diagnosis of cancer. Polymers (Basel) 9: 13.

Joseph Berkmans, A., Jagannatham, M., Priyanka, S. and Haridoss, P. 2014. Synthesis of branched, nano channeled, ultrafine and nano carbon tubes from PET wastes using the arc discharge method. Waste Manag. 34: 2139–2145.

Kamran, A., Chaudhry, M.N. and Batool, S.A. 2015. Effects of socio-economic status and seasonal variation on municipal solid waste composition: a baseline study for future planning and development. Environ. Sci. Eur. 27: 16.

Kariim, I., Abdulkareem, A.S., Tijani, J.O. and Abubakre, O.K. 2020. Development of MWCNTs/TiO_2 nanoadsorbent for simultaneous removal of phenol and cyanide from refinery wastewater. Sci. African 10: e00593.

Karthikeyan, S. and Mahalingam, P. 2010. Synthesis and characterization of multi-walled carbon nanotubes from biodiesel oil: green nanotechnology route. Int. J. Green Nanotechnol. Phys. Chem. 2: 39–46.

Konicki, W. and Pełech, I. 2019. Removing cationic dye from aqueous solutions using As-grown and modified multi-walled carbon nanotubes. Polish J. Environ. Stud. 28: 717–727.

Kowthaman, C.N. and Arul Mozhi Selvan, V. 2020. Synthesis and characterization of carbon nanotubes from engine soot and its application as an additive in schizochytrium biodiesel fuelled DICI engine. Energy Reports 6: 2126–2139.

Kumar, R., Singh, R.K. and Singh, D.P. 2016. Natural and waste hydrocarbon precursors for the synthesis of carbon based nanomaterials: graphene and CNTs. Renew. Sustain. Energy Rev. 58: 976–1006.

Kumar, R., Singh, R.K. and Tiwari, R.S. 2016. Growth analysis and high-yield synthesis of aligned-stacked branched nitrogen-doped carbon nanotubes using sesame oil as a natural botanical hydrocarbon precursor. Mater. Des. 94.

Kumar, R., Tiwari, R.S. and Srivastava, O.N. 2011. Scalable synthesis of aligned carbon nanotubes bundles using green natural precursor: neem oil. Nanoscale Res. Lett. 6: 92.

Kumar, R., Yadav, R.M., Awasthi, K., Shripathi, T., Sinha, A.S.K., Tiwari, R.S. and Srivastava, O.N. 2013. Synthesis of carbon and carbon–nitrogen nanotubes using green precursor: jatropha-derived biodiesel, J.Exp. Nanosci. 8: 606-620.

Lewandowski, W.M., Januszewicz, K. and Kosakowski, W. 2019. Efficiency and proportions of waste tyre pyrolysis products depending on the reactor type—a review. J. Anal. Appl. Pyrolysis 140: 25–53.

Li, Y., Liu, F., Xia, B., Du, Q., Zhang, P., Wang, D., Wang, Z. and Xia, Y. 2010. Removal of copper from aqueous solution by carbon nanotube/calcium alginate composites. J. Hazard. Mater. 177: 876–880.

Li, Z., Saini, V., Dervishi, E., Kunets, V.P., Zhang, J., Xu, Y., Biris, A.R., Salamo, G.T. and Biris, A.S. 2010. Polymer functionalized N-type single wall carbon nanotube photovoltaic devices. Appl. Phys. Lett. 96: 033110.

Lian, L., Wang, L., Yan, H., Ali, S., Wang, J., Zhao, L., Yang, C., Wu, R. and Ma, L. 2020. Non-mercury catalytic acetylene hydrochlorination over Bi/CNTs catalysts for vinyl chloride monomer production. J. Mater. Res. Technol. 9: 14961–14968.

Lotfy, V.F., Fathy, N.A. and Basta, A.H. 2018. Novel approach for synthesizing different shapes of carbon nanotubes from rice straw residue. J. Environ. Chem. Eng. 6: 6263–6274.

Luo, C., Xie, H., Wang, Q., Luo, G. and Liu, C. 2015. A review of the application and performance of carbon nanotubes in fuel cells. J. Nanomater. 2015.

Machado, B.F., Gomes, H.T., Serp, P., Kalck, P. and Faria, J.L. 2010. Liquid-phase hydrogenation of unsaturated aldehydes: enhancing selectivity of multiwalled carbon nanotube-supported catalysts by thermal activation. ChemCatChem 2: 190–197.

Malpani, S.K. and Rani, A. 2019. A greener route for synthesis of fly ash supported heterogeneous acid catalyst. Mater. Today Proc. 9: 551–559.

Mewada, A., Pandey, S., Shinde, S., Mishra, N., Oza, G., Thakur, M. et al. 2013. Green synthesis of biocompatible carbon dots using aqueous extract of trapa bispinosa peel. Mater. Sci. Eng. C 33: 2914–2917.

Mishra, N., Das, G., Ansaldo, A., Genovese, A., Malerba, M., Povia, M., Ricci, D., Fabrizio, E.D., Zitti, E.D., Sharon, M. and Sharon, M. 2012. Pyrolysis of waste polypropylene for the synthesis of carbon nanotubes. J. Anal. Appl. Pyrolysis 94: 91–98.

Moradi, F., Darvish Ganji, M. and Sarrafi, Y. 2018. Remediation of phenol-contaminated water by pristine and functionalized SWCNTs: Ab Initio van Der Waals DFT investigation. Diam. Relat. Mater. 82: 7–18.

Murali, A., Sarswat, P.K. and Free, M.L. 2020. Adsorption-coupled reduction mechanism in ZnO-functionalized MWCNTs nanocomposite for Cr (VI) removal and improved anti-photocorrosion for photocatalytic reduction. J. Alloys Compd. 843: 155835.

Murillo, R., Aylón, E., Navarro, M.V., Callén, M.S., Aranda, A. and Mastral, A.M. 2006. The application of thermal processes to valorise waste tyre. Fuel Process. Tech. 87: 143–147 Elsevier.

Mwafy, E.A. 2020. Eco-friendly approach for the synthesis of MWCNTs from waste tires via chemical vapor deposition. Environ. Nanotech., Monit. Manag. 14: 100342.

Nath, D.C.D. and Sahajwalla, V. 2012. Analysis of carbon nanotubes produced by pyrolysis of composite film of poly (vinyl alcohol) and modified fly ash. Mater. Sci. Appl. 03.

Qu, J., Cong, Q., Luo, C. and Yuan, X. 2013. Adsorption and photocatalytic degradation of bisphenol a by low-cost carbon nanotubes synthesized using fallen leaves of poplar. RSC Adv. 3: 961–965.

Qu, J., Luo, C., Cong, Q. and Yuan, X. 2012. Carbon nanotubes and Cu–Zn nanoparticles synthesis using hyperaccumulator plants. Environ. Chem. Lett. 10: 153–158.

Quan, C., Li, A. and Gao, N. 2010. Synthesis of carbon nanotubes and porous carbons from printed circuit board waste pyrolysis oil. J. Hazard. Mater. 179: 911–917.

Rashid, M.H.O. and Ralph, S.F. 2017. Carbon nanotube membranes: synthesis, properties, and future filtration applications. Nanomaterials 7: 99.

Ren, J., Li, F.-F., Lau, J., González-Urbina, L. and Licht, S. 2015. One-pot synthesis of carbon nanofibers from CO_2. Nano Lett. 15: 6142–6148.

Restivo, J., Orge, C.A., Guedes Gorito dos Santos, A.S., Gonçalves Pinto Soares, O.S. and Ribeiro Pereira, M.F. 2020. Influence of preparation methods on the activity of macro-structured ball-milled MWCNT catalysts in the ozonation of organic pollutants. J. Environ. Chem. Eng. 9: 104578.

Robaiah, M., Rusop, M., Abdullah, S., Khusaimi, Z., Azhan, H., Fadzlinatul, M.Y., Salifairus, M.J. and Asli N.A. 2018. Synthesis of carbon nanotubes from palm oil on stacking and non-stacking substrate by thermal-CVD method. *In:* AIP Conference Proceedings. 1963: 020027.

Salah, N., Al-Ghamdi, A.A., Memic, A., Habib, S.S. and Khan, Z.H. 2016. Formation of carbon nanotubes from carbon-rich fly ash: growth parameters and mechanism. Mater. Manuf. Process. 31: 146–156.

Santos, A.S.G.G., Orge, C.A., Soares, O.S.G.P. and Pereira, M.F.R. 2020. 4-nitrobenzaldehyde removal by catalytic ozonation in the presence of CNT. J. Water Process. Eng. 38: 101573.

Sengupta, A. and Gupta, N.K. 2017. MWCNTs based sorbents for nuclear waste management: a review. J. Environ. Chem. Eng. 5: 5099–5114.

Sharma, R., Sharma, A.K., Sharma, V. and Harkin-Jones, E. 2015. Synthesis of carbon nanotubes by arc-discharge and chemical vapor deposition method with analysis of its morphology, dispersion and functionalization characteristics. Cogent Eng. 2: 1094017.

Shi, K., Yan, J., Lester, E. and Wu, T. 2014. Catalyst-free synthesis of multiwalled carbon nanotubes via microwave-induced processing of biomass. Ind. Eng. Chem. Res. 53: 15012–15019.

Singh, V., Chatterjee, S., Palecha, M., Sen, P., Ateeq, B. and Verma, V. 2020. Chickpea peel waste as sustainable precursor for synthesis of fluorescent carbon nanotubes for bioimaging application. Carbon Lett. 31: 117–123.

Suriani, A.B., Asli, N.A., Salina, M., Mamat, M.H., Aziz, A.A., Falina, A.N., Maryam, M., Shamsudin, M.S., Nor, R.M. and Abdullah, S. 2013. Effect of iron and cobalt catalysts on the growth of carbon nanotubes from palm oil precursor. *In:* IOP Conference Series: Materials Science and Engineering. Vol. 46: 12014. IOP Publishing.

Suriani, A.B., Dalila, A.R., Mohamed, A., Mamat, M.H., Salina, M., Rosmi, M.S., Rosly, J., Nor, R.M. and Rusop, M. 2013. Vertically aligned carbon nanotubes synthesized from waste chicken fat. Mater. Lett. 101: 61–64.

Tang, J., Fan, G., Li, Z., Li, X., Xu, R., Li, Y., Zhang, D., Moon, W.J., Kaloshkin, S.D. and Churyukanova, M. 2013. Synthesis of carbon nanotube/aluminium composite powders by polymer pyrolysis chemical vapor deposition. Carbon N. Y. 55: 202–208.

TermehYousefi, A., Bagheri, S., Shinji, K., Rouhi, J., Rusop Mahmood, M. and Ikeda, S. 2014. Fast synthesis of multilayer carbon nanotubes from camphor oil as an energy storage material. Biomed Res. Int. 2014: 691537.

Titirici, M.M., White, R.J., Brun, N., Budarin, V.L., Su, D.S., Del Monte, F., Clark, J.H. and MacLachlan, M.J. 2015. Sustainable carbon materials. Chem. Soc. Rev. 44: 250–290.

Tripathi, P., Durbach, S. and Coville, N. 2017. Synthesis of multi-walled carbon nanotubes from plastic waste using a stainless-steel CVD reactor as catalyst. Nanomaterials 7: 284.

Wang, Y., Zhao, W., Qi, Z., Zhang, L., Zhang, Y., Huang, H., and Peng, Y. 2020. Designing ZIF-8/Hydroxylated MWCNT nanocomposites for phosphate adsorption from water: capability and mechanism. Chem. Eng. J. 394: 124992.

Wang, Z., Ogata, H., Morimoto, S., Ortiz-Medina, J., Fujishige, M., Takeuchi, K., Muramatsu, H., Hayashi, T., Terrones, M., Hashimoto, Y. and Endo, M. 2015. Nanocarbons from rice husk by microwave plasma irradiation: from graphene and carbon nanotubes to graphenated carbon nanotube hybrids. Carbon N. Y. 94: 479–484.

Wu, C., Wang, Z., Wang, L., Williams, P.T. and Huang, J. 2012. Sustainable processing of waste plastics to produce high yield hydrogen-rich synthesis gas and high quality carbon nanotubes. RSC Adv. 2: 4045–4047.

Wu, Z.L., Zhang, P., Gao, M.X., Liu, C.F., Wang, W., Leng, F. and Huang, C.Z. 2013. One-pot hydrothermal synthesis of highly luminescent nitrogen-doped amphoteric carbon dots for bioimaging from Bombyx Mori silk-natural proteins. J. Mater. Chem. B 1: 2868–2873.

Xu, Junqing, Yu, J., Xu, Jianglin, Sun, C., He, W., Huang, J. and Li, G. 2020. High-value utilization of waste tires: a review with focus on modified carbon black from pyrolysis. Sci. Total Environ. 742: 140235.

Yadav, R.K., Awasthi, R., Tiwari, K. and Shrivastav, R.M. 2011. Effect of nitrogen variation on the synthesis of vertically aligned bamboo-shaped C–N nanotubes using sunflower oil. Int. J. Nanosci. 10: 809–813.

Yan, Y., Miao, J., Yang, Z., Xiao, F.X., Bin Yang, H., Liu, B. and Yang, Y. 2015. Carbon nanotube catalysts: recent advances in synthesis, characterization and applications. Chem. Soc. Rev. 44: 3295–3346.

Yang, R.-X., Chuang, K.-H. and Wey. M.-Y. 2016. Carbon nanotube and hydrogen production from waste plastic gasification over Ni/Al–SBA-15 catalysts: effect of aluminum content. RSC Adv. 6: 40731–40740.

Yang, S.B., Kong, B.S., Jung, D.H., Baek, Y.K., Han, C.S., Oh, S.K. and Jung, H.T. 2011. Recent advances in hybrids of carbon nanotube network films and nanomaterials for their potential applications as transparent conducting films. Nanoscale 3: 1361–1373.

Yang, T., Tang, Y., Liu, L., Gao, Y. and Zhang, Y. 2020. Cu-anchored CNTs for effectively catalytic reduction of 4-nitrophenol. Chem. Phys. 533: 110738.

Yang, W., Sun, W.J., Chu, W., Jiang, C.F. and Wen, J. 2012. Synthesis of carbon nanotubes using scrap tyre rubber as carbon source. Chinese Chem. Lett. 23: 363–366.

Yin, Z., Cui, C., Chen, H., Duoni, Yu, X. and Qian, W. 2020. The application of carbon nanotube/graphene-based nanomaterials in wastewater treatment. Small 16: 1902301.

Zhang, Q., Bai, Z., Du, F. and Dai, L. 2019. Carbon nanotube energy applications. pp. 695–728. *In:* Nanotube Superfiber Materials: Science, Manufacturing, Commercialization. Elsevier.

Zhang, Y. and Williams, P.T. 2016. Carbon nanotubes and hydrogen production from the pyrolysis catalysis or catalytic-steam reforming of waste tyres. J. Anal. Appl. Pyrolysis 122: 490–501.

Zhao, S., Zhan, Y., Wan, X., He, S., Yang, X., Hu, J. and Zhang, G. 2020. Selective and efficient adsorption of anionic dyes by core/shell magnetic MWCNTs nano-hybrid constructed through facial polydopamine tailored graft polymerization: insight of adsorption mechanism, kinetic, isotherm and thermodynamic study. J. Mol. Liq. 319: 114289.

Zhou, S., Zhang, B., Liao, Z., Zhou, L. and Yuan, Y. 2020. Autochthonous N-doped carbon nanotube/activated carbon composites derived from industrial paper sludge for chromate (VI) reduction in microbial fuel cells. Sci. Total Environ. 712: 136513.

Zhu, J., Jia, J., Kwong, F.L., Ng, D.H.L. and Tjong, S.C. 2012. Synthesis of multiwalled carbon nanotubes from bamboo charcoal and the roles of minerals on their growth. Biomass and Bioenergy 36: 12–19.

Zhu, L., Deng, C., Chen, P., You, X.D., Su, H.B., Yuan, Y.H. et al. 2013. Glucose oxidase biosensors based on carbon nanotube non-woven fabrics. Xinxing Tan Cailiao/New Carbon Mater. 28: 342–348.

Zhuo, C., Hall, B., Richter, H. and Levendis, Y. 2010. Synthesis of carbon nanotubes by sequential pyrolysis and combustion of polyethylene. Carbon N. Y. 48: 4024–4034.

Zhuo, C. and Levendis, Y.A. 2014. Upcycling waste plastics into carbon nanomaterials: a review. J. Appl. Polym. Sci. 131: 39931.

Role of Carbon-Based Nanomaterials With Its Application for Wastewater Treatment

Parwathi Pillai and Swapnil Dharaskar*

Department of Chemical Engineering, School of Technology,
Pandit Deendayal Energy University, Gandhinagar - 382355, India

1. Introduction

The world population has increased environmental pollution because of rapid industrialization and urbanization. Developing countries and underdeveloped countries have a pollution threat. These have become significant issues all over the world. The major problem of the human race and aquatic life is water pollution. On the Earth's surface, 70% is covered with water, but only 3% of water can be consumed by a human being (Fuerhacker et al. 2012). Worldwide waterborne diseases are the primary cause of safe drinking water. Around 1.1 billion people are affected by water shortage (statistical data were done by the World Health Organization study). The organic and inorganic contaminants are the pollutants that cause a significant problem for producing freshwater (Sheet et al. 2014). The wastewater from industries is introduced in rivers, streams, and agriculture, which eventually contributes to overall pollution. Inorganic pollutants are nutrients and sediments coming from the industry of heavy metals. Maximum of inorganic contamination in the environment is because of mining industries, agricultural, and industrial activities. Organic pollutants are caused by pesticides, pharmaceuticals, dyes, and other industrial discharges (Ojekunle et al. 2016). Organic pollutants also come from the discharge of sewage, surface water runoff, and agricultural activities. These organic and inorganic pollutants can cause severe problems for human beings and animals. In human beings, the problem mainly occurs in heart failure, skin problems, lung infection, etc. To prevent these, poisonous contaminant removal is necessary from wastewater (Kumar and Chawla 2014). The maximum concentration level of heavy metals and their effects are illustrated in Table 1.

Corresponding author: swapnil.dharaskar@sot.pdpu.ac.in, swapnildharaskar11@gmail.com

Table 1: Heavy metals concentrations with health effects
(Yu et al. 2018, Ihsanullah et al. 2016)

Heavy Metals	Limit (mg/L)	Effects in Health
Arsenic (As)	10	Damage on the skin or respiratory problem
Cadmium (Cd)	5	Kidney damage
Chromium (Cr)	100	Allergic dermatitis
Copper (Cu)	1300	Kidney and liver damage
Lead (Pb)	15	Kidney problem and blood pressure
Mercury (Hg)	2	Kidney damage
Uranium (U)	30	Cancer risk

Nanotechnology has gained immense attention in wastewater treatment. Nanomaterials manipulation at the atomic level is involved in nano-related techniques. The range of nanomaterials is 1-100 nm, and its surface activity, specific affinity, and surface area properties are the reason for the attention. In addition to the large pore size, it also has unique features like high reactivity and catalytic potential (Qu et al. 2013). It is widely used in water and wastewater treatment. With the evolution of nano-related techniques, carbon-related techniques and their byproducts have occurred. Carbon materials have embarked on their importance in the field of water and wastewater treatment. In this current review, deep insight into carbon-based nanomaterials with applications in wastewater treatment is discussed. Their adsorption role in wastewater treatment is discussed (Wang et al. 2017). The upcoming sections give general information on different types of carbon nanomaterials.

2. Wastewater Treatment Processes

The wastewater treatment process is divided into four parts: physical, mechanical, biological and chemical.

2.1 Physical Method

This method includes removing the contamination with physical force. There are many types of involvement in a physical method, like In-flow equalizations and sedimentation. In-flow equalization helps to remove the secondary and basic things from the wastewater treatment process with optimized temperature and flow of pollutant levels. Sedimentation helps the suspended particle to settle with the help of gravitational force (Jia et al. 2010).

2.2 Mechanical Method

In this, screening and filter are included to help the pollutant to remove from water. Screening is the oldest method that removes gross contaminants from waste stream to protect downstream. This pollutant protects the equipment from damage and evades interference with plant operation from entering the primary tank. Filters can be classified in a biological method rather than a mechanical method. It is an aerobic attached growth biological treatment process used to remove organic matter (Guo et al. 2009).

2.3 Biological Method

It helps with finely divided and dissolved organic matter in wastewater. In this process, microorganisms, basically bacteria, adapt to the colloidal and dissolved organic matter in a mixture of gas present in a sedimentation tank (Sharma et al. 2019).

2.4 Chemical Method

Various methods are used, such as coagulation, photocatalysis, reverse osmosis, filtration, precipitation, oxidation, membrane technology, adsorption, etc., for wastewater treatment (Wu et al. 2013). Moreover, some of the methods are high cost, high amount of sludge for generation, and economically viable. Disposal of sludge has a significant problem in many ways. Amid the above conventional methods, nano-based technology like adsorption is used widely because of its low cost and simple to use (Park et al. 2014).

3. Preparation of Nanoparticles

There are many methods used for the development of nanoparticles, such as co-precipitation, sol-gel, spray pyrolysis, spark discharge, pulsed laser ablation, thermal plasma, mechano-chemical route, flame synthesis, and electrodeposition. The individual method has merits and demerits, but mostly sol-gel method was preferred by many researchers. The sol-gel has the benefits of getting high purity, offers a degree of control of composition and structure molecular level. This may be the reason scientists use this method more often than other methods (Hasan and Hasan 2015).

4. Carbon Derivatives

For the treatment of wastewater, carbon-based nanomaterials have established a lot of consideration in the present invention (Shan et al. 2017). Researchers are mainly using these nanomaterials because of their exclusive characteristics, like a smaller size, large surface area to volume ratio, fast reactivity, better thermal and chemical stability, more availability, and catalytic potential to the nanoscale (Madhura et al. 2019). In wastewater treatment, their pore size has more active sites that get attached to the pollutant with various chemical species. These are present in different allotropic forms like diamond, graphite, carbon nitrate, fullerenes, etc. Figure 1 shows the allotropic types of carbon (Jayaraman et al. 2018). Their contribution is in the generation of clean, renewable, and viable forms of energy from light-based water splitting and pollutant removal. In this review, carbon-based nanotubes (CNTs), activated carbon, graphite carbon nitrate (g-C_3N_4), graphene, and their derivatives are discussed in wastewater treatment.

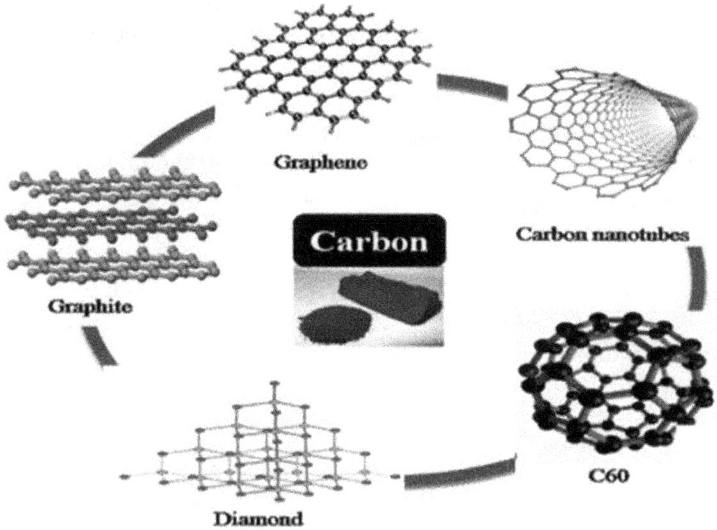

Figure 1: Allotropical arrangements of carbon nanomaterials (Jayaraman et al. 2018).

4.1 Carbon Nanotubes (CNTs)

CNTs are tube-like structures consisting of cylindrical carbon molecules. It was first developed by Iijima (Cha et al. 2013). It was divided into two parts, i.e., single-walled carbon nanotubes (SWCNTs) and multi-walled carbon nanotubes (MWCNTs), as illustrated in Figure 2 (Ihsanullah et al. 2016). Their names depend on the carbon layer attached to them. These are widely used in the application of sensors, membranes, and catalysts that have properties like large surface area, high porosity, and hallow structure, lightweights, layered structure, and also their strong interaction with pollutants (Das et al. 2018). Many poisonous contaminants were decreased by using carbon nanotubes, which consist of properties like structural, electronic, semiconductors, and optoelectronic. CNTs have been considered as their promising adsorbent for wastewater and dyes (Sadegh et al. 2015). MWCNTs were used to pass cadmium hydroxide nanowires-loaded AC with efficient removal of safranin O (SO) (M. Ghaedi et al. 2012). However, very few articles are there for dye removal using CNTs. These CNTs were used directly without further treatment (Ghaedi et al. 2011). Thus, CNTs' functionalization has produced different functional groups with new active sites in CNTs. To functionalize the active sites in CNTs, the oxidation method is an easy method to provide hydroxyl and carbonyl groups. The HNO_3, NaClO, and $KMnO_4$ solutions were used to oxidize CNTs to overcome such problems. These types of MWCNTs are used in methylene red (MB) and methylene blue (MB) removal from an aqueous solution (Mehrorang Ghaedi et al. 2012). Yao et al. (2010) performed on CNTs for adsorption capacity for MR removal. The result obtained at 333 K was 41.63 mg/g adsorption capacity (Yao et al. 2010). Shahyrari et al. reported the result for the same CNTs, which was used for Yao et al. (2010) got a better result with a surface of 280 m^2/g at 310 K with 132.6 mg/g adsorption capacity (Shahryari et al. 2010). Moreover, the MB adsorption was done on rod-like nanocrystals of soy protein isolate/hydroxyapatite with cellulose grafted on it with 454 mg/g adsorption capacity. The reusability decreases from the first cycle to the fourth cycle from 99% to 95%, as illustrated in Figure 3 (Salama 2017). The different heavy metals and their adsorbents with adsorption capacity are represented as chromium, and its adsorbent named MWCNTs and acid-modified MWCNTs with adsorbent capacity as 0.37mg/g. Heavy metal named arsenic, cadmium, mercury, copper, and its adsorbents named Fe-MWCNT, acid-modified CNTs, HNO_3-oxidized CNTs, CNT-COO, MWCNTs, COOH-MWCNT, CNT-CONH$_2$, HNO_3-modified CNTs, MWCNTs/Fe$_3$O$_4$, CNT-OH with adsorption capacity of 9.86 µg/g, 2.02 mg/g, 5.1 mg/g, 3.325 mmol/g, 0.49 mg/g, 81.57 mg/g, 1.658 mmol/g, 29 mg/g, 19 mg/g, 1.342 mmol/g (Santhosh et al. 2016).

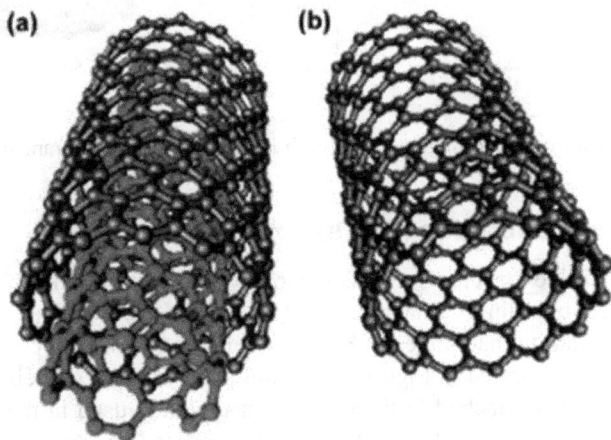

Figure 2: Structure representation: (a) MWCNT and (b) SWCNT (Ihsanullah et al. 2016).

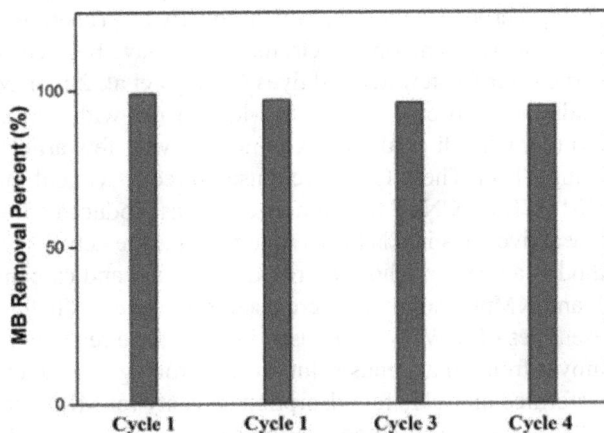

Figure 3: Reusability of cellulose grafted/SPI/hydroxyapatite for MB adsorption (Salama, 2017).

CNTs impregnated chitosan hydrogel beads (CSBs) have been experimented with the removal of congo red (CR). The maximum adsorption capacity was 450.4 mg/g with impregnated chitosan, while 200 mg/g for without impregnating Langmuir adsorption (Chatterjee et al. 2010). To improve the mechanical properties, new CSBs were prepared with sodium dodecyl sulfate (Chatterjee et al. 2011). The new CSBs have a high adsorption capacity of 375.94 mg/g for CR. MWCNTs and hybrid CNTs (HCNTs) were compared with SWCNTs, which have a high specific area that helps in better absorption of pollutants. SWCNTs are more efficient for the removal of benzene and toluene, which shows the adsorption capacity of 9.98 and 9.96 mg/g (Bina et al. 2012). The removal of reactive blue 29 (RB29) from the aqueous solution with an adsorption capacity of 496 mg/g.; the structure of RB29 is shown in Figure 4 (Nadafi et al. 2011). On the contrary, CNTs have high efficiency in divalent metal ions. The mechanism was studied, and it shows the ion exchange, electrochemical potential, and surface phenomenon which were essential factors for the adsorption of heavy metal using CNTs (Gao et al. 2009).

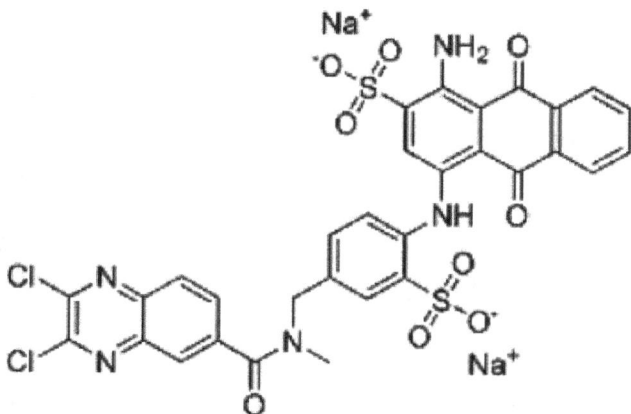

Figure 4: Structure of RB29 (Nadafi et al. 2011).

4.2 Activated Carbon (AC)

AC was used for adsorbent for removal of heavy metals and dye, but it is complicated to remove the adsorbent at ppb level. To overcome this problem CNTs, fullerenes and graphene were used for nano-adsorbent (Chatterjee et al. 2011). The advantages of AC are high porosity, active sites, and can be easily available and made from wood, coal, coconut shell, and agricultural waste. AC is used to remove organic and inorganic pollutants in wastewater and water stream. The granular activated carbon (GAC) for arsenic removal was studied for pentavalent arsenic (Rodríguez et al. 2010). The Cr(VI) was removed using AC from coconut tree sawdust from synthetic water (Machado et al. 2011). The mercury removal for emission control was done using AC for stability and adsorption (Chatterjee et al. 2011). The removal of Cu(II) and Pb(II) shows the adsorption capacity of 0.85 and 0.89 mmol/g at 60 °C using powdered activated carbon (PAC) prepared from *Eucalyptus camaldulensis* Dehn bark (Patnukao et al. 2008). A unique adsorbent was made from gamma radiation by sodium polyacrylate grafted AC for metal removal. The properties of high adsorption capacity and low cost make AC promising material for dye and heavy metals removal (Bina et al. 2012). Some of the adsorbents are illustrated in Table 2.

Table 2: Adsorption capacity of heavy metal ions with activated carbon

Adsorbent	q_m	Surface Area (m^2/g)	Removal Percentage (%)	pH	Reference
Granular AC	As(V):2.5	950	NA	2-11.5	(Di Natale et al. 2008)
AC	Cr(VI):3.46	486	99	10	(Selvi et al. 2001)
BPL AC	Hg(II):110.8	1026	NA	2.88	(Graydon et al. 2009)
HGR AC	Hg(II):101.6	482	NA	2.88	(Patnukao et al. 2008)
BPL-S AC	Hg(II):95.4	1142	NA	2.88	(Ewecharoen et al. 2009)
FC-S AC	Hg(II):89.5	160	NA	2.88	(Iijima 1991)
Powdered AC	Cu(II):0.85	1.239	NA	3-6	(Pyrzyńska and Bystrzejewski 2010)

4.3 Graphene-Based Nanomaterials

Graphene is an advanced carbon nanomaterial, which gained a lot of attention from 2004, It consists of a two-dimensional (2D) monolayer of sp^2-hybrid carbon atoms arranged in a hexagonal crystalline structure. Their properties are a high specific area, high thermal conductivity, and rapid heterogeneous electron transfer. Additionally, mechanical strength, extremely hydrophilic properties, and high negative charge density are some essential features of graphene (Sitko et al. 2013). Graphene nanomaterials are divided into three types: graphene (G), graphene oxide (GO), and reduced graphene oxide (RGO). G and GO are shown in Figure 5. G, GO, and RGO was made through the chemical exfoliation method. G can be prepared through Hammers method. GO was synthesized from a simple and low-cost method using chemical oxidation through exfoliation in ultrasonication. Its hydrophilic nature, high active sites, and the large functional group are an adsorbent and catalyst for wastewater (Chowdhury and Balasubramanian 2014). The graphene-based nanomaterials were used in wastewater, but its drawback is poor separation and reusability. These drawbacks can be overcome by modifying surface and hybridization. Because of the oxidation reaction, nanomaterials are usually bonded with zero bandgaps. These will increase the photocatalytic activities of nanoparticles by accepting electrons and also as a transporter between the particles (Shan et al. 2017). The Cd(II), Co(II), Pb(II), and U(VI) ions from aqueous solution have been experimented with using few-layered GONSs (graphene oxide nanosheet). The vital role in GONSs was the abundant oxygen-containing functional group. Zhao et al. have reported that GONSs have a good agreement in metal adsorption (Zhao et al. 2011a). The Cd(II) and Co(II) adsorption on GONSs have better adsorptions on pH and are weakly dependent on ionic strength. The better adsorption is at pH < 8 of humic acid. The maximum adsorption capacity of Cd(II) and Co(II) is at 6.0 pH with 106.3 and 68.2 mg/g, respectively, at 303 K. For Pb (II), maximum adsorption at 293, 313, and 333 K with 842, 1,150, and 1,850 mg/g (Zhao et al. 2011b). At 293 K, the removal of U(VI) maximum adsorption was at 97.5 mg/g (Zhao et al. 2012). Table 3 shows the different carbon-based nanomaterials using different heavy metals. The mechanism is shown in Figure 5.

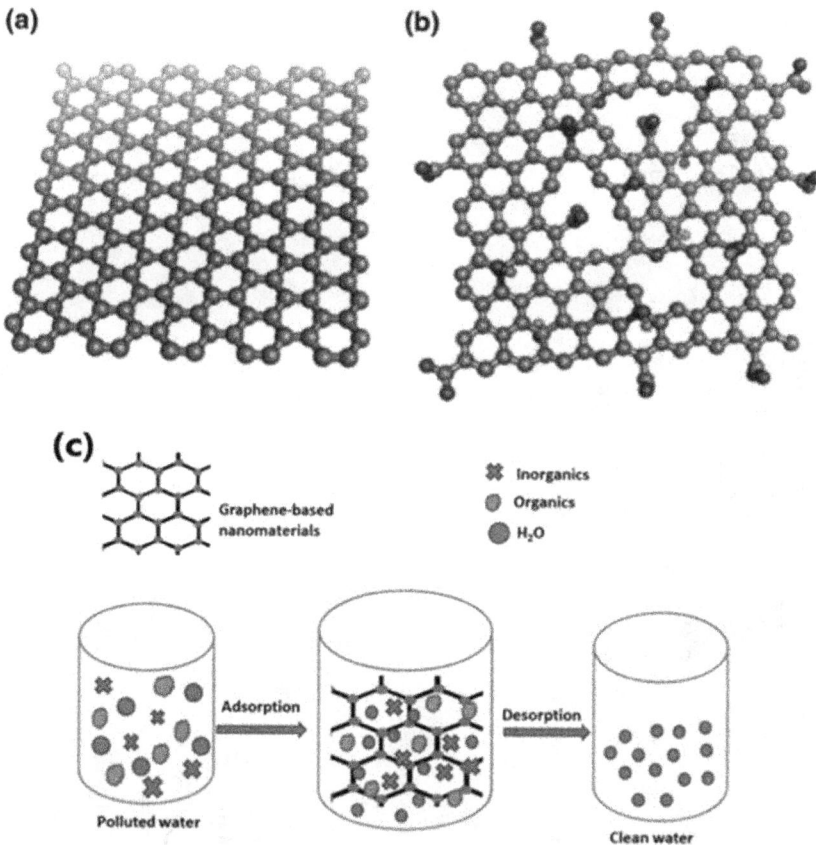

Figure 5: Schematic diagram of (a) Graphene, (b) Graphene oxide (Sadegh et al. 2017) and (c) Graphene-based nanomaterials mechanism for organic and inorganic pollutants (Madima et al. 2020).

5. Summary of the Mechanism of Pollutants

The adsorption process helps to remove organic and inorganic contaminants. The adsorption rates and adsorption capacity are most useful for adsorption. The removal

Table 3: Comparison of different carbon based nanomaterials using heavy metals.

Adsorbent	Adsorbate	Adsorption capacity (mg/g)	Rate constant (k_i, h^{-1})	Reference
AC	Reactive red M-2BE	260.7	1.503	(Machado et al. 2011)
PAC	Nitrofurazone	50.8	0.1129	(Ying-Ying and Zhen-Hu, 2016)
SWCNTs	Reactive blue 29 Cr(VI)	496	-	(Nadafi et al. 2011)
	Acid Red 18	166.7	21.12	(Shirmardi et al. 2013)
	Reactive red 120	426.49		(Bazrafshan et al. 2013)
	Cr(VI)	1.26		(Dehghani et al. 2015)
MWCNTs	Reactive red M-2	335.7	-	(Machado et al. 2011)
	Nitrofurazone	59.9	-	(Ying-Ying and Zhen-Hu, 2016)
	MB	95.3	2.860	(Selen et al. 2016)
	CR	352.1	-	(Zare et al. 2015)
	Maxilon blue	260.7	3.18	(Alkaim et al. 2015)
	Cr(VI)	2.35	-	(Dehghani et al. 2015)
Oxidized MWCNTs	Bromothymol blue (BTB)	55	0.42	(M. Ghaedi et al. 2012)
Diethylenetriamine-MWCNTs	Pb(II)	58.26	0.042	(Vuković et al. 2011)
	Cd(II)	31.45	-	
GO	MB	714	-	(Yang et al. 2011)
rGONSs	Chlorpyrifos	1200	-	(Maliyekkal et al. 2013)
	Endosulfan	1100	-	(Maliyekkal et al. 2013)
	Malathion	800	-	(Maliyekkal et al. 2013)
Graphane	Cd(II)	106.3	-	(Zhao et al. 2011a)
	Co(II)	68.2	-	
GONSs	Pb(II)	842	-	(Zhao, Ren, et al. 2011)

efficiency of an organic and inorganic pollutant from wastewater is only due to the mechanism involved between the adsorbent and adsorbate. Many different factors are there for adsorption. The factors include in the adsorption removal due to hydrophobic interaction, II-II bond, electrostatic attraction, and hydrogen bond. In adsorption, the surface phenomenon plays a significant role in properties, and the functional group is associated with the pollutant. The elements contain the functional group available on the surface of an adsorbent, such as amine, hydroxyl group, and carboxyl group. Moreover, electrostatic attraction is significant and with nonpolar hydrocarbons, hydrophobic interaction plays an important role in the material attached with the functional group (Tong et al. 2019).

Carbon derivatives have higher wastewater pollutants removal efficiency. Carbon nanotubes show high productivity because of the high surface area and higher interaction among carbon-based contaminants. The reason behind the adsorption of organic contaminants on carbon nanotube is due to van der Waals force, electrostatic force, Π-Π stacking, electrostatic attraction, hydrogen bonding, and hydrophobic interaction among themselves or in groups. The interaction between carbon nanotubes and inorganic pollutants is because of the carboxyl, hydroxyl and phenol functional group at the surface of the carbon (Gupta et al. 2013).

Figure 6 shows the mechanism of untreated carbon nanotubes and exterior oxidation to occur on the materials using the adsorption process. The adsorption takes place through a functional group that is present on the pollutant. This active group has adsorption sites in organic and inorganic pollutants with electrostatic attraction and chemical bonding. The adsorption of organic contaminants with graphene nanoparticles is because of the physisorption adsorption and has a large surface area with higher adsorption capacity (Ihsanullah et al. 2016).

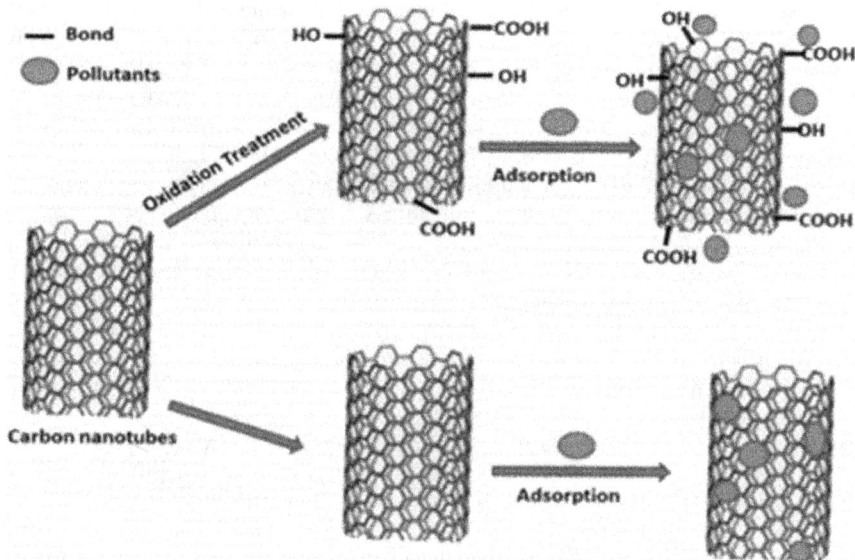

Figure 6: Mechanism of carbon nanotube and oxidized carbon nanotube (Madima et al. 2020).

6. Conclusion

The contamination present in water is one of the significant issues globally, which should be addressed seriously. These effects on living organisms and ecosystems. Therefore, the removal of toxic pollutants is necessary from water and wastewater systems. Many methods are used for removal purposes, but adsorption is a secure and widely used method. Adsorption method with nanotechnology holds promising approaches to buy any other conventional methods. The usage of carbon-based nanomaterials has gained a lot of attention as a catalyst for wastewater treatment. This review highlights the carbon-based nanoparticles and their derivatives to remove pollutants present in wastewater. Many researchers have done experiments with an adsorbent, which gives the result according to the efficiency of the catalyst. The pore sites available on the surface of the material and strong interaction among the pores and contamination bring the adsorption capacity better and faster. Though, to make an effective nano adsorbent in the market, much-needed modification is to be done to its properties, characteristics, and cost-effectiveness.

Conflict of Interests

The authors have no conflict of interests.

References

Alkaim, A.F., Sadik, Z., Mahdi, D.K., Alshrefi, S.M., Al-Sammarraie, A.M., Alamgir, F.M., Singh, P.M. and Aljeboree, A.M. 2015. Preparation, structure and adsorption properties of synthesized multiwall carbon nanotubes for highly effective removal of maxilon blue dye. Korean J. Chem. Eng. 32: 2456–2462. https://doi.org/10.1007/s11814-015-0078-y

Bazrafshan, E., Mostafapour, F.K., Hosseini, A.R., Raksh Khorshid, A. and Mahvi, A.H. 2013. Decolorisation of reactive red 120 dye by using single-walled carbon nanotubes in aqueous solutions. J. Chem. https://doi.org/10.1155/2013/938374

Bina, B., Amin, M.M., Rashidi, A. and Pourzamani, A. 2012. Benzene and toluene removal by carbon nanotubes. Arch. Environ. Prot. 38: 3–25. https://doi.org/10.2478/v10265-012-0001-0

Cha, C., Shin, S.R., Annabi, N. and Dokmeci, M.R. 2013. Carbon-based nanomaterials: multifunctional materials for. ACS Nano 7: 2891–2897.

Chatterjee, S., Chatterjee, T., Lim, S.R. and Woo, S.H. 2011. Effect of the addition mode of carbon nanotubes for the production of chitosan hydrogel core-shell beads on adsorption of congo red from aqueous solution. Bioresour. Technol. 102: 4402–4409. https://doi.org/10.1016/j.biortech.2010.12.117

Chatterjee, S., Lee, M.W. and Wooa, S.H. 2010. Adsorption of congo red by chitosan hydrogel beads impregnated with carbon nanotubes. Bioresour. Technol. 101: 1800–1806. https://doi.org/10.1016/j.biortech.2009.10.051

Chowdhury, S. and Balasubramanian, R. 2014. Recent advances in the use of graphene-family nanoadsorbents for removal of toxic pollutants from wastewater. Adv. Colloid Interface Sci. 204: 35–56. https://doi.org/10.1016/j.cis.2013.12.005

Das, R., Leo, B.F. and Murphy, F. 2018. The toxic truth about carbon nanotubes in water purification: a perspective view. Nanoscale Res. Lett. 13. https://doi.org/10.1186/s11671-018-2589-z

Dehghani, M.H., Taher, M.M., Bajpai, A.K., Heibati, B., Tyagi, I., Asif, M., Agarwal, S. and Gupta, V.K. 2015. Removal of noxious Cr (VI) ions using single-walled carbon nanotubes and multiwalled carbon nanotubes. Chem. Eng. J. 279: 344–352. https://doi.org/10.1016/j.cej.2015.04.151

Di Natale, F., Erto, A., Lancia, A. and Musmarra, D. 2008. Experimental and modelling analysis of As(V) ions adsorption on granular activated carbon. Water Res. 42: 2007–2016. https://doi.org/10.1016/j.watres.2007.12.008

Ewecharoen, A., Thiravetyan, P., Wendel, E. and Bertagnolli, H. 2009. Nickel adsorption by sodium polyacrylate-grafted activated carbon. J. Hazard. Mater. 171: 335–339. https://doi.org/10.1016/j.jhazmat.2009.06.008

Fuerhacker, M., Haile, T.M., Kogelnig, D., Stojanovic, A. and Keppler, B. 2012. Application of ionic liquids for the removal of heavy metals from wastewater and activated sludge. Water Sci. Technol. 65: 1765–1773. https://doi.org/10.2166/wst.2012.907

Gao, Z., Bandosz, T.J., Zhao, Z., Han, M. and Qiu, J. 2009. Investigation of factors affecting adsorption of transition metals on oxidized carbon nanotubes. J. Hazard. Mater. 167: 357–365. https://doi.org/10.1016/j.jhazmat.2009.01.050

Ghaedi, M., Haghdoust, S., Kokhdan, S.N., Mihandoost, A., Sahraie, R. and Daneshfar, A. 2012. Comparison of activated carbon, multiwalled carbon nanotubes, and cadmium hydroxide nanowire loaded on activated carbon as adsorbents for kinetic and equilibrium study of removal of safranine O. Spectrosc. Lett. 45: 500–510. https://doi.org/10.1080/00387010.2011.641058

Ghaedi, Mehrorang, Khajehsharifi, H., Yadkuri, A.H., Roosta, M. and Asghari, A. 2012. Oxidized multiwalled carbon nanotubes as efficient adsorbent for bromothymol blue. Toxicol. Environ. Chem. 94: 873–883. https://doi.org/10.1080/02772248.2012.678999

Ghaedi, M., Shokrollahi, A., Tavallali, H., Shojaiepoor, F., Keshavarz, B., Hossainian, H., Soylak, M. and Purkait, M.K. 2011. Activated carbon and multiwalled carbon nanotubes as efficient adsorbents for removal of arsenazo(III) and methyl red from waste water. Toxicol. Environ. Chem. 93: 438–449. https://doi.org/10.1080/02772248.2010.540244

Graydon, J.W., Zhang, X., Kirk, D.W. and Jia, C.Q. 2009. Sorption and stability of mercury on activated carbon for emission control. J. Hazard. Mater. 168: 978–982. https://doi.org/10.1016/j.jhazmat.2009.02.118

Guo, Z., Shin, K., Karki, A.B., Young, D.P., Kaner, R.B. and Hahn, H.T. 2009. Fabrication and characterization of iron oxide nanoparticles filled polypyrrole nanocomposites. J. Nanoparticle Res. 11: 1441–1452. https://doi.org/10.1007/s11051-008-9531-8

Gupta, V.K., Kumar, R., Nayak, A., Saleh, T.A. and Barakat, M.A. 2013. Adsorptive removal of dyes from aqueous solution onto carbon nanotubes: a review. Adv. Colloid Interface Sci. 193–194: 24–34. https://doi.org/10.1016/j.cis.2013.03.003

Hasan, S. and Hasan, S. 2015. A review on nanoparticles: their synthesis and types incom 7–10.

Ihsanullah, Abbas, A., Al-Amer, A.M., Laoui, T., Al-Marri, M.J., Nasser, M.S., Khraisheh, M. and Atieh, M.A. 2016. Heavy metal removal from aqueous solution by advanced carbon nanotubes: critical review of adsorption applications. Sep. Purif. Technol. 157: 141–161. https://doi.org/10.1016/j.seppur.2015.11.039

Jayaraman, T., Murthy, A.P., Elakkiya, V., Chandrasekaran, S., Nithyadharseni, P., Khan, Z., Senthil, R.A., Shanker, R., Raghavender, M., Kuppusami, P., Jagannathan, M. and Ashokkumar, M. 2018 Recent development on carbon based heterostructures for their applications in energy and environment: a review. J. Ind. Eng. Chem. 64: 16–59. https://doi.org/10.1016/j.jiec.2018.02.029

Jia, A., Liang, X., Su, Z., Zhu, T. and Liu, S. 2010. Synthesis and the effect of calcination temperature on the physical-chemical properties and photocatalytic activities of Ni, La codoped SrTiO$_3$. J. Hazard. Mater. 178: 233–242. https://doi.org/10.1016/j.jhazmat.2010.01.068

Kumar, R. and Chawla, J. 2014. Removal of cadmium ion from water/wastewater by nano-metal oxides: a review. Water Qual. Expo. Heal. 5: 215–226. https://doi.org/10.1007/s12403-013-0100-8

Machado, F.M., Bergmann, C.P., Fernandes, T.H.M., Lima, E.C., Royer, B., Calvete, T. and Fagan, S.B. 2011. Adsorption of reactive red M-2BE dye from water solutions by multiwalled carbon nanotubes and activated carbon. J. Hazard. Mater. 192: 1122–1131. https://doi.org/10.1016/j.jhazmat.2011.06.020

Madhura, L., Singh, S., Kanchi, S., Sabela, M., Bisetty, K. and Inamuddin, I. 2019. Nanotechnology-based water quality management for wastewater treatment. Environ. Chem. Lett. Springer International Publishing. https://doi.org/10.1007/s10311-018-0778-8

Madima, N., Mishra, S.B., Inamuddin, I. and Mishra, A.K. 2020. Carbon-based nanomaterials for remediation of organic and inorganic pollutants from wastewater: a review. Environ. Chem. Lett. https://doi.org/10.1007/s10311-020-01001-0

Maliyekkal, S.M., Sreeprasad, T.S., Krishnan, D., Kouser, S., Mishra, A.K., Waghmare, U.V. and Pradeep, T. 2013. Graphene: a reusable substrate for unprecedented adsorption of pesticides. Small 9: 273–283. https://doi.org/10.1002/smll.201201125

Nadafi, K., Mesdaghinia, A., Nabizadeh, R., Younesian, M. and Rad, M.J. 2011 The combination and optimization study on RB29 dye removal from water by peroxy acid and single-wall carbon nanotubes. Desalin. Water Treat. 27: 237–242. https://doi.org/10.5004/dwt.2011.1980

Ojekunle, O.Z., Ojekunle, O.V., Adeyemi, A.A., Taiwo, A.G., Sangowusi, O.R., Taiwo, A.M. and Adekitan, A.A. 2016. Evaluation of surface water quality indices and ecological risk assessment for heavy metals in scrap yard neighbourhood. Springerplus 5. https://doi.org/10.1186/s40064-016-2158-9

Park, S., Yang, J., Kim, Y., Park, M. and Baek, K. 2014. Removal of As(III) and As(V) using iron-rich sludge produced from coal mine drainage treatment plant. 18: 10878-89. https://doi.org/10.1007/s11356-014-3023-4

Patnukao, P., Kongsuwan, A. and Pavasant, P. 2008. Batch studies of adsorption of copper and lead on activated carbon from Eucalyptus camaldulensis Dehn. bark. J. Environ. Sci. 20: 1028–1034. https://doi.org/10.1016/S1001-0742(08)62145-2

Pyrzyńska, K. and Bystrzejewski, M. 2010. Comparative study of heavy metal ions sorption onto activated carbon, carbon nanotubes, and carbon-encapsulated magnetic nanoparticles. Colloids Surfaces A Physicochem. Eng. Asp. 362: 102–109. https://doi.org/10.1016/j.colsurfa.2010.03.047

Qu, X., Alvarez, P.J.J. and Li, Q. 2013. Applications of nanotechnology in water and wastewater treatment. Water Res. 47: 3931–3946. https://doi.org/10.1016/j.watres.2012.09.058

Rodríguez, A., Ovejero, G., Sotelo, J.L., Mestanza, M. and García, J. 2010. Adsorption of dyes on carbon nanomaterials from aqueous solutions. J. Environ. Sci. Heal. – Part A Toxic/Hazardous Subst. Environ. Eng. 45: 1642–1653. https://doi.org/10.1080/10934529.2010.506137

Sadegh, H., Ali, G.A.M., Gupta, V.K., Makhlouf, A.S.H., Shahryari-ghoshekandi, R., Nadagouda, M.N., Sillanpää, M. and Megiel, E. 2017. The role of nanomaterials as effective adsorbents and their applications in wastewater treatment. J. Nanostructure Chem. 7: 1–14. https://doi.org/10.1007/s40097-017-0219-4

Sadegh, H., Shahryari-Ghoshekandi, R., Agarwal, S., Tyagi, I., Asif, M. and Gupta, V.K. 2015. Microwave-assisted removal of malachite green by carboxylate functionalized multiwalled carbon nanotubes: kinetics and equilibrium study. J. Mol. Liq. 206: 151–158. https://doi.org/10.1016/j.molliq.2015.02.007

Salama, A. 2017. New sustainable hybrid material as adsorbent for dye removal from aqueous solutions. J. Colloid Interface Sci. 487: 348–353. https://doi.org/10.1016/j.jcis.2016.10.034

Santhosh, C., Velmurugan, V., Jacob, G., Jeong, S.K., Grace, A.N. and Bhatnagar, A. 2016. Role of nanomaterials in water treatment applications: a review. Chem. Eng. J. 306: 1116–1137. https://doi.org/10.1016/j.cej.2016.08.053

Selen, V., Güler, Ö., Özer, D. and Evin, E. 2016. Synthesized multiwalled carbon nanotubes as a potential adsorbent for the removal of methylene blue dye: kinetics, isotherms, and thermodynamics. Desalin. Water Treat. 57: 8826–8838. https://doi.org/10.1080/1944399 4.2015.1025851

Selvi, K., Pattabhi, S. and Kadirvelu, K. 2001. Removal of Cr(VI) from aqueous solution by adsorption onto activated carbon. Bioresour. Technol. 80: 87–89. https://doi.org/10.1016/S0960-8524(01)00068-2

Shahryari, Z., Goharrizi, A.S. and Azadi, M. 2010. Experimental study of methylene blue adsorption from aqueous solutions onto carbon nano tubes. Int. J. Water Resour. Environ. Eng. 2: 16–028.

Shan, S.J., Zhao, Y., Tang, H. and Cui, F.Y. 2017. A mini-review of carbonaceous nanomaterials for removal of contaminants from wastewater. IOP Conf. Ser. Earth Environ. Sci. 68. https://doi.org/10.1088/1755-1315/68/1/012003

Sharma, P., Sen, K., Thakur, P., Chauhan, M. and Chauhan, K. 2019. Spherically shaped pectin-g-poly(amidoxime)-Fe complex: a promising innovative pathway to tailor a new material in high amidoxime functionalization for fluoride adsorption. Int. J. Biol. Macromol. 140: 78–90. https://doi.org/10.1016/j.ijbiomac.2019.08.098

Sheet, I., Kabbani, A. and Holail, H. 2014. Removal of heavy metals using nanostructured graphite oxide, silica nanoparticles and silica/graphite oxide composite. Energy Procedia 50: 130–138. https://doi.org/10.1016/j.egypro.2014.06.016

Shirmardi, M., Mahvi, A.H., Mesdaghinia, A., Nasseri, S. and Nabizadeh, R. 2013. Adsorption of acid red18 dye from aqueous solution using single-wall carbon nanotubes: kinetic and equilibrium. Desalin. Water Treat. 51: 6507–6516. https://doi.org/10.1080/19443994.201 3.793915

Sitko, R., Turek, E., Zawisza, B., Malicka, E., Talik, E., Heimann, J., Gagor, A., Feist, B. and Wrzalik, R. 2013. Adsorption of divalent metal ions from aqueous solutions using graphene oxide. Dalt. Trans. 42: 5682–5689. https://doi.org/10.1039/c3dt33097d

Tong, Y., McNamara, P.J. and Mayer, B.K. 2019. Adsorption of organic micropollutants onto biochar: a review of relevant kinetics, mechanisms and equilibrium. Environ. Sci. Water Res. Technol. 5: 821–838. https://doi.org/10.1039/c8ew00938d

Vuković, G.D., Marinković, A.D., Škapin, S.D., Ristić, M.T., Aleksić, R., Perić-Grujić, A.A. and Uskoković, P.S. 2011. Removal of lead from water by amino modified multiwalled carbon nanotubes. Chem. Eng. J. 173: 855–865. https://doi.org/10.1016/j.cej.2011.08.036

Wang, S., Ke, X., Zhong, S., Lai, Y., Qian, D., Wang, Y., Wang, Q. and Jiang, W. 2017. Bimetallic zeolitic imidazolate frameworks-derived porous carbon-based materials with efficient synergistic microwave absorption properties: the role of calcining temperature. RSC Adv. 7: 46436–46444. https://doi.org/10.1039/c7ra08882e

Wu, K., Liu, R., Li, T., Liu, H., Peng, J. and Qu, J. 2013. Removal of arsenic (III) from aqueous solution using a low-cost byproduct in Fe-removal plants—Fe-based backwashing sludge. Chem. Eng. J. 226: 393–401. https://doi.org/10.1016/j.cej.2013.04.076

Yang, S.T., Chen, S., Chang, Y., Cao, A., Liu, Y. and Wang, H. 2011. Removal of methylene blue from aqueous solution by graphene oxide. J. Colloid Interface Sci. 359: 24–29. https://doi.org/10.1016/j.jcis.2011.02.064

Yao, Y., Xu, F., Chen, M., Xu, Z. and Zhu, Z. 2010. Adsorption of cationic methyl violet and methylene blue dyes onto carbon nanotubes. 2010 IEEE 5th Int. Conf. Nano/Micro Eng. Mol. Syst. NEMS 2010: 1083–1087. https://doi.org/10.1109/NEMS.2010.5592561

Ying-Ying, W. and Zhen-Hu, X. 2016. Multiwalled carbon nanotubes and powder-activated carbon adsorbents for the removal of nitrofurazone from aqueous solution. J. Dispers. Sci. Technol. 37: 613–624. https://doi.org/10.1080/01932691.2014.981337

Yu, G., Lu, Y., Guo, J., Patel, M., Bafana, A., Wang, X., Qiu, B., Jeffryes, C., Wei, S., Guo, Z. and Wujcik, E.K. 2018. Carbon nanotubes, graphene, and their derivatives for heavy metal removal. Adv. Compos. Hybrid Mater. 1: 56–78. https://doi.org/10.1007/s42114-017-0004-3

Zare, K., Sadegh, H., Shahryari-Ghoshekandi, R., Maazinejad, B., Ali, V., Tyagi, I., Agarwal, S. and Gupta, V.K. 2015. Enhanced removal of toxic Congo red dye using multi walled carbon nanotubes: kinetic, equilibrium studies and its comparison with other adsorbents. J. Mol. Liq. 212: 266–271. https://doi.org/10.1016/j.molliq.2015.09.027

Zhao, G., Li, J., Ren, X., Chen, C. and Wang, X. 2011a. Few-layered graphene oxide nanosheets as superior sorbents for heavy metal ion pollution management. Environ. Sci. Technol. 45: 10454–10462. https://doi.org/10.1021/es203439v

Zhao, G., Ren, X., Gao, X., Tan, X., Li, J., Chen, C., Huang, Y. and Wang, X. 2011b. Removal of Pb(II) ions from aqueous solutions on few-layered graphene oxide nanosheets. Dalt. Trans. 40: 10945–10952. https://doi.org/10.1039/c1dt11005e

Zhao, G., Wen, T., Yang, X., Yang, S., Liao, J., Hu, J., Shao, D. and Wang, X. 2012. Preconcentration of U(VI) ions on few-layered graphene oxide nanosheets from aqueous solutions. Dalt. Trans. 41: 6182–6188. https://doi.org/10.1039/c2dt00054g

The Impact of Carbon-Based Nanomaterials in Biological Systems

Leirika Ngangom[1], Kunal Sharma[1], Pankaj Bhatt[2], Nilay Singh[3] and Neha Pandey[3]*

[1] Department of Life Sciences, Graphic Era Deemed to be University,
Dehradun - 248002, Uttarakhand, India
[2] State Key Laboratory for Conservation and Utilization of Subtropical Agro-bioresources,
Guangong Province Key Laboratory of Microbial Signals and Disease Control,
Integrative Microbiology Research Centre, South China Agricultural University,
Guangzhou 510642, China
[3] Department of Biotechnology, Graphic Era Deemed to be University,
Dehradun - 248002, Uttarakhand, India

1. Introduction

Carbon is the sole and imperative element on the Earth and is ranked as the second component that is present in a human body comprising about 18% of the individual's weight. The remarkable character of the carbon element is that its wide range of metastable stages can be arranged or formed around the intermediate surroundings. In addition to this, the function of carbon is vital as it forms a bond with the enclosing light components and also with itself despite the scarce amount available around the Earth's crust, which is about 0.032% from the overall mass of the planet (Zhang et al. 2012, Marty et al. 2013). Therefore, the functional capability of carbon elements has bought a wide expansion in the field of biology and chemistry. In the current era, the study of carbon science is very contemporary, and in some research areas, such as engineering and technology, materials science, nanoscience, and carbon nanostructures, have carbon elements of various low dimensions comprising carbon nanotubes, graphene, and activated carbon (Geim et al. 2007, Titirici et al. 2015, Deng et al. 2016). There are different allotropic forms of carbon, namely graphite, buckminsterfullerene (smallest fullerene molecule), and diamond. Among the allotropes of carbon mentioned, the thermodynamically stable allotrope is graphite. It has a high thermal and electrical conductivity that makes it suited for use in various applications that demands high temperatures. Graphite is a crystalline form of carbon molecule and its atoms configured in a hexagonal

Corresponding author: neha.pandey@geu.ac.in

pattern. Another allotrope form of carbon is graphene that comprises a single-layered carbon atom in a two-dimensional honeycomb network. Graphene is mainly used as a precursor for synthesizing various carbon nanoparticles. It undergoes oxidation to form graphene oxide and several other graphene derivatives are obtained, namely graphene quantum dots (GQDs) and reduced graphene oxide (rGO) (Smith et al. 2019). Each carbon derivatives shows incomparable characteristics and is broadly used in numerous biological applications, like tissue engineering, drug delivery, bioimaging, cancer therapy, as well as biosensing (Hong et al. 2015, Bhattacharya et al. 2016). The carbon nanostructures are classified into two sets based on their orbital hybridization; CNMs (Carbon nanomaterials) like carbon nanotubes (SWNTs and MWNTs), graphene, fullerene, graphite, graphene oxide (GO), graphite carbon nitride (gCN) is composed of sp^2 hybridized carbon atoms and nanodiamonds have sp^3 hybridized carbon atoms, whereas carbon dots (CDs) are composed of both sp^2 and sp^3 hybridized carbon atoms (Georgakilas et al. 2015).

Over the years, nanotechnology had lured considerable interest through its immediate application to engender relevant materials with elite properties. Various nanostructure features, like flexibility, high surface area, and superior directionality, make it appropriate for a broad array of applications. Mainly because of this reason, researchers from science-based fields are much more interested in such applications considering the important role they have in advanced technologies. Under the category of nanomaterials, carbon nanomaterials exhibit an essential property. Various carbon nanomaterials, namely carbon nanotubes (CNTs), carbon dots, nanodiamonds, and graphene, are applied widely in the industrial field. The primary reason for the utilization of CNMs in medical and commercial fields is the unusual chemical features of carbon elements. The pattern and chemical features of carbon are arranged in sp, sp^2, and sp^3 hybridizations (Tucek et al. 2018). Over the years, we have seen immense attention rising in the utilization of CNMs employed in biological applications. Moreover, the present study has emphasized the synthesis of carbon-based nanomaterials (CBNs) and their effective applications in various biological systems. The reduced cytotoxicity and elevated biocompatibility of CNMs were outlined in addition to antioxidant and immunological activities. Such beneficial features are applicable for bioimaging, diagnostic study, and treatment of distinct health disorder applications. (Giordani et al. 2019).

The CNMs comprise several carbon allotropes that have been reported in detail in the account of their significant properties, like thermal, optical, mechanical, and electronic. CNMs are beneficial in the field of biomedical, industrial, plant systems, and the environment (Yang and Westerhoff 2014). The production and utilization of CNMs will ascertain elevate the risk to mankind, and concern more about their biosafety. Furthermore, there is still some lag in the research field in concern to the safety measure resulting in the restrictions of CNMs applications in the biomedicine area. There is numerous research work carried out to increase the potentials of CNMs in the biological systems for antibacterial, bioimaging, disease diagnostics, therapy, and drug delivery (Lin et al. 2018). Therefore, there must be a thorough knowledge about nano-bio interactions concerning its potency and biosafety measures.

In the past few decades, on the discovery of various low-dimensional carbon derivatives, the well-known carbon allotropes have widened. Such novel CNMs have aroused great interest in the science discipline as these carbon allotropes comprise new features and are applicable in several biological applications. According to the findings of novel properties of CNMs, they are often called 'wonder materials' in the fields of science (Ko and Grigoropoulos 2014, Peng et al. 2014, Kah and Hofmann 2015). The current review emphasizes the role of CBNs, namely CNTs, graphene, and its derivatives (graphite, graphene oxide, reduced graphene oxide, and graphene quantum dots), carbon dots, and nanodiamonds in biological applications.

2. Types of Carbon-Based Nanomaterials

2.1 Carbon Nanotubes (CNTs)

CNTs consist of barrel-shaped graphene sheets covered with fullerenes at the ends. In the nanotechnology world, the limelight is on CNTs mainly because of their industrial applications. CNTs are categorized into two types, namely single-walled carbon nanotubes (SWNTs) and multi-walled carbon nanotubes (MWNTs) (Russier et al. 2011) as described in Figure 1. First and foremost, CNTs were reported by Iijima in 1991. In the world of nanotechnology, CNTs have acquired a great amount of attention because of their unexceptional properties and applications in biomedicine. CNTs are one-dimensional nanostructures that exhibit optical and electronic properties that vary from other carbon elements and nanoparticles (Gupta et al. 2018). In CNTs, the chemical bonding within carbon atoms is known to be the strongest bond among the other allotropes of carbon and imparts high mechanical strength (Falvo et al. 1997). They are formed by various methods, like chemical vapor deposition (CVD), arc-discharge, and their electronic characteristics, which rely merely on geometric parameters like chiral angle and diameter. The essential properties of CNTs are used in optics, microelectronics, and biomedicine applications, such as tissue engineering and using CNTs-based nanoelectrodes for stimulating neural activity.

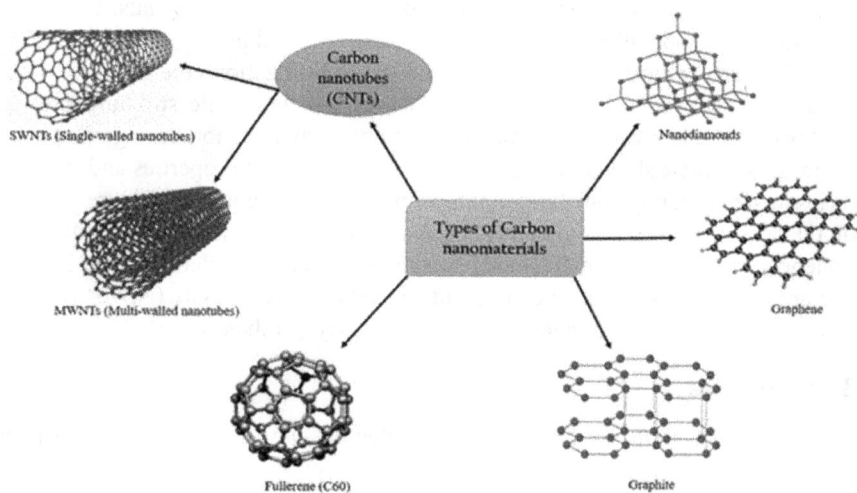

Figure 1: Different types of carbon nanomaterials.

2.1.1 Single-Walled Nanotubes (SWNTs)

SWNTs are a peculiar form of CNTs and are commonly characterized as one-dimensional. They comprise a monolayer of graphene, which is folded to form a cylindrical structure. The diameter of the cylindrical structure ranges from 0.4-2 nm (He et al. 2013). SWNTs are carbon allotropes that are transitional between graphene and fullerene. They are categorized into three different types: armchair, zigzag, and chiral.

2.1.2 Multi-Walled Nanotubes (MWNTs)

MWNTs are another special type of CNTs after SWNTs in which collective single-walled nanotubes are embedded one after another. The diameter of the outer side of MWNTs is 2-100 nm and the inner side is 1-3 nm (He et al. 2013). They can be differentiated from SWNTs based on the rigidity and multi-walled Russian-doll structure.

The special characteristics possessed by CNTs that are one-dimensional nanostructure have shown alluring properties such as optical and electronic that differentiate them from other carbon nanomaterials (Gupta et al. 2018). Besides this, CNTs are biocompatible, have a small size, great surface function, and their reactivity is high. Consequently, CNTs are broadly used in the area of biomedicine (Mehra et al. 2018), energy (Luo et al. 2018), and photoelectricity (Yi et al. 2018). There are quite a few reports that demonstrate that CNTs have an antagonistic effect on the cellular compartments of the human body (Gholizadeh et al. 2017). CNTs and their derivatives enhance the growth and development of the plant. Such nanomaterials can be broadly utilized in the nanosensor field for detecting pathogens that are toxic to human health or biological systems.

2.2 Graphene

Graphene is reported to be an advanced type of carbon nanomaterials. It is a two-dimensional allotrope of carbon elements and is composed of a monolayer of carbon atoms. The carbon atoms in graphene show sp^2-hybridization connected by σ- and π-bonds and packed in a hexagonal crystal pattern. Its atomic structure provides a platform for other allotropic carbon, namely carbon nanotubes, graphite, and fullerenes. Graphene shows similarities to CNTs in terms of properties and structure. It varies in size ranging from nanometers to microns. Consequently, Figure 2 focuses upon the potential of graphene to form derivatives exhibiting numerous applications. Graphene was discovered in 2004 sing scotch tape peeling (Guler et al. 2020). Some important properties of graphene are its ideal thermal conductivity (Novoselov et al. 2004), mechanical strength, and optical transparency (Mohajeri et al. 2019).

2.3 Fullerene

A fullerene is an allotrope form of carbon element, and it was discovered in the year 1985 by Harry Kroto and colleagues (Kroto et al. 1985). Fullerene displays a series of carbon atoms arranged either in a cylindrical shape or in a buckyballs pattern. The carbon atoms in fullerenes are mainly in the sp^2-hybridization and

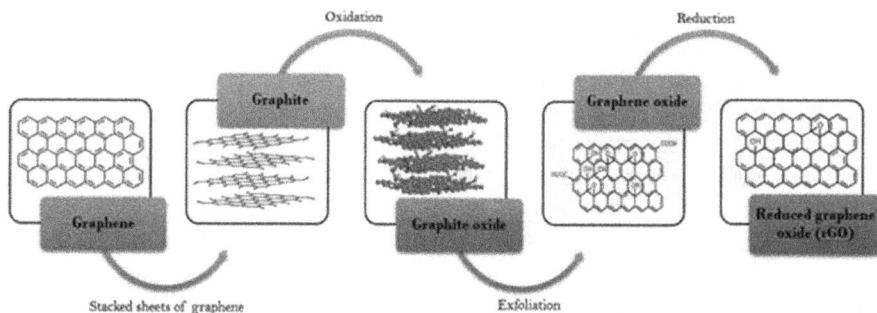

Figure 2: Role of graphene as a precursor to form different types of graphene derivatives.

connected by covalent bonds. The commonly known fullerene is C60 and is a three-dimensional compound. Some peculiar properties of fullerenes are that it acts as a radical scavenger and can be altered to adapt properties as a derivative. The vital physiochemical characteristic of fullerenes is that various solutions, such as water at a relevant pH can form crystalline nanoparticles ranging from 25-500 nm in diameter in stable conditions (Fortner et al. 2005, Chen et al. 2006).

2.4 Nanodiamonds (NDs)

The carbon atoms of nanodiamonds are in sp^3 hybridization. Diamonds exhibit unique properties, like low friction coefficient, extremely high thermal conductivity, rigidity, and ideal mobility of electrical charge carriers. The optical features of NDs are excellent; it is colorless and transparent in pure diamond form and has a colored appearance in the presence of lattice impurities (Aharonovich et al. 2011). Nanodiamonds are allotropic carbon with a nanoscale size, and they can be obtained from an oxygen-deprived TNT-hexogen combination in inert conditions (Shenderova et al. 2002). Synthesis of NDs can be done by various methods, such as pulsed laser ablation (Amans et al. 2009), detonation (Kovalenko et al. 2010), and ion irradiation of graphite can produce nanodiamonds at optimum temperature (Daulton et al. 2001).

3. Biomedical Applications of CBNs

Numerous biological applications based on carbon nanomaterials have received excellent attention because of their unusual physical and chemical features, including mechanical, electrical, thermal, and synthetic variations. Due to their fundamental properties, derivatives of carbon nanomaterials, such as graphene oxide, carbon nanotubes, graphene quantum dots, and nanodiamonds are broadly examined for biomedical applications. Until now, it is acknowledged that CNMs can penetrate inside organisms through various routes, like digestion, inhalation, and injection, and move forward to other parts of the body, such as organs or tissues via the circulatory or cardiovascular system (Mu et al. 2014). According to some research investigations, it has been manifested that disclosure to CNMs shows detrimental effects on the natural physiological activities of the circulatory, respiratory, and immune systems at a voluminous level (Braakhuis et al. 2014, Wen et al. 2015).

The bioavailability of CNMs determines the nature and toxic level based on the biological processes of cellular uptake and exocytosis, which eventually depends on CNMs interacting with the biomolecules. Therefore, to ensure the biosafety of CNMs toward living organisms, a thorough understanding should be approached for nano-bio interactions at a micro-level. Most of the biological applications mentioned earlier (Figure 3) generally start with the interactions of CNM derivatives with the targeted biomolecule or cells. However, applications of various types of CBNs in cancer treatment are shown in Table 1. Subsequently, there is a need for a comprehensive study to figure out the impact of CNMs at a cellular level as to how the CNMs intrude on the cellular components and the cell itself. At present, there is a huge number of research work that demonstrates the direct interactions of CNMs with cellular components, like proteins, lipids, or the cell membrane (Li et al. 2013, Calvaresi et al. 2014).

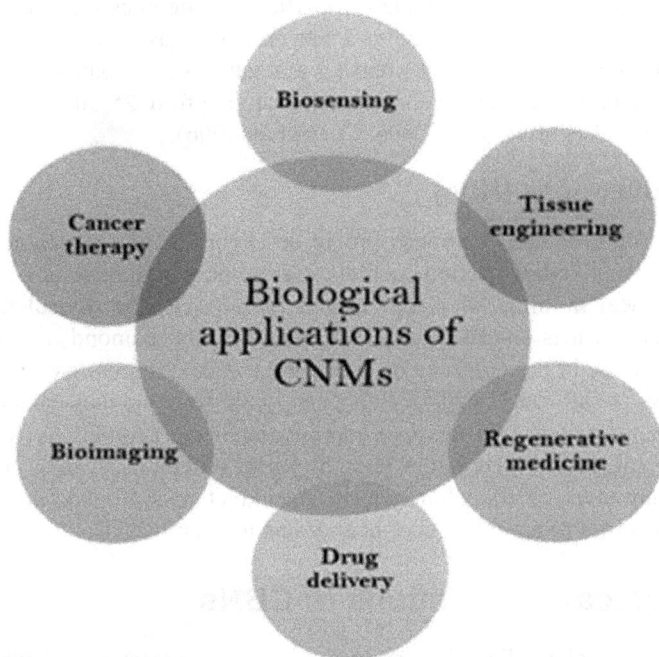

Figure 3: CNMs role in biological applications (Hong et al. 2015, Bhattacharya et al. 2016).

3.1 Carbon Nanotubes (CNTs)

CNTs possess peculiar optical, electronic, and mechanical characteristics and are examined as nanoprobes in the new era (Tilmaciu et al. 2015). The essential feature of CNT as biosensors is the degradation of biomolecules present on the surface, which improves recognition and signal transfer activity. Depending on such features, these CNTs derived biosensors are classified as optical biosensors and electronic-based biosensors (Maiti et al. 2019). For detection of nitric oxide and detecting epinephrine, electrochemical biosensors based on CNTs were designed (Ulissi et al. 2014). For

Table 1: Applications of carbon nanomaterials as theragnostic in cancer

Carbon Nanomaterials	Theragnostic Implications	References
Graphene		
Graphene oxide	Targeting cancer stem cells	Marco et al. 2015
Graphene quantum dots	Improving photodynamic therapy in cancer treatment	Hua-Yang et al. 2019
Carbon Nanotubes and Dots		
Single-walled carbon nanotubes	Killing cancer cells by thermal effect	Nadine Wong Shi et al. 2005
Mesoporous carbon nanoparticles	Multi-functional carrier for cancer treatment	Meiting et al. 2020
Carbon dots	Combination carbon dots and doxorubicin resulted in inhibition of tumor cells	Jufeng et al. 2019
Nanodiamonds and Fullerenes		
Nanodiamonds	Chemotherapeutic	Charu et al. 2017
Fullerenes (C_{60})	Modulation of anticancer drug	Larysa M et al. 2018

identifying human blood proteins, about 20 discrete SWCNTs corona phases were set up (Bisker et al. 2016), and this investigation showed that the distinct corona phase can recognize fibrinogen. To detect arginase-1, MWCNTs were efficiently designed (Baldo et al. 2016). Single-stranded DNA SWCNTs were designed to check the lipid substance in the endosomal lumen of cellular compartments (Jena et al. 2017).

3.2 Graphene Oxide (GO)

Over the decade graphene oxides are being used for their potential, mechanical, and structural abilities with less biotoxicity in wide biomedical applications, like in drug delivery and sensing-bioimaging techniques with some antibacterial activities. Latest studies showed that nanocomposite combined structures of graphene and hydrogels provide a much effective and improved targeted drug delivery approach with high mechanical strength rather than the conventional hydrogels (Yi et al. 2020). Previously it was also shown that double networked structured graphene oxides carrying nanogel and drug were effective photothermal agents with a cytotoxic effect in colorectal cancer cells (Fiorica et al. 2020). In a recent review, it was deduced that a bionic optimal size range of hydrogel electrode prepared from graphene oxide incorporating fish sperm DNA was applied to detect ultrasensitive mutation in mitochondrial DNA (Khajouei et al. 2020).

3.3 Graphene Quantum Dots (GQDs)

Recent studies showed that improved and advanced photodynamic therapy based on non-composite graphene quantum dots can be much efficient in cancer treatment (Fan et al. 2019). A literature study also showed the efficient synthesis of GQDs can be useful in targeted drug delivery (Zhao et al. 2020). Some studies focus their ideas on the bioimaging-based application of GQDs and have shown that these GQDs have a much efficient role in bioimaging and visualization of cellular integrity along with biomolecules and can substitute other imaging techniques. These GQDs based techniques include two-photon imaging, flurosense imaging, dual-modal imaging, and MRI-based imaging (Younis et al. 2020). Using biodegradable nano-transporters based on their therapeutical approach with GQDs was assessed and proved to be effective in drug delivery in certain pancreatic cancer cells (Yang et al. 2019). In 2017, a well-described review of researchers showed the *in-vitro* and *in-vivo* imaging and biosensing applications along with the improved analytical and systematic performance of GQDs-based nanoparticles, like immunological and nucleic acid assays (Chen et al. 2018). GQDs also proved to be a perfect biocompatible for cisplatin drug delivery mode with a much low biotoxic effect (Nasrollahi et al. 2019).

4. Role of CNMs on the Plant System

The interaction of CNMs on plants causes various physiological changes in the plant, primarily based on the CNMs features such as types, size, shape, chemical composition, dose concentration, and surface covering. The potency of CNMs differs from one plant to another and the interaction between CNMs and plants is mainly dependent on the dosage concentration. CNMs play multiple roles in the plant system as shown in Figure 4. Currently, various types of CNMs, namely CNTs, fullerenes, and fullerols have acquired attention because of their beneficial utilization in the growth and development of the plant. Several research investigations have manifested the potential and also toxicity effect of different types of CNMs on plants.

Numerous research study has been done over the years demonstrating the effect of CNMs both in positive and negative aspects. Some of the recent investigations

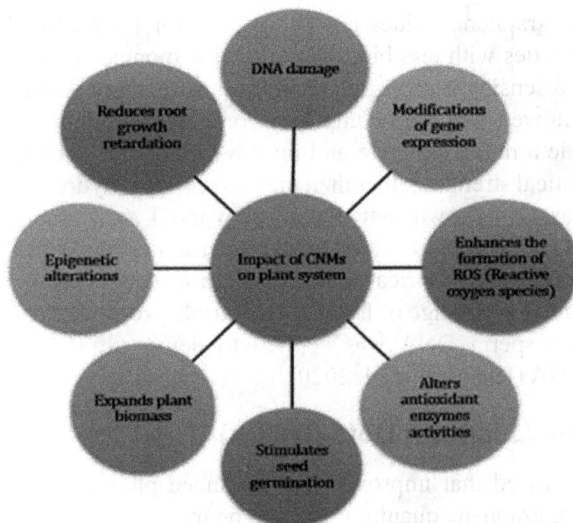

Figure 4: Impact of CNMs on the plant system on various levels.

carried out have been documented in this section. In *Arabidopsis thaliana,* it was reported that CNPs exhibited effect during the time of flowering and on photomorphogenesis (Kumar et al. 2018). It was also reported that the concentration of C60 (most common fullerene) was higher in comparison to the metal nanoparticles (CuO) in the stems and roots of the rice plant, *Oryza sativa,* but the concentration of metal NPs was higher than C60 in panicles of the rice plant (Liang et al. 2018). The exogenous fullerols enhances photosynthesis, seed germination, plant growth and elevate antioxidant activity in *Brassica nupus* under drought condition (Xiong et al. 2018). FNPs (Fullerenol nanoparticles) serve as an intracellular binding molecule of water, resulting in hygroscopic activity under water stress. Such FNPs can easily penetrate the leaves and root tissues of the plant (Borišev et al. 2016). Besides this, FNPs promote root elongation as a high concentration of FNPs is found in the roots as compared to the leaves and stems (Wang et al. 2016). A detailed investigation carried out demonstrates that graphene promotes root elongation but there are no changes in the growth of leaves and also inhibits the photosynthesis process and caused oxidative stress in roots (Zhang et al. 2016). In another study, graphene oxide was labeled with steady isotope 13C to analyze graphene concentration in *Triticum aestivum*, and it was concluded that 13C-GO content was low in the roots whereas 13C-GO NPs with larger size interrupts the movement of the NPs to leaves and stems (Chen et al. 2017). It was demonstrated that in *Avena sativa,* GO inhibits photosynthesis and seedling length (Chen et al. 2018). A research study concluded that exposure of two plant species namely *Nicotiana tabacum* and *Corylus avellana* to high dos concentrations of GO hinders pollen activity in *in vitro* conditions (Carniel et al. 2018). Detailed investigations of CNPs in plant systems under *in vitro* and *in vivo* conditions could expand novel ideas for improving commercial and valuable green plants. In view of this, some important positive effects of CBNs in diverse crops have been discussed in Table 2.

5. Environmental Applications of CBNs

In some studies, the use of carbon-based nanomaterials with their remarkable properties shows that these carbon nanomaterials can endow in a wide extent of environmental applications, like as absorbents materials, dialysis units, depth filters, disinfectant agents, sensors units, and also in controlling environment-related pollutions (Mauter et al. 2008). Studies also reflect the properties of activated carbon that its absorbent nature with the large surface area, and it can be employed in the standard treatment of wastewater by absorbing organic and inorganic emissions. However, this activated carbon sometimes shows imprecise absorbent and less efficacy toward microorganisms (Zaytseva et al. 2016). Many deep literature studies also showed and revealed that carbon-based nanomaterials have advantageous capabilities in improvising filtration approaches in wastewater treatments and purification (Qu et al. 2013, Liu et al. 2013, Saleem et al. 2020). With some disadvantages of activated carbon in adsorption kinetics, carbon nanotubes show much effective adsorption capacity for some microbial toxins like microcystins (Yan et al. 2006), lead (II) (Rahbari et al. 2009), and copper (III) (Dichiara et al. 2015). Many studies have also shown that multi-walled nanotubes were much efficient in

Table 2: Some important effect of carbon-based nanomaterials in crop improvement

Types	Positive Effects	Crop Types	References
Carbon nanotubes	Elongation and enhanced root growth	Onion, cucumber, and rye-grass	Serag et al. 2015
SWCNTs	Increased expression of associated genes of seminal root whereas decreased root hairs gene expression.	Maize	Yan et al. 2013
	Improves seedling growth.	Fig plants, tomato	Khodakovskaya et al. 2011, 2013
MWCNTs	Doubling of flowering time and increased yield	Tomato	Mariya et al. 2013 Lahiani et al. 2013
	Germination and development of seedling	Soybean, corn, and barley	
Graphene	Promotes growth in adventitious roots	Rice	Shangjie et al. 2015
Hydrated graphene ribbon	Increased seed germination and resistance to oxidative stress	Wheat	Xiangang, and Qixing Zhou 2014
Graphene oxide	Ripeness, increased fringe and carbohydrates content of the fruit	Watermelon	Park et al. 2020
Fullerol	Increased biomass yield, water content with enhanced anticancer antidiabetic properties	Bitter melon	Chittaranjan et al. 2013
Nano-carbon sol	Increased root and shoot biomass, ameliorated potassium absorption	Tobacco	J. et al. 2015
Fullerene C60	Increased level of phytomedicinal content with increased biomass	Bitter melon	Husen and Siddiqi 2014

the absorption of toxic nitrogenous waste of herbicides (Duman et al. 2019) as well as antibiotics like tetracycline (Babaei et al. 2016). In contrast, carbon nanotubes and fullerenes display a mobilization capacity for different organic-based pollutants, like lindane an agricultural insecticide (Srivastava et al. 2011) and sustained polychlorinated biphenyls (Wang et al. 2013). Recycling of filter-based carbon nanotubes can be done by desorbing contaminants (Wang et al. 2014). Carbon nanomaterials also have antibacterial activity, even though the action mode of mechanism is still being investigated. Other research studies reveal the feasible use of CNTs for disinfection reasons or to cover the outer surface of the antimicrobials. In particular, silver nanoparticles (AgNPs) coated with silver exhibit antimicrobial properties and materials showing such activity can be introduced in the biomedical tools and antibacterial control systems. Moreover, some studies also suggested the decontamination of water with the help of integrative functioning of carbon nanotubes filters with microwaves (Jung et al. 2011). Water-based purification technology uses CNTs, including disinfection in water outlined in the effective reviews (Upadhyayula et al. 2009, Das et al. 2018).

6. Toxic Effects of Carbon Nanomaterials

The broad attention received by carbon nanomaterials due to their uncommon chemical and physical properties is remarkable. The extensive utilization of CNMs led some scientists to study the toxicity impacts of CNMs against the environment, humans, plants, and animals. In this literature, we have listed down some of the toxicity effects of CNMs in Table 3. In some previous investigations, it was demonstrated that multi-walled carbon nanotubes (MWCNTs) show cytotoxicity against neutrophil-like differentiated HL-60 cells (Tabei et al. 2019). An investigation on aquatic larvae of *Chironomus riparius* revealed that MWCNTs hinder the gene transcription that is involved in apoptosis (Martinez-Paz et al. 2019). In a study, it was stated that graphene can form graphene accumulates, and it exhibits toxicity to CRL-2522 cells (Chng et al. 2015). Two investigations reported that graphene and graphene oxide exhibit a toxic effect on the human skin and erythrocytes (Chng et al. 2013, Liao et al. 2011). Graphene and its derivatives showed a toxic effect on various levels, like membrane integrity, metabolic activity, macrophages, and lysosomal action of fish cells (Kalman et al. 2019). The surface charge on fullerene (C_{60}) imparts a toxic effect on the biological activity of proteins and cell structure (Gieldon et al. 2017). C_{60} is aquatic pollutants and is capable of DNA damage despite being insoluble in water (Matsuda et al. 2011). The size of CNMs is quite small which allows them to enter into an immune cell directly and causes an inflammatory reaction (Vasyukova et al. 2015) and demonstrated that CNTs can impair bacterial cell membrane causing leakage of nucleic acid and declination of the metabolism activity.

7. Conclusion

In the present study, various reasons were laid down as to how CBNs have acquired great attention in the past years in biomedical applications, plant systems, and environmental applications. In the last two decades, researchers have put terrific efforts to understand CBNs which are one of the broadly used nanomaterials. The interesting fact about CBNs is the occurrence of both organic π-π stacking characteristics and inorganic semiconducting properties. Concurrently, it reacts to light and interacts with other biological molecules. Such features bring a huge advantage for the future. In terms of the toxic impact on the biological system, various chemical modification approaches are being drafted and effectively employed in biological applications, like cancer therapy, drug delivery, and recognition of biomolecules. The reviewed literature of CBNs has focused on a few advancements in environmental applications, biomedical applications, and green plants. Therefore, more scientific investigations are required to know thoroughly about the toxicity impact of CBNs on humans, animals, plants, and the environment.

Table 3: Implications of carbon-based nanomaterials on cellular cytotoxicity

Structure Types	Particle Size	Toxicity Dose Concentration	Cytotoxicity	Target Cells	References
Water-soluble fullerene	10.18 Å	8×10^{-5} M	Release of Lactate dehydrogenase, cellular membrane disruption, and lipid peroxidation	Human dermal blasts, HepG2, neuronal astrocytes	Pulskamp, Karin et al. 2007
Carbon nanotubes	1.5 µm; 9.5 nm	120 µg/ml	Release of LDH, increased TNF-α production, ROS production	Macrophage	Agathe et al. 2015
Graphene oxides (GOs)	205-500 nm	10~50 µg/ml	Decrease in autophagic degradation, destabilization of lysosomal membrane	Serosa macrophages	Bin et al. 2013
Multi-walled carbon nanotube	>50 nm	100-120 µg/ml	Bronchioalveolar cells, 3T3 cells, Macrophage, A_{549} cells	Mitochondrial membrane destabilization, ROS generation, lysosomal destabilization	Bo et al. 2011, Syed K et al. 2010
Carbon nanohorns	>70 nm	100 µg/ml	Macrophage	Generation of ROS, lysosomal membrane permeabilization, and degradation	Mei et al. 2014
SWCNTs	>530 nm	8 and 32 mg/L	Mesothelium cells of normal and malignant tissues	Increased cell apoptosis, generation of ROS, increased signaling of PARP, AP-1, NF-κB, p38, and Akt pathways	Maricica et al. 2008, Tao et al. 2020

Material	Size	Concentration	Cell type	Effect	Reference
Aminated MWCNT; Carboxylated-MWCNT	652 nm	1.6 or 4 mg/kg	Pulmonary cells	Acute and chronic pulmonary toxicity	Alexia J et al. 2020
MWCNTs	30-100 nm	$0.625 \sim 10 \ \mu g/cm^2$	Alveolar Macrophage	Reduced mitochondrial activity, LDH release	D van Berlo et al. 2014
Carbon black nanoparticles	>50 nm	50-800 μg/mL	THP-1 cells	Inducing cytotoxicity, inflammation, and altered phagocytosis in human monocytes	Devashri et al. 2014

Conflicts of Interest

The authors state to have no conflict of interest.

References

Aharonovich, I., Castelletto, S., Simpson, D.A., Su, C.H., Greentree, A.D. and Prawer, S. 2011. Diamond-based single-photon emitters. Rep. Prog. Phys. 74(7): 076501. DOI: 10.1088/0034-4885/74/7/076501

Amans, D., Chenus, A.C., Ledoux, G., Dujardin, C., Reynaud, C., Sublemontier, O. and Guillois, O. 2009. Nanodiamond synthesis by pulsed laser ablation in liquids. DRM 18(2-3): 177-180. DOI: 10.1016/j.diamond.2008.10.035.

Babaei, A.A., Lima, E.C., Takdastan, A., Alavi, N., Goudarzi, G., Vosoughi, M., Hassani, G. and Shirmardi, M. 2016. Removal of tetracycline antibiotic from contaminated water media by multi-walled carbon nanotubes: operational variables, kinetics, and equilibrium studies. Water Sci Technol 74(5): 1202-1216. DOI: 10.2166/wst.2016.301

Baldo, S., Buccheri, S., Ballo, A., Camarda, M., La Magna, A., Castagna, M.E. and Scalese, S. 2016. Carbon nanotube-based sensing devices for human Arginase-1 detection. Sens. Bio-Sens. Res. 7: 168-173. DOI: 10.1016/j.sbsr.2015.11.011

Bhattacharya, K., Mukherjee, S.P., Gallud, A., Burkert, S.C., Bistarelli, S. and Bellucci, S. 2016. Biological interactions of carbon-based nanomaterials: from coronation to degradation. Nanomedicine 12: 333-351. DOI: 10.1016/j.nano.2015.11.011

Bisker, G., Dong, J., Park, H.D., Iverson, N.M., Ahn, J., Nelson, J.T. and Strano, M.S. 2016. Protein-targeted corona phase molecular recognition. Nature Commun. 7(1): 1-14. DOI: 10.1038/ncomms10241

Borišev, M., Borišev, I., Župunski, M., Arsenov, D., Pajević, S., Ćurčić, Ž. and Djordjevic, A. 2016. Drought impact is alleviated in sugar beets (Beta vulgaris L.) by foliar application of fullerenol nanoparticles. PLoS One, 11(11): e0166248. DOI: 10.1371/journal.pone.0166248

Braakhuis, H.M., Gosens, I., Krystek, P., Boere, J.A., Cassee, F.R., Fokkens, P.H. and Park, M.V. 2014. Particle size dependent deposition and pulmonary inflammation after short-term inhalation of silver nanoparticles. Part. Fibre Toxicol. 11(1): 1-16. DOI: 10.1186/s12989-014-0049-1

Calvaresi, M., Arnesano, F., Bonacchi, S., Bottoni, A., Calo, V., Conte, S. and Zerbetto, F. 2014. C60@ Lysozyme: direct observation by nuclear magnetic resonance of a 1:1 fullerene protein adduct. ACS Nano 8(2): 1871-1877. DOI: 10.1021/nn4063374

Carniel, F.C., Gorelli, D., Flahaut, E., Fortuna, L., Del Casino, C., Cai, G. and Tretiach, M. 2018. Graphene oxide impairs the pollen performance of Nicotiana tabacum and Corylus avellana suggesting potential negative effects on the sexual reproduction of seed plants. Environ. Sci.: Nano 5(7): 1608-1617. DOI: 10.1039/C8EN00052B

Chen, K.L. and Elimelech, M. 2006. Aggregation and deposition kinetics of fullerene (C60) nanoparticles. Langmuir 22(26): 10994-11001. DOI: 10.1021/la062072v

Chen, B., Liu, Y., Song, W.M., Hayashi, Y., Ding, X.C. and Li, W.H. 2011. In vitro evaluation of cytotoxicity and oxidative stress induced by multiwalled carbon nanotubes in murine RAW 264.7 macrophages and human A549 lung cells. BES 24(6): 593-601.DOI: 10.3967/0895-3988.2011.06.002

Chen, L., Wang, C., Li, H., Qu, X., Yang, S.T. and Chang, X.L. 2017. Bioaccumulation and toxicity of 13C-skeleton labeled graphene oxide in wheat. ES&T 51(17): 10146-10153. DOI: 10.1021/acs.est.7b00822

Chen, Fei, Gao, Weiyin, Qiu, Xiaopei, Zhang, Hong, Liu, Lianhua, Liao, Pu, Fu, Weiling, Luo, Yang. 2018. Graphene quantum dots in biomedical applications: recent advances and future challenges. FILM 1. DOI: 10.1016/j.flm.2017.12.006.

Chen, L., Yang, S., Liu, Y., Mo, M., Guan, X., Huang, L. and Chang, X.L. 2018. Toxicity of graphene oxide to naked oats (Avena sativa L.) in hydroponic and soil cultures. RSC Adv. 8(28): 15336-15343. DOI: 10.1039/C8RA01753K

Chng, E.L.K. and Pumera, M. 2013. The toxicity of graphene oxides: dependence on the oxidative methods used. Chemistry – A European Journal 19(25): 8227-8235. DOI: 10.1002/chem.201300824

Chng, E.L.K. and Pumera, M. 2015. Toxicity of graphene related materials and transition metal dichalcogenides. RSC Adv. 5(4): 3074-3080. DOI: 10.1039/C4RA12624F

Das, R., Leo, B.F. and Murphy, F. 2018. The toxic truth about carbon nanotubes in water purification: a perspective view. Nanoscale Res. Lett. 13(1): 183. DOI: 10.1186/s11671-018-2589-z

Daulton, T.L., Kirk, M.A., Lewis, R.S. and Rehn, L.E. 2001. Production of nanodiamonds by high-energy ion irradiation of graphite at room temperature. Nucl Instrum Methods P. 175: 12-20. DOI: 10.1016/S0168-583X(00)00603-0

Deng, J., You, Y., Sahajwalla, V. and Joshi, R.K. 2016. Transforming waste into carbon-based nanomaterials. Carbon 96: 105-115. DOI: 10.1016/j.carbon.2015.09.033

Dichiara, A.B., Webber, M.R., Gorman, W.R. and Rogers, R.E. 2015. Removal of copper ions from aqueous solutions via adsorption on carbon nanocomposites. ACS Appl. Mater. Interfaces 7(28): 15674-15680. DOI: 10.1021/acsami.5b04974

Duman, O., Ozcan, C., Gürkan Polat, T. and Tunc, S. 2019. Carbon nanotube-based magnetic and non-magnetic adsorbents for the high-efficiency removal of diquat dibromide herbicide from water: OMWCNT, OMWCNT-Fe_3O_4 and OMWCNT-κ-carrageenan-Fe_3O_4 nanocomposites. Environ. Pollution 244: 723-732. DOI: 10.1016/j.envpol.2018.10.071

Falvo, M.R., Clary, G.J., Taylor, R.M., Chi, V., Brooks, F.P., Jr, Washburn, S. and Superfine, R. 1997. Bending and buckling of carbon nanotubes under large strain. Nature 389(6651): 582-584. DOI: 10.1038/39282

Fan, H.Y., Yu, X.H., Wang, K., Yin, Y.J., Tang, Y.J., Tang, Y.L. and Liang, X.H. 2019. Graphene quantum dots (GQDs)-based nanomaterials for improving photodynamic therapy in cancer treatment. Eur. J. Med. Chem. 182: 111620. DOI: 10.1016/j.ejmech.2019.111620

Figarol, A., Pourchez, J., Boudard, D., Forest, V., Akono, C., Tulliani, J.M., Lecompte, J.P., Cottier, M., Bernache-Assollant, D. and Grosseau, P. 2015. In vitro toxicity of carbon nanotubes, nano-graphite and carbon black, similar impacts of acid functionalization. Toxicol. In. Vitro. 30(1 Pt B): 476-485. DOI: 10.1016/j.tiv.2015.09.014

Fiorica, C., Mauro, N., Pitarresi, G., Scialabba, C., Palumbo, F.S. and Giammona, G. 2017. Double-network-structured graphene oxide-containing nanogels as photothermal agents for the treatment of colorectal cancer. Biomacromolecules 18(3): 1010-1018. DOI: 10.1021/acs.biomac.6b01897

Fiorillo, M., Verre, A.F., Iliut, M., Peiris-Pagés, M., Ozsvari, B., Gandara, R. and Lisanti, M.P. 2015. Graphene oxide selectively targets cancer stem cells, across multiple tumor types: implications for non-toxic cancer treatment, via "differentiation-based nano-therapy". Oncotarget 6(6): 3553. DOI: 10.18632/oncotarget.3348

Flores, D., Chaves, J.S., Chacón, R. and Schmidt, A. 2013. A novel technique using SWCNTs to enhanced development and root growth of fig plants (Ficus carica). In: Technical Proceedings of the NSTI Nanotechnology Conference and Expo (NSTI-Nanotech'13), Vol. 3, pp. 167-170.

Fortner, J.D., Lyon, D.Y., Sayes, C.M., Boyd, A.M., Falkner, J.C., Hotze, E.M., Alemany, L.B., Tao, Y.J., Guo, W., Ausman, K.D., Colvin, V.L. and Hughes, J.B. 2005. C60 in water:

nanocrystal formation and microbial response. Environ. Sci. Tech. 39(11): 4307-4316. DOI: 10.1021/es048099n

Geim, A.K. and Novoselov, K.S. 2007. The rise of graphene. Nat. Mater. 6: 183-191. DOI: 10.1038/nmat1849

Georgakilas, V., Perman, J.A., Tucek, J. and Zboril, R. 2015. Broad family of carbon nanoallotropes: classification, chemistry, and applications of fullerenes, carbon dots, nanotubes, graphene, nanodiamonds, and combined superstructures. Chemical Reviews 115(11): 4744-4822. DOI: 10.1021/cr500304f

Gholizadeh, S., Moztarzadeh, F., Haghighipour, N., Ghazizadeh, L., Baghbani, F., Shokrgozar, M.A. and Allahyari, Z. 2017. Preparation and characterization of novel functionalized multiwalled carbon nanotubes/chitosan/β-Glycerophosphate scaffolds for bone tissue engineering. Int. J. Biol. Macromol. 97: 365-372. DOI: 10.1016/j.ijbiomac.2016.12.086

Gieldon, A., Witt, M.M., Gajewicz, A. and Puzyn, T. 2017. Rapid insight into C60 influence on biological functions of proteins. Structural Chem. 28(6): 1775-1788. DOI: 10.1007/s11224-017-0957-4

Giordani, S., Camisasca, A, and Maffeis, V. 2019. Carbon nano-onions: a valuable class of carbon nanomaterials in biomedicine. Current Med. Chem. 26(38): 6915-6929. DOI: 10.2174/0929867326666181126113957

Guler, O. and Katmer, H. 2020. Investigating the synergistic effect of CNT+ MLG hybrid structure on copper matrix and electrical contact properties of the composite. Eur. Phys. J. Plus. 135(3): 308. DOI: 10.1140/epjp/s13360-020-00315-w

Gupta, C., Prakash, D. and Gupta, S. 2017. Cancer treatment with nano-diamonds. Front Biosci (Schol Ed) (Scholar edition), 9: 62-70. DOI: 10.2741/s473

Gupta, S., Murthy, C.N. and Prabha, C.R. 2018. Recent advances in carbon nanotube based electrochemical biosensors. Int. J. Biol. Macromol. 108: 687-703. DOI: 10.1016/j.ijbiomac.2017.12.038

He, H., Pham-Huy, L.A., Dramou, P., Xiao, D., Zuo, P. and Pham-Huy, C. 2013. Carbon nanotubes: applications in pharmacy and medicine. BioMed Res. Inter. 2013. DOI: 10.1155/2013/578290

Hong, G., Diao, S., Antaris, A.L. and Dai, H. 2015. Carbon nanomaterials for biological imaging and nanomedicinal therapy. Chem. Rev. 115(19): 10816-10906. DOI: 10.1021/acs.chemrev.5b00008

Husen, A. and Siddiqi, K.S. 2014. Carbon and fullerene nanomaterials in plant system. J. nanobiotechnology 12: 16. https://doi.org/10.1186/1477-3155-12-16

Iijima, S. 1991. Helical microtubules of graphitic carbon. Nature 354: 56-58. DOI: 10.1038/354056a0

Jena, P.V., Roxbury, D., Galassi, T.V., Akkari, L., Horoszko, C.P., Iaea, D.B. and Heller, D.A. 2017. A carbon nanotube optical reporter maps endolysosomal lipid flux. ACS Nano 11(11): 10689-10703. DOI: 10.1021/acsnano.7b04743

Jiang, T., Amadei, C.A., Gou, N., Lin, Y., Lan, J., Vecitis, C.D. and Gu, A.Z. 2020. Toxicity of single-walled carbon nanotubes (SWCNTs): effect of lengths, functional groups and electronic structures revealed by a quantitative toxicogenomics assay. Environ. Sci. Nano 7(5): 1348-1364. DOI: 10.1039/d0en00230e

Jung, J.H., Hwang, G.B., Lee, J.E. and Bae, G.N. 2011. Preparation of airborne Ag/CNT hybrid nanoparticles using an aerosol process and their application to antimicrobial air filtration. Langmuir 27(16): 10256-10264. DOI: 10.1021/la201851r

Kah, M. and Hofmann, T. 2015. The challenge: carbon nanomaterials in the environment – new threats or wonder materials? Environ. Toxicol. Chem. 34(5): 954-954. DOI: 10.1002/etc.2898

Kalman, J., Merino, C., Fernández-Cruz, M.L. and Navas, J.M. 2019. Usefulness of fish cell lines for the initial characterization of toxicity and cellular fate of graphene-related materials (carbon nanofibers and graphene oxide). Chemosphere 218: 347-358. DOI: 10.1016/j.chemosphere.2018.11.130

Kam, N.W.S., O'Connell, M., Wisdom, J.A. and Dai, H. 2005. Carbon nanotubes as multifunctional biological transporters and near-infrared agents for selective cancer cell destruction. PNAS 102(33): 11600-11605. DOI: 10.1073/pnas.0502680102

Khajouei, S., Ravan, H. and Ebrahimi, A. 2020. DNA hydrogel-empowered biosensing. Adv. Colloid Interface Sci. 275: 102060. DOI: 10.1016/j.cis.2019.102060

Khodakovskaya, M.V., de Silva, K., Nedosekin, D.A., Dervishi, E., Biris, A.S., Shashkov, E.V. and Zharov, V.P. 2011. Complex genetic, photothermal, and photoacoustic analysis of nanoparticle-plant interactions. PNAS 108(3): 1028-1033. DOI: 10.1073/pnas.1008856108

Khodakovskaya, M.V., Kim, B.S., Kim, J.N., Alimohammadi, M., Dervishi, E., Mustafa, T. and Cernigla, C.E. 2013. Carbon nanotubes as plant growth regulators: effects on tomato growth, reproductive system, and soil microbial community. Small (Weinheim an der Bergstrasse, Germany), 9(1): 115-123. DOI: 10.1002/smll.201201225

Ko, S.H. and Grigoropoulos, C.P. (Eds.). 2014. Hierarchical nanostructures for energy devices. Royal Society of Chemistry. DOI: 10.1039/9781849737500

Kole, C., Kole, P., Randunu, K.M., Choudhary, P., Podila, R., Ke, P.C., Rao, A.M. and Marcus, R.K. 2013. Nanobiotechnology can boost crop production and quality: first evidence from increased plant biomass, fruit yield and phytomedicine content in bitter melon (Momordica charantia). BMC Biotech. 13: 37. DOI: 10.1186/1472-6750-13-37

Kovalenko, I., Bucknall, D.G. and Yushin, G. 2010. Detonation nanodiamond and onion-like-carbon-embedded polyaniline for supercapacitors. Adv. Func. Mat. 20(22): 3979-3986. DOI: 10.1002/adfm.201000906

Kroto, H.W., Heath, J.R., O'Brien, S.C., Curl, R.F. and Smalley, R.E. 1985. C60: buckminsterfullerene. Nature 318(6042): 162-163. DOI: 10.1038/318162a0

Kumar, A., Singh, A., Panigrahy, M., Sahoo, P.K. and Panigrahi, K.C. 2018. Carbon nanoparticles influence photomorphogenesis and flowering time in Arabidopsis thaliana. Plant Cell Rep. 37(6): 901-912. DOI: 10.1007/s00299-018-2277-6

Lahiani, M.H., Dervishi, E., Chen, J., Nima, Z., Gaume, A., Biris, A.S. and Khodakovskaya, M.V. 2013. Impact of carbon nanotube exposure to seeds of valuable crops. ACS Appl. Mater. Interfaces 5(16): 7965-7973. DOI: 10.1021/am402052x

Li, Y., Yuan, H., von Dem Bussche, A., Creighton, M., Hurt, R.H., Kane, A.B. and Gao, H. 2013. Graphene microsheets enter cells through spontaneous membrane penetration at edge asperities and corner sites. PNAS 110(30): 12295-12300. DOI: 10.1073/pnas.1222276110

Liang, C., Xiao, H., Hu, Z., Zhang, X. and Hu, J. 2018. Uptake, transportation, and accumulation of C60 fullerene and heavy metal ions (Cd, Cu, and Pb) in rice plants grown in an agricultural soil. Environ. Pollution 235: 330-338. DOI: 10.1016/j.envpol.2017.12.062

Liao, K.H., Lin, Y.S., Macosko, C.W. and Haynes, C.L. 2011. Cytotoxicity of graphene oxide and graphene in human erythrocytes and skin fibroblasts. ACS Appl. Mater. Interfaces 3(7): 2607-2615. DOI: 10.1021/am200428v

Lin, J., Huang, Y. and Huang, P. 2018. Graphene-based nanomaterials in bioimaging. BAFN 247-287. DOI: 10.1016/B978-0-323-50878-0.00009-4

Liu, X., Wang, M., Zhang, S. and Pan, B. 2013. Application potential of carbon nanotubes in water treatment: a review. J. Environ. Sci. (China), 25(7): 1263-1280. DOI: 10.1016/S1001-0742(12)60161-2

Luo, Q., Ma, H., Hou, Q., Li, Y., Ren, J., Dai, X. and Guo, Z. 2018. All-carbon-electrode-based endurable flexible perovskite solar cells. Adv. Func. Mat. 28(11): 1706777. DOI: 10.1002/adfm.201706777

Maiti, D., Tong, X., Mou, X. and Yang, K. 2019. Carbon-based nanomaterials for biomedical applications: a recent study. Front. Pharmacol. 9: 1401. DOI: 10.3389/fphar.2018.01401

Martinez-Paz, P., Negri, V., Esteban-Arranz, A., Martínez-Guitarte, J.L., Ballesteros, P. and Morales, M. 2019. Effects at molecular level of multi-walled carbon nanotubes (MWCNT) in Chironomus riparius (DIPTERA) aquatic larvae. Aquatic Toxicology 209: 42-48. DOI: 10.1016/j.aquatox.2019.01.017

Marty, B., Alexander, C.M.O. and Raymond, S.N. 2013. Primordial origins of earth's carbon. Rev. Mineral. Geochem. 75: 149-181. DOI: 10.2138/rmg.2013.75.6

Matsuda, S., Matsui, S., Shimizu, Y. and Matsuda, T. 2011. Genotoxicity of colloidal fullerene C60. Environ. Sci. Tech. 45(9): 4133-4138. DOI: 10.1021/es1036942

Mauter, M.S. and Elimelech, M. 2008. Environmental applications of carbon-based nanomaterials. Environ. Sci. Tech. 42(16): 5843-5859. DOI: 10.1021/es8006904

Mehra, N.K., Jain, A.K. and Nahar, M. 2018. Carbon nanomaterials in oncology: an expanding horizon. Drug Discov. Today 23(5): 1016-1025. DOI: 10.1016/j.drudis.2017.09.013

Mohajeri, M., Behnam, B. and Sahebkar, A. 2019. Biomedical applications of carbon nanomaterials: drug and gene delivery potentials. J. Cell. Physiol. 234(1): 298-319. DOI: 10.1002/jcp.26899

Mu, Q., Jiang, G., Chen, L., Zhou, H., Fourches, D., Tropsha, A. and Yan, B. 2014. Chemical basis of interactions between engineered nanoparticles and biological systems. Chemical Reviews 114(15): 7740-7781. DOI: 10.1021/cr400295a

Nasrollahi, F., Koh, Y.R., Chen, P., Varshosaz, J., Khodadadi, A.A. and Lim, S. 2019. Targeting graphene quantum dots to epidermal growth factor receptor for delivery of cisplatin and cellular imaging. Mater. Sci. Eng. C. 94: 247-257. https://doi.org/10.1016/j.msec.2018.09.020

Novoselov, K.S., Geim, A.K., Morozov, S.V., Jiang, D., Zhang, Y., Dubonos, S.V. and Firsov, A.A. 2004. Electric field effect in atomically thin carbon films. Science 306(5696): 666-669. DOI: 10.1126/science.1102896

Pace, N.R. 2001. The universal nature of biochemistry. Proc. Natl. Acad. Sci. USA 98(3): 805-808. DOI: 10.1073/pnas.98.3.805

Park, S., Choi, K. S., Kim, S., Gwon, Y., & Kim, J. 2020. oxide-assisted promotion of plant growth and stability. Nanomaterials 10(4): 758. 10.3390/nano10040758. DOI: 10.3390/nano10040758

Peng, Q., Dearden, A.K., Crean, J., Han, L., Liu, S., Wen, X. and De, S. 2014. New materials graphyne, graphdiyne, graphone, and graphane: review of properties, synthesis, and application in nanotechnology. Nanotechnol. Sci. Appl. 7: 1. DOI: 10.2147/NSA.S40324

Pulskamp, K., Diabaté, S. and Krug, H.F. 2007. Carbon nanotubes show no sign of acute toxicity but induce intracellular reactive oxygen species in dependence on contaminants. Toxicol. Lett. 168(1): 58-74. DOI: 10.1016/j.toxlet.2006.11.001

Qu, X., Alvarez, P.J. and Li, Q. 2013. Applications of nanotechnology in water and wastewater treatment. Water Res. 47(12): 3931-3946. DOI: 10.1016/j.watres.2012.09.058

Rahbari, M. and Goharrizi, A.S. 2009. Adsorption of lead (II) from water by carbon nanotubes: equilibrium, kinetics, and thermodynamics. Water Environ. Res. 81(6): 598-607. DOI: 10.2175/106143008x370511

Russier, J., Ménard-Moyon, C., Venturelli, E., Gravel, E., Marcolongo, G., Meneghetti, M., Doris, E. and Bianco, A. 2011. Oxidative biodegradation of single- and multi-walled carbon nanotubes. Nanoscale 3(3): 893-896. DOI: 10.1039/C0NR00779J

Sahu, D., Kannan, G.M. and Vijayaraghavan, R. 2014. Carbon black particle exhibits size dependent toxicity in human monocytes. Int J. Inflam. 2014: 827019. DOI: 10.1155/2014/827019

Saleem, H. and Zaidi, S.J. 2020. Developments in the application of nanomaterials for water treatment and their impact on the environment. Nanomaterials 10(9): 1764. DOI: 10.3390/nano10091764

Serag, M.F., Kaji, N., Tokeshi, M. and Baba, Y. 2015. Carbon nanotubes and modern nanoagriculture. *In:* Siddiqui, M., Al-Whaibi, M., Mohammad, F. (eds.). Nanotechnology and Plant Sciences. Springer, Cham. https://doi.org/10.1007/978-3-319-14502-0_10

Shenderova, O.A., Zhirnov, V.V. and Brenner, D.W. 2002. Carbon nanostructures. Crit. Rev. Solid State Mater. Sci. 27(3-4): 227-356. DOI: 10.1080/10408430208500497

Skivka, L.M., Prylutska, S.V., Rudyk, M.P., Khranovska, N.M., Opeida, I.V., Hurmach, V.V., Prylutskyy, Y.I., Sukhodub, L.F. and Ritter, U. 2018. C60 fullerene and its nanocomplexes with anticancer drugs modulate circulating phagocyte functions and dramatically increase ROS generation in transformed monocytes. Cancer Nanotechnol. 9(1): 8. DOI: 10.1186/s12645-017-0034-0

Smith, A.T., Lachance, A.M., Zeng, S., Liu, B. and Sun, L. 2019. Synthesis, properties, and applications of graphene oxide/reduced graphene oxide and their nanocomposites, Nano Mater. Sci. 1(1): 31-47. DOI: 10.1016/j.nanoms.2019.02.004

Sohaebuddin, S.K., Thevenot, P.T., Baker, D., Eaton, J.W. and Tang, L. 2010. Nanomaterial cytotoxicity is composition, size, and cell type dependent. Part. Fibre Toxicol. 7: 22. DOI: 10.1186/1743-8977-7-22

Srivastava, M., Abhilash, P.C. and Singh, N. 2011. Remediation of lindane using engineered nanoparticles. J. Biomed. Nanotechnol. 7(1): 172-174. DOI: 10.1166/jbn.2011.1255

Tabei, Y., Fukui, H. and Nishioka, A. 2019. Effect of iron overload from multi walled carbon nanotubes on neutrophil-like differentiated HL-60 cells. Sci. Rep. 9: 2224. DOI: 10.1038/s41598-019-38598-4

Taylor-Just, A.J., Ihrie, M.D., Duke, K.S., Lee, H.Y., You, D.J., Hussain, S., Kodali, V.K., Ziemann, C., Creutzenberg, O., Vulpoi, A., Turcu, F., Potara, M., Todea, M., van den Brule, S., Lison, D. and Bonner, J.C. 2020. The pulmonary toxicity of carboxylated or aminated multi-walled carbon nanotubes in mice is determined by the prior purification method. Part. Fibre Toxicol. 17(1): 60. DOI: 10.1186/s12989-020-00390-y

Tilmaciu, C.M. and Morris, M.C. 2015. Carbon nanotube biosensors. Front. Chem. 3: 59. DOI: 10.3389/fchem.2015.00059

Titirici, M.M., White, R.J., Brun, N., Budarin, V.L., Su, D.S., Del Monte, F., Clark, J.H. and MacLachlan, M.J. 2015. Sustainable carbon materials. Chem. Soc. Rev. 44, 250-290. DOI: 10.1039/C4CS00232F

Tucek, J., Błoński, P., Ugolotti, J., Swain, A.K., Enoki, T. and Zbořil, R. 2018. Emerging chemical strategies for imprinting magnetism in graphene and related 2D materials for spintronic and biomedical applications. Chem. Soc. Rev. 47(11): 3899-3990. DOI: 10.1039/C7CS00288B

Ulissi, Z.W., Sen, F., Gong, X., Sen, S., Iverson, N., Boghossian, A.A. and Strano, M.S. 2014. Spatiotemporal intracellular nitric oxide signaling captured using internalized, near-infrared fluorescent carbon nanotube nanosensors. Nano Lett. 14(8): 4887-4894. DOI: 10.1021/nl502338y

Upadhyayula, V.K., Deng, S., Mitchell, M.C. and Smith, G.B. 2009. Application of carbon nanotube technology for removal of contaminants in drinking water: a review. Sci. Total Environ. 408(1): 1-13. DOI: 10.1016/j.scitotenv.2009.09.027

van Berlo, D., Wilhelmi, V., Boots, A.W., Hullmann, M., Kuhlbusch, T.A., Bast, A., Schins, R. P. and Albrecht, C. 2014. Apoptotic, inflammatory, and fibrogenic effects of two different types of multi-walled carbon nanotubes in mouse lung. Arch. Toxicol. 88(9): 1725-1737. DOI: 10.1007/s00204-014-1220-z

Vasyukova, I., Gusev, A. and Tkachev, A. 2015. Reproductive toxicity of carbon nanomaterials: a review. *In:* IOP Conference Series: Materials Science and Engineering. Vol. 98(1): 012001. DOI: 10.1088/1757-899X/98/1/012001

Wan, B., Wang, Z.X., Lv, Q.Y., Dong, P.X., Zhao, L.X., Yang, Y. and Guo, L.H. 2013. Single-walled carbon nanotubes and graphene oxides induce autophagosome accumulation and lysosome impairment in primarily cultured murine peritoneal macrophages. Toxicol. Lett. 221(2): 118-127. DOI: 10.1016/j.toxlet.2013.06.208

Wang, L., Fortner, J.D., Hou, L., Zhang, C., Kan, A.T., Tomson, M.B. and Chen, W. 2013. Contaminant-mobilizing capability of fullerene nanoparticles (nC60): effect of solvent-exchange process in nC60 formation. Environ. Toxicol. Chem. 32(2): 329-336. DOI: 10.1002/etc.2074

Wang, H., Ma, H., Zheng, W., An, D. and Na, C. 2014. Multifunctional and recollectable carbon nanotube ponytails for water purification. ACS Appl. Mater. Interfaces 6(12): 9426-9434. DOI: 10.1021/am501810f

Wang, C., Zhang, H., Ruan, L., Chen, L., Li, H., Chang, X.L. and Yang, S.T. 2016. Bioaccumulation of 13 C-fullerenol nanomaterials in wheat. Environ. Sci.: Nano 3(4): 799-805. DOI: 10.1039/C5EN00276A

Wen, K.P., Chen, Y.C., Chuang, C.H., Chang, H.Y., Lee, C.Y. and Tai, N.H. 2015. Accumulation and toxicity of intravenously-injected functionalized graphene oxide in mice. J. Appl. Toxicol. 35(10): 1211-1218. DOI: 10.1002/jat.3187

Xia, J., Kawamura, Y., Suehiro, T., Chen, Y. and Sato, K. 2019. Carbon dots have antitumor action as monotherapy or combination therapy. Drug Discov. Ther. 13(2): 114-117. DOI: 10.5582/ddt.2019.01013

Xiong, J.L., Li, J., Wang, H.C., Zhang, C.L. and Naeem, M.S. 2018. Fullerol improves seed germination, biomass accumulation, photosynthesis and antioxidant system in Brassica napus L. under water stress. Plant Physiol. Biochem. 129: 130-140. DOI: 10.1016/j.plaphy.2018.05.026

Yan, H., Gong, A., He, H., Zhou, J., Wei, Y. and Lv, L. 2006. Adsorption of microcystins by carbon nanotubes. Chemosphere, 62(1): 142-148. DOI: 10.1016/j.chemosphere.2005.03.075

Yan, S., Zhao, L., Li, H., Zhang, Q., Tan, J., Huang, M., He, S. and Li, L. 2013. Single-walled carbon nanotubes selectively influence maize root tissue development accompanied by the change in the related gene expression. J. Hazard. Mater. 246-247: 110-118. DOI: 10.1016/j.jhazmat.2012.12.013

Yang, M., Zhang, M., Tahara, Y., Chechetka, S., Miyako, E., Iijima, S. and Yudasaka, M. 2014. Lysosomal membrane permeabilization: carbon nanohorn-induced reactive oxygen species generation and toxicity by this neglected mechanism. Toxicol. Appl. Pharmacol. 280(1): 117–126. DOI: 10.1016/j.taap.2014.07.022

Yang, Y. and Westerhoff, P. 2014. Presence in, and release of, nanomaterials from consumer products. Nanomaterial 1-17. DOI: 10.1007/978-94-017-8739-0_1

Yang, C., Chan, K.K., Xu, G., Yin, M., Lin, G., Wang, X., Lin, W.J., Birowosuto, M.D., Zeng, S., Ogi, T., Okuyama, K., Permatasari, F.A., Iskandar, F., Chen, C.K. and Yong, K.T. 2019. Biodegradable polymer-coated multifunctional graphene quantum dots for light-triggered synergetic therapy of pancreatic cancer. ACS Appl. Mater. Interfaces 11(3): 2768-2781. DOI: 10.1021/acsami.8b16168

Yi, H., Huang, D., Qin, L., Zeng, G., Lai, C., Cheng, M. and Guo, X. 2018. Selective prepared carbon nanomaterials for advanced photocatalytic application in environmental pollutant treatment and hydrogen production. Applied Catalysis B: Environmental 239: 408-424. DOI: 10.1016/j.apcatb.2018.07.068

Yi, J., Choe, G. and Park, J. 2020. Graphene oxide-incorporated hydrogels for biomedical applications. Polym. J. 52: 823-837. DOI: 10.1038/s41428-020-0350-9

Younis, M.R., He, G., Lin, J. and Huang, P. 2020. Recent advances on graphene quantum dots for bioimaging applications. Front. Chem. 8: 424. DOI: 10.3389/fchem.2020.00424

Zaytseva, O. and Neumann, G. 2016. Carbon nanomaterials: production, impact on plant development, agricultural and environmental applications. Chem. Biol. Technol. Agric. 3: 17. DOI: 10.1186/s40538-016-0070-8

Zhang, Y. and Yin, Q.Z. 2012. Carbon and other light element contents in the Earth's core based on first-principles molecular dynamics. Proc. Natl. Acad. Sci. USA 109: 19579-19583. DOI: 10.1073/pnas.120382610

Zhang, Y., Bai, W., Cheng, X., Ren, J., Weng, W., Chen, P. and Peng, H. 2014. Flexible and stretchable lithium-ion batteries and supercapacitors based on electrically conducting carbon nanotube fiber springs. Angewandte Chemie International Edition, 53(52): 14564-14568. DOI: 10.1002/anie.201409366

Zhang, P., Zhang, R., Fang, X., Song, T., Cai, X., Liu, H. and Du, S. 2016. Toxic effects of graphene on the growth and nutritional levels of wheat (Triticum aestivum L.): short-and long-term exposure studies. J. Hazard. Mater. 317: 543-551. DOI: 10.1016/j.jhazmat.2016.06.019

Zhao, C., Song, X., Liu, Y., Fu, Y., Ye, L., Wang, N., Wang, F., Li, L., Mohammadniaei, M., Zhang, M., Zhang, Q. and Liu, J. 2020. Synthesis of graphene quantum dots and their applications in drug delivery. J. Nanobiotechnology 18(1): 142. DOI: 10.1186/s12951-020-00698-z

Zhou, M., Zhao, Q., Wu, Y., Feng, S., Wang, D., Zhang, Y. and Wang, S. 2020. Mesoporous carbon nanoparticles as multi-functional carriers for cancer therapy compared with mesoporous silica nanoparticles. Aaps Pharmscitech 21(2): 1-12. DOI: 10.1208/s12249-019-1604-8

Carbon-Based Nanomaterials for Application in Different Fields – A Short Review

Debasish Malik[1], Indrani Bhattacharyya[2] and Sabina Yeasmin[2]*

[1] Department of Molecular Biology & Biotechnology, University of Kalyani, West Bengal, India
[2] Department of Microbiology, CNMC, Kolkata, West Bengal, India

1. Introduction

Carbon nanotubes are unique nanostructures with remarkable electronic and mechanical properties, some stemming from the close relation between carbon nanotubes and graphite while some from their one-dimensional aspects. Initially, carbon nanotubes aroused great interest in the research community because of their exotic electronic structure. Researchers used CNTs as fillers in polymer and got a nanocomposite containing the properties of CNTs (Wagner et al. 2000). In the field of science for the last few decades nanotechnology was a well-known field (Franks 1987). Scientists investigated that nanotechnology mainly consists of the processing steps of the separation, consolidation, and deformation of materials by one atom or one molecule (Taniguchi 1974). CNTs are also known as buckytubes and have a wide range of applications due to their unique properties and cylindrical carbon structure. CNTs have a variety of applications in optics, materials, and nanoelectronics. Besides these, they also have mechanical, thermal, and electrical properties. Scientists discovered CNTs as a member of the fullerene family (Kroto et al. 1985). The name of CNTs is derived from its size, as the diameter is in the order of a few nanometers. In the arc-discharge method, scientists first prepared multi-walled carbon nanotubes (Iijima 1991). Worm-like carbon was discovered by scientists (Radushkevich and Lukyanovich 1952). At 600°C, the CO decomposed on iron particles, and the soot formed was observed. Due to the lack of technologies in ancient times, it was difficult to study its characteristics in the field of nanotechnology (Lau and Hui 2002). CNTs, also known as tubular fullerenes, are cylindrical graphene sheets of sp2-bonded carbon atoms. Different allotropes form carbon that can be prepared by graphene

Corresponding author: sabina.kolkata21@gmail.com

sheet (Iijima and Ichihashi 1993). On the basis of tubes present on CNTs, it can be divided into three categories: (1) single-walled carbon nanotubes (SWCNTs), (2) double-walled carbon nanotubes, and (3) multi-walled carbon nanotubes (MWCNT) (Geim and Novoselov 2007). Single-walled carbon nanotubes with a diameter of 1 to 2 nm were prepared by graphene sheet (Sheshmania et al. 2013). Depending on the preparation method, the length of SWNTs varies (Saether et al. 2003). Double-walled carbon nanotubes prepared by double tubes are enclosed with each other (Pichler 2007). MWNTs of diameter 2 to 50 nm were made by multiple tubes. The distance between tubes in MWNTs was approximately 0.34 nm (Ajayan 2004).

2. Synthesis

Gas-phase processes are the main processes involved in the synthesis of carbon nanotubes. Carbon nanotubes can be prepared by three main processes. One of them was the laser-ablation technique (Taniguchi 1974, Geim and Novoselov 2007), another was chemical vapor deposition (CVD) (Pichler 2007, Ajayan 2004), and the other was the carbon-arc discharge technique (Chico et al. 1996, Ajayan and Ebbesen 1997, Abbasi et al. 2014). In the previous time, high-temperature techniques, like laser ablation and arc discharge, were used for the preparation of carbon nanotubes. Nowadays, these processes are substituted by low temperature CVD or chemical vapor deposition method. In low temperature CVD or chemical vapor deposition method length, diameter, alignment, purity, orientation, and density of carbon nanotubes can be controlled (Jose-Yacaman et al. 1993).

2.1 Laser Ablation Technique

In an argon atmosphere, maintaining the temperature of the furnace at $1,200 \pm C$, a quartz-tube containing a block of pure graphite is heated inside a furnace at a high power laser vaporization temperature (Pichler 2007). When graphite gets heated inside the quartz using a laser, it gets vaporized. In the laser ablation technique, graphite was heated with catalyst (metal particles), and SWNTs were produced. The diameter of the nanotubes is controlled by laser power. The diameter of the nanotubes decreases with an increase in laser pulse power (Ajayan 2004). From different studies, it was revealed that ultra-fast (subpicosecond) laser pulse was responsible for the production of large amounts of SWNTs (Chico et al. 1996). More than one gram per hour of carbon nanotube can be produced by using the laser ablation technique. A lot of parameters like the structural and chemical composition of the target material can determine the properties of carbon nanotubes. Different laser properties like repetition rate, cw versus pulse, oscillation wavelength, peak power, energy fluence, chemical composition, flow and pressure of the buffer gas, the pressure of the chamber, the pressure of the buffer gas, the distance between the substrate and target, and ambient temperature can determine the properties of carbon nanotubes. In the laser ablation technique, SWNTs were produced of high quality and purity. Generally, the laser ablation technique had a similar technique to the arc discharge technique. The only dissimilarity was the laser hit the graphite pellet with cobalt or nickel-metal catalyst. High yield and greater purity were the main advantages of this technique over others.

Once the tube was closed the metal got evaporated from the end of the tube. The carbon nanotubes formed in this method have branching than straight structure, and this was the main disadvantage of this structure. The nanotube formed in this method was not large as in the arc discharge method. Also, in this method, a high-power laser was required and high purity graphite rods were expensive.

2.2 Chemical Vapor Deposition

Carbon nanotubes can also be prepared by chemical vapor deposition or CVD technique. There are many different types of CVDs, such as catalytic chemical vapor deposition (CCVD), either thermal (Iijima et al. 1992) or plasma-enhanced (PE) oxygen assisted CVD (Iijima 1991), microwave plasma (MPECVD) (Bernholc et al. 1997), water-assisted CVD (Journet et al. 1997, He et al. 2010, Ebbesen and Ajayan 1992), hot-filament (HFCVD) (Ajayan et al. 1999, Terrones 1997) or radiofrequency CVD (RF-CVD) (Dervishi et al. 2009). The standard technique for the synthesis of carbon nanotubes is catalytic chemical vapor deposition (CCVD). In this method, hydrocarbons materials break down, chemically spread, and carbon nanotubes form. Both arc discharge method and chemical vapor deposition method have the same technique. In both cases, carbon got excited and attached with metallic catalyst particles. The bottom of the tube in which carbon nanotubes formed was

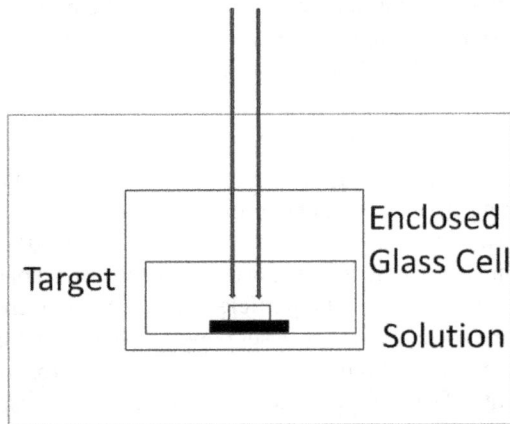

Figure 1: Carbon nanotubes preparation by laser ablation method.

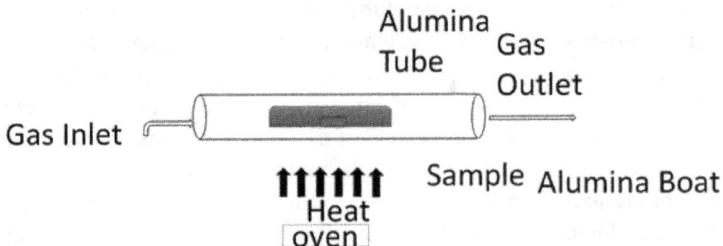

Figure 2: Carbon nanotubes synthesis by chemical vapor deposition method.

covered with iron nanoparticles, and it was drilled into silicon. Then a hydrocarbon like acetylene was decomposed on the surface of the substrate. In the tunnel when carbon came in contact with the metal catalyst inside the wholes, it started the production of carbon nanotubes. These carbon nanotubes initially act as a template for the production of carbon nanotubes. In the angle of the tunnel by maintaining the alignment carbon nanotubes grow very long. At above 700°C, the metal catalyst carbon nanotubes were formed. Nickel, cobalt, iron, or in a combination of all these are used as catalysts (Landi et al. 2005, Eklund et al. 2002). Catalyst support such as MgO or Al_2O_3 in combination with metal nanoparticles provides a great production of carbon nanotubes over the metal-catalyst surface. In the fluidized bed reactor, two types of fluid gases were used as fuel in the reactor and the nanotube formation started expansion (Steiner et al. 2009, Choudhary et al. 2014). In this type of reactor process gas (such as hydrogen, nitrogen, or ammonia) and gas-containing carbon compounds (acetylene, ethylene, methane, or ethanol) were used. The carbon-containing gas breakdown over the catalyst surface and at the edges of the catalytic surface formation of nanotubes was clear. The mechanism of the formation of carbon nanotubes was discussed in this literature (Tempel et al. 2010). The conventional models were discussed in this literature where scientists focus on a different type of growth (Smajda et al. 2009). After the production of carbon nanotubes over the catalyst, it reacts as a template during its growth and expansion (Patole et al. 2008). If we compare laser ablation and chemical vapor deposition technique, it gets priority for its high purity and yield over the CVD technique. So, controlled reaction, high yield, and purity are the different advantages of the CVD process (Grobert 2007).

2.3 Electric Arc Discharge

In the arc-discharge technique, high temperatures (above 1,700°C) were used for the synthesis of carbon nanotubes. In the arc discharge method graphite electrode of high purity and water-cooled electrodes were separated by a few millimeters in a gas chamber. The gas contains helium at high pressure or other gases like methane or hydrogen (Sheshmania et al. 2013). The chamber contains nickel or cobalt catalyst with two graphite electrodes surrounded by gaseous carbon compounds. It is known as the arc method since direct current passes through the chamber. In this method in high pressure and high temperature, the chamber was heated. In this procedure, evaporated carbon gets solidified over the cathode (negative electrode) tip. It was deposited at a specific rate known as 'cylindrical hard deposit or cigar-like structure'. A hard grey shell of the remaining carbon-coated into the 'chamber soot' near the walls of the chamber and 'cathode soot' on the cathode. The soft and dark 'chamber soot' and 'cathode soot' form single-walled or multi-walled carbon nanotubes. Depending on the use of catalyst carbon nanotubes can be prepared by two different methods: (1) Synthesis of CNTs in presence of a catalyst; and (2) Synthesis in absence of a catalyst.

Synthesis of MWNTs can be done in absence of a catalyst, and SWNTs can be prepared in presence of a catalyst. The anode in the arc discharge chamber was made up of metal and graphite; for example, Gd (Saether et al. 2003), Co, Ni, Fe, Ag, Pt, Pd, etc., or mixtures of Co, Ni, and Fe with other elements, like Co-Pt,

Co-Ru (Thess et al. 1996), Ni-Y, Fe-Ni, Co-Ni, Co-Cu, Ni-Cu, Fe-No, Ni-Ti, etc. SWNTs are produced in a high percentage with an average diameter of little less than two nanometers can be prepared in Ni-Y-graphite mixtures (Vander et al. 2003). Nowadays, SWNTs on a larger scale can be prepared in different institutes. The main advantage of the arc discharge method was the preparation of large-scale nanotubes. The main disadvantage of this method was very little control over the alignment of the nanotubes. As metallic catalysts are necessary for the reaction, so purification is essential.

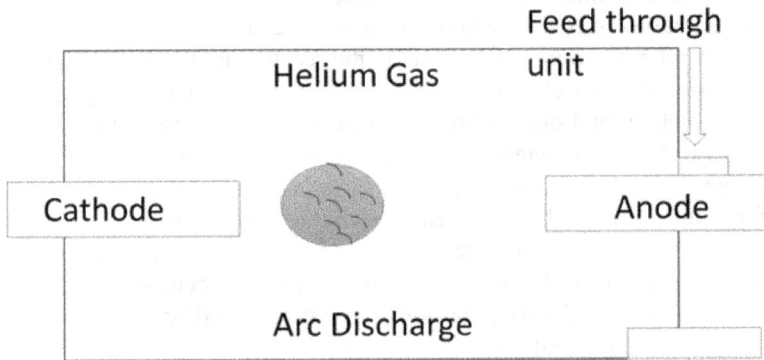

Figure 3: Carbon nanotube synthesis by electric arc discharge method.

3. Nanotube Purification

Different purification methods were used depending on the synthesis of carbon nanotubes. The main steps of purification were discussed here. The large graphite aggregation was deleted followed by filtration, using concentrated acid as solvent prepared carbon nanotubes dissolved in it to remove catalyst particles. Fullerenes can be removed by using organic solvents. Carbon clusters can be removed and nanotubes of different sizes can be separated by using microfiltrations and chromatography techniques (Grobert 2007). In the arc-discharge technique, the MWNTs prepared can be purified by using oxidation techniques. In this method, MWNTs and polyhedral graphite-like particles got separated (Sheshmania et al. 2013). More than 95% of starting materials were destroyed in this process, low purity, and the existence of unsatisfied valence bond of remaining carbon nanotubes were the main disadvantages of this process (Vander Wal 2003). A high annealing temperature of almost three thousand degrees centigrade is required to remove such dangling bonds. Surfactants, other colloidal particles, and polymers were used to prevent aggregation in the non-destructive method of CNTs separation (Askeland and Phule 2003). Size exclusion chromatography was used to separate CNTs depending on the size and porous filters, and microfiltration can be used to separate SWNTs and catalytic particles, and amorphous carbon (Askeland and Phule 2003, Saito et al. 1998). Nitric acid (Banerjee et al. 2008) and hydrofluoric acid (Brown et al. 2011) solutions were used to separate SWNTs and metal catalyst particles and amorphous carbon. In the carbon nanotubes separation method, it will dissolve in different media followed by sonication at 470°C thermal oxidation of SWNTS and hydrochloric acid treatments (Xu et al. 2011). In thermal oxidation methods, SWNTs at under 400 °C can be purified in presence of gold clusters which the CNTs (unsatisfied valency) got oxidized (Prasek et al. 2011). Using size exclusion chromatography (SEC) of DNA-dispersed carbon nanotubes (DNA-SWNT), semiconducting and metallic SWNTs can be separated (Varshney et al. 2010). Depending on the diameter, SWNTs can be separated by using density-gradient ultracentrifugation (Inami et al. 2007). CNTs of different chiralities can be separated by ion-exchange chromatography (IEC) and DNA-SWNT (IEC-DNA-SWNT). Short DNA oligomers were used to purify SWNT chiral compounds. MWNT and SWNT can be purified, cut, and suspended in organic solvents by bromination, fluorination, and acid treatments (Ishigami et al. 2008, Pinilla et al. 2007). Depending on the synthesis, there were also different methods of purification of CNTs. In this purification process, it should be considered that the properties remain unchanged.

4. Applications

4.1 Artificial Implants

Nanotubes have attractive physical and chemical properties and play a very important role in regenerative medicine (Muradov 2001). Nanotubes were attached with proteins and amino acids to avoid rejection of post-administration pain. In artificial joints and other implants, both SWNTs and MWNTs were used. Carbon nanotubes were filled with calcium and shaped-like bone structures to implant bone with artificial bones (Naha and Puri 2008, Fotopoulos and Xanthakis 2010). Cellular adhesion and proliferation can be accelerated with both CNTs. Inside natural and synthetic materials, these CNTs can be incorporated to increase cell adhesion and proliferation. Some nanotube applications as artificial implants are discussed here. Porous SWCNT polycarbonate membrane of natural or synthetic type osteoblast-like cells increase lamellipodia (cytoskeletal) extensions and lamellipodia extensions (Calvert 1999). SWCNT-incorporated of natural or synthetic type chitosan scaffolds C2Cl2 cells/C2 myogenic cell line help in cell growth improvement (Marquis 2003). MWCNT collagen sponge of natural or synthetic type honeycomb scaffold MC3T3-E1 cells, a mouse osteoblast-like cell line increases cellular adhesion and proliferation (Bian et al. 2004). MWCNT prepared by polyurethane acts on fibroblast cells to enhance interactions between the cells and the polyurethane surface (Flahaut et al. 2004). SWCNT of alginate acts on rat heart endothelial cells and enhance cellular adhesion and proliferation (Yanagi et al. 2006). MWCNT of poly(acrylic acid) acts on human embryonic stem cells and increases cellular differentiation toward neurons (Baughman et al. 1999). SWCNT of propylene fumarate help in rabbit tibia support cell attachment and proliferation (Niu et al. 1997).

4.2 Tissue Engineering

Tissue engineering was used to replace damaged or disease tissue or retain its normal and original function. Material science and engineering have advantages, like progress in tissue regenerative medicine and engineering. In cell tracking and labeling, sensing

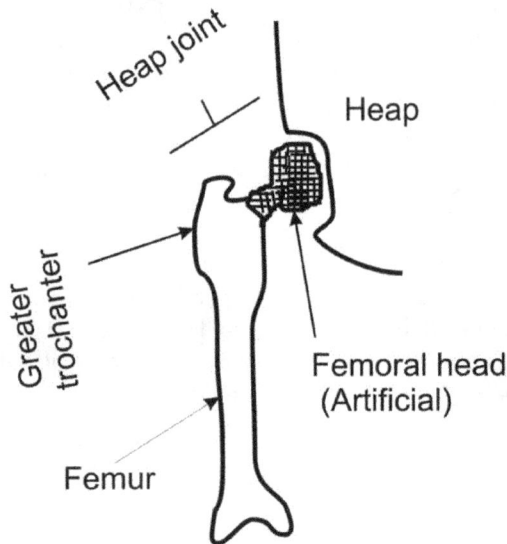

Figure 4: Artificial implants by nanoparticles.

cellular behavior, augmenting cellular behavior, and enhancing tissue matrices are the areas where tissue engineering can be used (Dai et al. 1996). *In vivo* and non-invasive tissue, the formation can be observed by cell tracking and labeling of the implanted cells. Implanted cells are labeled not only to evaluate the viable cells but also to understand the biodistribution, migration, relocation, and movement pathways of transplanted cells. Flow cytometry and non-invasive methods got priority over traditional methods as it was traditional and difficult to handle. In magnetic resonance, optical, and radiotracer modalities carbon nanotubes act as imaging contrast agents. One more application of tissue engineering is biodistribution modified radiotracers for gamma scintigraphy (Tibbetts et al. 2001). The biodistribution of the nanotube can be examined on BALB/c mice (Wei et al. 2004). Enzyme/cofactor interactions, protein and metabolite secretion, cellular behavior, and ion transport are included in monitoring better cellular physiology and help in designing engineered tissues. The performance of the engineered tissues is constantly monitored by the utilization of nanosensors. Carbon nanotubes for their large surface area, electrical properties, and capacity to immobilize DNA or other proteins behave as ideal elements for nanosensors. Carbon nanotube has a unique electronic structure which makes it an electrochemical sensor. This sensor investigates engineered tissue amino acids and redox-active proteins. In rat presence of glutathione and L-cysteine thiol, amino acids can be detected by using conjugation of MWNTs and platinum nanoparticles (Wang et al. 2004). In tissue engineering, the matrix of cells plays a vital role. In comparison to carbon nanotubes synthetic polymers, PLGA and PLA cannot be singly used in tissue engineering due to lack of mechanical strength. Carbon nanotubes were used in tissue scaffolds and provided strength to the structure. Polymer and nanotubes combined and increased the mechanical strength of the composites. Instead of using chitosan, if we use chitosan-MWNTs composite, the development in mechanical properties occurred (Chen et al. 2009). To gain smooth muscle cell growth, SWNTs-collagen were used (Lacerda et al. 2006, Zhang and Webster 2009, Kam et al. 2005, Ding et al. 2001, Aoki et al. 2005, Abarrategi et al. 2008, Hirata et al. 2009).

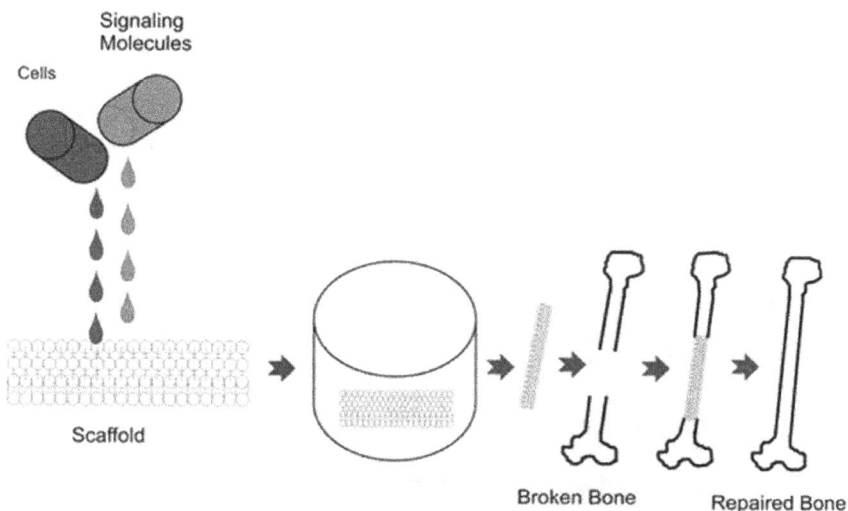

Figure 5: Nanoparticles help in tissue engineering.

4.3 Cancer Cell Identification

Cancer treatment, detection, and diagnosis can be done by nanodevices. Nanostructures less than 100 nm can be quickly cleared from the body. After taking entry inside the cells, it reacts with proteins and DNA and therefore imaging and detection can be easily studied. A peptide nanotube-folic acid-modified graphene electrode can be used to detect cervical cancer cells that overexpressed using folate receptors (Meng et al. 2009, Yildirim et al. 2008, Chao et al. 2009, Shi et al. 2008, Harrison and Atala 2007, Singh et al. 2006, Wang et al. 2005). In most of the cancers, morphologic modifications were absent and neoplastic disorders in an early stage cannot be detected by different cancer imaging methods, like CT, X-ray, and MRI. In the early stage, due to a lack of adequate spatial resolution, the disease cannot be detected. In the past few decades, imaging studies have been done with SWNTs. To study molecular imaging through SWNTs in the MRI field in combined Gd3 + functionalized SWCNTs high resolution of good tissue was observed (MacDonald et al. 2005). PET and SPECT (labeled SWCNTs + radionuclide) imaging techniques can be used to study the resolution of medium, sensitivity, and tissue penetration. Protein biomarkers were overexpressed in cancer cells. In advanced disease early diagnosis, prognosis maintaining surveillance followed by curative surgery and monitoring therapy used to predict and treat disease.

P-type carbon nanotubes and multi-label secondary antibody-nanotube bioconjugates containing biomarker prostate-specific antigen (PSA) are used to detect prostate cancer (Castillo et al. 2013, Eatemadi et al. 2014). Microelectrode arrays modified with single-walled carbon nanotubes (SWNTs) also contain total prostate-

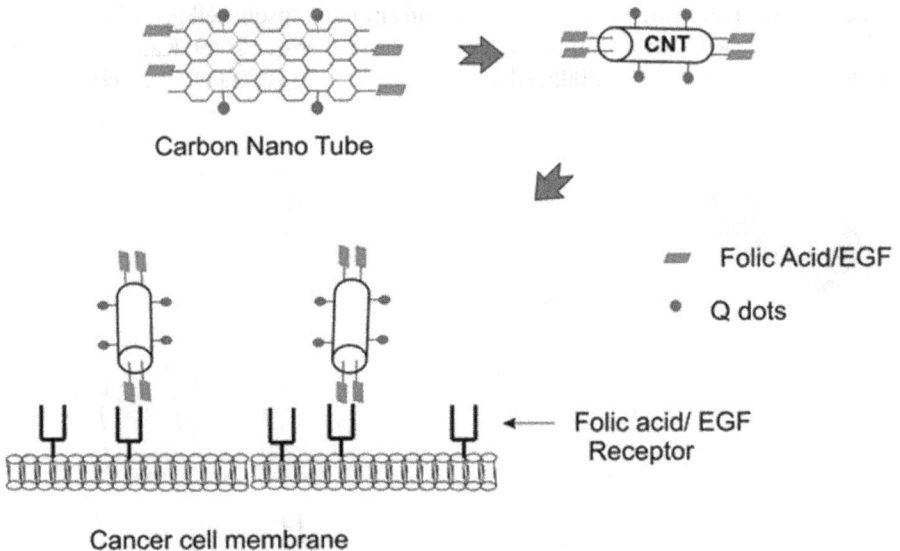

Carbon Nano Tube

Folic Acid/EGF

Q dots

Folic acid/ EGF Receptor

Cancer cell membrane

Figure 6: Cancer cell identification.

specific antigen (T-PSA) to detect prostate cancer (Eatemadi et al. 2014). Multi-walled carbon nanotubes-thionine-chitosan (MWCNTs-THI-CHIT) nanocomposite film containing biomarkers chlorpyrifos residues to detect many forms of cancer (Hong et al. 2009). Carbon nanomaterial, MWCNT-platinum nanoparticle-doped chitosan (CHIT), poly-L-lysine/hydroxyapatite/carbon nanotube (PLL/HA/CNT) hybrid nanoparticles, MWCN-polysulfone (PSf) polymer, multi-walled carbon nanotube-chitosan matrix, MWCNT-glassy carbon electrode (GCE) contained the following biomarkers carcinoma antigen-125 (CA125), AFP, carbohydrate antigen 19–9 (CA19-9), and human chorionic gonadotropin (hCG), prostate-specific antigen (PSA) to detect carcinoma, many forms of cancer and prostate cancer (Li et al. 2005, Yu et al. 2006, Okuno et al. 2007, Ou et al. 2007, Wu et al. 2007, Ding et al. 2008).

4.4 Drug and Gene Delivery by CNTs

The use of conventional chemotherapeutic agents has a lot of drawbacks, like systemic toxicity, poor distribution among cells, lack of selectivity, limited solubility, the inability of drugs to cross cellular barriers, and lack of clinical procedures for overcoming multidrug-resistant (MDR) cancer (Sánchez et al. 2008, Li et al. 2008). Polymers, liposomes, quantum dots, micelles, silica nanoparticles, dendrimers, molecular conjugates, and emulsions were the different systems used to overcome multidrug-resistant cancer (Panini et al. 2008). Due to the ultra-high surface area of carbon nanotubes (CNTs), it becomes a potential agent for drug delivery, nucleic acids, and peptides. In drug delivery, drugs conjugated with carbon nanotubes recognize cancer on the cell surface. By the process of endocytosis or other mechanisms, it crosses the mammalian cell line (Heister et al. 2009). Carbon nanotubes can easily transport drugs in the target cells (Jabr-Milane et al. 2008).

Scientists have discovered the SWNT-based tumor-targeted drug delivery system (DDS) that contains anticancer drugs, tumor-targeting ligands, and functionalized SWNTs. The drug delivery system reacts with the receptor on the cancer cells by the process of receptor-mediated endocytosis.

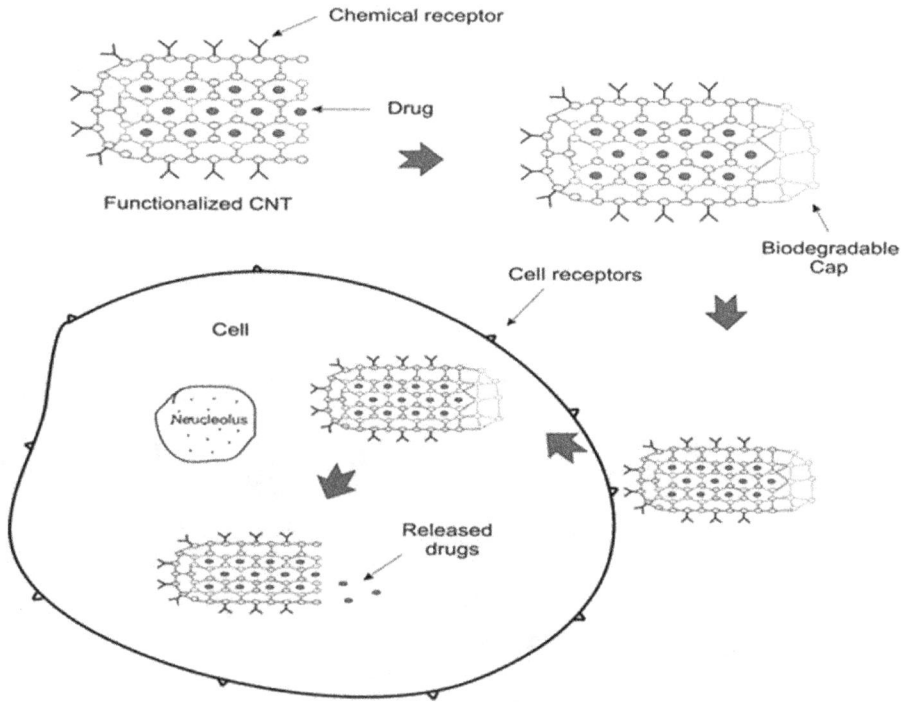

Figure 7: Drug delivery by nanoparticles.

4.5 Sensors

The important detecting devices sensors can be widely used in different fields. If CNTs were attached to bio- or molecular sensors, the efficiency of these sensors increase. Functional chemical groups attached to the ends of CNTs can be sensed by chemical force microscopy techniques. Sensors containing nanotube composite pellets, which were sensitive to gases, can be used to check leaks in gas chambers (Goldstein et al. 2005). SWNTs were extremely sensitive to air and other gases observed high variations in the electrical resistance of the samples. For NH_3, H_2O, CO_2, and CO, MWNTs act as efficient sensors (Zhang et al. 2009). With a slightly modified environment, scientists observed that the capacitance and resistance of CNTs also change (Chen et al. 2008). Using SWNTs or MWNTs highly sensitive and fast-responsive microwave-resonant sensors were prepared to detect NH_3 (Garcia-Gutierrez et al. 2006). Environmental pressure can also be measured by sensors prepared by CNTs and its composites [195]. CNTs were deformed in liquid mediums, like liquid immersion or polymer-embedding processes (Bhattacharyya et al. 2004).

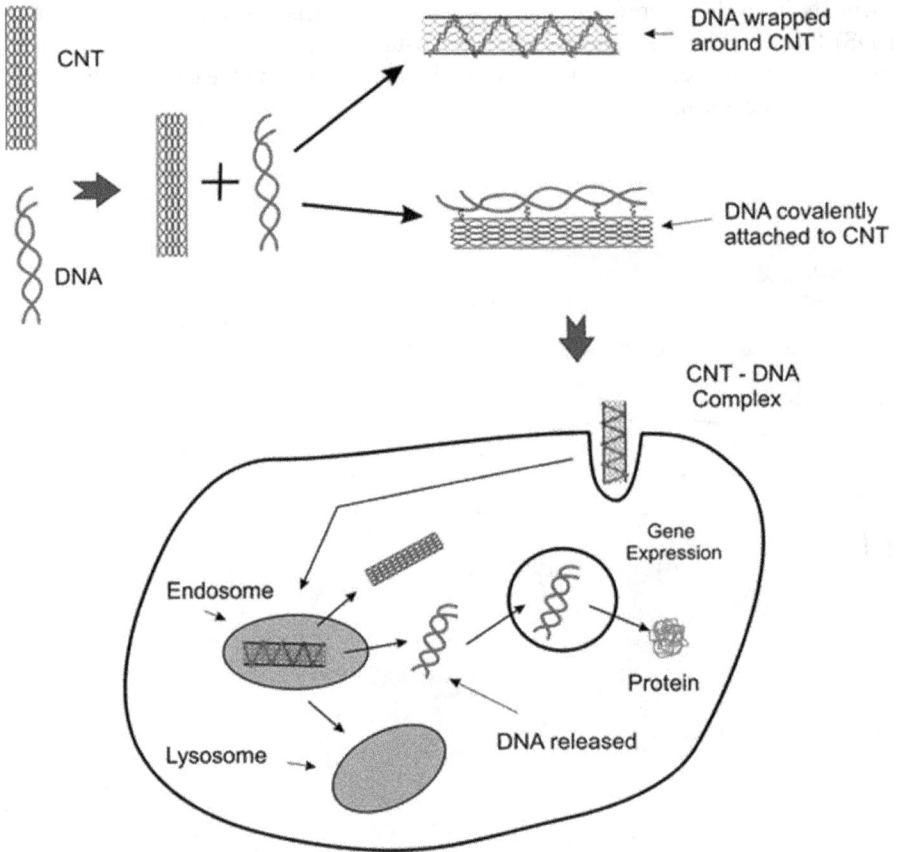

Figure 8: Gene delivery by nanoparticles.

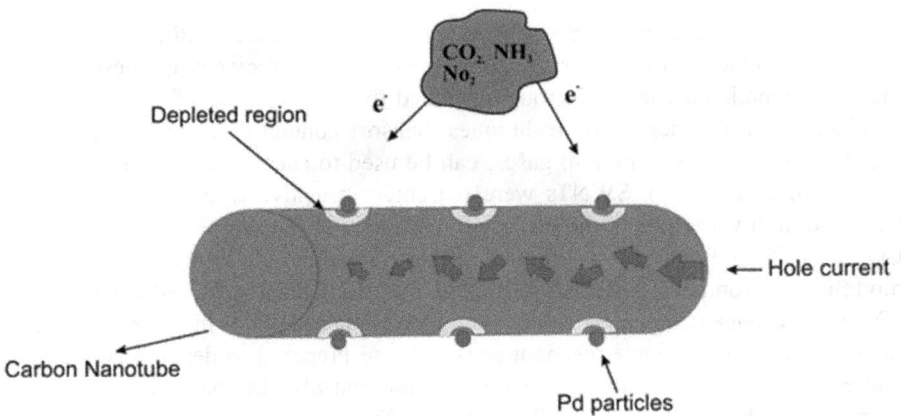

Figure 9: Carbon nanotubes used in bio-censors.

Biomedical Imaging

Biomedical imaging is a powerful tool that provides a clear and high-resolution image of cells, tissues, and body parts. Carbon nanotube (CNT) has been used in different biomedical imaging to analyze their functions in the respective environment (Gong et al. 2013). The fluorescence properties of some CNTs have been used as label-free imaging in a biomedical application (Tong et al. 2012). With photoacoustic imaging (it allows dense tissue to be imaged) and magnetic resonance imaging, CNTs can be guided to specific organs with an external magnetic field (Zerda et al. 2008). CNTs modifications along with the addition of new elements make them more efficient in biomedical imaging. Gold nanostructures with CNTs enhance the fluorescence intensity and provide ferritin receptor-mediated biomedical imaging. CNTs with other different nanoparticles include gold nanoparticles, quantum dots, iron oxide nanoparticles, and up-conversion nanoparticles (Li et al. 2014). CNTs in biomedical imaging are themselves providing a platform for easy-to-use technology and enhanced access to difficult imaging techniques.

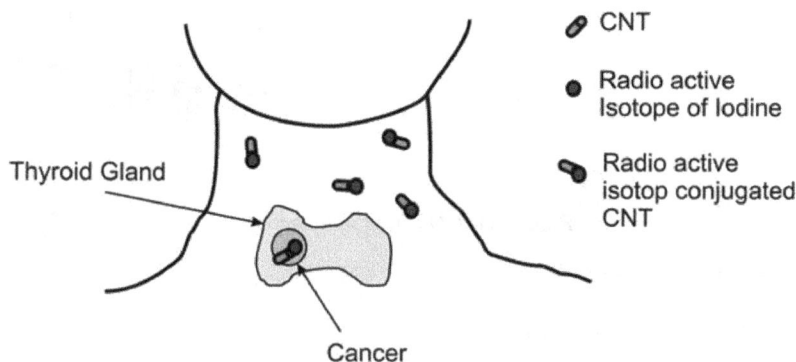

Figure 10: CNT help in biomedical imaging.

CNTs Used for Diagnostics

Hydrogels are a cross-linked polymeric structure that has unique properties of water absorption due to hydrophilic groups. Hydrogels are classified into two types depending on their cross linking interaction: chemical hydrogel and physical hydrogel. In chemical hydrogel, cross-linking is irreversible and stable whereas in physical hydrogel cross-linking is electrostatic and reversible. CNT-based hydrogel biosensors are mainly focused on glucose detection, majority of them measure the electron transfer of redox couple glucose oxidase (GOx) glucose. Kim et al. developed a bacterial cellulose-based biocompatible film that was directly used as an electrode (Kim et al. 2013). A long-term stable and very reliable glucose sensor has been developed by Comba et al. that is composed of albumin, Gox, and carbon nanotube mucin composite cross-linked with glutaraldehyde (Comba et al. 2018). Barone et al. developed an optical glucose sensor where apo-glucose oxidase (apoGOx) and glucose interaction induced by a shift of photoluminescence emission of CNT embedded in polyvinyl alcohol hydrogel (Barone et al. 2009).

Glucose oxidase

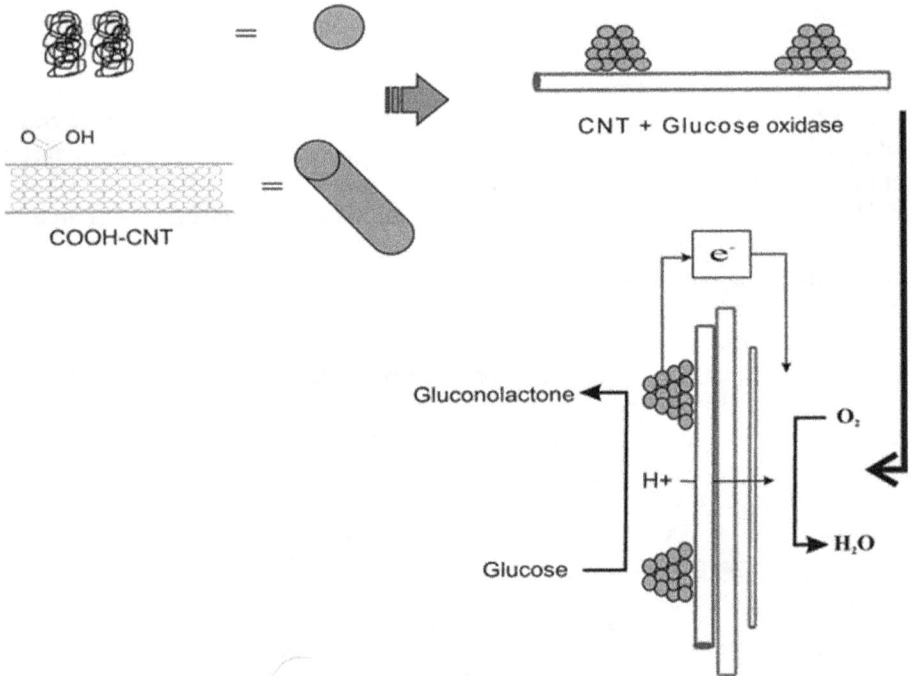

Figure 11: CNT used in diagnosis

Conflict of Interests

There is no conflict of interests.

References

Abarrategi, A., Gutierrez, M.C., Moreno-Vicente, C., Ramos, V., Lopez-Lacomba, J.L., Ferrer, M.L. and Monte, del. F. 2008. Multiwall carbon nanotube scaffolds for tissue engineering purposes. Biomaterials 29: 94–102.

Abbasi, E.A., Sedigheh, F., Abolfazl, A., Morteza, M., Hamid, T.N., Younes, H., Kazem, N.K. and Roghiyeh, P.-A. 2014. Dendrimers: synthesis, applications, and properties. Nanoscale Res. Lett. 9: 247–255.

Ajayan, P.M. and Ebbesen, T.W. 1997. Nanometre-size tubes of carbon. Rep. Prog. Phys. 60: 1025–1062.

Ajayan, P.M., Charlier, J.C. and Rinzler, A.G. 1999. Carbon nanotubes: from macromolecules to nanotechnology. Proc. Natl. Acad. Sci. 96: 14199–14200.

Ajayan, P.M. 2004. Bulk metal and ceramics nanocomposites. In nanocomposite science and technology. doi.org/10.1002/3527602127.ch1.

Aoki, N., Yokoyama, A., Nodasaka, Y., Akasaka, T., Uo, M., Sato, Y., Tohji, K. and Watari, F. 2005. Cell culture on a carbon nanotube scaffold. J. Biomed. Nanotechnol. 1: 402–405.

Askeland, D.R. and Phule, P.P. 2003. The science and engineering of materials. 4th Edition,

Brooks/Cole Publishing/Thompson Learning, USA.

Banerjee, S., Naha, S. and Puri, I.K. 2008. Molecular simulation of the carbon nanotube growth mode during catalytic synthesis. Appl. Phys. Lett. 92: 233121.

Barone, P.W., Yoon, H., Ortiz-García, R., Zhang, J., Ahn, J.-H., Kim, J.-H. and Strano, M.S. 2009. Modulation of single-walled carbon nanotube photoluminescence by hydrogel swelling. ACS Nano 3: 3869–3877.

Baughman, R.H., Cui, C., Zakhidov, A.A., Iqbal, Z., Barisci, J.N., Spinks, G.M., Wallace, G.G., Mazzoldi, A., Rossi, De. D. and Rinzler, A.G. 1999. Carbon nanotube actuators. Science 284: 1340–1344.

Bernholc, J.C., Roland, B.I. and Yakobson. 1997. Nanotubes. Curr. Opinion Solid State Mater. Sci. 2: 706–715.

Bhattacharyya, A.R., Potschke, P., Abdel-Goad, M. and Fischer, D. 2004. Effect of encapsulated SWNT on the mechanical properties of melt mixed PA12/SWNT composites. Chem. Phys. Lett. 392: 28-33.

Bian, Z., Wang, R.J., Wang, W.H., Zhang, T. and Inoue, A. 2004 Carbon-nanotube-reinforced Zr-based bulk metallic glass composites and their properties. Adv. Funct. Mater. 14: 55–63.

Brown, B., Parker, C.B., Stoner, B.R. and Glass, J.T. 2011. Growth of vertically aligned bamboo-like carbon nanotubes from ammonia/methane precursors using a platinum catalyst. Carbon 49: 266–274.

Calvert, P., 1999. Nanotube composites: a recipe for strength. Nature 399: 210–211.

Castillo, J.J., Svendsen, W.E., Rozlosnik, N. and Escobar, P. 2013. Detection of cancer cells using a peptide nanotube-folic acid modified grapheme electrode. Analyst 138: 1026–1031.

Chao, T.I., Xiang, S., Chen, C.S., Chin, W.C., Nelson, A.J., Wang, C. and Lu, J. 2009. Carbon nanotubes promote neuron differentiation from human embryonic stem cells. Biochem. Biophys. Res. Commun. 384: 426–430.

Chen, J., Chen, S., Zhao, X., Kuznetsova, L.V., Wong, S.S. and Ojima, I. 2008. Functionalized single-walled carbon nanotubes as rationally designed vehicles for tumor-targeted drug delivery. J. Am. Chem. Soc. 130: 16778–16785.

Chen, S., Yuan, R., Chai, Y., Min, L., Li, W. and Xu, Y. 2009. Electrochemical sensing platform based on tris (2,2'-bipyridyl) cobalt (III) and multiwall carbon nanotubes-Nafion composite for immunoassay of carcinoma antigen-125. Electro. Chim. Acta 54: 7242–7247.

Choudhary, N., Hwang, S. and Choi, W. 2014. Carbon nanomaterials: a review. pp. 709-730. *In:* Handbook of Nanomaterials Properties. USA: Springer.

Comba, F.N., Romero, M.R., Garay, F.S. and Baruzzi, A.M. 2018. Mucin and carbon nanotube-based biosensor for detection of glucose in human plasma. Anal. Biochem. 550: 34–40.

Dai, H., Hafner, J.H., Rinzler, A.G., Colbert, D.T. and Smalley, R.E. 1996. Nanotubes as nanoprobes in scanning probe microscopy. Nature 384: 147–150.

Dervishi, E., Li, Z., Xu, Y., Saini, V., Biris, A.R., Lupu, D. and Biris, A.S. 2009. Carbon nanotubes: synthesis, properties, and applications. Part Sci. Technol. 27: 107–125.

Ding, R.G., Lu, G.Q., Yan, Z.F. and Wilson, M.A. 2001. Recent advances in the preparation and utilization of carbon nanotubes for hydrogen storage. J. Nanosci. Nano Technol. 1: 7–29.

Ding, Y., Liu, J., Jin, X., Lu, H., Shen, G. and Yu, R. 2008. Poly-L lysine/hydroxyapatite/carbon nanotube hybrid nanocomposite applied for piezoelectric immunoassay of carbohydrate antigen. Analyst 133: 184–190.

Eatemadi, A., Daraee, H., Zarghami, N., Hassan Melat, Y. and Abolfazl, A. 2014. Nanofiber: synthesis and biomedical applications, Artif. Cells Nanomed. Biotechnol. 43: 1–11.

Ebbesen, T.W. and Ajayan, P.M. 1992. Large-scale synthesis of carbon nanotubes. Nature 358: 220–222.

Eklund, P.C., Pradhan, B.K., Kim, U.J., JE, Q., JE. Xiong, Fischer, A.D., Friedman, B.C., Holloway, K. Jordan and Smith, M.W. 2002. Large-scale production of single-walled carbon nanotubes using ultrafast pulses from a free electron laser. Nano Lett. 2: 561–566.

Flahaut, E., Rul, S., Laurent, C. and Peigney, A. 2004. Carbon nanotubes-ceramic composites. Ceramic Nanomat. Nanotech. II 148: 69–82.

Fotopoulos, N. and Xanthakis, J.P. 2010. A molecular level model for the nucleation of a single-wall carbon nanotube cap over a transition metal catalytic particle. Diam. Relat. Mater. 19: 557–561.

Franks, A. 1987. Nanotechnology. J. Phys. E 20: 1442–1451.

Garcia-Gutierrez, M.C., Nogales, A., Rueda, D.R., Domingo, C., Garcia-Ramos, J.V., Broza, G., Roslaniec, Z., Schulte, K., Davies, R.J. and Ezquerra, T.A. 2006. Templating of crystallization and shear-induced self-assembly of single-wall carbon nanotubes in a polymer-nanocomposite. Polymer 47: 341–345.

Geim, AK. and Novoselov, K.S. 2007. The rise of graphene. Nature Mater. 6: 183–191.

Goldstein, D., Nassar, T., Lambert, G., Kadouche, J. and Benita, S. 2005. The design and evaluation of a novel targeted drug delivery system using cationicemulsion-antibody conjugates. J. Control Release 108: 418–432.

Gong, H., Peng, R. and Liu, Z. 2013. Carbon nanotubes for biomedical imaging: the recent advances. Adv. Drug Delivery Rev. 65: 1951–1963.

Grobert, N. 2007. Carbon nanotubes—becoming clean. Mater. Today 10: 28–35.

Harrison, B.S. and Atala, A. 2007. Carbon nanotube applications for tissue engineering. Biomaterials 28: 344–353.

He, Z.B., Maurice, J.L., Lee, C.S., Cojocaru, C.S. and Pribat, D. 2010. Nickel catalyst faceting in plasma-enhanced direct current chemical vapor deposition of carbon nanofibers. Arab. J. Sci. Eng. 35(1C): 11–19.

Heister, E., Neves, V., Lipert, K., Coley, H.M., Silva, S.R. and McFadden, J. 2009. Triple functionalisation of single-walled carbon nanotubes with doxorubicin, a monoclonal antibody, and a fluorescent marker for targeted cancer therapy. Carbon 47: 2152–2160.

Hirata, E., Uo, M., Takita, H., Akasaka, T., Watari, F. and Yokoyama, A. 2009. Development of a 3D collagen scaffold coated with multiwalled carbon nanotubes. J. Biomed. Mater. Res. B Appl. Biomater. 90: 629–634.

Hong, H., Gao, T. and Cai, W. 2009. Molecular imaging with single-walled carbon nanotubes. Nano Today 4: 252–261.

Iijima, S. 1991. Helical microtubules of graphitic carbon. Nature 354: 56–58.

Iijima, S. and Ichihashi, T. 1993. Single-shell carbon nanotubes of 1-nm diameter. Nature 363: 603-605.

Iijima, S., Ajayan, P.M. and Ichihashi, T. 1992. Growth model for carbon nanotubes. Phys. Rev. Lett. 69: 3100–3103.

Inami, N., Mohamed, A.M., Shikoh, E. and Fujiwara, A. 2007. Synthesis-condition dependence of carbon nanotube growth by alcohol catalytic chemical vapor deposition method. Sci. Technol. Adv. Mater. 8: 292–295.

Ishigami, N., Ago, H., Imamoto, K., Tsuji, M., Iakoubovskii, K. and Minami, N. 2008. Crystal plane dependent growth of aligned single-walled carbon nanotubes on sapphire. J. Am. Chem. Soc. 130: 9918–9924.

Jabr-Milane, L.S., van Vlerken, L.E., Yadav, S. and Amiji, M.M. 2008. Multi-functional nano carriers to overcome tumor drug resistance. Cancer Treat. Rev. 34: 592–602.

Jose-Yacaman, M.Y., Miki, M., Rendon, L. and Santiesteban, J.G. 1993. Catalytic growth of carbon microtubules with fullerene structure. Appl. Phys. Lett. 62: 202–204.

Journet, C., Maser, K.W., Bernier. P., Loiseau, A., La, De., Chapelle, M.L., Lefrant, D., Deniard, P., Lee, R. and Fischer, J.E. 1997. Large-scale production of single-walled carbon nanotubes by the electric-arc technique. Nature 388: 756–758.

Kam, N.W.S., Connell, O.'M., Wisdom, J.A. and Dai, H. 2005. Carbon nanotubes as multifunctional biological transporters and near-infrared agents for selective cancer cell destruction. Proc. Natl. Acad. Sci. USA 102: 11600–11605.

Kim, Y.-H., Park, S., Won, K., Kim, H.J. and Lee, S.H. 2013. Bacterial cellulose-carbon nanotube composite as a biocompatible electrode for the direct electron transfer of glucose oxidase. J. Chem. Technol. Biotechnol. 88: 1067–1070.

Kroto, H.W., Heath, J.R., Brien, O'S.C., Curl, R.F. and Smalley, R.E. 1985. C60: Buckminster-fullerene. Nature 318: 162–163.

Lacerda, L., Bianco, A., Prato, M. and Kostarelos, K. 2006. Carbon nanotubes as nanomedicines: from toxicology to pharmacology. Adv. Drug Deliv. Rev. 58: 1460–1470.

Landi, B.J., Raffaelle, R.P., Castro, S.L. and Bailey, S.G. 2005. Single-wall carbon nanotube—polymer solar cells. Prog. Photovolt Res. Appl. 13: 165–172.

Lau, A.K.T. and Hui, D. 2002. The revolutionary creation of new advanced materials—carbon nanotube composites. Composites B 33: 263–277.

Li, J., Chang, X., Chen, X., Gu, Z., Zhao, F., Chai, Z. and Zhao, Y. 2014. Toxicity of inorganic nanomaterials in biomedical imaging. Biotechnol. Adv. 32: 727–743.

Li, N., Yuan, R., Chai, Y., Chen, S. and An, H. 2008. Sensitive immunoassay of human chorionic gonadotrophin based on multi-walled carbon nanotube-chitosan matrix. Bioprocess. Biosyst. Eng. 31: 551–558.

Li. C., Currelli, M., Lin, H., Lei, B., Ishikawa, F.N., Datar, R., Cote, R.J., Thompson, M.E. and Zhou, C. 2005. Complementary detection of prostate-specific antigen using In_2O_3 nanowires and carbon nanotubes. J. Am. Chem. Soc. 127: 12484–12485.

MacDonald, R.A., Laurenzi, B.F., Viswanathan, G., Ajayan, P.M. and Stegemann, J.P. 2005. Collagen-carbon nanotube composite materials as scaffolds in tissue engineering. J. Biomed. Mater. Res. A 74: 489–496.

Marquis, F.D. 2003. Fully integrated hybrid polymeric carbon nanotube composites. Trans. Tech. Publ. 100: 85–88.

Meng, J., Kong, H., Han, Z., Wang, C., Zhu, G., Xie, S. and Xu, H. 2009. Enhancement of nano fibrous scaffold of multiwalled carbon nanotubes/polyurethane composite to the fibroblasts growth and biosynthesis. J. Biomed. Mater. Res. A 88: 105–116.

Muradov, N. 2001. Hydrogen via methane decomposition: an application for decarbonization of fossil fuels. Int. J. Hydrog. Energy 26: 1165–1175.

Naha, S. and Puri, I.K. 2008. A model for catalytic growth of carbon nanotubes. J. Phys. D Appl. Phys. 41: 065304.

Niu, C., Sichel, E.K., Hoch, R., Moy, D. and Tennent, H. 1997. High power electrochemical capacitors based on carbon nanotube electrodes. Appl. Phys. Lett. 70: 1480–1482.

Okuno, J., Maehashi, K., Kerman, K., Takamura, Y., Matsumoto, K. and Tamiya, E. 2007. Label-free immunosensor for prostate-specific antigen based on single-walled carbon nanotube array-modified microelectrodes. Biosens. Bioelectron. 22: 2377–2381.

Ou, C., Yuan, R., Chai, Y., Tang, M., Chai, R. and He, X. 2007 A novel amperometric immunosensor based on layer-by-layer assembly of gold nanoparticles-multi-walled carbon nanotubes-thionine multilayer films on polyelectrolyte surface. Anal. ChimActa 205–213.

Panini, N.V., Messina, G.A., Salinas, E. and Raba, J. 2008. Integrated microfluidic systems with an immune sensor modified with carbon nanotubes for detection of prostate specific antigen (PSA) in human serum samples. Biosens. Bioelectron. 23: 1145–1151.

Patole, S.P., Alegaonkar, P.S., Lee, H.C. and Yoo, J.B. 2008. Optimization of water assisted chemical vapor deposition parameters for super growth of carbon nanotubes. Carbon 46: 1987–1993.

Pichler, T. 2007. Molecular nanostructures: carbon ahead. Nature Mater. 6: 332–333.

Pinilla, J.L., Moliner, R., Suelves, I., Lízaro, M.J., Echegoyen, Y. and Palacios, J.M. 2007. Production of hydrogen and carbon nanofibers by thermal decomposition of methane using metal catalysts in a fluidized bed reactor. Int. J. Hydrog. Energy 32: 4821–4829.

Prasek, J., Drbohlavova, J., Chomoucka, J., Hubalek, J., Jasek, O., Adam, V. and Kizek, R. 2011. Methods for carbon nanotubes synthesis—review. J. Mater. Chem. 21: 15872–15884.

Radushkevich, L.V. and Lukyanovich, V.M. 1952. The structure of carbon forming in thermal decomposition of carbon monoxide on an iron catalyst. Russ. J. Phys. Chem. 26: 88–95.

Saether, E., Frankland, S.J.V. and Pipes, R.B. 2003. Transverse mechanical properties of single walled carbon nanotube crystals. Part I: determination of elastic moduli. Compos. Sci. Technol. 63: 1543–1550.

Saito, R., Dresselhaus, G. and Dresselhaus, M.S. 1998. Physical properties of carbon nanotubes. 4th edition. USA: World Scientific.

Sánchez, S., Roldán, M., Pérez, S. and Fàbregas, E. 2008. Toward a fast, easy, and versatile immobilization of biomolecules into carbon nanotube/ polysulfone-based biosensors for the detection of hCGhormone. Anal. Chem. 80: 6508–6514.

Sheshmania, S., Ashorib, A. and Fashapoyeha, M.A. 2013. Wood plastic composite using grapheme nanoplatelets. Int. J. Biol. Macromol. 58: 1–6.

Shi, X., Sitharaman, B., Pham, Q.P., Spicer, P.P., Hudson, J.L., Wilson, L.J., Tour, J.M., Raphael, R.M. and Mikos, A.G. 2008. In vitro cytotoxicity of single-walled carbon nanotube/biodegradable polymer nanocomposites. J. Biomed. Mater. Res. A 86: 813–823.

Singh, R., Pantarotto, D., Lacerda, L., Pastorin, G., Klumpp, C., Prato, M., Bianco, A. and Kostarelos, K. 2006. Tissue bio distribution and blood clearance rates of intravenously administered carbon nanotube radiotracers. Proc. Natl. Acad. Sci. USA 103: 3357–3362.

Siochi, E.J., Working, D.C., Park, C., Lillehei, P.T., Rouse, J.H., Topping, C.C., Bhattacharyya, A.R. and Kumar, S. 2004. Melt processing of SWCNT-polyimide nanocomposite fibers. Composites B 35: 439–446.

Smajda, R., Andresen, J.C., Duchamp, M., Meunier, R., Casimirius, S., Hernadi, K.F.L. and Magrez, A. 2009. Synthesis and mechanical properties of carbon nanotubes produced by the water assisted CVD process. Physica Statussolidi (b) 246: 2457–2460.

Steiner, S.A., Baumann, T.F., Bayer, B.C., Blume, R., Worsley, M.A., Moberly, C.W.J. and Shaw, E.L. 2009. Nanoscale zirconia as a nonmetallic catalyst for graphitization of carbon and growth of single- and multiwall carbon nanotubes. J. Am. ChemSoc. 131: 12144–12154.

Taniguchi, N. 1974. On the basic concept of 'nano-technology'. Proceedings of the International Conference on Production Engineering, Tokyo, Japan, Part II, 18-23.

Tempel, H., Schneider, J.J. and Joshi, R. 2010. Ink jet printing of ferritin as method for selective catalyst patterning and growth of multi walled carbon nanotubes. Mater. Chem. Phys. 121: 178–183.

Terrones, M. 1997. Production and characterization of novel fullerene related materials, nanotubes, nanofibres and giant fullerenes, https://ethos.bl.uk/OrderDetails. do?uin=uk .bl. ethos.361404 (Thesis work).

Thess, A., Lee, R., Nikolaev, P., Dai, H., Petit, P., Robert, J., Xu, C., Lee, Y.H., Kim, S.G. and Rinzler, A.G. 1996. Crystalline ropes of metallic carbon nanotubes. Science-AAAS-Weekly Paper Edition 273: 483–487.

Tibbetts, G.G., Meisner, G.P. and Olk, C.H. 2001. Hydrogen storage capacity of carbon nanotubes, filaments, and vapor-grown fibers. Carbon 39: 2291–2301.

Tong, L., Liu, Y., Dolash, B.D., Jung, Y., Slipchenko, M.N., Bergstrom, D.E. and Cheng, J.-X. 2012. Label-free imaging of semiconducting and metallic carbon nanotubes in cells and mice using transient absorption microscopy. Nat. Nanotechnol. 7: 56–61.

Vander, Wal. R.L., Berger, G.M. and Ticich, T.M. 2003. Carbon nanotube synthesis in a flame using laser ablation for in situ catalyst generation. Applied Physics A 77: 885–889.

Varshney, D., Weiner, B.R. and Morell, G. 2010. Growth and field emission study of a monolithic carbon nanotube/diamond composite. Carbon 48: 3353–3358.

Wagner, F.E., Haslbeck, S., Stievano, L., Calogero, S., Pankhurst, Q.A. and Martinek, K.P. 2000. Before striking gold in gold-ruby glass. Nature 407: 691–692.

Wang, S.F., Shen, L., Zhang, W.D. and Tong, Y.J. 2005. Preparation and mechanical properties of chitosan/carbon nanotubes composites. Biomacromolecules 6: 3067–3072.

Wang, Y., Da, S., Kim, M.J., Kelly, K.F., Guo, W., Kittrell, C., Hauge, R.H. and Smalley, R.E. 2004. Ultrathin "bed-of-nails" membranes of single-wall carbon nanotubes. J. Am. Chem. Soc. 126: 9502–9503.

Wei, J., Zhu, H., Wu, D. and Wei, B. 2004. Carbon nanotube filaments in household lightbulbs. Appl. Phys. Lett. 84: 4869–4871.

Wu, L., Yan, F. and Ju, H. 2007. An amperometric immunosensor for separation-free immunoassay of CA125 based on its covalent immobilization coupled with thionine on carbon nanofiber. J. Immunol. Methods 322: 12–19.

Xu, Y., Dervishi, E., Biris, E.A.R. and Biris, A.S. 2011. Chirality-enriched semiconducting carbon nanotubes synthesized on high surface area MgO-supported catalyst. Mater. Lett. 65: 1878–1881.

Yanagi, H., Kawai, Y., Kita, T., Fujii, S., Hayashi, Y., Magario, A. and Noguchi, T. 2006. Carbonnanotube/aluminum composites as a novel field electron emitter. Jpn. J. Appl. Phys. 45(7L): L650.

Yildirim, E.D., Yin, X., Nair, K. and Sun, W. 2008. Fabrication, characterization, and biocompatibility of single-walled carbon nanotube-reinforced alginate composite scaffolds manufactured using free form fabrication technique. J. Biomed. Mater. Res. B. Appl. Biomater. 87: 406–414.

Yu, X., Munge, B., Patel, V., Jensen, G., Bhirde, A., Gong, J.D., Kim, S.N., Gillespie, J., Gutkind, J.S. and Papadimitrakopoulos, F. 2006. Carbon nanotube amplification strategies for highly sensitive immunodetection of cancer biomarkers. J. Am. Chem. Soc. 128: 11199–11205.

Zerda, De., La., Zavaleta, C., Keren, S., Vaithilingam, S., Bodapati, S., Liu, Z., Levi, J., Smith, B.R., Ma, T.-J. and Oralkan, O. 2008. Carbon nanotubes as photo acoustic molecular imaging agents in living mice. Nat. Nanotechnol. 3: 557–562.

Zhang, L. and Webster, T.J. 2009. Nanotechnology and nanomaterials: promises for improved tissue regeneration. Nano Today 4: 66–80.

Zhang, X., Meng, L., Lu, Q., Fei, Z. and Dyson, P.J. 2009. Targeted delivery and controlled release of doxorubicin to cancer cells using modified single wall carbon nanotubes. Biomaterials 30: 6041–6047.

Biomass-Derived Carbonaceous Materials: Synthesis and Photocatalytic Applications

Kezia Sasitharan[1] and Anitha Varghese[2]*

[1] Department of Chemistry, The University of Sheffield, S3 7HF Sheffield, United Kingdom
[2] Department of Chemistry, CHRIST (Deemed to be University), Hosur Road, Bengaluru - 560029, India

1. Introduction

Population growth explosion and the growing energy demands of the world require a continuous need for improved and sustainable energy technologies (de Almeida Ribeiro et al. 2020). Many researchers envisage that novel nanomaterials and carbon-based or carbonaceous materials can provide improved environmental applications (Ramis et al. 2020). A range of carbonaceous nanomaterials has so far been employed to revolutionize the fields of energy conversion, storage, fuel cells, CO_2 capture, catalysis and photocatalysis (Colmenares et al. 2013, de Almeida Ribeiro et al. 2020, Koe et al. 2020, Ramis et al. 2020). Graphene and graphitic materials, Buckminster fullerenes, carbon nanotubes, carbon fibers and carbon nitrides have emerged as rising stars in the direction of the quest for novel materials (Liu et al. 2018, Hong et al. 2014). However, technology can be considered truly 'sustainable' only if its preparation procedures involve simple methods, reduced carbon footprint and low cost (De et al. 2015). The various methods that exist for the preparation of carbonaceous materials are chemical-vapor deposition, arc discharge, laser ablation, mechanical exfoliation and template-casting to name a few (McKendry 2002). Although these techniques give high-quality products, they end up being time-consuming, require the use of harsher precursors, are expensive and have a resultant problem of disposal of leftover residuals. Hence, not all these techniques can be truly classed as an environmentally friendly approach for the creation of sustainable technologies.

Biomass often refers to plants or plant-based materials derived from nature and consisting of carbohydrates as their building blocks (Luo et al. 2004, Zhang et al. 2007a, De et al. 2015). They can be synthesized through biological photosynthesis

Corresponding author: anitha.varghese@christuniversity.in

using CO_2 and water as raw materials in the presence of sunlight (Jazaeri and Tsuzuki 2013, Varanasi et al. 2013). Through controlled combustion, biomass can be converted into carbonaceous materials of varying functionalities and properties, making them a highly desirable research direction in recent years (Hamelinck et al. 2005, Patrick et al. 2013, H. Li et al. 2016, Jain et al. 2016, Lee et al. 2016). Combustion is associated with the release of toxic gases and is justifiably one of the most prominent disadvantages of fossil fuels. In that case, how does the combustion of biomass differ from that of fossil fuels and how can a technology that involves combustion be termed sustainable? These might be some of the valid questions arising in your mind at the moment. Biomass is primarily made up of carbon and very few quantities of heteroatoms like sulfur and nitrogen. Therefore, the release of SO_x and NO_x emissions from biomass combustion will be much lower as compared to that of the burning of fossil fuels (Zhang et al. 2007b, Cha et al. 2016, Deng et al. 2016, Tripathi et al. 2016). In addition, the CO_2 released during the combustion of biomass is from the CO_2 that initially got absorbed by the plant during its photosynthesis, which means there is no new additional CO_2 being released into the environment and thereby making biomass combustion a carbon-neutral process (McKendry 2002, Lackner 2010, Titirici and Antonietti 2010, Yang et al. 2019a). Biomass-derived carbonaceous materials have the distinct advantage of being inexpensive and can be formed using easy preparation techniques (Liu et al. n.d.). They also offer the capability of being functionalized with different groups to tailor their properties for a particular application (Hong et al. 2014, de Almeida Ribeiro et al. 2020, Ramis et al. 2020). In this chapter, we aim to highlight the key properties of these materials, their preparation methods and their role and progress in the field of photocatalysis.

2. Synthetic Strategies

The common techniques of preparing biomass-derived porous carbonaceous materials (outline in Figure 1) include biomass pyrolysis, hydrothermal carbonization, ionothermal carbonization and molten salt carbonization. The properties of the final product are hugely dependent on the preparation conditions used.

2.1 Biomass Pyrolysis

Pyrolysis occupies a central position in the techniques employed for the conversion of biomass to any value-added materials/product (Nzihou et al. 2019). This technique involves the thermal degradation of biomass in an inert environment at elevated temperatures. This is done either in the limited or complete absence of oxygen. As implied, this is a physically and chemically irreversible procedure. This technique is preferred for biomass sources with low moisture content (Biswal et al. 2013). Varying the conditions in which pyrolysis is carried out can yield different products (Falco et al. 2011). For example, slow pyrolysis wherein the biomass is subjected to a slow heating rate of <10°C/minutes for a wider temperature range (300-800°C) results in less volatile products. On the other hand, fast pyrolysis, which involves heating the biomass at a higher rate (<300°C/minutes), over a narrow temperature range (400-600°C) results in higher production of bio-oil and syn-gas and only

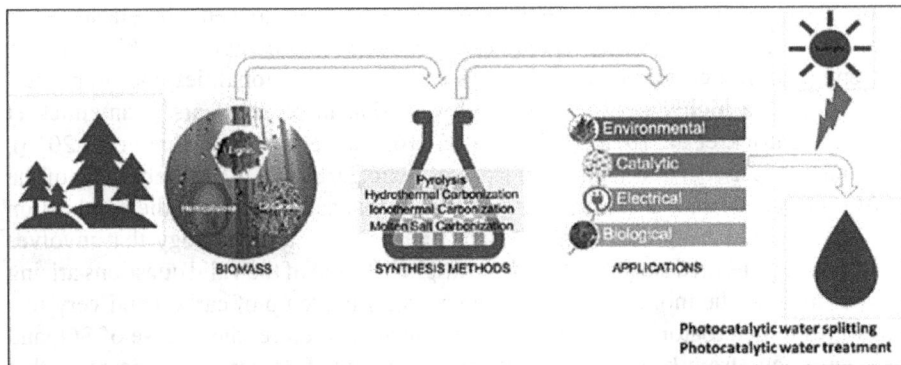

Figure 1: Schematic representation of the synthetic methods of biomass-derived materials discussed in this work and the relevant applications which include photocatalytic water splitting and water treatment. Adapted with permission from reference (Yang et al. 2019b) (© 2019 American Chemical Society).

about 10-12% of biochar (Ge et al. 2019). Thus, low-temperature pyrolysis offers the highest yield and is useful for the large-scale preparation of biochar. The high-temperature pyrolytic conditions are useful to produce biochemicals for the energy industry (Biswal et al. 2013).

Biomass is primarily composed of cellulose, hemicellulose and lignin. Each of these components upon calcination under inert gas can follow varying chemical pathways (Varma 2019). At the temperature range <100°C, the main step involved is the moisture release. At higher temperatures of 220-315°C hemicellulose pyrolyzes, and at 315-400°C cellulose begins to pyrolyze. All volatile materials are eliminated during pyrolysis and the residual material is mostly carbon.

The composition of organic precursors within the source biomass determines the properties of the final product. Inorganic materials present naturally within the source biomass can sometimes catalyze and alter the nature of the product (Nzihou et al. 2019). The processing conditions such as heating rate, residence time and final temperature are some of the physical factors that can be optimized for controlling the morphology, surface area and porosity of the resulting product. This technique is very simple and straightforward; however, the deduction of mechanism is complicated owing to the wide range of available reaction pathways (Liang et al. 2020). Some of the limitations of this method are that the process is time-consuming and leads to polluting emissions.

2.2 Hydrothermal Carbonization

Hydrothermal carbonization (HTC) of biomass is widely employed for biomass with higher moisture content (Titirici et al. 2012). This technique has been a promising approach for the synthesis of novel carbon-based materials. This is a thermochemical decomposition process where biomass and water are made to react at lower temperatures (150-350 °C). Here, water plays the dual role of a solvent and a catalyst, facilitating hydrolysis and cleavage of lignocellulose (Libra et al. 2011, Möller et al. 2011). During the hydrothermal carbonization process, a number of organic acids are formed such as acetic acid, formic acid, levulinic acid and lactic acid, which promote hydrolysis and formation of smaller fragments (Baccile et al. 2010, Berge et al. 2011).

Higher temperature leads to the formation of higher yields of gaseous products due to extensive dehydration and the solid carbonaceous materials produced are known to be highly porous. On the other hand, low-temperature HTC yields colloidal carbonaceous materials (Hoekman et al. 2011, Falco et al. 2013). Unlike temperature, the influence of pressure is insignificant on the nature of products. If the initial composition of the biomass starting material is known, then the hydrothermal temperature and residence time can be tuned for higher oxygenated functional groups in the product (Sevilla and Fuertes 2009, Falco et al. 2011, Jain et al. 2016). Catalysts have also been used for increasing the proton concentration, enhancing the reaction rate or tailoring a reaction path. Acids, oxidants and metal salts are some of the commonly employed catalysts that aid in reducing the activation energy for hydrolysis of biomass and yield high oxygenated functional groups in the product (Rabenau 1985, Makowski et al. 2008, Sevilla and Fuertes 2011, Donar et al. 2018).

The higher oxygenated functional groups in the products prepared using this technique offer functionalization routes. The products are also hydrophilic, hence more suitable for applications such as adsorption and catalysis (Shi et al. 2016, Zhu et al. 2020). The hydrochar produced in this technique is a good precursor for activated carbon. The hydrochar products also possess a lower degree of aromaticity and are more suitable for chemical activation. The temperature-dependent tunability of morphology and the use of less toxic reagents make this technique very useful. The resultant particle sizes are usually larger, which may be disadvantageous in applications and require a higher surface area of the material (Baccile et al. 2010, Libra et al. 2011, Möller et al. 2011, Titirici et al. 2012).

2.3 Ionothermal Carbonization

This technique involves the use of ionic liquids as solvents for carbonization instead of water. Ionic liquids offer good solubility for biomass materials. Their low volatility and high stability make them ideal as solvents for biomass decomposition (Bagheri et al. 2015). They can form hydrogen bonds with solutes, thus enhancing the solubility of carbohydrate chains that biomass materials are made up of. Ionothermal carbonization is known to produce a higher yield of carbonaceous materials with improved surface area as compared to hydrothermal carbonization (Varma 2019). Metal-containing ionic liquids are also available and can act as a solvent as well as a catalyst in carbonization processes. Although the ionic liquid can be recovered and reused after the end of the reaction, its high price is yet a disadvantage for the wider applicability of this technique.

2.4 Molten Salt Carbonization

In this method, biomass is carbonized in molten salt. Molten salts possess high heat transfer, high thermal stability and good dissolution. In the previously described methods, post-synthetic activation is required to enable high conductivity and porosity of the resultant products. Molten salt carbonization is in itself a high-temperature preparation technique, thus negating the need for any post-treatment step. This makes the method cost-effective and easily scalable (Yin et al. 2014, Pi et al. 2015, Deng et al. 2016, Yang et al. 2019a, 2019b).

3. Biomass-Derived Carbon Materials for Photocatalysis: Diversity and Evolution

Photocatalysis is a unique chemical transformation, inspired by photosynthesis (Yang and Wang 2018). A photocatalyst upon light absorption can produce chemical transformations of the reactants. The concept of photocatalysis has been used in literature for two distinctive processes:

a. Chemical transformations such as water splitting and CO_2 reduction where the photocatalyst uses light energy to drive thermodynamically uphill reactions (Figure 2a)
b. Chemical transformations such as dye degradations where light energy is used to drive thermodynamically downhill reactions (Figure 2b) (Berardi et al. 2014, Zhu and Wang 2017, Yang and Wang 2018).

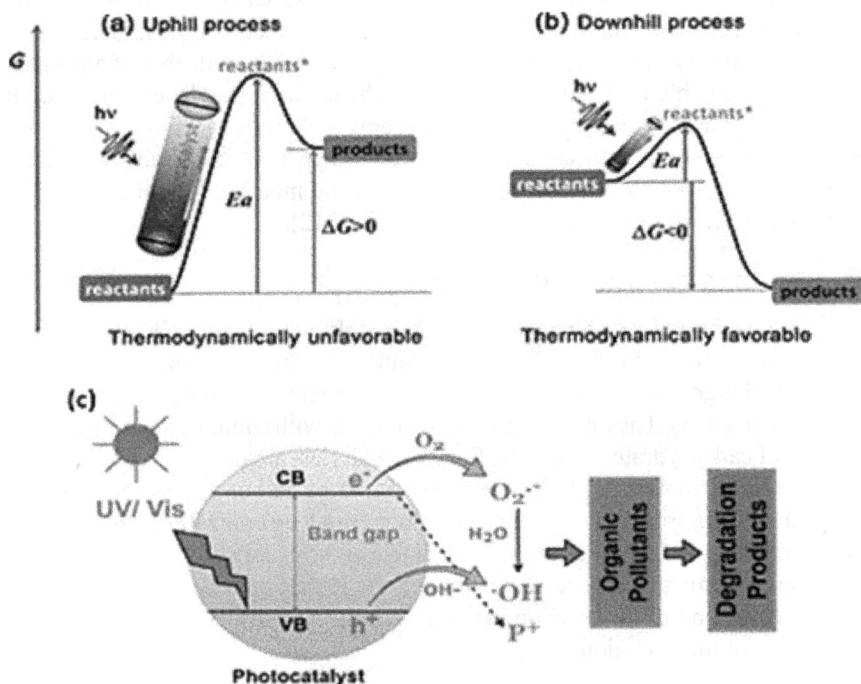

Figure 2: Thermodynamics of uphill and downhill photocatalysis: (a) uphill process; (b) downhill process. Adapted with permission from reference (Yang and Wang 2018). Copyright © 2018 American Chemical Society. (c) Fundamentals of organic pollutant degradation by photocatalysis, adapted with permission from reference (Mian and Liu 2018) (© The Royal Society of Chemistry 2018).

However, about the materials that can be used as photocatalysts for both of these reaction types, the desired properties remain the same – light absorption, charge separation and charge transfer (Fujishima and Honda 1972, Nocera 2012, Tachibana et al. 2012).

In a broader sense of description, the photocatalytic process can be described as an advanced oxidation process, which is driven by light energy (Figure 2c) (Fujishima and Honda 1972, Berardi et al. 2014, Zhu and Wang 2017, Yang and Wang 2018). For the choice of a photocatalyst, the material is expected to have good light sensitivity and have the correct band structure i.e. filled valence band and empty conduction bands. Upon irradiation with photons, a part of it is absorbed by the material and if the absorbed light consists of photons with energy greater than the bandgap of the photocatalyst, then the exciton (electron-hole pair) generation takes place where the electron jumps to the conduction band and the hole remains in the valence band. This is followed by exciton migration to the surface of the photocatalyst for redox reactions. Depending on the category of photocatalytic application, a series of redox reactions will take place on the surface of the photocatalyst. This can include oxidation that is induced by the photogenerated holes and reduction caused by the photogenerated electrons. The efficiency of the photocatalytic process depends on the nature of the photocatalyst. For the choice of the semiconductor as the photocatalyst, the desired features are:

a. Improved number of active sites to reduce premature electron-hole recombination.
b. The decreased bandgap of the photocatalyst enables visible light absorption.
c. Higher absorption ability/coefficient of the photocatalyst to increase the generated electron-hole pair.
d. Higher surface area and stability.
e. Higher quantum efficiency for suppressed electron-hole pair recombination.

TiO_2 has been a well-known photocatalyst for solar energy conversion and water purification. However, TiO_2 cannot absorb visible light due to its large bandgap of 3.2 Ev (Li et al. 2017, Osterloh 2017). Thus, about 45% of the solar spectrum remains unutilized in TiO_2 based photocatalysis. Therefore, in the quest for visible-light photocatalysts, there has been a recent shift toward non-TiO_2 based and more carbon-based or carbonaceous photocatalysts due to their unique electronic structures, coupling and synergistic effects (Fernando et al. 2015). Biomass and waste-derived materials have been considered as a potentially attractive option for photocatalytic applications with their highly sustainable design, tunable texture and morphology and synthetic conditions that can lead to increased active sites and surface area (Huang et al. 2019, Yang et al. 2019b).

Some of the desirable features of biomass-derived carbonaceous materials are:

a. Good electron conductivity because of the continuous connectivity in the carbon structure.
b. The improved surface area is due to the high-temperature activation techniques.
c. Biomass-derived carbon materials have shown excellent physicochemical stability.
d. Facile and well-established synthetic procedures.

e. The versatility of this technique is broadly applicable to any biomass source.
f. The vast variety of carbon materials that can be prepared using this technique is an advantage.

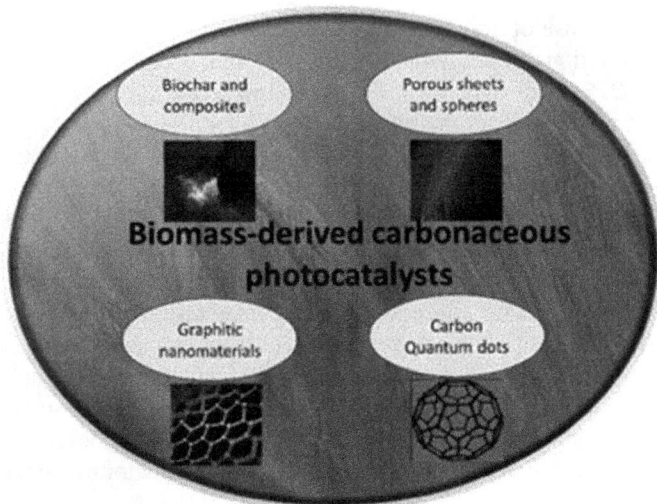

Figure 3: Schematic representation of the different biomass-derived carbonaceous photocatalysts.

4. Surface Tuning to Enhance the Photocatalytic Performance

Functionalization of the carbonaceous surface is important for fine-tuning its properties for photocatalytic applications. This can be done in two ways:

a. Pre-synthetically – where the functional groups are introduced *in-situ* during the material synthesis.
b. Post-synthetically – where the functional groups are introduced after the preparation of the carbonaceous material.

Pre-synthetic functionalization being a single-step process is often preferred for ease and scale-up. Post-synthetic functionalization is preferred for controlled functionalization of nanomaterials, and it introduces the functional group before the formation of the nanoparticles can distort the size and shape.

5. A Comparative Study of the Surface Morphology and Photocatalytic Parameters of Different Biomass-Derived Carbon Materials

5.1 Biochar from Biomass

Biochar is a porous carbon residue that is prepared by anaerobic pyrolysis of biomass. It is a low-cost and sustainable material. Biochar can be used as an excellent platform for hosting catalytic nanoparticles with contributed improvements in visible light sensitivity, stability and photocatalytic degradation. Commonly employed synthesis methods include sol-gel, ultrasound, thermal polycondensation, solvothermal and hydrothermal carbonizations. As photocatalyst supports, biomass-derived biochar is of high importance because the carbonization process gives them high oxygen functionalities that act as nucleation sites for the deposition or anchoring of the photocatalyst. One of the successful investigations of biomass-derived carbonaceous materials as catalyst supports involves the Starbon family—a new generation of porous carbon-based materials obtained from cheap biomass feedstock (Doi et al. 2002, Milkowski et al. 2004, Budarin et al. 2005). This material was developed as a highly porous and stable carbon-based material. Their surface properties and texture can be tuned depending on the carbonization temperature and the polysaccharide precursor. Colmenares and co-workers have shown for the first time the potential of these materials in heterogeneous photocatalysis (Colmenares et al. 2013). They showed that Starbon-800 can serve as solid support for TiO_2 photocatalysts with enhanced ordering and crystallinity. This leads to improved photocatalytic activity with reduced electron-hole recombination. The specific photocatalytic application studied in this work is phenol degradation. Starbon being hydrophobic effectively anchors the hydrophilic TiO_2 particles and modulates their charge transfer properties.

In another work highlighting the importance of biomass-derived materials in photocatalysis, titania-based wood from softwood pellets and straw-derived pellets were used (Lisowski et al. 2017). The TiO_2/biochar composite formed a heterojunction (Figure 4). The authors claim that the interfacial chemistry was controlled by sonication (ultrasonic wet impregnation method) and the charge transfer between TiO_2 and biochar was greatly improved leading to more efficient phenol photodegradation. A photo-responsive graphitic C_3N_4/leaf-derived biochar composite exhibited a high decontamination capability toward methylene blue (Pi et al. 2015). The conductive carbon material serves as efficient electron-transfer channels and also improves the charge separation of the electron-hole pairs. Li et al.

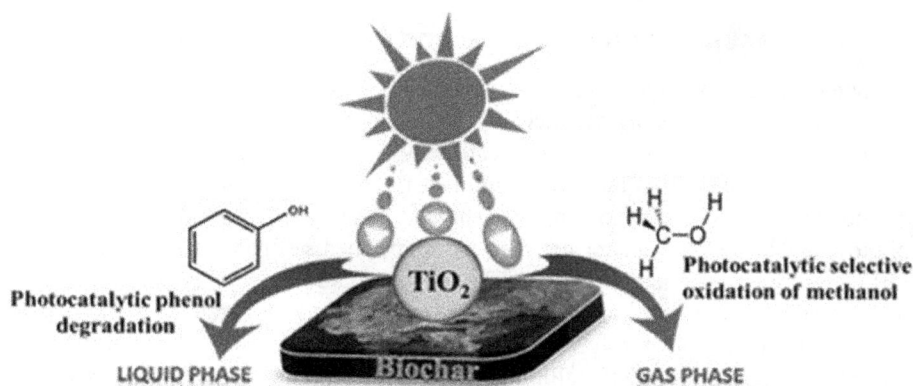

Figure 4: Representation of a titania/biochar photocatalytic composite, which enables efficient photocatalytic phenol degradation in the liquid phase, and photocatalytic selective oxidation of methanol in the gas phase. Adapted with permission from reference (Lisowski et al. 2017) (© 2017 American Chemical Society).

investigated the influence of biochar/BiOX composites on photocatalytic degradation of methyl orange (M. Li et al. 2016). Biochar was shown to act as a benign carrier that facilitated the uniform distribution of the BiOX particles, reduced the recombination of electrons/holes and promoted the photocatalytic reaction. TiO_2 supported on acid-treated reed straw biochar is another interesting composite evaluated for the photocatalytic degradation of antibiotics (Zhang et al. 2017). In the study by Zhang et al., (2017) the composites were immobilized by a sol-gel approach and exhibited a 91.27% efficiency of mineralization of the antibiotic sulfamethoxazole.

Khraisheh et al. developed a novel TiO_2/coconut shell composite for photocatalytic removal of carbamazepine, an organic pollutant increasingly found in wastewater streams (Khraisheh et al. 2013). The preparation method was sol-gel followed by thermal activation. The resulting highly crystalline photocatalytic absorbent material exhibited 98% removal efficiency. A titania-rice husk silica composite prepared by the sol-gel method yielded micro-structured particles which when calcined at 500°C, resulted in photodegradation of methylene blue (Manurung et al. 2015). In another work, Salamone et al. use activated carbon synthesized by carbonization of sugarcane bagasse waste as support for nano-TiO_2 (El-Salamony et al. 2017). This photocatalytic composite showed impressive dye degradation when tested with methylene blue with up to 96% catalytic activity. Overall, the anchoring effects and enhanced wettability of the biomass-derived carbonaceous materials enable uniform location of the active species and superior photocatalytic performance.

Therefore, the incorporation of transition metals, catalytic nanoparticles and non-metals with biochar templates are effective strategies for producing high-performance semiconductor nanoparticles. Biochar can host a large number of catalytic particles, provide an enhanced number of active sites, act as an electron reservoir, shuttle electrons like graphene, improve charge separation and reduce the bandgap via metal doping. Simultaneous adsorption and photodegradation lead to a higher percentage of organic-pollutant degradation (attack of reactive oxygen species) as shown in Figure 5.

5.2 Biomass-Derived Graphene and Graphitic Materials

Graphene with its sp^2 hybridized atoms, ordered in a two-dimensional honeycomb assembly offers interesting prospects for photocatalytic applications. N-doping of graphene has been evaluated as a good strategy for enhanced photocatalytic activity due to the increase in conductivity, light absorption and visible light sensitivity. N-doped graphene/CdS nanocomposite showed remarkable H_2 generation due to the photo-induced charge transfer between the components (Figure 5 a-b) (Peng et al. 2012). N-doping of graphene using chitosan aerogels via *in-situ* pyrolysis at 900°C imparts photoactivity for hydrogen generation from methanol/water (Lavorato et al. 2014). An increase in pyrolytic temperature is observed to decrease the nitrogen content and improve the crystallinity of the resulting N-graphene, which leads to improved photocatalytic performance.

Recently graphite-like carbon nitride (g-C_3N_4) has emerged as a photocatalytic material. For waste-derived g-C_3N_4, nitrogen-rich precursors are preferred as starting materials. This includes urea, thiourea, melamine, cyanamide, dicyanamide, etc.

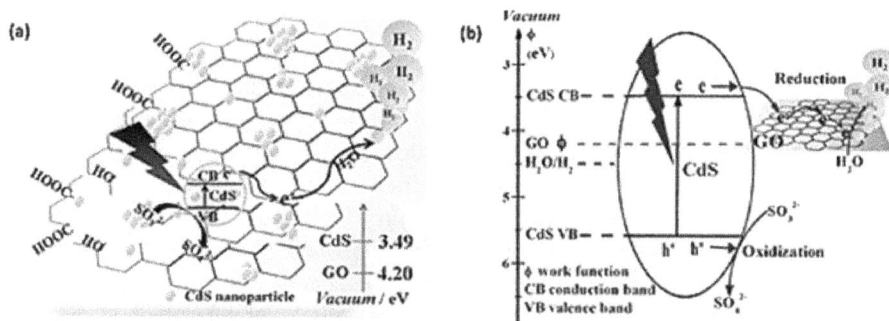

Figure 5: (a) Representation of the reduced GO/CdS composite and (b) the proposed mechanism for the photocatalytic hydrogen production in the system. Adapted with permission from ref (Peng et al. 2012) (© 2012 American Chemical Society).

These are easily obtained from agricultural wastes—a good step toward a circular economy (Antonetti et al. 2017). Pure g-C_3N_4 is not very desirable for photocatalysis owing to its high recombination rate and reduced charge mobility. When used in combination with metallic materials, g-C_3N_4 shows a promising photocatalytic response. Recently calcination of waste toner powder at 600°C was used to produce magnetic Fe_3O_4 powder. This when subjected to the second calcination in the presence of thiourea led to the formation of g-C_3N_4-Fe_2O_3 composite (Babar et al. 2019). There is a heterojunction formation between the g-C_3N_4 and Fe_2O_3 components, leading to enhanced visible light absorption and high photoluminescent yield. Their photocatalytic behavior was tested with methyl orange with a high degree of success accompanied by good recyclability as the particles could be efficiently recovered with the help of an external magnetic field. Various other N-containing carbonaceous materials have also been reported from biomass sources (Tuck et al. 2012, Lin et al. 2013). Carbonization of rabbit skin residues at 180-600°C resulted in well visible light absorption and improved photocatalytic response during phenol photooxidation (Colmenares et al. 2014).

5.3 Biomass-Derived Porous Carbon Sheets and Spheres

Biomass-derived porous carbon materials with a graphitic structure have also been explored for photocatalytic remediation of organic pollutants. Liu et al. fabricated highly porous BPGCs from fish waste and used the resulting material as a highly efficient adsorbent for methyl orange (Liu et al. 2016). These graphitic carbon systems are known to possess macro- and mesopores which promote mass transportation, higher surface area and enough active sites for heteroatom doping. The mechanism by which these materials aid in efficient extraction of the organic pollutants is reported to be π-π interaction. The graphene-like structure on the surface can form π-π interaction with the aromatic ring of the organic dyes/pollutants being removed. In another unique example of a biochar-graphene-type system, Zhong et al. prepared magnetic biochar via microwave-assisted pyrolysis of rice husk and mixed it with $FeSO_4$ (Zhong et al. 2018). The resultant product was reported to be magnetic biochar with a graphitic structure and has high conductivity. The system was used to facilitate electron transfer for adsorption and reduction of hexavalent chromium [Cr(VI)].

As supports for photocatalytic materials, these systems can increase the efficiency of photocatalytic degradation by imparting improved stability and reducing the reaction distance. The photogenerated electrons from the photocatalyst can be extracted and transported by the porous carbon, thus reducing the recombination rate. Li et al. used hydrothermal carbonization of Camellia oleifera that shells to prepare biomass-derived porous carbon microspheres coupled with g-C_3N_4 (Li et al. 2019). The synergistic effect between the two systems enabled substantial removal of Cr(VI) from aqueous solutions. The generated Cr(III) was subsequently fixed and recycled. Ye et al. (2019) demonstrated the production of porous graphitic carbon sheets through synchronous carbonization of straw with potassium ferrate as a catalyst. The resulting graphitic carbon sheets were then loaded with MoS_2 to form g-MoS_2/BPGC composite that showed a high removal efficiency toward tetracycline hydrochloride via photocatalysis under visible light. There is still scope for work in this field to further investigate the binding mechanisms and optimization of the composites.

5.4 Carbon Quantum Dots Derived from Biomass

Quantum dots are multicolor emissive materials with unique size-dependent optoelectronic properties. Carbon quantum dots are gaining attention because of their non-toxicity and bio-compatibility. There has been a steady rise in their application as photocatalytic materials for dye degradation, photochemical water splitting and solar fuel generation. The specific properties of interest include high solubility, stability, photobleaching resistance and routes for functionalization. Biomass-derived quantum dots BCDs are particularly interesting because of their low cost. The various techniques that have been used for BCD preparation from biomass include HTC and microwave treatments.

Biomass-derived carbon dots (BCDs) have unique photoluminescence and photoelectron transfer properties and therefore can play the role of a photosensitizer, a photocatalyst or a spectral converter (Meng et al. 2019). By doping with heteroatoms, the bandgap and quantum yield of BCDs can be tuned for a particular photocatalytic application. Qin et al. show the preparation of water-soluble BCDs by carbonization of willow barks (Qin et al. 2013). The photoluminescent BCDs were demonstrated as excellent photocatalysts for the reduction of Au complexes and reduced graphene oxide to form Au-decorated rGO composites (Figure 6a). BCDs synthesized from lemon peel waste by HTC when immobilized onto electrospun TiO_2, demonstrated good photocatalytic degradation behavior toward methylene blue (about 2.5 times higher than pristine TiO_2) (Tyagi et al. 2016). In another similar study, Arun et al. show the photocatalytic activity of Hylocereus undatus derived BCDs in the successful reduction of methylene blue (Arul et al. 2017). Wang et al. demonstrate the preparation of fluorescent BCDs embedded in a carbon matrix by HTC of glucose (Wang et al. 2015). The BCDs act as photocatalysts for degradation of 4-nitrophenol shown in Figure 6b.

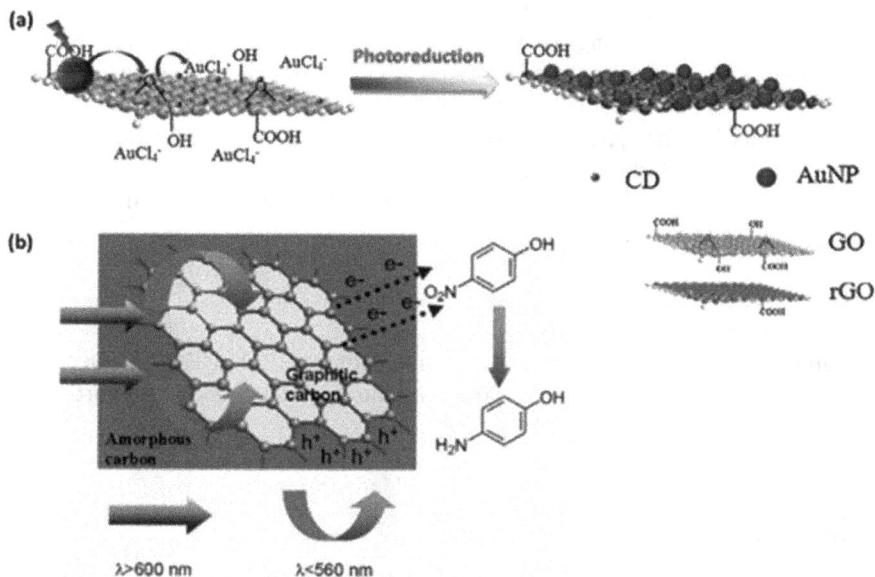

Figure 6: (a) A schematic diagram illustrating the proposed mechanism for the photochemical synthesis of AuNPs-rGO nanocomposites; adapted with permission from reference (Qin et al. 2013), copyright ©The Royal Society of Chemistry 2013. (b) Schematic photocatalytic mechanism of the reduction of 4-nitrophenol to 4-aminophenol over the CD-based carbocatalysts. Adapted with permission from reference (Wang et al. 2015) (© 2015 American Chemical Society).

6. Conclusions and Perspective

In this chapter, we have reviewed the different synthetic methods and properties of biomass-derived carbonaceous photocatalysts. We have systematically reviewed the processes involved in converting biomass into carbonaceous materials and the functionalities that enable their use in photocatalysis. Biomass as a carbon source is renewable, inexpensive, abundantly available, has good biocompatibility and can be scaled up for production to industrially relevant quantities. The method of preparation imparts high surface area and active sites to the resultant carbon materials. Numerous functionalities and dopants can be easily introduced into these materials to enable the formation of composites, heterostructures and heterojunctions, all of which are impressive strategies to improve photocatalytic efficiency. While a lot of research has been focused on developing new synthetic routes and exploring novel biomass sources, some challenges include tuning the absorption/emission properties, mechanistic investigations and detailed electron/charge transfer studies. Finding out more information about the mechanism of enhancement of photocatalytic efficiency will help in developing better performing and highly scalable systems and can expedite the field from lab-scale research to industrially and commercially relevant applications.

In addition to plant-based biomass various hazardous wastes, such as municipal waste, sewage and manure sludge, can also be useful in the synthesis of biomass-derived carbon materials. This is a very impressive and much-desired step in waste management and in assisting the global efforts for achieving a circular economy. While most existing applications focus on photocatalytic degradation of organic pollutants in the field of wastewater treatment, there is still a large scope for exploring these materials in other photocatalytic applications, such as CO_2 reduction, water splitting and artificial photosynthesis. We anticipate that this chapter has provided the readers with a good idea of what has already been explored in this field and invokes useful directions for future work.

Acknowledgements

The authors express their gratitude to the Center for Research, CHRIST (Deemed to be University), Bangalore, India for providing the required financial support.

References

Antonetti, E., Iaquaniello, G., Salladini, A., Spadaccini, L., Perathoner, S. and Centi, G. 2017. Waste-to-chemicals for a circular economy: the case of urea production (waste-to-urea). ChemSusChem. 10(5): 912–920. doi: 10.1002/cssc.201601555.

Arul, V., Edison, T.N.J.I., Lee, Y.R. and Sethuraman, M.G. 2017. Biological and catalytic applications of green synthesized fluorescent N-doped carbon dots using Hylocereus undatus. J. Photochem. Photobiol. B. Biol. 168: 142–148. https://doi.org/10.1016/j.jphotobiol.2017.02.007.

Babar, S., Gavade, N., Shinde, H., Gore, A., Mahajan, P., Lee, K. H., … Garadkar, K. 2019. An innovative transformation of waste toner powder into magnetic g-C_3N_4-Fe_2O_3 photocatalyst: sustainable e-waste management. J. Environ. Chem. Eng. 7(2): 103041. https://doi.org/10.1016/j.jece.2019.103041

Baccile, N., Antonietti, M. and Titirici, M.M. 2010. One-step hydrothermal synthesis of nitrogen-doped nanocarbons: albumine directing the carbonization of glucose. ChemSusChem. Wiley-VCH Verlag, 3(2): 246–253. doi: 10.1002/cssc.200900124.

Bagheri, S., Muhd Julkapli, N. and Bee Abd Hamid, S. 2015. Functionalized activated carbon derived from biomass for photocatalysis applications perspective. Int. J. Photoenergy, 2015(June). https://doi.org/10.1155/2015/218743.

Berardi, S., Drouet, S., Francàs, L., Gimbert-Suriñach, C., Guttentag, M., Richmond, C., … Llobet, A. (2014, November 21). Molecular artificial photosynthesis. Chem. Soc. Rev. Royal Society of Chemistry. https://doi.org/10.1039/c3cs60405e.

Berge, N.D., Ro, K.S., Mao, J., Flora, J.R.V., Chappell, M.A. and Bae, S. 2011. Hydrothermal carbonization of municipal waste streams. Environ. Sci. Technol. 45(13): 5696–5703. https://doi.org/10.1021/es2004528.

Biswal, M., Banerjee, A., Deo, M. and Ogale, S. 2013. From dead leaves to high energy density supercapacitors. Energy Environ. Sci. 6(4): 1249–1259. https://doi.org/10.1039/c3ee22325f.

Budarin, V., Clark, J.H., Deswarte, F.E.I., Hardy, J.J.E., Hunt, A.J. and Kerton, F.M. 2005. Delicious not siliceous: expanded carbohydrates as renewable separation media for column chromatography. Chem. Commun. (23): 2903–2905. https://doi.org/10.1039/b502330k.

Cha, J.S. Park, S.H., Jung, S.C., Ryu, C., Jeon, J.K., Shin, M.C. and Park, Y.K. 2016. Production and utilization of biochar: a review. J. Ind. Eng. Chem. Korean Society of Industrial Engineering Chemistry. https://doi.org/10.1016/j.jiec.2016.06.002.

Colmenares, J.C., Juan C., Lisowski, P., Bermudez, J.M., Cot, J. and Luque, R. 2014. Unprecedented photocatalytic activity of carbonized leather skin residues containing chromium oxide phases. Appl. Catal. B. Environ. 150–151: 432–437. https://doi.org/10.1016/j.apcatb.2013.12.038.

Colmenares, J.C., Lisowski, P. and Łomot, D. 2013. A novel biomass-based support (Starbon) for TiO_2 hybrid photocatalysts: a versatile green tool for water purification. RSC Adv. The Royal Society of Chemistry 3(43): 20186–20192. doi: 10.1039/c3ra43673j.

de Almeida Ribeiro, R.S., Monteiro Ferreira, L.E., Rossa, V., Lima, C.G.S., Paixão, M.W., Varma, R.S. and de Melo Lima, T. 2020. Graphitic carbon nitride-based materials as catalysts for the upgrading of lignocellulosic biomass-derived molecules. ChemSusChem (1): 3992–4004. https://doi.org/10.1002/cssc.202001017.

De, S., Balu, A.M., Van Der Waal, J.C. and Luque, R. 2015. Biomass-derived porous carbon materials: synthesis and catalytic applications. ChemCatChem 7(11): 1608–1629. https://doi.org/10.1002/cctc.201500081.

Deng, J., Li, M. and Wang, Y. 2016. Biomass-derived carbon: synthesis and applications in energy storage and conversion. Green Chemistry. Royal Society of Chemistry. 4824–4854. https://doi.org/10.1039/c6gc01172a.

Doi, S., Clark, J.H., Macquarrie, D.J. and Milkowski, K. 2002. New materials based on renewable resources: chemically modified expanded corn starches as catalysts for liquid phase organic reactions. Chemical Communications. The Royal Society of Chemistry, 2(22): 2632–2633. https://doi.org/10.1039/b207780a.

Donar, Y.O., Bilge, S., Sınağ, A. and Pliekhov, O. 2018. TiO_2/carbon materials derived from hydrothermal carbonization of waste biomass: a highly efficient, low-cost visible-light-driven photocatalyst. ChemCatChem. 10(5): 1134–1139. https://doi.org/10.1002/cctc.201701405.

El-Salamony, R.A., Amdeha, E., Ghoneim, S.A., Badawy, N.A., Salem, K.M. and Al-Sabagh, A.M. 2017. Environmental Technology Titania modified activated carbon prepared from sugarcane bagasse: adsorption and photocatalytic degradation of methylene blue under visible light irradiation. Titania modified activated carbon prepared from sugarcane bagasse: adsorption and photocatalytic degradation of methylene blue under visible light irradiation. https://doi.org/10.1080/21622515.2017.1290148.

Falco, C., Sieben, J.M., Brun, N., Sevilla, M., Van Der Mauelen, T., Morallón, E., ... Titirici, M.M. 2013. Hydrothermal carbons from hemicellulose-derived aqueous hydrolysis products as electrode materials for supercapacitors. ChemSusChem 6(2): 374–382. https://doi.org/10.1002/cssc.201200817.

Falco, C., Baccile, N. and Titirici, M.M. 2011. Morphological and structural differences between glucose, cellulose and lignocellulosic biomass derived hydrothermal carbons. Green Chemistry 13(11): 3273–3281. doi: 10.1039/c1gc15742f.

Fernando, K.A.S., Sahu, S., Liu, Y., Lewis, W.K., Guliants, E.A., Jafariyan, A. ... Sun, Y.P. 2015. Carbon quantum dots and applications in photocatalytic energy conversion. ACS Appl. Mater. Interfaces. American Chemical Society 7(16): 8363–8376. https://doi.org/10.1021/acsami.5b00448.

Fujishima, A. and Honda, K. 1972. Electrochemical photolysis of water at a semiconductor electrode. Nature, 238(5358): 37–38. doi: 10.1038/238037a0.

Ge, J., Zhang, Y. and Park, S.J. 2019. Recent advances in carbonaceous photocatalysts with enhanced photocatalytic performances: a mini review. Materials 12(12). doi: 10.3390/ma12121916.

Hamelinck, C.N., Van Hooijdonk, G. and Faaij, A.P.C. 2005. Ethanol from lignocellulosic biomass: techno-economic performance in short-, middle- and long-term. Biomass and Bioenergy. Elsevier Ltd, 28(4): 384–410. doi: 10.1016/j.biombioe.2004.09.002.

Hoekman, S.K., Broch, A. and Robbins, C. 2011. Hydrothermal carbonization (HTC) of lignocellulosic biomass. Energy and Fuels 25(4): 1802–1810. doi: 10.1021/ef101745n.

Hong, C., Jin, X., Totleben, J., Lohrman, J., Harak, E., Subramaniam, B. … Ren, S. 2014. Graphene oxide stabilized Cu_2O for shape selective nanocatalysis. J. Mater. Chem. A 2(20): 7147–7151. doi: 10.1039/c4ta00599f.

Huang, Q., Song, S., Chen, Z., Hu, B., Chen, J. and Wang, X. 2019. Biochar-based materials and their applications in removal of organic contaminants from wastewater: state-of-the-art review. Biochar. 1(1): 45–73. https://doi.org/10.1007/s42773-019-00006-5.

Jain, A., Balasubramanian, R. and Srinivasan, M.P. 2016. Hydrothermal conversion of biomass waste to activated carbon with high porosity: a review. Chem. Eng. J. Elsevier, pp. 789–805. doi: 10.1016/j.cej.2015.08.014.

Jazaeri, E. and Tsuzuki, T. 2013. Effect of pyrolysis conditions on the properties of carbonaceous nanofibers obtained from freeze-dried cellulose nanofibers. Cellulose 20(2): 707–716. doi: 10.1007/s10570-012-9858-2.

Khraisheh, M., Kim, J., Campos, L., Al-Muhtaseb, A.H., Walker, G.M. and AlGhouti, M. 2013. Removal of carbamazepine from water by a novel TiO_2–coconut shell powder/UV process: composite preparation and photocatalytic activity. Environ. Eng. Sci. 30(9): 515–526. https://doi.org/10.1089/ees.2012.0056.

Koe, W.S., Lee, J.W., Chong, W.C., Pang, Y.L. and Sim, L.C. 2020. An overview of photocatalytic degradation: photocatalysts, mechanisms, and development of photocatalytic membrane. Environ. Sci. Pollut. Res. Springer. https://doi.org/10.1007/s11356-019-07193-5.

Lackner, K.S. 2010. Comparative impacts of fossil fuels and alternative energy sources. *In:* Carbon Capture. Royal Society of Chemistry, pp. 1–40. doi: 10.1039/9781847559715-00001.

Lavorato, C., Primo, A., Molinari, R. and Garcia, H. 2014. N-doped graphene derived from biomass as a visible-light photocatalyst for hydrogen generation from water/methanol mixtures. Chemistry - A European Journal. 20(1): 187–194. https://doi.org/10.1002/chem.201303689.

Lee, S., Lee, M.E., Song, M.Y., Cho, S.Y., Yun, Y.S. and Jin, H.J. 2016. Morphologies and surface properties of cellulose-based activated carbon nanoplates. Carbon Letters. 20(1): 32–38. https://doi.org/10.5714/CL.2016.20.032.

Li, H., Yuan, D., Tang, C., Wang, S., Sun, J., Li, Z. … He, C. 2016. Lignin-derived interconnected hierarchical porous carbon monolith with large areal/volumetric capacitances for supercapacitor. Carbon. 100, 151–157. https://doi.org/10.1016/j.carbon.2015.12.075.

Li, K., Huang, Z., Zhu, S., Luo, S., Yan, L., Dai, Y. … Yang, Y. 2019. Removal of Cr(VI) from water by a biochar-coupled g-C_3N_4 nanosheets composite and performance of a recycled photocatalyst in single and combined pollution systems. Appl. Catal. B Environ. 243: 386–396. https://doi.org/10.1016/j.apcatb.2018.10.052.

Li, M., Huang, H., Yu, S., Tian, N., Dong, F., Du, X. and Zhang, Y. 2016. Simultaneously promoting charge separation and photoabsorption of BiOX (X = Cl, Br) for efficient visible-light photocatalysis and photosensitization by compositing low-cost biochar. Appl. Surf. Sci. 386, 285–295. https://doi.org/10.1016/j.apsusc.2016.05.171.

Li, Z., Wang, W., Ding, C., Wang, Z., Liao, S. and Li, C. 2017. Biomimetic electron transport via multiredox shuttles from photosystem II to a photoelectrochemical cell for solar water splitting. Energy Environ. Sci. 10(3): 765–771. https://doi.org/10.1039/c6ee03401b.

Liang, W., Pan, J., Duan, X., Tang, H., Xu, J. and Tang, G. 2020. Biomass carbon modified flower-like Bi_2WO_6 hierarchical architecture with improved photocatalytic performance. Ceramics International, 46(3): 3623–3630. https://doi.org/10.1016/j. ceramint.2019.10.081.

Libra, J.A., Ro, K.S., Kammann, C., Funke, A., Berge, N.D., Neubauer, Y. ... Emmerich, K.H. 2011. Hydrothermal carbonization of biomass residuals: a comparative review of the chemistry, processes and applications of wet and dry pyrolysis. Biofuels 2(1): 71–106. https://doi.org/10.4155/bfs.10.81.

Lin, C.S.K., Pfaltzgraff, L.A., Herrero-Davila, L., Mubofu, E.B., Abderrahim, S., Clark, J.H. ... Luque, R. 2013. Food waste as a valuable resource for the production of chemicals, materials and fuels. Current situation and global perspective. Energy and Enviro. Sci. Royal Society of Chemistry, https://doi.org/10.1039/c2ee23440h 426–464. doi: 10.1039/c2ee23440h.

Lisowski, P., Colmenares, J.C., Mašek, O., Lisowski, W., Lisovytskiy, D., Kamińska, A. and Łomot, D. 2017. Dual functionality of TiO_2/biochar hybrid materials: photocatalytic phenol degradation in the liquid phase and selective oxidation of methanol in the gas phase. ACS Sustain. Chem. Eng. 5(7): 6274–6287. https://doi.org/10.1021/acssuschemeng.7b01251.

Liu, Y., Chen, J., Cui, B., Yin, P. and Zhang, C. (n.d.). Design and preparation of biomass-derived carbon materials for supercapacitors: a review. J. Carbon Res., 2018; 4(4): 53.

Liu, Z., Zhang, F., Liu, T., Peng, N. and Gai, C. 2016. Removal of azo dye by a highly graphitized and heteroatom doped carbon derived from fish waste: adsorption equilibrium and kinetics. J. Environ. Manage., 182: 446–454. https://doi.org/10.1016/j.jenvman.2016.08.008.

Luo, Z., Z., Wang, S., Liao, Y., Zhou, J., Gu, Y. and Cen, K. 2004. Research on biomass fast pyrolysis for liquid fuel. Biomass and Bioenergy. 26(5): 455–462. https://doi.org/10.1016/j.biombioe.2003.04.001.

Makowski, P., Demir Cakan, R., Antonietti, M., Goettmann, F. and Titirici, M.M. 2008. Selective partial hydrogenation of hydroxy aromatic derivatives with palladium nanoparticles supported on hydrophilic carbon. Chemical Commun. Royal Society of Chemistry, 8: 999–1001. https://doi.org/10.1039/b717928f.

Manurung, P., Situmeang, R., Ginting, E. and Pardede, I. 2015. Synthesis and characterization of titania-rice husk silica composites as photocatalyst. Indones. J. Chem. 15(1): 36–42. https://doi.org/10.22146/ijc.21221.

McKendry, P. 2002. Energy production from biomass (part 1): overview of biomass. Biores. Tech. 83(1): 37–46. https://doi.org/10.1016/S0960-8524(01)00118-3.

Meng, W., Bai, X., Wang, B., Liu, Z., Lu, S. and Yang, B. 2019. Biomass-derived carbon dots and their applications. Energy Environ. Mater. 2(3): 172–192. https://doi.org/10.1002/eem2.12038.

Mian, M.M. and Liu, G. 2018. Recent progress in biochar-supported photocatalysts: synthesis, role of biochar, and applications. RSC Adv. Royal Society of Chemistry, 8(26): 14237–14248. doi: 10.1039/c8ra02258e.

Milkowski, K., Clark, J.H. and Doi, S. 2004. New materials based on renewable resources: chemically modified highly porous starches and their composites with synthetic monomers. Green Chemistry. Royal Society of Chemistry, 6(4): 189–190. doi: 10.1039/b316322a.

Möller, M., Nilges, P., Harnisch, F. and Schröder, U. 2011. Subcritical water as reaction environment: fundamentals of hydrothermal biomass transformation. ChemSusChem. Wiley-VCH Verlag. https://doi.org/10.1002/cssc.201000341.

Nocera, D.G. 2012. The artificial leaf. Acc. Chem. Res. 45(5): 767–776. doi: 10.1021/ar2003013.

Nzihou, A., Stanmore, B., Lyczko, N. and Minh, D.P. 2019. The catalytic effect of inherent and adsorbed metals on the fast/flash pyrolysis of biomass: a review. Energy. 170: 326–337. https://doi.org/10.1016/j.energy.2018.12.174.

Osterloh, F.E. 2017. Photocatalysis versus photosynthesis: a sensitivity analysis of devices for solar energy conversion and chemical transformations. ACS Energy Lett. American Chemical Society, pp. 445–453. doi: 10.1021/acsenergylett.6b00665.

Patrick, J.W., Botha, F.C. and Birch, R.G. 2013. Metabolic engineering of sugars and simple sugar derivatives in plants. Plant Biotechnol. J. pp. 142–156. doi: 10.1111/pbi.12002.

Peng, T., Li, K., Zeng, P., Zhang, Q. and Zhang, X. 2012. Enhanced photocatalytic hydrogen production over graphene oxide-cadmium sulfide nanocomposite under visible light irradiation. J. Phys. Chem. C 116(43): 22720–22726. https://doi.org/10.1021/jp306947d.

Pi, L., Jiang, R., Zhou, W., Zhu, H., Xiao, W., Wang, D. and Mao, X. 2015. G-C_3N_4 modified biochar as an adsorptive and photocatalytic material for decontamination of aqueous organic pollutants. Appl. Surf. Sci. 358: 231–239. Elsevier B.V. https://doi.org/10.1016/j.apsusc.2015.08.176.

Qin, X., Lu, W., Asiri, A.M., Al-Youbi, A.O. and Sun, X. 2013. Green, low-cost synthesis of photoluminescent carbon dots by hydrothermal treatment of willow bark and their application as an effective photocatalyst for fabricating Au nanoparticles-reduced graphene oxide nanocomposites for glucose detection. Catal. Sci. Technol. 3(4): 1027–1035. https://doi.org/10.1039/c2cy20635h.

Rabenau, A. 1985. The role of hydrothermal synthesis in preparative chemistry. Angewandte Chemie International Edition in English, pp. 1026–1040. doi: 10.1002/anie.198510261.

Ramis, G., Bahadori, E. and Rossetti, I. 2020. Design of efficient photocatalytic processes for the production of hydrogen from biomass derived substrates. Int. J. Hydrogen Energy, (xxxx). https://doi.org/10.1016/j.ijhydene.2020.02.192.

Sevilla, M. and Fuertes, A.B. 2009. Chemical and structural properties of carbonaceous products obtained by hydrothermal carbonization of saccharides. Chem. A Eur. J. 15(16): 4195–4203. doi: 10.1002/chem.200802097.

Sevilla, M. and Fuertes, A.B. 2011. Sustainable porous carbons with a superior performance for CO_2 capture. Energy Environ. Sci. 4(5): 1765–1771. doi: 10.1039/c0ee00784f.

Shi, M., Wei, W., Jiang, Z., Han, H., Gao, J. and Xie, J. 2016. Biomass-derived multifunctional TiO_2/carbonaceous aerogel composite as a highly efficient photocatalyst. RSC Advances, 6(30), 25255–25266. https://doi.org/10.1039/c5ra28116d.

Tachibana, Y., Vayssieres, L. and Durrant, J.R. 2012. Artificial photosynthesis for solar water-splitting. Nature Photonics 511–518. doi: 10.1038/nphoton.2012.175.

Titirici, M.M., White, R.J., Falco, C. and Sevilla, M. 2012. Black perspectives for a green future: hydrothermal carbons for environment protection and energy storage. Energy Environ. Sci. https://doi.org/10.1039/c2ee21166a.

Titirici, M.M. and Antonietti, M. 2010. Chemistry and materials options of sustainable carbon materials made by hydrothermal carbonization. Chem. Soc. Rev. 39(1): 103–116. doi: 10.1039/b819318p.

Tripathi, M., Sahu, J.N. and Ganesan, P. 2016. Effect of process parameters on production of biochar from biomass waste through pyrolysis: a review. Renew. Sustain. Energy Rev. Elsevier Ltd, pp. 467–481. doi: 10.1016/j.rser.2015.10.122.

Tuck, C.O., Pérez, E., Horváth, I.T., Sheldon, R.A. and Poliakoff, M. 2012. Valorization of biomass: deriving more value from waste. Science. American Association for the Advancement of Science. https://doi.org/10.1126/science.1218930.

Tyagi, A., Tripathi, K.M., Singh, N., Choudhary, S. and Gupta, R.K. 2016. Green synthesis of carbon quantum dots from lemon peel waste: applications in sensing and photocatalysis. RSC Advances, 6(76), 72423–72432. https://doi.org/10.1039/c6ra10488f.

Varanasi, P., Singh, P., Auer, M., Adams, P.D., Simmons, B.A. and Singh, S. 2013. Survey of renewable chemicals produced from lignocellulosic biomass during ionic liquid pretreatment. Biotechnology for Biofuels. https://doi.org/10.1186/1754-6834-6-14.

Varma, R.S. 2019. Biomass-derived renewable carbonaceous materials for sustainable chemical and environmental applications. ACS Sustain. Chem. Eng. American Chemical Society, 7(7): 6458–6470. doi: 10.1021/acssuschemeng.8b06550.

Wang, H., Zhuang, J., Velado, D., Wei, Z., Matsui, H. and Zhou, S. 2015. Near-infrared- and visible-light-enhanced metal-free catalytic degradation of organic pollutants over carbon-dot-based carbocatalysts synthesized from biomass. ACS Appl. Mater. Interfaces 7(50): 27703–27712. https://doi.org/10.1021/acsami.5b08443.

Yang, D.P., Li, Z., Liu, M., Zhang, X., Chen, Y., Xue, H. ... Luque, R. 2019a. Biomass-derived carbonaceous materials: recent progress in synthetic approaches, advantages, and applications. ACS Sustain. Chem. Eng. American Chemical Society. https://doi.org/10.1021/acssuschemeng.8b06030.

Yang, X. and Wang, D. 2018. Photocatalysis: from fundamental principles to materials and applications. ACS Appl. Energy. Mater. American Chemical Society, pp. 6657–6693. doi: 10.1021/acsaem.8b01345.

Yin, H., Lu, B., Xu, Y., Tang, D., Mao, X., Xiao, W. ... Alshawabkeh, A.N. 2014. Harvesting capacitive carbon by carbonization of waste biomass in molten salts. Environ. Sci. Tech. 48(14): 8101–8108. https://doi.org/10.1021/es501739v.

Zhang, H., Wang, Z., Li, R., Guo, J., Li, Y., Zhu, J. and Xie, X. 2017. TiO_2 supported on reed straw biochar as an adsorptive and photocatalytic composite for the efficient degradation of sulfamethoxazole in aqueous matrices. Chemosphere. Elsevier Ltd, 185, pp. 351–360. https://doi.org/10.1016/j.chemosphere.2017.07.025.

Zhang, Q., Chang, J., Wang, T. and Xu, Y. 2007a. Review of biomass pyrolysis oil properties and upgrading research. Energy Convers. Manag. Pergamon, 48(1): 87–92. doi: 10.1016/j.enconman.2006.05.010.

Zhong, D., Zhang, Y., Wang, L., Chen, J., Jiang, Y., Tsang, D.C.W. ... Crittenden, J.C. 2018. Mechanistic insights into adsorption and reduction of hexavalent chromium from water using magnetic biochar composite: key roles of Fe_3O_4 and persistent free radicals. Environ. Poll. 243: 1302–1309. https://doi.org/10.1016/j.envpol.2018.08.093.

Zhu, S. and Wang, D. 2017. Photocatalysis: basic principles, diverse forms of implementations and emerging scientific opportunities. Adv. Energy Mat. Wiley-VCH Verlag. doi: 10.1002/aenm.201700841.

Zhu, Z., Yang, P., Li, X., Luo, M., Zhang, W., Chen, M. and Zhou, X. 2020. Green preparation of palm powder-derived carbon dots co-doped with sulfur/chlorine and their application in visible-light photocatalysis. Spectrochim. Acta. A 227: 117659. https://doi.org/10.1016/j.saa.2019.117659.

Production of Multifunctional Carbon Nanotubes for Sensor and Water Treatment Applications

Kingsley I. John[1,2], Aderemi Timothy Adeleye[3,4]*, Abesa Solomon[5] and A.A. Audu[6]

[1] Lab of Department of Pure and Applied Chemistry, College of Natural Sciences,
   Veritas University Abuja, PMB 5171, Abuja, Nigeria
[2] State Key Laboratory of Catalysis & Division of Solar Energy, Dalian National Laboratory
   for Clean Energy, Dalian Institute of Chemical Physics, Chinese Academy of Sciences,
   Dalian, 116023, China
[3] CAS Key Laboratory of Science and Technology on Applied Catalysis, Dalian Institute of
   Chemical Physics, Chinese Academy of Sciences, Dalian 116023, China
[4] Organization of African Academic Doctor (OAAD), Off Kamiti Road,
   P.O. Box 25305000100, Nairobi, Kenya
[5] Chemistry Department, Federal University of Agriculture, Makurdi, Nigeria
[6] Department of Pure and Industrial Chemistry, Bayero University, PMB 3011,
   Kano, Nigeria

1. Introduction

Nanomaterials (NMs) are functional materials comprising particulates with at least one dimension below 100 nm (Saito et al. 2005). The technological applicability of nanomaterials is termed 'nanotechnology'. Owing to the unique characteristic properties they possess, NMs have a vast application in devices for energy, catalytic activities, adsorption of pollutants as electrochemical sensors, in medicine as biomedical sensors, wastewater treatment, fuel cell, biodiesel production, and many others (Luxembourg et al. 2005, Chao et al. 2018, Oliveira et al. 2018, Shaw et al. 2017, Li et al. 2012). The unique properties include various inherent electronic, biocompatibility, magnetic and optical properties, excellent activity, versatility, and ease of handling (Lehman et al. 2011). CNTs are made of a tubular structure of carbon, and the dimension is calculated in nanometers (Li et al. 2020). CNTs are carbon allotropes made of a nanostructure with a length ratio to a diameter of more than 1,000,000 (Clegg et al. 2019). CNTs are formulated from a graphite sheet; the

---

Corresponding author: aderemi4crown@yahoo.com

substance is like a rolled mesh of a continuous chain of a hexagonal structure. CNTs have a different range of conductivity features that may be useful for sensor devices. The rolled graphene sheet's nature, which determines the species' chirality and hinges on these properties (Lehman et al. 2011). Some unique properties of CNTs, which make them superior in a technological application, include high elasticity, surface area, high thermal conductivity, stiffness, low density, and toughness.

2. A Brief History of Carbon Nanotubes

In the 1980s, the only type of carbon was amorphous, diamond, and graphite. Other variants of carbon, such as fullerene (C_{60}), were discovered later. There are over 30 forms of fullerene (C_{32}, C_{50}, C_{70}, C_{72} and C_{84}) and CNTs, which are part of such linear molecules. As shown in Figure 1 above, fullerene is a spherical molecule arranged in a soccer ball shape consisting of carbon atoms. It consists of a fused system of five and six-membered rings (Ahamad et al. 2019). In 1952, the Soviet Journal of Physical Chemistry published candid photographs of carbon tubes with a diameter of about 50 nm as investigated by Radushkevich and Lukyanovich. A researcher from Hyperion Catalysis named Howard G. Tennent obtained a US patent to produce 'cylindrical discrete carbon fibrils' in 1987. John Abrahamson, in 1979, also presented evidence of carbon nanotubes (Clegg et al. 2019). Though research on carbon nanotubes has been carried out previously, a significant proportion of scholarly and mainstream literature commonly credits discovering hollow, nanometer-sized graphite carbon tubes to the Nippon Electric Company's Sumio Iijima in 1991 (Li et al. 2020). He accidentally discovered it from his desire to ascertain fullerene's molecular structure and examine their crystals' growth (Amjadi et al. 2017). In 1993, Iijimi and Donald Bethune found the single-walled nanotubes known as buckytubes.

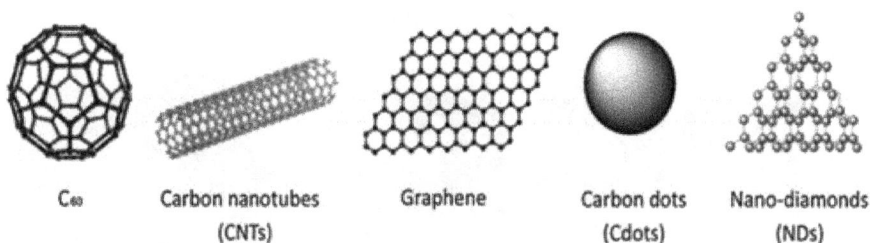

| C_{60} | Carbon nanotubes (CNTs) | Graphene | Carbon dots (Cdots) | Nano-diamonds (NDs) |

Figure 1: Some carbon nanostructures (Song et al. 2020, Lehman et al. 2011).

3. Structure and Morphology of Carbon Nanotubes (CNTs)

Carbon nanotubes are made of graphite sheets; these sheets are rolled up into cylindrical shapes that give rise to what we now call carbon nanotubes. Graphite is a buildup of flat two-dimension sheets of carbon atoms, and each sheet is a honeycomb or a hexagonal net of carbon atoms held together by a relatively weak Van der Waal force (Wang et al. 2020). The bonding in carbon nanotubes is of sp² hybridisation. The bonding structure of graphite is more formidable than the sp³ bonds present in diamond (Wang et al. 2020). Graphite and its derivatives, of which carbon nanotubes are a chief member, have attracted tremendous interest in biomedical sensors, electronics, energy sensors, catalytic activities, drug delivery activities, and wastewater treatment. The vast application of CNTs may be ascribed to their porosity, reactivity, remarkable aspect ratios, a large surface area, mechanical stability, electrical, and chemical properties (Amjadi et al. 2017).

4. Classification of Carbon Nanotubes

Generally, according to the number of carbon layers, CNTs are categorised into two, namely single-walled carbon nanotubes (SWCNTs) and multi-walled carbon nanotubes (MWCNTs). SWCNTs are hexagonal one-dimensional (1D) carbon nanomaterials consisting of a hollow tube with a one-atom-thick wall possessing a diameter within the range of 0.4-2 nm, as shown in Figure 2a (Li et al. 2020, Filip et al. 2018). As the name implies, carbon nanotubes with multi-rolled graphite layers are called multi-wall carbon nanotubes. They usually have two or more cylinders arranged in the hollow, and the diameter varies between 1-3 nm and as seen in Figure 2b and Figure 2c which show the hollow of double-walled carbon nanotubes (DWCNTs) and multi-walled carbon nanotubes, respectively (Filip et al. 2018). Carbon nanotubes can be prepared via various methods; plasma arcing, a laser method, catalysed chemical vapour deposition, and the ball milling method to mention a few. The purification could be done via gas phase, liquid phase, intercalation, and plasma CNTs purification methods.

Apart from the structural difference between SWCNTs and MWCNTs, many other differences exist between these two forms of CNTs (see Table 1).

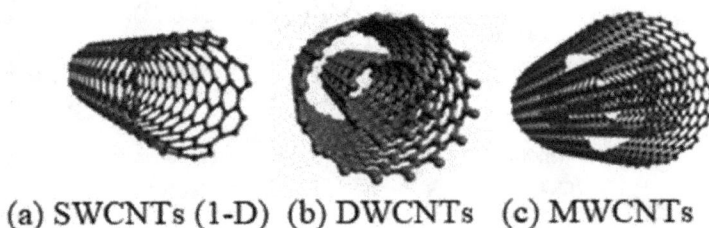

(a) SWCNTs (1-D) (b) DWCNTs (c) MWCNTs

Figure 2: Building block and various forms of CNTs (Filip et al. 2018).

5. Synthesis Methods for Carbon Nanotubes (CNTs)

After finding CNTs in 1991, a lot of consideration has been paid to basic and applied research on this material (Rajisha et al. 2011). After their discovery, carbon nanotubes (CNTs), a fourth carbon allotrope, were widely studied worldwide. In many applications, CNTs have distinctive properties that make them useful. Chemical

Table 1: Differences between single-walled carbon nanotubes (SWCNTs) and multi-walled carbon nanotubes (MWCNTs) (Chongin et al. 2019, Yang et al. 2021)

SWCNTs	MWCNTs
SWCNTs possess a single 1D graphene sheet spunned into a tube	MWCNTs comprise multiple graphene panes rolled into cylinders
They usually involve the use of the catalyst for their synthesis	They can be synthesised in the absence of a catalyst
Bulk synthesis comes with challenges of how to manage overgrowth and atmospheric condition	Bulk synthesis poses no issues
SWCNTs are susceptible to defect when functionalised	They are less susceptible to defect when produced by an arc-discharged technique
Purity is poor	Good purity
They can be twisted with ease and as well more pliable	They cannot be easily twisted
Characterisation and evaluation are not difficult	They have a very complex structure

vapour deposition, arc discharge, spray pyrolysis, and laser ablation among others are applicable in the synthesis of CNTs. The preparation conditions are crucial to determining the nature of CNTs.

5.1 Synthesis Methods for Single-Walled Carbon Nanotubes (SWCNTs)

For many decades, because of their wide range of applications in microelectronic, biomedical, polymer, energy, and environmental devices, SWCNTs' outstanding properties have attracted a lot of interest. Nevertheless, for many applications, the enormous potential of SWCNTs involves scaling up the techniques of synthesis. This move always poses a challenge. Efforts have been devoted to developing economic methods for the production of SWCNTs. The techniques applied for the synthesis of SWCNTs include combustion, laser ablation, non-equilibrium plasma, CVD, arc discharge, and arc-jet plasmas. A brief description of solar and CVD techniques, respectively, representing a traditional and a contemporary synthesis method of SWCNTs is presented below:

5.1.1 Solar Synthesis of SWCNTs

This technique involves the vapourisation of carbon precursors, such as graphite, in the presence of a sufficient energy source from the solar system. The objective diameter determines vapourisation rate. Luxembourg and colleagues reported quantifying the vapourisation rate of pure graphite set at the 1-megawatt solar furnace's focus (Anzar et al. 2020). For instance, a diameter target of 6 cm, conforming to 50 kW usable solar power, yielded carbon black of approximately 10–20 g/h which was obtained from the kinetic law of vapourisation, proposed to be (Anzar et al. 2020):

$$r_v = A \exp\left(\frac{-B}{T_{target}}\right)$$

where $A = 1.63 \times 10^9$ kg m^2/s, $B = 97259$ K, for P $= 240$ hPa and 3000 K $< T_{target} <$ 3700 K.

Figure 3 displays the experimental setup. Essential subunits include generator, gas circulation scheme, calculation system, records procurement, and filter. A detailed description of the solar synthesis of SWCNTs is reported in the literature (Anzar et al. 2020).

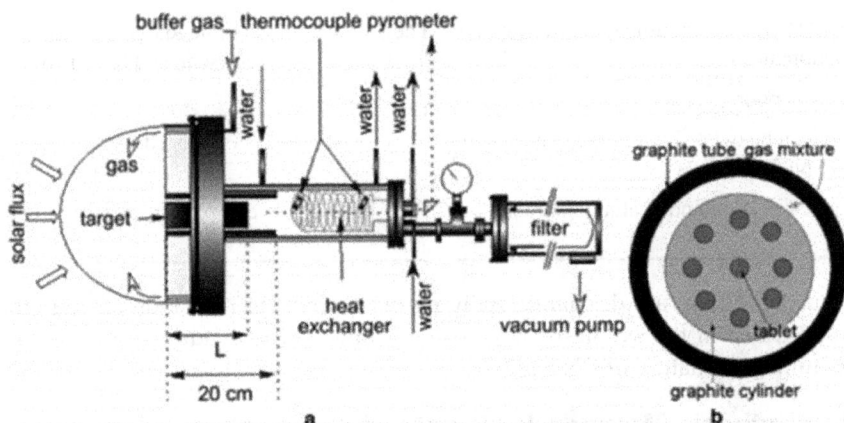

Figure 3: Experimental setup set at the centre of the PROMES-CNRS Lab 1 MW solar furnace: (a) side view of the entire setup and (b) a top view of the goal rod (paraphrased by authors) (Anzar et al. 2020).

5.1.2 CVD Synthesis of SWCNTs

The CVD technique is among the most advanced methods for the synthesis of SWCNTs. Hussain and coworkers demonstrated the preparation of SWCNTs using the CVD setup shown in Figure 4 (Baptista et al. 2015). As a carbon source, a minimum flow of ethylene is fed into the reactor. Hydrogen is applied to the surface of the catalyst particle to control ethylene breakdown. A mixture of reactants and nitrogen as a carrier gas is introduced into the reactor. As shown in Figure 4a, there are different ways of introducing the carrier gases. The main flow, Figure 4a, is ethylene, hydrogen, and N$_2$ transitory via the ferrocene container. The centreline temperature profile, shown in Figure 4c, is calculated using a thermocouple in a quartz tube at a constant temperature. The influence of feeding points on the synthesis of SWCNT can be examined by separately incorporating the gaseous reactants at high feed point H and low feed point L. In a high-temperature zone, nanoparticles are formed from the decomposition of the vapour phase ferrocene. At elevated temperature, the gaseous ethylene fragments into carbon over the surface of the catalyst NP. The fragmented carbons in the gas phase are transformed into carbon nanotubes over the surface of a catalyst.

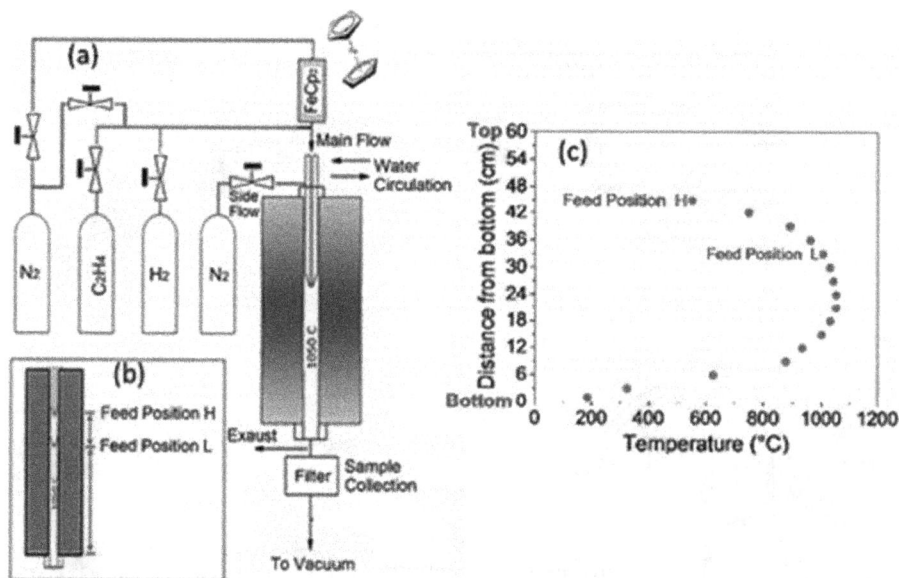

Figure 4: Schemes for (a) an FC-CVD reactor for the synthesis of SWCNTs, (b) feeding the reactant at two different positions inside the reactor, and (c) temperature profile inside the quartz tube measured in N_2 flow with the aid of thermocouple. Feed position H is at a comparatively lower temperature than feed position L (paraphrased by the author) (Baptista et al. 2015).

5.2 Synthesis of Multi-walled Carbon Nanotubes (MWCNTs)

MWCNTs can be synthesised using similar techniques outlined for SWCNTs. An overview of the layout and the process involved in the synthesis of MWCNTs is shown in Figure 5. Although the techniques mentioned above efficiency is encouraging under appropriate operating conditions, it is still unclear which method is best due to questions related to material properties, technicality, and production economy (Power et al. 2011). The electrical arch-discharge technique, shown in Figure 5a, is comparatively cheaper, gives better output quality; however, its operation involves a high temperature. A plasma comprising carbon and metallic vapour catalysts is produced by electric arc discharge between two electrodes under a rare gas atmospheric condition.

The vapour content causes corrosion as a result of the energy transmitted to the catalyst-doped anode. At the same time, MWCNTs are produced at the cathode as a consistent and robust deposit.

The laser ablation method, displayed in Figure 5b, is also a promising alternative to the synthesis at room temperature of excellent carbon nanotubes. However, it is expensive (Power et al. 2011, Kan et al. 2016). Herein, a laser beam is directed on a catalyst comprising graphite pellets, positioned at the centre of a quartz tube containing a torpid gas stored at a specific temperature. The radiant energy sublimates the carbon substrate, swept by the gas flow towards the conical water-cooled copper collector. MWCNTs get deposited in large quantities in the same collector, quartz tube walls, and the graphite pellet's backside.

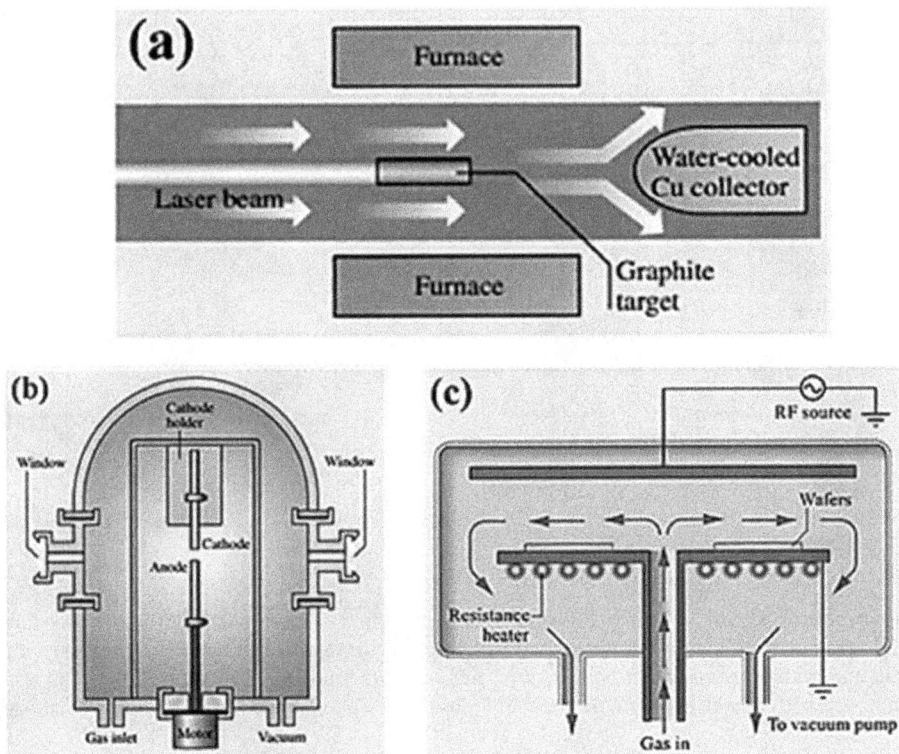

Figure 5: The general layout of (a) an electrical arch-discharge, (b) laser ablation, and (c) chemical vapour deposition apparatus used to synthesise multi-walled carbon nanotubes (paraphrased by the author) (Power et al. 2011).

The synthesis carried out by a CVD method (Figure 3c) overcomes the obstacles mentioned above, guarantees large-scale output, and has become more commonly used to obtain MWCNTs for these reasons. This approach can be performed in a heterogeneous or homogeneous system (Power et al. 2011, Kan et al. 2016). The critical synthetic pathway explores the catalytic disintegration of a source containing carbon on substrates made up of transition metals and their aggregates, enabling aligned and dense nanotube arrays to expand. It is also possible to combine optical excitation plasma to synthesise MWCNTs at lower temperatures in EAD and LA (Power et al. 2011, Kan et al. 2016). In contrast to the others, it is possible to manipulate the homogeneity, selectivity, and nanotubes' size using the CVD method. However, these features are mainly contingent on the nature and configuration of the catalysts and synthesis conditions.

6. Properties and Characterisation Techniques of CNTs

The most prominent properties of CNTs include size and shape, optical properties, surface area, and thermal stability. Properties of CNTs can be investigated using different techniques such as scanning electron microscopy (SEM), transmission electron microscopy (TEM), high-resolution transmission electron microscopy (HRTEM), selected area electron diffraction (SAED), fast Fourier transform (FFT), Fourier-transform infrared spectroscopy (FTIR), BET (Brunauer-Emmett-Teller), Raman spectroscopy, and thermogravimetric analysis (TGA) among others. No single technique has the robustness of measuring all the properties of CNTs.

6.1 Microscopy

Microscopy is a branch of science that involves using microscopes to examine structures and parts of objects that are not visible to the naked eye (applicable to objects that are not within the normal eye's resolution range). Microscopy is divided into electron, optical, and scanning probe microscopy alongside the new X-ray microscopy field. SEM and TEM are the most common microscopy techniques for investigating the morphology of CNTs. While SEM displays the sample's outer size and shape, as shown in Figure 6a, TEM reveals such materials' internal structures (see Figures 6b and 6c) (Li et al. 2020). In contrast to other techniques, SEM is facile and can be performed frequently. For a better view of the inner structure of CNTs, a high-resolution TEM (HRTEM) was utilised (Kim et al. 2019).

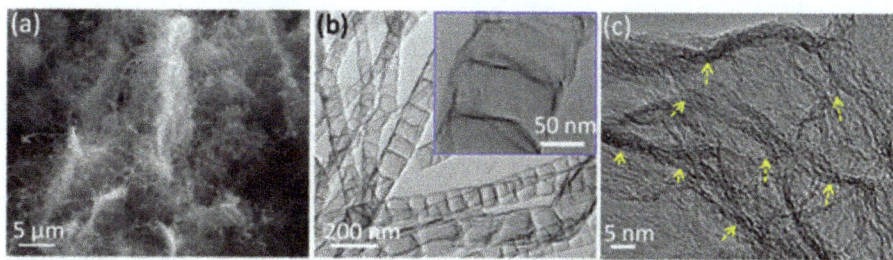

Figure 6: (a) SEM image, and (b, c) TEM images of N,S-doped CNTs (Li et al. 2020).

6.2 Diffraction Technique

Diffraction methods are the most efficient and information-rich tool for determining chemical structure. Single-crystal X-ray diffraction is the most commonly used of these, but the technique's use is broadened by using neutrons and microcrystalline powders instead of single-crystal samples. The method, which has been in use for over a century and is well-known, is still being developed (Chongin et al. 2019). Diffraction techniques are employed basically to study the crystallinity of CNTs. X-ray diffraction (XRD) and selected area electron diffraction (SAED) are frequently used for this purpose. However, the crystallinity and disorderliness of CNTs can also be examined using the HRTEM. The degree of crystallinity can be calculated by measuring one segment of a tube under HRTEM's fast Fourier transform (FFT). The substance is exceptionally crystalline if the FFT is made up of sharp spots (narrow peaks).

Nonetheless, the material is not incredibly crystalline if the FFT contains faint marks. Using selected area electron diffraction (SAED), the measure of crystalline nature and chirality of the tubes can be ascertained. Ahamad and colleagues investigated the crystalline nature of multi-walled functionalized CNTs using XRD and SAED (Ahamad et al. 2019). They reported a consistent result using both techniques, as presented in Figure 7.

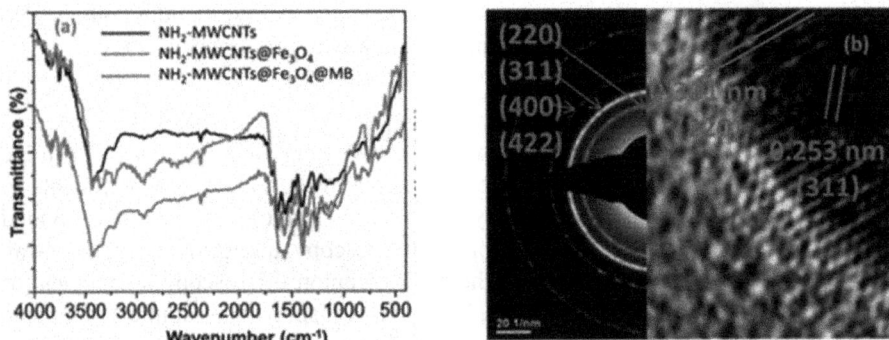

Figure 7: (a) XRD and (b) SAED of the NH_2-MWCNTs@Fe_3O_4 with rings marked (220), (311), (400), and (422) (Ahamad et al. 2019).

6.3 Spectroscopy Technique

Spectroscopic techniques use light to interact with matter, allowing them to probe particular sample characteristics to learn more about its quality or structure. Light is electromagnetic radiation, which has various energies and can probe various molecular features depending on that energy (Amjadi et al. 2017). Spectroscopy techniques like ultraviolet-visible spectroscopy and Raman spectroscopy, and Fourier-transform infrared spectroscopy (FTIR) are helpful for colour, purity, and functional groups analyses, respectively, of CNTs. The colour characteristics or wavelength reflectance CNTs depend on the morphology and electronic configuration, which form the basis for the difference between MWCNTs and SWCNTs. SWCNTs have structures active in the infrared, red and visible region of the solar spectrum. Raman spectroscopy provides evidence on the purity, imperfections, and orientation and helps to discern the existence of MWCNTs from other carbon allotropes. In explaining the structural properties of SWCNTs, the method has been remarkably effective. Unfortunately, the Raman spectra analysis from an MWCNT is always very complex and is yet to exhibit the same level of activity as SWCNTs. The functionalisation of CNTs is typically probed using FTIR. The level of functionalization in different surfactants will alter the nanotubes' wettability and, therefore, alter the toxicity. Recently, Wang et al. reported the synthesis and FTIR and Raman characterizations of reduced graphene/carbon nanotube composites (RGO/CNTs) as shown in Figures 8a and 8b, respectively (Wang et al. 2020).

6.4 BET (Brunauer-Emmett-Teller)

BET (Brunauer-Emmett-Teller) is a model established in 1938 for isothermal N_2 adsorption. A comprehensive explanation of BET theory's derivations can be accessed in the literature (Al-Husseini et al. 2018). The characterisation of the surface

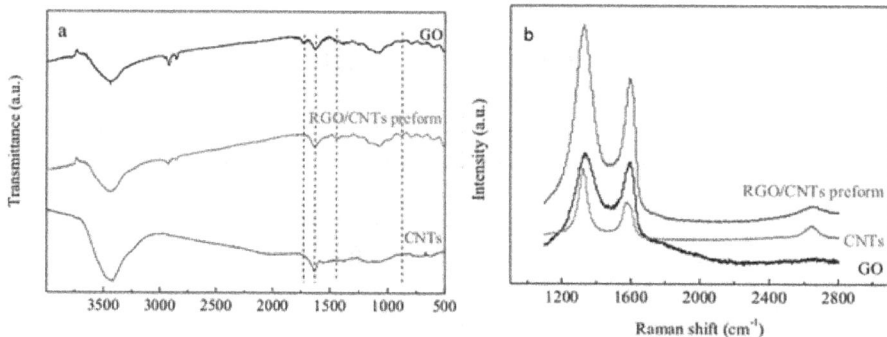

Figure 8: (a) FTIR and (b) Raman characterization of RGO/CNTs preform (Wang et al. 2020).

region of the nanotube of CNTs supports the unravelling of material behaviour and molecular interactions. Most generally, the adsorption of N_2 gas is the basis for determining the surface area of CNTs. In general, because of micropore filling or more significant pressure due to capillary condensation, lower pressure adsorption may not mean coverage is a monolayer. The BET model has a variety of critiques. Uniform adsorption sites over a surface are the basis of the theory. For MWCNTs, due to bundling/aggregation and faults, the adsorption sites are not similar.

Moreover, sufficient elucidation of the adsorbed molecule interaction can hardly be obtained using the BET theory. Nevertheless, BET is still a traditional technique for evaluating the surface area of CNTs, regardless of its limitations. BET is a widely employed technique by researchers for CNTs surface area, pore-volume, and pore size measurement. For instance, Yang and coworkers reported the use of the BET technique for the elucidation of the surface area and pore size of nitrogen-doped carbon nanotube network CP@NCNT, cobalt phosphide (CP) nanoparticles, and nitrogen-doped CNTs (NCNT) (Yang et al. 2021). CP@NCNT was reported to show the highest surface area and pore size, as shown in Figure 9, due to nitrogen and CNT's synergetic presence.

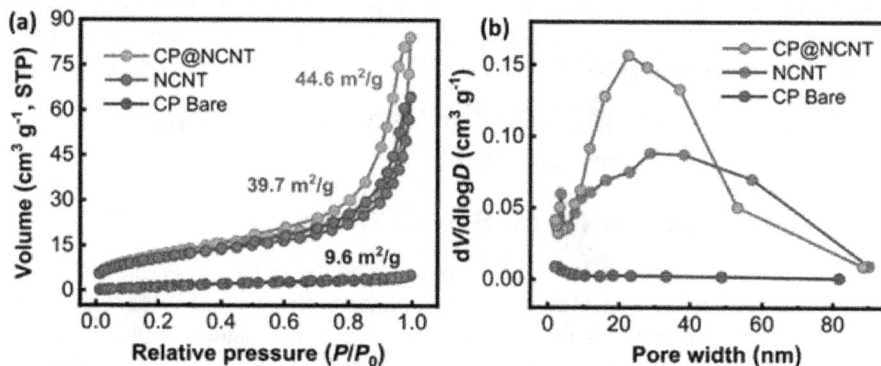

Figure 9: (a) Nitrogen adsorption-desorption isotherm plots and (b) corresponding pore-size distribution plots of CP@NCNT, NCNT, and bare CP (Yang et al. 2021).

6.5 Thermogravimetric Analysis (TGA)

Thermogravimetric analysis (TGA) is an analytical technique that monitors the weight shift when a sample is heated at a constant rate to assess a material's thermal stability and a fraction of volatile components (Rajisha et al. 2011). Thermogravimetric analysis (TGA) is used to probe the purity and resilience of CNTs under high-temperature conditions. From the curve, residual mass, and oxidation, and initiation temperature can be estimated. The initiation temperature is simply the point where the material starts decomposing. The oxidation temperature is the maximum weight loss point, which correlates with the material's thermal stability. The residual mass of material is subject to its consistency, aromatic bondings, and homogeneity. Figure 10 shows a TGA curve obtained in research executed by Liu et al. (2019). The incorporation of CNT to Fe_2O_3 was demonstrated to enhance the composite's thermal stability, suggesting the significant role of CNTs.

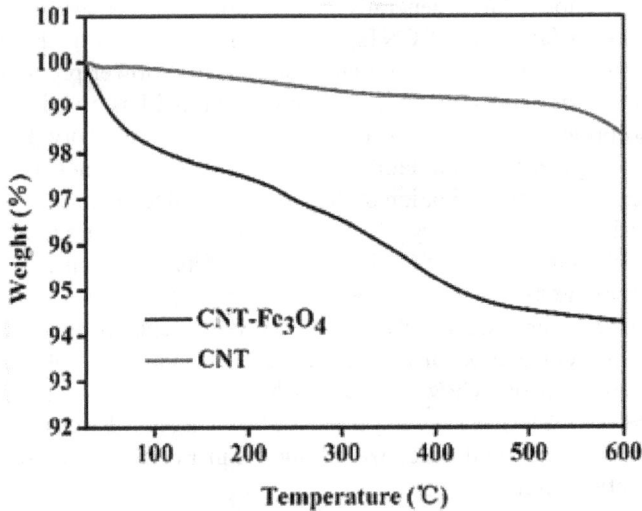

Figure 10: TGA curves of CNTs and CNT-Fe_3O_4 composites. The different weight loss behaviours between the CNTs and CNT-Fe_3O_4 composites indicated that functional groups were introduced on the surface of CNTs during the oxidation procedure (Liu et al. 2019).

7. Applications of CNTs for Sensor and Water Treatment Technologies

7.1 Application of CNTs for Sensor Technology

In the broadest sense, a sensor is a computer, module, machine, or subsystem that detects events or changes in its environment and transmits the data to other electronics. Manufacturing and equipment, airplanes and aerospace, automobiles, medicine, robots, and many other aspects of our daily lives use sensors. CNTs are exceptional to the carbon family. They possess a unique atomic structure and mechanical deformations, making them helpful in developing miniaturised sensors sensitive to a chemical known as electrochemical sensors (Li et al. 2020). Electrochemical sensors consist of an electrochemical cell containing at least a pair of electrodes and a transducer where the transfer of charges occurs in a closed circuit.

In contrast, the carriage of charges can be ionically or electronically in the analyte sample. The distortion of electron clouds makes the CNT electrochemically active [22]. Besides, CNTs store essential biomolecules, which helps increase the probes' selectivity and sensitivity. It is crucial that CNTs are correctly functionalised and immobilised for applications in electrochemical sensing devices (Wang et al. 2018).

In 2016, Kan et al. fabricated an organised layer of mercapto-β-cyclodextrin matrix attached to MWCNTs to detect quercetin. The sensor exhibited good sensitivity, activity, selectivity, and a low detection limit (Ali et al. 2019). For humidity level detection, Kim et al. (2019) developed a sensor based on CNTs ornamented with gold electrodes. The synthesised nanosensor showed high efficacy without relying on voltage and frequency (Yan et al. 2019). Figure 11 shows the SEM and TEM of the CNT composites as well images of the humidity sensor. According to their findings, the fabricated chitosan (CS)-MWCNT nanohybrid humidity sensor had good electrical efficiency without voltage or frequency dependence and a low-temperature dependency with a temperature coefficient (RH/°C) of 0.187 percent. In 2018, Husseini and colleagues fabricated an OH-MWCNT sensor for the discriminatory recognition of ammonia. The sensor checks the functionalised MWCNTs to detect ammonia gas selectively at room temperature (Al-Husseini et al. 2018).

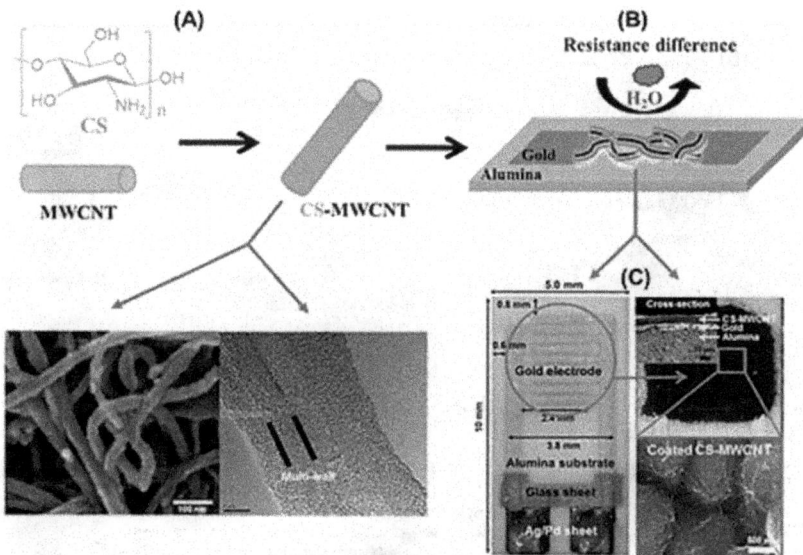

Figure 11: (a) Schematic illustration and SEM and TEM images of CS-shell/MWCNT-core nanohybrids structure. (b) Photograph of the commercial alumina-based humidity sensor electrode. (c) Photograph of the humidity sensor electrode coated by CS-MWCNT nanohybrids and SEM image showing electrodes interconnected by CS-MWCNT nanohybrids (Yan et al. 2019).

7.2 Application of CNTs for Water Treatment Technology

Wastewater pollution is a major challenge to humans and the environment. Several strategies and new materials have been employed to decontaminate wastewater. So far, among the materials used, CNTs seem very promising due to their large surface area, mechanical stability, conductivity, high porosity, small size, and easy functionalisation. Generally, nanocomposites have higher adsorption capacity than non-functionalised CNTs. For instance, Alijani and coworkers reported higher efficiency (99.56%) for SWCNT-magnetite cobalt sulfide-based nanocomposite and 45.39% for pristine SWCNTs applied for the decontamination of mercury in wastewater (Power et al. 2011). The pollutant removal performance of CNTs has been reported to depend on contact time, the initial concentration of a solution, and pH. For SWCNTs, adsorption is said to follow the Langmuir isotherm model. However, unlike SWCNTs, the adsorption features of MWCNTs fit the Freundlich model of adsorption (Power et al. 2011). The oxidised form of MWCNTs is known to show enhanced water ion adsorption ability and proficiency. Researches have reported better adsorption ability using plasma-oxidised MWCNTs than the chemically oxidised counterpart, which may be likened to a larger number of oxygen-functionalised CNTs surface (Power et al. 2011). Compared to other organic oxide products, the adsorption performance of functionalised MWCNTs is higher, and functionalised MWCNTs are also expected to be remarkably more efficient in metal ion removal than non-functionalised MWCNTs. The sorption's key mechanism is

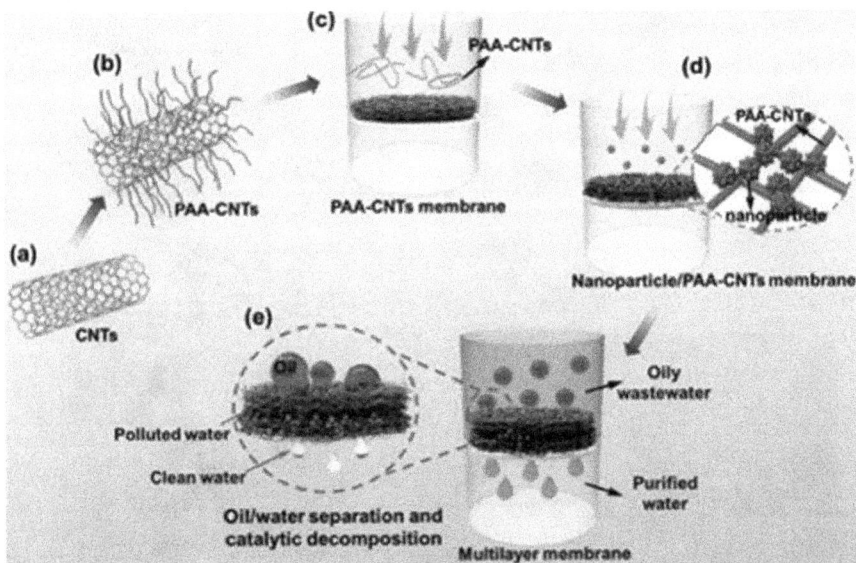

Figure 12: Fabrication of multifunctional CNTs composite separation membrane by filtration. (a,b) PAA modified CNTs (PAA-CNTs). (c) PAA-CNTs membrane prepared by vacuum filtration. (d) The deposition of catalytic nanoparticles in PAA-CNTs membrane. (e) The resultant multilayer PAA-CNTs/Pd@Pt/PAA-CNTs composite membrane is employed for oil/water separation and catalytic decomposition (Yan et al., 2019).

commonly thought to be the interaction amongst ions and the surface of CNTs (Ali et al. 2019). Figure 12 shows a scheme for a robust and straightforward strategy for the fabrication of multifunctional hybrid CNTs by Yan et al. 2019. The nanocomposites were demonstrated to be effective for oil-in-water emulsion separation and catalytic decontamination of various types of organic pollutants in water (Yan et al. 2020).

8. Conclusions

Recently, CNTs have gained more attention due to their distinctive characteristic features. They can be synthesised via various methods. Although all approaches reported yielded more than 75% under suitable operating conditions, it is still unclear which method is best due to puzzles related to the properties of CNTs, technicality, and cost of production. However, compared to the other techniques, the CVD method gives room for greater control of the nature of the resultant CNTs. CNTs are widely applicable in electrochemical sensing devices and wastewater pollution control. More so, the functionalised forms of CNTs have been demonstrated to be more effective and showed potentials for industrial applications.

References

Ahamad, T., Naushad, M., Eldesoky, G.E., Al-Saeedi, S.I., Nafady, A., Al-Kadhi, N.S., Al-Muhtaseb, A.H., Khan, A.A. and Khan, A. 2019. Effective and fast adsorptive removal of toxic cationic dye (MB) from aqueous medium using amino-functionalized magnetic multiwall carbon nanotubes. J. Mol. Liq. 282: 154–161. https://doi.org/10.1016/j.molliq.2019.02.128.

Al-Husseini, A.H., Al-Sammarraie, A.M.A. and Saleh, W.R. 2018. Specific NH_3 gas sensor worked at room temperature based on MWCNTs-OH network. Nano Hybrids Compos. 23: 8–16. https://doi.org/10.4028/www.scientific.net/nhc.23.8.

Ali, S., Rehman, S.A.U., Luan, H.Y., Farid, M.U. and Huang, H. 2019. Challenges and opportunities in functional carbon nanotubes for membrane-based water treatment and desalination. Sci. Total Environ. 646: 1126–1139. https://doi.org/10.1016/j.scitotenv.2018.07.348.

Amjadi, M., Hallaja, T., Asadollahia, H., Song, Z., Marta de Frutos and Hildebrand, N. 2017. Facile synthesis of carbon quantum dot/silver nanocomposite and its application for colorimetric detection of methimazol. Sensors and Actuators B 244: 425–432. http://dx.doi.org/10.1016/j.snb.2017.01.003

Anzar, N., Hasan, R., Tyagi, M., Yadav, N. and Narang, J. 2020. Carbon nanotube – a review on synthesis, properties and plethora of applications in the field of biomedical science. Sensors Int. https://doi.org/10.1016/j.sintl.2020.100003.

Baptista, F.R., Belhout, S.A., Giordani, S. and Quinn, S.J. 2015. Recent developments in carbon nanomaterial sensors. Chem. Soc. Rev. 44: 4433–4453. https://doi.org/10.1039/c4cs00379a.

Chao, W., Vignesh, M., Jie, K., Zhenfeng, H., Xianmin, M., Qian, S., Yanjun, C., Li, G., Chuntai, L., Subramania, A. and Zhanhu, G. 2018. Overview of carbon nanostructures and nanocomposites for electromagnetic wave shielding. Carbon (2018), doi:10.1016/j.carbon.2018.09.006.

Chongin, P.V., Ovidiu, G., Vincent, Y., Huibo, C., Alimamy, F.B., and Michael, S. 2019. Na2Mn3Se4: Strongly Frustrated Antiferromagnetic Semiconductor with Complex Magnetic Structure. Inorg. Chem. 2019, 58, 9, 5799–5806. https://doi.org/10.1021/acs.inorgchem.9b00134.

Clegg, W., Tyne, N. and Kingdom, U. 2019. X-Ray and Neutron Diffraction. Elsevier Inc.. https://doi.org/10.1016/B978-0-12-409547-2.14635-9.

Filip, A., Thomas, J.M., Vladimir, M. and Ivan, P.P. 2018. Evaluation of the BET theory for the characterization of meso and microporous MOFs. Small Methods 2018, 1800173 1800173. https://doi.org/10.1002/smtd.201800173.

Kan, X., Zhang, T., Zhong, M. and Lu, X. 2016. Biosensors and bioelectronics CD/AuNPs/MWCNTs based electrochemical sensor for quercetin dual-signal detection. Biosens. Bioelectron. 77: 638–643. https://doi.org/10.1016/j.bios.2015.10.033.

Kim, H.S., Kim, J.H., Park, S.Y., Kang, J.H., Kim, S.J., Choi, Y.B. and Shin, U.S. 2019. Carbon nanotubes immobilized on gold electrode as an electrochemical humidity sensor. Sensors Actuators, B Chem. 300: 127049. https://doi.org/10.1016/j.snb.2019.127049.

Lehman, J.H., Terrones, M., Meunier, V., Mansfield, E. and Hurst, K.E. 2011. Evaluating the characteristics of multiwall carbon. Carbon N.Y. 49: 2581–2602. https://doi.org/10.1016/j.carbon.2011.03.028.

Li, B., Liu, T., Wang, Y. and Wang, Z. 2012. ZnO/graphene-oxide nanocomposite with remarkably enhanced visible-light-driven photocatalytic performance. J. Colloid Interface Sci. 377: 114–121. https://doi.org/10.1016/j.jcis.2012.03.060.

Luxembourg, D., Flamant, G. and Laplaze, D. 2005. Solar synthesis of single-walled carbon nanotubes at medium scale. Carbon N.Y. 43: 2302–2310. https://doi.org/10.1016/j.carbon.2005.04.010.

Oliveira, T.M.B.F. and Morais, S. 2018. New generation of electrochemical sensors based on multi-walled carbon nanotubes. Appl. Sci. 8: 5–7. https://doi.org/10.3390/app8101925.

Power, A., Gorey, B. and Chandra, J.C.S. 2011. Carbon nanomaterials and their application to electrochemical sensors. Nanotechnology Reviews 'Just Accepted' paper ISSN (online) 2191-9097 DOI: 10.1515/ntrev-2017-0160.

Rajisha, K.R., Deepa, B., Pothan, L.A. and Thomas, S. 2011. Thermomechanical and spectroscopic characterization of natural fibre composites. Interface Eng. Nat. Fibre Compos. Maximum Perform, 241–274. https://doi.org/10.1533/9780857092281.2.241.

Saito, T., Ohshima, S., Xu, W.C., Ago, H., Yumura, M. and Iijima, S. 2005. Size control of metal nanoparticle catalysts for the gas-phase synthesis of single-walled carbon nanotubes. J. Phys. Chem. B. 109: 10647–10652. https://doi.org/10.1021/jp044200z.

Shaw, J.M., Satyro, M.A. and Yarranton, H.W. 2017. Phase behavior and properties of heavy oils. https://doi.org/10.1007/978-3-319-49347-3_8.

Song, L., Hu, C., Xiao, Y., He, J., Lin, Y., Connell, J.W. and Dai, L. 2020. An ultra-long life, high-performance, flexible Li-CO$_2$ battery based on multifunctional carbon electrocatalysts. Nano Energy. https://doi.org/10.1016/j.nanoen.2020.104595.

Wang, B., Dou, S., Li, W. and Gao, Y. 2020. Multifunctional reduced graphene oxide/carbon nanotubes/epoxy resin nanocomposites based on carbon nanohybrid preform. Soft Mater. 18: 89–100. https://doi.org/10.1080/1539445X.2019.1688833.

Yan, L., Zhang, G., Zhang, L., Zhang, W., Gu, J., Huang, Y., Zhang, J. and Chen, T. 2019. Robust construction of underwater superoleophobic CNTs/nanoparticles multifunctional hybrid membranes via interception effect for oily wastewater purification. J. Memb. Sci. 569: 32–40. https://doi.org/10.1016/j.memsci.2018.09.060.

Yang, D., Hou, W., Lu, Y., Zhang, W. and Chen, Y. 2021. Cobalt phosphide nanoparticles supported within network of N-doped carbon nanotubes as a multifunctional and scalable electrocatalyst for water splitting. J. Energy Chem. 52: 130–138. https://doi.org/10.1016/j.jechem.2020.04.005.

Index

For Product Safety Concerns and Information please contact our EU
representative GPSR@taylorandfrancis.com
Taylor & Francis Verlag GmbH, Kaufingerstraße 24, 80331 München, Germany

www.ingramcontent.com/pod-product-compliance
Lightning Source LLC
Chambersburg PA
CBHW060744220326
41598CB00022B/2317